641.36
Mus

Muscle as Food

421873   1000

W9-BSD-218

G 25⁰⁰
294274

WITHDRAWN BY
ORANGE COUNTY LIBRARY SYSTEM

ORANGE COUNTY LIBRARY SYSTEM

UNIVERSTIY OF KENTUCKY
College of Agriculture
Food Science
Date-

Hesham Elgaali

MUSCLE AS FOOD

UNIVERSTIY OF KENTUCKY
College of Agriculture
Food Science
Date-

FOOD SCIENCE AND TECHNOLOGY
A SERIES OF MONOGRAPHS

Series Editor

Bernard S. Schweigert
University of California, Davis

Advisory Board

S. Arai
University of Tokyo, Japan

C. O. Chichester
Nutrition Foundation, Washington, D.C.

J. H. B. Christian
CSIRO, Australia

Larry Merson
University of California, Davis

Emil Mrak
University of California, Davis

Harry Nursten
University of Reading, England

Louis R. Rockland
Chapman College, Orange, California

Kent K. Stewart
Virginia Polytechnic Institute
and State University, Blacksburg

A list of the books in this series is available from
the publisher on request.

MUSCLE AS FOOD

Edited by

Peter J. Bechtel

Meat Science Laboratory
University of Illinois at Urbana-Champaign
Urbana, Illinois

1986

ACADEMIC PRESS, INC.
Harcourt Brace Jovanovich, Publishers

Orlando   San Diego   New York   Austin
London   Montreal   Sydney   Tokyo   Toronto

COPYRIGHT © 1986 BY ACADEMIC PRESS, INC.
ALL RIGHTS RESERVED.
NO PART OF THIS PUBLICATION MAY BE REPRODUCED OR
TRANSMITTED IN ANY FORM OR BY ANY MEANS, ELECTRONIC
OR MECHANICAL, INCLUDING PHOTOCOPY, RECORDING, OR
ANY INFORMATION STORAGE AND RETRIEVAL SYSTEM, WITHOUT
PERMISSION IN WRITING FROM THE PUBLISHER.

ACADEMIC PRESS, INC.
Orlando, Florida 32887

United Kingdom Edition published by
ACADEMIC PRESS INC. (LONDON) LTD.
24-28 Oval Road, London NW1 7DX

LIBRARY OF CONGRESS CATALOGING-IN-PUBLICATION DATA

Main entry under title:

Muscle as food.

    (Food science and technology)
    Includes index.
    1. Meat—Addresses, essays, lectures.   I. Bechtel,
Peter J.   II. Series.
TX556.M4M87   1986   641.3'6   85-19951
ISBN 0-12-084190-8 (alk. paper)
ISBN 0-12-084191-6 (paperback)

PRINTED IN THE UNITED STATES OF AMERICA

86 87 88 89   9 8 7 6 5 4 3 2 1

Contents

8. Nutritional Composition and Value of Meat and Meat Products

C. E. Bodwell and B. A. Anderson

9. Poultry Muscle as Food

P. B. Addis

10. Fish Muscle as Food

W. D. Brown

Contributors

Numbers in parentheses indicate the pages on which the authors' contributions begin.

P. B. *Addis* (371), Department of Food Science and Nutrition, University of Minnesota, St. Paul, Minnesota 55108

B. A. *Anderson* (321), Nutrition Monitoring Division, Human Nutrition Information Service, U.S. Department of Agriculture, Hyattsville, Maryland 20782

P. J. *Bechtel* (1), Meat Science Laboratory, University of Illinois at Urbana–Champaign, Urbana, Illinois 61801

C. E. *Bodwell* (321), Energy and Protein Nutrition Laboratory, Beltsville Human Nutrition Research Center, Agricultural Research Service, U.S. Department of Agriculture, Beltsville, Maryland 20705

W. D. *Brown*[1] (405), Institute of Marine Resources, Department of Food Science and Technology, University of California, Davis, Davis, California 95616

H. R. *Cross*[2] (279), USDA–ARS Roman L. Hruska U.S. Meat Animal Research Center, Clay Center, Nebraska 68933

P. R. *Durland* (279), USDA–ARS Roman L. Hruska U.S. Meat Animal Research Center, Clay Center, Nebraska 68933

M. L. *Greaser* (37), Muscle Biology Laboratory, University of Wisconsin, Madison, Wisconsin 53706

R. *Hamm* (135), Bundesanstalt für Fleischforschung, 8650 Kulmbach, Federal Republic of Germany

A. A. *Kraft* (239), Department of Food Technology, Iowa State University, Ames, Iowa 50011

A. M. *Pearson* (103), Department of Food Science and Human Nutrition, Michigan State University, East Lansing, Michigan 48824

---

[1] Deceased.

[2] Present address: Department of Animal Science, Texas A & M University, College Station, Texas 77843.

G. R. *Schmidt* (201), Department of Animal Sciences, Colorado State University, Fort Collins, Colorado 80523

S. C. *Seideman* (279), USDA–ARS Roman L. Hruska U.S. Meat Animal Research Center, Clay Center, Nebraska 68933

Preface

Muscle as Food provides a broad-based coverage of the conversion of muscle to meat and meat properties. Advanced undergraduate or beginning graduate students in animal science or food science will find this book a valuable and useful reference. Also, individuals involved in research and development of meat systems will find the methods and references useful. Emphasis has been placed on understanding the properties of muscle proteins. The properties of the contractile proteins are included in Chapter 1, and the biochemical changes that occur as muscle is converted to meat are described in detail in Chapter 2. Physical and biochemical changes that occur in muscle during storage and preservation are described in Chapter 3, and the functional properties of myofibrillar systems are the topic of Chapter 4. These four chapters plus sections of Chapters 9 and 10 on poultry and fish muscle provide a basis for understanding the basic biochemical properties of meat systems. An overview of meat processing and fabrication is given in Chapter 5, and the unique aspects of meat microbiology are elaborated on in Chapter 6. Sensory qualities of meat and sensory methods are discussed in Chapter 7. A detailed listing of the nutritional composition of meat and meat products constitutes Chapter 8. Too often both poultry and fish muscle are not covered in texts focusing on meat science and technology. In this book, two chapters have been devoted to these topics to emphasize the unique properties of these tissues.

Each chapter is written by an expert in his or her respective area. The subject matter of each chapter has been arranged to avoid overlaps and to provide as broad a coverage as possible of the immense subject area of meat science and technology.

Peter J. Bechtel

1

Muscle Development and Contractile Proteins

P. J. BECHTEL

Meat Science Laboratory
University of Illinois at Urbana-Champaign
Urbana, Illinois

Copyright © 1986 by Academic Press, Inc.
All rights of reproduction in any form reserved.

I. SKELETAL MUSCLE STRUCTURE

Muscle tissue is a major component of the body: it has been estimated that the human body contains no fewer than 434 muscles, which constitute ~25% of the body mass at birth and ~40% in the adult man (Adams, 1975).

A. Gross Structure and Muscle Bundles

Skeletal muscles are composed of bundles of muscle fibers embedded in connective tissue. The major cell component is called the muscle cell or muscle fiber; however, other types of cells found in muscle include those of vascular, connective, adipose, and nerve tissues. The gross organization of a muscle is shown in Fig. 1. The muscle is encased in a layer of connective tissue called the epimysium, which is composed predominately of collagen. Individual muscle cells are surrounded by connective tissue called the endomysium, which forms a continuous connective tissue network from the muscle cell surface to the epimysium. Muscle cells are grouped into bundles that are surrounded by a layer of connective tissue called the preimysium. Small bundles are grouped into larger bundles, which eventually form the muscle. As a generality, the vascular system and nerves are found between muscle bundles, although capillaries and a nerve ending are found in each endomysial layer.

A single motor nerve axon innervates a group of muscle cells and sends a single axonal filament to each cell. Large muscles such as the gluteus maximus may contain up to 200 muscle cells per group; the small eye muscles, only five muscle cells per group (Lockart, 1973).

Most muscles attach to bones. The muscle connective tissue (type III collagen) becomes continuous with the tendon connective tissue. Tendons contain type I collagen and are lubricated to facilitate the gliding action of the tendon (Ramachandran and Reddi, 1976).

B. Muscle Cell

The muscle cell is the cellular contractile unit of skeletal muscle. The number of total muscle cells in large meat animals is unknown but is probably in the range of hundreds of millions. Muscle cells are large multinucleated cells that vary in both width and length. The diameter of muscle cells ranges from 10 to 100 μm and their length can vary from several millimeters to more than 30 cm (Lockhart, 1972). The diameter of individual muscle cells varies over their length and often is smaller at the ends. Muscle cell diameters vary with exercise, animal age, animal sex, nutritional state, breed, species, and type of muscle. Individual muscle cells can span the length of small muscles but

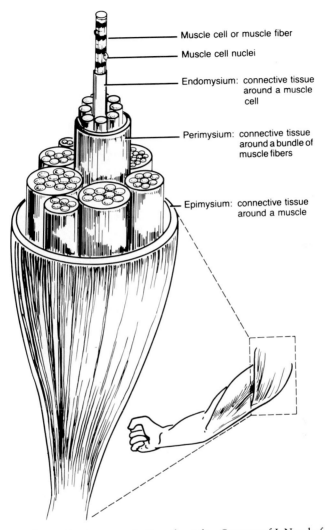

Muscle cell or muscle fiber

Muscle cell nuclei

Endomysium: connective tissue around a muscle cell

Perimysium: connective tissue around a bundle of muscle fibers

Epimysium: connective tissue around a muscle

Fig. 1 Connective tissue organization of muscle. Courtesy of J. Novakofski.

generally span only a fraction of the muscle length in large animals. Fibers can run parallel to the length of the muscle but often are arranged at an angle.

1. SARCOLEMMA AND T-TUBULE SYSTEM

The sarcolemma is the plasma membrane of the muscle cell and is similar in structure to other plasma membranes. The membrane has a thickness of ~7 nm and has the same trilaminate structure as other cell membranes (Adams,

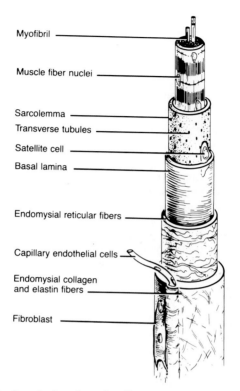

Myofibril —————————

Muscle fiber nuclei —————————

Sarcolemma —————————
Transverse tubules —————————
Satellite cell —————————
Basal lamina —————————

Endomysial reticular fibers —————————

Capillary endothelial cells —————————

Endomysial collagen
and elastin fibers —————————

Fibroblast —————————

Fig. 2 Organization of muscle cell. Courtesy of J. Novakofski.

1975). The structure is formed from a bimolecular leaflet of phospholipids in which the nonpolar portions of the molecule are oriented inward and the polar moieties are on the surface. Carbohydrate moieties and proteins are integral components of the membrane. Some proteins are located only on the outer or inner membrane surface while others span the membrane. Many of the proteins on the outer membrane surface are glycoproteins.

The muscle cell membrane is surrounded by a basement membrane structure which contains reticular fibers and has a thickness > 20 nm. The basal lamina is enclosed by the endomysial collagen sheath (Fig. 2). The endomysial collagen sheath is attached to the basement membrane, which in turn is attached to the sarcolemma.

The sarcolemma has a number of functions, the first of which is to compartmentalize the contractile units. Another major function is regulation of uptake and release of some molecules by the muscle cell. Probably the major distinguishing feature of the sarcolemma is its ability to depolarize in response to a nerve impulse. The site at which the nerve axon is attached to the muscle membrane is the motor endplate. The muscle membrane under the motor

endplate region contains an increased number of acetylcholine receptors, mitochondria, and other components. Once the nerve is stimulated and releases
acetylcholine, receptors on the sarcolemma bind the acetylcholine and go
through a series of events ultimately resulting in the depolarization of the
sarcolemma.

The transverse tubule (T-tubule) system is a series of small (20-nm diameter)
invaginations of the sarcolemma. There are two T-tubule systems per sarcomere, which surround the myofibrils and are located at the site at which the A
and I bands meet (A–I junction) in mammalian skeletal muscle (Fig. 3). The
function of the T-tubule system is to bring the membrane depolarization event
from the sarcolemma to the inner regions of the muscle cell. T-Tubule membrane is thought to have electrical excitability properties similar to the sarcolemma. There is nine times as much T-tubule membrane as sarcolemma membrane.

The structure of the sarcoplasmic reticulum is shown in Fig. 3. This membrane structure has a biochemical composition different from both the sarcolemma and T-tubule system. Its main function is to release and sequester
calcium in the muscle cell. Sarcoplasmic reticulum forms a matrix around the
myofibril between the T-tubules. The sarcoplasmic reticulum is not continuous with the T-tubule system; however, a series of small structures called feet
bridge the gap between the T-tubule membrane and the sarcoplasmic reticulum. When the muscle cell is stimulated, the sarcoplasmic reticulum releases
calcium; however, the mechanism of transmitting the signal from the T-tubule
system to the sarcoplasmic reticulum is not fully understood. The sarcoplasmic reticulum has the ability to sequester calcium against a concentration
gradient, a process that requires energy. On a relative scale, there is one part
sarcolemma membrane to nine parts T-tubule membrane to one hundred and
forty-four parts sarcoplasmic reticulum membrane.

There is a correlation between the amount of sarcoplasmic reticulum and the
speed of contraction (Peachy and Porter, 1959). Muscles with faster contraction speeds have a more developed sarcoplasmic reticular system. A more
developed sarcoplasmic reticulum may allow for a faster, more uniform release
and uptake of calcium.

2. MYOFIBRIL

Myofibrils are the long thin contractile elements inside the muscle cell that
give the characteristic striated pattern (Fig. 4). Vertebrate myofibrils are 0.5 –
1.0 μm in width and have a fairly uniform diameter throughout their length
(Adams, 1975). The sarcomere is the unit of muscle structure between the two
Z lines (Bloom and Fawcett, 1975). Other bands that can be observed with the
light microscope include the A band, I band, and Z line (Fig. 4). Areas that
appear darkest are the Z line and the regions of the A band where thick and thin

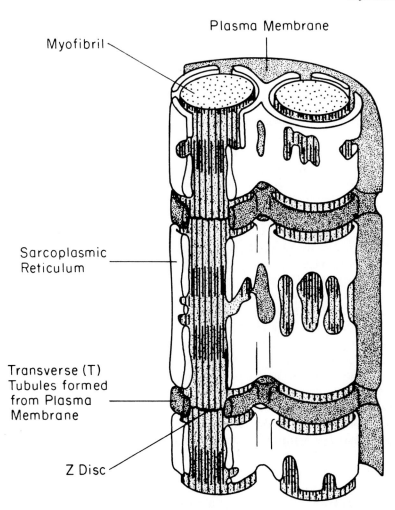

Fig. 3   Organization of the sarcoplasmic reticulum and transverse tubule systems.

filaments overlap.   Sarcomere length changes depending on the contractile state of the muscle and a resting mammalian psoas muscle will have a sarcomere length of approximately 2.6 μm (Huxley, 1973).

The sarcomere, depicted in Fig. 4, is constructed of thick filaments, thin filaments, and Z lines.   Mammalian thick filaments have diameters of 10 to 12 nm and lengths of 1.5 μm and thin filaments have diameters of 5 to 7 nm and lengths of 0.375 μm (Huxley, 1973).

The sliding filament model for contraction has the thick and thin filaments

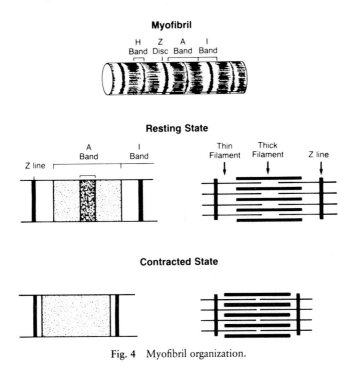

Fig. 4   Myofibril organization.

sliding past one another, resulting in an increase or decrease in sarcomere length. The thick and thin filaments do not change length, but the degree of overlap between thick and thin filaments changes. A cross-sectional examination of the myofibril reveals a hexagonal array of six thin filaments around a thick filament.

The Z line maintains the spacing between the thin filaments and is ~50–65 nm in diameter depending on the muscle type. Structural analysis of the Z line has not been completed, but the protein α-actinin is present. When the Z-line is examined at high magnification, it appears as a zigzag structure that anchors the thin filaments. Thin filaments do not span the Z line (Knappeis and Carlsen, 1962).

Two other myofibril structures are the M line and N_2 band. The M line appears as a line running parallel to the Z line that connects the centers of the thick filaments together. The M line is located in the center of the H zone. Knappeis and Carlsen (1968) described a complex model for the M line structure that consisted of both M filaments and M bridges in conjunction with the thick filaments. The N_2 line is a structure located in the I band that runs parallel to the Z lines. This structure may be composed of a high molecular weight protein (Wang and Williamson, 1980).

Poorly understood myofibrillar structures are the 2-nm filaments observed in stretched myofibrils (Locker and Leet, 1975). Maruyama *et al.* (1977b) and Wang *et al.* (1979) described the purification of new proteins that could form the 2-nm filaments. Much work remains to be done on the identification and characterization of the thin 2-nm filaments and associated proteins.

Other types of filaments that can be found in skeletal muscle include small numbers of microtubules, especially in developing muscle, and the 10-nm intermediate filaments. The intermediate filaments in skeletal muscle are thought to connect adjacent myofibrils together at the level of the Z line (Robson *et al.*, 1981).

3. OTHER CELL ORGANELLES

The mature skeletal muscle cell is multinucleated and can contain several thousand nuclei per cell. In mammalian muscle, the nuclei are located directly beneath the sarcolemma and have a width of 1 to 3 μm and a length of 5 to 12 μm (Adams, 1975). These small nuclei are surrounded by a nuclear envelope and the nuclei inside mature muscle cells do not divide. The primary function of skeletal muscle nuclei is to provide the DNA template for RNA synthesis.

Skeletal muscle contains lysosomes, which are small spherical particles 0.2 – 2 μm in diameter surrounded by a membrane and located in the interior of the muscle cell (Bird *et al.*, 1980). Lysosomes are thought to be involved in intracellular digestion and turnover of biochemical components. Lysosomes contain a variety of enzymes including proteases, nucleases, glycosidases, lipases, phospholipases, and phosphatases. As a generality, muscle contains far less lysosomal enzyme activity than most other tissues such as liver and spleen. There is more lysosomal enzyme activity in slow-twitch than fast-twitch muscles. Bird *et al.* (1980) have shown that the lysosomal enzymes cathepsins B and D are found inside the muscle cell.

Muscle mitochondria are found in the sarcoplasma and are aligned with the contractile fibers of skeletal muscle. These mitochondria are usually cigar-shaped and consist of two separate membrane sacs. The inner membrane has infoldings called cristae, which are studded with small protruding structures called inner membrane particles. Mitochondria contain enzymes involved in the citric acid cycle, electron transport assemblage, fatty acid oxidation, other lipid-metabolizing enzymes, some nitrogen-metabolizing enzymes, and some of the enzymes involved in urea synthesis (White *et al.*, 1978). Slow-twitch oxidative muscles contain more mitochondria than fast-twitch glycolytic muscle. There does not appear to be anything unique about muscle mitochondria except that they are located in close proximity to the myofibrils.

A unique feature of muscle is the large number of glycogen particles found in the sarcoplasm. The glycogen particle is not enclosed in a membrane and isolated particles have variable diameters, often in the range 20–30 nm (Preiss and Walsh, 1981). The glycogen particle is associated with a number of the enzymes involved in glycogen metabolism. The particle size will change as the glycogen is being synthesized or degraded to supply glucose units for energy production.

Muscle cells have a number of small Golgi bodies located near one pole of each muscle nucleus (Bloom and Fawcett, 1975). The Golgi organelle plays an important role in the secretion of proteins from many types of cells. Skeletal muscle does not secrete as many proteins as many other cell types and the Golgi apparatus of muscle cells is not a major cellular component.

C. Other Cell Types in Muscle

There are a number of different cell types in muscle including nerve cells, cells involved in the synthesis of collagen, fat cells, and the different types of cells that constitute the vascular system. There is a large variation in the number of mature fat cells in muscle. Differences in fat cell number can be attributed to species, breed, animal maturity, nutrition state, and a large number of other variables (Allen et al., 1976). Fat cells are derived from mesenchyme; when they start to accumulate lipid, they are called adipoblasts. The mature lipid-filled fat cells are called adipocytes. The mature adipocyte is a large spherical cell that can have a diameter $> 100 \mu$m; the small precursor cells have diameters of $\sim 5 \mu$m. The potential of changing the diameter of adipose tissue cells 50-fold through dietary manipulation is an indication of the plasticity of this tissue.

In skeletal muscle, fat cells are usually located next to small vessels and often associated with the perimysial connective tissue. The lipid inside the fat cell is in the form of a large droplet; other cellular components, such as the nucleus, are located next to the cell membrane. There are two major types of adipose tissue, commonly called white and brown fat. White fat is found in skeletal muscle and constitutes the bulk of the adipose tissue. Brown fat is found in some fat pads and has been associated with thermal regulation.

Connective tissue is a major component of skeletal muscle and is composed of extracellular fibers of collagen, elastin, and reticulin, ground substance mucopolysaccharides, and fibroblast cells. The protein and ground substance are synthesized by fibroblasts and form the fibrous connective tissue matrix that surrounds the muscle. During embryonic development, it is difficult to distinguish between cell types; however, the connective tissue in muscle begins to acquire a recognizable form at about the time muscle cells are formed. At this stage, the connective tissue will divide the muscle into primary and secondary groups of muscle cells. Endomysial connective tissue will extend between individual muscle fibers late in fetal development (Adams, 1975). The amount of connective tissue between the muscle fibers increases with age. There is a higher content of connective tissue in muscles used for locomotion such as those in the leg.

D. Cardiac and Smooth Muscles

Muscles can be classified as either striated or nonstriated. Striated muscles exhibit regularly spaced transverse bands along the length of the cell and can be further subdivided into skeletal and cardiac muscle. Nonstriated or smooth muscle is composed of cellular units that are not subject to voluntary control.

Smooth muscle is found in the walls of the digestive tract, ducts of glands, respiratory passages, walls of blood vessels, and many other places. Smooth muscle cells have one nucleus per cell and are long spindle-shaped cells, having lengths up to 0.5 mm (Bloom and Fawcett, 1975). The cells are often found in bundles or sheets. Contraction of smooth muscle can be initiated by nerve impulses, hormonal stimulation, or changes in the muscle cell such as stretching. The major contractile proteins in smooth muscle are myosin, actin, tropomyosin, α-actinin, and filamin; however, troponin is not present. As the name implies, nonstriated muscle lacks the filamentous organization found in cardiac and skeletal muscle.

Cardiac muscle cells are $50-120$ μm in length and have an intercalated disc that connects adjacent cardiac cells. Cardiac muscle cells contain only one or two nuclei, which are located in the cell interior. Cardiac muscle contains more cytoplasm and mitochondria than skeletal muscle. Differences between atrial and ventrical cardiac muscles include smaller cell size, fewer T tubules in atrial muscle, and different isoforms of some contractile proteins such as myosin. The intercalated discs of cardiac muscle are located on the opposing ends of cardiac muscle cells and have complex interdigitating structures that maintain cell–cell cohesion. The thin filament of the adjacent I bands terminate in the matrix of the intercalated disc, resulting in continuity from cell to cell. In the mature meat animal, the heart constitutes $<1\%$ of the live animal's weight.

II. MUSCLE DEVELOPMENT

A. Muscle Embryology

Cells forming limb muscle are derived from the mesoderm. Striated voluntary muscle is derived from paired somites (limb muscles) and from mesenchyme or the branchial arches (head and neck muscles). The portion of a somite left after emigration of the schlerotome mass to form a vertebrae is the myotome or

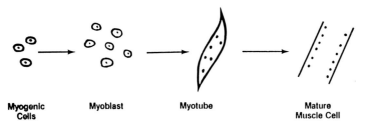

| Myogenic Cells | Myoblast | Myotube | Mature Muscle Cell |

Fig. 5   Myotube formation.

muscle plate (Arey, 1974). This plate thickens and the cells eventually differentiate into myoblasts. These spindle-shaped cells arrange themselves parallel to the long axis of the body and will form into skeletal muscle fibers.

The mesodermal somites first appear in very small embryos. In humans, by the fifth week, myotomes form muscles that rapidly develop and become capable of contraction. The muscle fibers aggregate in groups constituting individual muscles and become enclosed by connective tissue. As a generality, muscles retain their original innervation throughout life. The skeletal musculature is attached primarily to the skeleton.

B. Myotube Formation

The process of myotube development starts with the accumulation of mononucleated mitotically active cells in the areas destined to form muscle such as somites and limb rudiments. Some of these cells within the somite become spindle-shaped. The spindle-shaped cells fuse forming multinucleated myotubes. Cell fusion is followed by the assembly of myofibrils and tubular systems, mitochondrial proliferation, glycogen accumulation, and, eventually, innervation.

The formation of a myotube is shown in Fig. 5. Dividing cells are referred to as presumptive myoblasts. These cells do not fuse to form myotubes and do not synthesize large amounts of contractile proteins. Eventually, the presumptive myoblasts stop dividing and are then called myoblasts. Myoblasts are often elongated, spindle-shaped cells with one nucleus. Myoblasts fuse to form myotubes or fuse with existing myotubes. Myoblasts will not fuse with nonmuscle cells but will fuse with myoblasts from other species. There is a large volume of literature on the events that take place prior to and during the fusion process (Fischmann, 1972; Pearson and Epstein, 1982).

The myotube is a multinucleated syncitium that arises from myoblast fusion. The immature myotube contains myofibrils and the nuclei occupies the core of the cell.   As the myotube matures, the myofibrils are distributed more evenly and the nuclei are located toward the cell periphery.   The mature myotube is called a muscle fiber or muscle cell.

C. Determinants of Muscle Fiber Number

Approximately 80% of the muscle mass is from muscle cells.   When the size of a skeletal muscle increases, there is either an increase in muscle cell number, an increase in the muscle cell diameter, or both (Goldspink and Williams, 1980).   In most muscles, the number of muscle cells become fixed around the time of birth and does not change dramatically throughout the life of the animal (Gollnick et al., 1981) or may show a slight decrease in number during aging (Layman et al., 1980).

The number of muscle cells per muscle has been shown to be different in separate strains of mice (Aberle and Doolittle, 1976).   The large muscles from double-muscled cattle have an increased number of muscle cells per muscle (Ashmore et al., 1974).   These studies as well as others indicate there is a high degree of heritability for muscle cell number.   Although the maximum number of muscle cells is present at about the time of birth, it is possible to reduce the number of muscle cells at birth by severe nutritional restriction during gestation.

One potential mechanism for increasing the number of muscle cells after birth is by muscle cell splitting.   After a large increase in diameter, it may be possible for muscle cells to split (Hall-Craggs, 1970).   This phenomena does not appear to be a major factor in muscle growth of meat animals.

D. Muscle Protein/DNA Ratios

The ratio of protein to DNA from different muscle types varies (Waterlow et al., 1978; Layman et al., 1981).   There are greater than twofold differences in the ratio of protein to DNA between predominately slow-twitch and fast-twitch skeletal muscles from avian and mammalian species.   It is of interest to note that genetic selection for increased muscle mass in meat animals often results in an increased number of fast-twitch muscle fibers, which contain more protein per unit DNA than is found in slow-twitch muscle cells.

The relationship between protein and DNA ratios in strains of animals with different growth rates was examined by Millward (Waterlow et al., 1978).   Rat strains with markedly different growth rates had similar protein/DNA ratios in comparable muscles.

E. Muscle Nuclei Number

The number of muscle nuclei increases more than 20-fold from birth to adulthood (Moss, 1968). There is a high correlation between the increase in nuclei number and the increase in muscle weight. It is possible that the addition of a unit of muscle DNA will result in the addition of a unit of muscle protein in a normal animal.

Nuclei that are in a muscle cell do not divide (Stockdale and Holtzer, 1961). The increase in muscle nuclei number during muscle growth results from fusion of satellite cells with the muscle cell. Satellite cells are mononucleated cells located between the muscle cell and its surrounding basal lamina (Mauro, 1961). Satellite cells can divide, and some of the resulting cells will fuse with the muscle cells. This results in an increased number of nuclei in the muscle cell.

The cessation of muscle growth or a reduced growth rate has been correlated with a reduction in nuclear proliferation (Enesco and Puddy, 1964). The number of satellite cells per muscle cell is reduced during aging (Beerman *et al.*, 1983).

F. Muscle Fiber Type

Some muscles from the same animal have differences in color, like those between turkey breast and thigh muscles. The muscle cells from dark red muscle are usually slow-contracting, smaller, and more granular in appearance than those from white muscles. Different types of muscle cells can be distinguished by histochemical methods such as staining for the mitochondrial enzyme succinate dehydrogenase (Fig. 6) and myosin ATPase.

Initially there were two types of muscle fibers, often referred to as red and white; however, with the use of histochemical stains and the detection of different isoforms of contractile proteins, a number of fiber types intermediate between red and white have been reported. The most universally used system was developed by Ashmore and Doerr (1971) and Peters *et al.* (1972). This system segregates the fibers into three categories: slow-twitch oxidative fibers (βR), fast-twitch oxidative–glycolytic (αR) fibers, and fast-twitch glycolytic (αW) fibers. The distinctive properties associated with each fiber type are described in a later chapter.

Different muscles will contain varying percentages of the different fiber types. Some muscles such as the chicken pectoralis muscle contain predominately αW fibers while other muscles such as the rat soleus muscle are βR. The majority of muscles in large animals contain a mixture of all three fiber types. Often within an individual muscle there will be an increased percentage of βR fibers in the deep areas close to bone, and an increased percentage of αR and

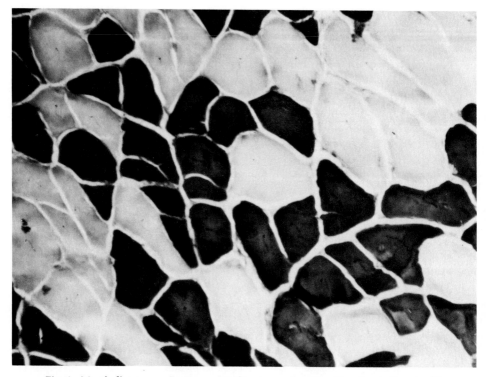

Fig. 6 Muscle fiber types. Equine semitendinosis muscle section histochemically stained for succinate dehydrogenase. Dark fibers are βR plus R types and the unstained fibers are the W type.

αW in the outer areas of the muscle. Fetal muscles from a number of species contain reduced numbers of slow-twitch fibers (Ashmore *et al.*, 1973). In the growing pig, the percentage of slow-twitch fibers steadily increased at the expense of fast-twitch fibers for about the first 8 weeks of life and then remained constant (Suzuki and Cassens, 1980).

III. FACTORS AFFECTING MUSCLE GROWTH AND DEVELOPMENT

A. Growth Factors

Endocrine factors have pronounced effects on muscle growth and development and in maintaining muscle mass. The muscle mass in mature animals can be decreased by the action of catabolic hormones or increased with some anabolic hormones.

The action of several hormones regulating protein degradation and synthesis

Table I
Effects of Hormones on Skeletal Muscle[a]

	Protein synthesis	Protein degradation
Insulin	↑	↓
Growth hormone	↑	No change
Thyroid hormone	↑	↑
Glucocorticoids (fasted state)	↓	↑

[a] Data from Goldberg et al. (1980).

in skeletal muscle are shown in Table I. Insulin is an important hormone for regulating protein turnover in skeletal muscle (Goldberg et al., 1980). This hormone increases the net uptake of amino acids, increases protein synthesis, and inhibits protein degradation in skeletal muscle. The thyroid hormones can have both positive and negative effects depending on the level present. A minimum level of a thyroid hormone is essential for normal muscle growth, but an excess can result in muscle wasting. During this severe muscle wasting, the rate of protein breakdown is increased, possibly through a mechanism involving lysosomal changes. The glucocorticoid hormones have a marked catabolic effect on muscle size. The ability of glucocorticoids to promote the breakdown of muscle protein is thought to be an important factor in the regulation of blood glucose levels during fasting.

The growth factors that effect myogenic cells have been the subject of a review by Allen (1978). Guerriero and Florini (1978) examined the effect of glucocorticoids on myogenic cell proliferation and found that some glucocorticoids would stimulate proliferation while other closely related steroids had no effect. Testosterone does not enhance myogenic cell proliferation (Gospodarowicz et al., 1976). Recently, attention has focused on protein factors that stimulate myogenic cell proliferation. At physiological concentration, growth hormone does not stimulate myogenic cell proliferation but the small protein growth factor called multiplication-stimulating activity (MSA) does stimulate proliferation (Florini et al., 1977). MSA is in a class with other small polypeptides known as insulinlike growth factors, which could be involved in myogenic cell proliferation.

B. Muscle Stretching

Possibly the most important physiological mechanism for increasing the DNA content in growing muscle is stretching of the muscle over a prolonged period of time, such as would happen with bone growth. Different models have been used to investigate the mechanisms of stretch-induced muscle growth including use of weights attached to an adult chicken wing, which cause the wing to droop and stretch several muscles (Laurent and Sparrow, 1977). Barnett *et al.* (1980) developed a method for stretching a wing skeletal muscle by use of a spring clip. Muscles that are mechanically stretched increase in weight, protein, and DNA content. The protein/DNA ratio in the muscle after stretching was similar to that of unstretched muscle. On a cellular level, the growth mechanisms are poorly understood.

C. Muscle Regeneration

The mechanisms that underlie skeletal muscle regeneration have been extensively investigated (Mauro, 1979). This section will deal only with muscle regeneration in mammalian muscle, regeneration that is different in a number of major respects from amphibian limb regeneration (Carlson, 1979a,b).

The course of events involved in mammalian regeneration are shown in Table II. On day 0, the injury occurs and within 24–48 hours the damaged muscle in areas close to blood vessels is slowly degraded by necrotic processes. During this time, large oval nuclei appear between the basement membrane and degenerating fiber. After 2 to 3 days, myotubes begin to form beneath basement membranes. The formation of myotubes and their enlargement continues throughout the early phases of the regeneration process.

Regeneration events requiring more time include enlargement of muscle fibers and the formation, when necessary, of new basement membrane. After ~3 weeks, the formation of motor end plates has been observed (Zhenevskaya, 1962) and by 30 to 40 days, different fiber types are present. The presence of a functional nerve has a stimulatory effect on muscle regeneration (Peterson and Crain, 1972) but early regeneration events do not require nerves (Mong, 1977).

If the regenerated muscle is to regain its function, the newly formed fibers must have a similar orientation to the original muscle. If the basal membrane

Table II
Events Occurring during Skeletal Muscle Regeneration

Day 0: Injury
Days 1–2:
Infiltration of muscle cell by phagocytic cells
Appearance of cells with large oval nuclei
Days 2–3:
Maximum number of myoblastic cells
Start formation of myotubes
Days 3–5: Formation of myofilaments in myotubes
Day ~14: Formation of new basement membrane
Day ~21: Formation of motor end plate
Days 30–40: Differentiation of muscle fiber type

has been severely disrupted, the regenerated muscle fibers grow in a chaotic manner. If tension is applied to the muscle, the regenerating tissue tends to become more organized. The origin of the myogenic cells responsible for the regenerative process is thought to be from dormant presumptive myoblasts activated by the injury (Snow, 1977). There are many common features between normal growth and development of skeletal muscle and the regeneration process.

D. Nutritional Status

The nutritional status of an animal can greatly effect skeletal muscle growth. The DNA content of muscle in growing rats is sensitive to both energy restriction and protein restriction (Trenkle, 1974). The status of the protein/DNA ratio in skeletal muscle during "catch-up" growth following diet restriction is complicated due to the large number of variables, which include the degree and type of restriction, duration of restriction, and age of the animal during restriction. If an animal is allowed to catch up after a nutritional restriction, there is little evidence that it will have a greater protein to DNA ratio than the restricted control at maturity. Millward *et al.* (1975) reported that muscles from animals that had their growth rate altered by early malnourishment had the same ratio of protein to DNA as muscles from well-fed controls. Many nutrition studies indicate that protein/DNA ratios can be reduced during restriction, but few indicate that protein/DNA ratios greater than normal can be obtained through dietary manipulations.

E. Environmental and Genetic Factors

Environmental factors affecting animal growth, including temperature, humidity, light intensity, and many others, have been extensively reviewed by Curtis (1981). Most environmental factors that influence growth and development of whole animals generally have the same effect on the skeletal muscle component. An important environmental factor is temperature, and it is common knowledge that different breeds of meat animals thrive better in different climates. When compared to other body organs and tissues, muscle is not uniquely effected by environmental factors; however, muscle will respond to environmental insults more slowly than many other tissues.

Genetic factors that alter muscle characteristics are numerous. Development of the large turkey breast muscle is an excellent example of how muscle size can be altered through genetic selection. Different breeds of animals often have different muscle/fat ratios and the cross-sectional area of muscle is a highly heritable trait. Large differences in the number of skeletal muscle cells in a muscle can be found between breeds; for example, the number of cells in the muscle of a Welsh pony is less than the number of cells in the same muscle of a Belgian work horse. The percentage of the different fiber types within a muscle is also under some degree of genetic control and Gunn (1978) has shown that in dogs and horses breed differences exist in the distribution of fiber type.

F. Abnormalities of Skeletal Muscle

There are a number of factors causing abnormal growth and development of muscle (Bourne, 1974). A number of these abnormal growth and development patterns are due to genetic problems such as dwarfism, some muscular dystrophies, and glycogen storage diseases. As a generality, the state of nutrition can influence the size of muscle tissue; however, the lack of specific nutrients will rarely show a specific effect on muscle. One exception is Vitamin E or selenium deficiency in cattle, sheep, and pigs, which results in a dystrophic muscle that is superficially white in appearance.

An interesting abnormality in the muscles from meat animals is the double-muscling of cattle. The large bulging muscles of double-muscled cattle are the result of a genetic abnormality that results in more muscle tissue (although double-muscled cattle have a normal number of muscles). The larger muscles have more muscle cells and there is an increased percentage of fast-twitch glycolytic cells. Double-muscled animals tend to have fertility and calving problems.

IV. MUSCLE PROTEINS

A. Thick Filament Proteins

1. MYOSIN

The myosin molecule from mammalian skeletal muscle is ~ 150 nm in length and consists of a long α-helical rod-shaped region that has two globular regions attached to one end (Lowey et al., 1969). The two globular regions are referred to as the myosin heads. Myosin heads contain both actin and ATP binding sites. The rod regions of myosin molecules bind to one another resulting in the formation of the backbone of thick filaments.

Skeletal muscle myosin has a molecular weight of $\sim 480,000$ and is composed of two large subunits called heavy chains and four small subunits called light chains (Fig. 7). Each heavy chain has a molecular weight of $\sim 200,000$. The two heavy chains form the rod portion and a large part of the myosin head. The light chains have molecular weights that range from approximately 16,000 to 27,500 depending on the type of subunit and source of the myosin (Table III).

Heavy chains   Light chains

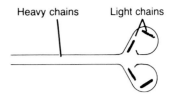

Fig. 7 Myosin subunit organization. There are two large subunits (heavy chains) and four small subunits (light chains).

Two light chains are located in each of the myosin heads. There are two types of myosin light chains, often referred to as the 5,5′-dithiobis-nitrobenzoic acid (DTNB) (LC2) and alkali light chains (LC1 and LC3) (Mannherz and Goody, 1976). The DTNB light chains can be removed from myosin with 5,5′-dithiobis-nitrobenzoic acid without the loss of ATPase activity. These light chains can undergo phosphorylation by light chain kinase (Bárány and Bárány, 1980; Adelstein and Eisenberg, 1980). The alkali subunits can be removed from myosin but this usually results in loss of ATPase activity; however, it has been reported that the alkali subunits can be removed without loss of ATPase activity (Wagner and Ginger, 1981). Each myosin head contains one DTNB subunit and one alkali subunit.

Different types of myosin are found in skeletal muscle cells having the βR and αW fiber type. Histochemical assays for myosin ATPase have shown that these fiber types stain differently (Guth and Samaha, 1969). Myosin ATPase activity has been correlated with maximal shortening speed (Bárány, 1967). Myosin heavy chains from βR and W fiber types are different and although complete comparative primary sequence studies have not been completed, differences have been reported in the 3-methylhistidine content (Huszar, 1972). Also, heavy chain differences have been documented by immunochemical criteria (Arndt and Pepe, 1975), peptide mapping, and sodium dodecyl sulfate gel electrophoresis (Rushbrook and Stracher, 1979).

Myosin light chains from fast- and slow-twitch muscles have been extensively

Table III
Myosin Subunit Molecular Weights Determined on Sodium Dodecyl Sulfate–Polyacrylamide Gels[a]

	Fast-twitch myosin	Slow-twitch myosin
Heavy chains	200,000	200,000
Light chains		
LC1	25,000	27,500
(alkali 1)		26,500
LC2 (DTNB)	18,000	19,000
LC3 (alkali 2)	16,000	—

[a] Data from Weeds (1980).

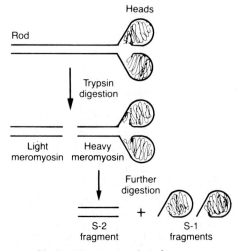

Fig. 8   Myosin proteolytic fragments.

studied (Weeds, 1980).   There are three light chains found in most myosins from mature fast-twitch muscles.   These light chains have been extensively studied and the complete amino acid sequences determined.   The fast-twitch LC1 and LC3 light chains are similar and have a common 141-amino acid sequence at the C-terminus; however, LC1 is larger and has an additional 41 residues at the N-terminus.   Chemically related alkali chains have been identified from cardiac and slow muscle.   Myosin from fast-twitch muscle consists of two myosin heavy chains, two LC2 myosin light chains and LC1 and LC3 light chains.   The fast-twitch myosin can contain either two LC1 or two LC3 or one of each.

   Myosin from embryonic skeletal muscles contains unique heavy chains and also a unique light chain (Whalen, 1980).   During skeletal muscle development the embryonic form of myosin is replaced by the fast- or slow-twitch form of myosin found in the adult skeletal muscle.   The study of the expression of different myosin genes during development of skeletal muscle is a research area that is very active.   Major questions remain with regard to the number of myosin isoforms expressed in muscles, the number of myosin genes, and the mechanisms that regulate expression of the different myosin genes.

   A number of useful myosin proteolytic fragments can easily be made.   Because myosin is a large salt-soluble protein that has a tendency to aggregate, it is difficult to use for certain types of studies.   One method of circumventing these problems has been to make proteolytic fragments of myosin with trypsin, papain, or chymotrypsin (Lowey et al., 1969; Weeds and Pope, 1977).   When myosin is digested with trypsin, two major fragments are formed.   The largest fragment is called heavy meromyosin; it contains both of the myosin heads and a portion of the myosin rod (Fig. 8).   Heavy meromyosin has a molecular weight of ~350,000, is soluble at low ionic strength, and contains both the actin binding and the ATPase activity. The other major fragment from the trypsin digestion is called light meromyosin; it has a molecular weight of ~150,000, is not soluble at low ionic strength, and comes from the α-helical rod portion of the myosin molecule. Using the appropriate conditions and proteolytic enzyme, heavy meromyosin can be further digested to form the S-1 and S-2 fragments. The S-1 fragments have a molecular weight of ~115,000 and are the individual myosin heads that contain the actin binding and ATPase sites. The S-2 fragment is a small portion of the myosin rod but is soluble at low ionic strength. Papain digestion of myosin can be used to produce the entire myosin rod region plus S-1 fragments.

2. C PROTEIN

C protein is part of the thick filaments of both skeletal and cardiac muscle. This protein has a molecular weight of ~140,000 by SDS gel electrophoresis and binds to myosin at low ionic strengths (Moos et al., 1975). C protein forms a series of seven transverse stripes on each side of the thick filament. The stripes are 7 nm in width and are spaced at 43-nm intervals. The function of C protein has not been determined although a number of theories have been proposed. Isoforms of C protein have been identified in fast- and slow-twitch muscles (Callaway and Bechtel, 1981).

3. M-LINE PROTEINS

The M line of skeletal muscle is where the enzyme creatine kinase (Turner et al., 1973) and M-line protein (Mani and Kay, 1978) are located. Creatine kinase is a dimer with a subunit molecular weight of 42,000. Creatine kinase could have two functional roles, one as a structural component of the M line and another as the enzyme involved in the reaction that forms ATP and creatine from creatine phosphate and ADP. The other protein associated with the M line has a molecular weight of 165,000 by SDS gel electrophoresis. This high molecular weight protein binds to both myosin and creatine kinase; however, its function is not fully understood. The M line in skeletal muscle is thought to join together the thick filament of the myofibril.

4. THICK FILAMENT STRUCTURE

Thick filaments from mammalian skeletal muscle have a length of 1.5 μm and are shaped like a cigar. The diameter of the thick filament is ~14 nm. The M line is located in the middle of the filament and is surrounded on either side by the H zone (Fig. 9). The myosin rod forms the backbone of the thick filament and the myosin heads protrude from the filaments. Myosin molecules are packed in the thick filament such that there are ~12 rods forming any given

Myosin head arrangement

H-zone structure

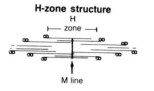

Fig. 9 Structure of the thick filaments. Upper panel depicts the protrusion of the myosin heads from the thick filament. Lower panel depicts the organization of myosin molecules in the H zone.

section of the thick filament. The myosin heads project from the filament with an axial spacing of 14.3 nm and have a threefold rotational symmetry (Squire, 1975).

B. Thin Filament Proteins

1. ACTIN

Actin is the main constituent of the thin filament. This protein has a molecular weight of ~ 42,000; the complete 376-amino acid sequence of rabbit skeletal muscle actin has been determined (Elzinga *et al.*, 1973). An unusual chemical feature of actin is that it contains a single 3-methylhistidine at position 73. Like that of most contractile proteins, the N-terminus of actin is acetylated. One molecule of actin has a diameter of ~ 5.5 nm. Actins from a number of different tissues share a great deal of sequence homology (Vandekerckhove and Weber, 1979). Using isoelectric focusing techniques on polyacrylamide gel, Zechel and Weber (1978) found three major actin isoforms, which have been labeled α, β, and γ. The α form is the most abundant form in skeletal muscle and the β and γ isoforms are found in smooth muscle and other tissues.

When actin is in its monomer form (42,000), it is called G actin. Actin molecules polymerize together, forming actin filaments referred to as F actin. Actin binds ATP and the following scheme for ATP hydrolysis during polymerization has been demonstrated, although a complete understanding of this phenomenon is not available.

$$n(\text{G-Actin-ATP}) \rightarrow \text{F-actin-ADP} + n\text{PO}_4$$

The F actin forms the backbone of the thin filament and also provides binding sites for both tropomyosin and troponin. One end of the F-actin filament is bound to the Z line. In muscle, when calcium is present, F actin comes into contact with the myosin heads of the thick filaments and there is a rapid breakdown of ATP, ultimately resulting in muscle contraction.

2. TROPOMYOSIN

Tropomyosin is a rod-shaped protein that is ~400Å in length and nearly 100% α helix. The protein is composed of two subunits, each with a molecular weight of 34,000. The two subunits twist around one another to form a coiled structure (Cohen and Holmes, 1963). The complete amino acid sequence of tropomyosin has been determined (Stone and Smillie, 1978) and the protein can be easily crystallized. From the sequence and X-ray diffraction studies, the actin and troponin binding sites of tropomyosin have been determined.

There are two tropomyosin subunit isoforms found in most skeletal muscles, which are referred to as α and β chain. The two isoforms are similar except that the α chain contains one cysteine and the β chain, two. Skeletal muscle tropomyosin is formed from either α-α or α-β chains (Lehrer, 1975).

Tropomyosin is involved in regulating skeletal muscle contraction in conjunction with troponin. When calcium concentrations are elevated in muscle, tropomyosin may move such that F actin can bind to myosin, resulting in muscle contraction.

3. TROPONIN

Troponin is a thin filament protein found in skeletal and cardiac muscle. The protein has a molecular weight of 76,000 and is composed of three nonidentical subunits (TnC, TnT, and TnI). Troponin is the contractile protein that binds calcium and in concert with tropomyosin regulates the interaction of actin and myosin (Mannherz and Goody, 1976).

The calcium-binding subunit of troponin is called TnC and has a molecular weight of 18,000. This subunit has been crystallized (Mercola *et al.*, 1975) and the complete amino acid sequence has been determined. Several models for binding of metal ions to TnC have been proposed. One model has a total of six metal ion–binding sites, which include two magnesium ion–binding sites, two low-affinity calcium ion–binding sites and two high-affinity binding sites for which both magnesium and calcium ions compete. Both the magnesium ion and the TnI subunit affect the affinity of calcium for TnC. Because of the high millimolar concentrations of magnesium in muscle, it is probable that the two low-affinity calcium binding sites are the calcium regulatory sites. The two low-affinity calcium binding sites bind calcium in the physiological 1- to 10-μM range. TnC does not bind actin or tropomyosin but binds the TnT and TnI subunits.

The troponin inhibitory subunit (TnI) has a molecular weight of ~21,000 and is insoluble at low ionic strength. The amino acid sequence of TnT from different striated muscles has been determined (Wilkinson and Grand, 1978). TnI with the other troponin subunits and tropomyosin is involved in the inhibition of the actin–myosin interaction. High concentrations of TnI can inhibit

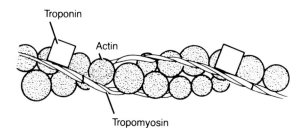

Fig. 10 Structure of a segment of the thin filament. The major proteins include troponin (TnT, TnI, TnC), tropomyosin (dimer) and actin. One end of the thin filament is bound to the Z line.

the interaction of actin with myosin. Lower concentrations of TnI plus tropomyosin result in complete inhibition of the actin–myosin interaction. TnI binds to TnC in the presence of calcium, and the TnI–TnC complex is soluble at low ionic strength.

The troponin subunit that binds tropomyosin is called TnT. This is the largest troponin subunit and has a molecular weight that ranges from 37,000 to 45,000 depending on the source of the skeletal muscle (Greaser et al., 1972). This subunit has an isoelectric point of ~8.8 and is insoluble at low ionic strength. TnT binds to tropomyosin and actin–tropomyosin complex. The function of TnT is to connect the TnI–TnC complex to the thin filament.

The mechanism of calcium regulation of muscle contraction by troponin is not totally understood; however, it is thought that when calcium binds to TnC it results in stronger interactions between the subunits and weakens the TnI–actin interaction. This results in movement of the tropomyosin, which then allows the actin–myosin interaction to occur.

4. THIN FILAMENT STRUCTURE

The thin filaments are ~1 μm in length and 7–8 nm in diameter (Huxley, 1972). The three most abundant proteins are actin, tropomyosin, and troponin, which are present in a 7:1:1 molar ratio (Murray and Weber, 1974; Potter, 1976). The F-actin helix has a repeat of 37 nm. This helix appears as two chains of actin that are wound around one another (Fig. 10). The long rod-shaped tropomyosin molecules are located in the grooves between the two actin strands and are arranged in the grooves such that there is a very small amount of overlap from the end of one tropomyosin to the start of the next. Troponin is attached to both actin and tropomyosin and is bound to a region of the tropomyosin molecule approximately one-third of the distance from the C-terminus (Squire, 1975).

C. Z line and α-Actinin

The Z line is a very prominent structure and the distance from Z line to Z line is the sarcomere. The width of the Z line is about 50 nm in fast-twitch muscle and 65 nm in a slow-twitch muscle. The Z-line provides the framework for connecting the thin filaments of the myofibril and maintains the correct spacing between these filaments (Knappeis and Carlsen, 1962). The protein α-actinin is located in the Z line. This protein has a molecular weight of 206,000 and is composed of two 100,000 molecular weight subunits. The protein has an α-helix content of ~74% and has a 40- to 50-nm length and 4-nm diameter (Suzuki et al., 1976). The function of α-actinin is not fully understood but it could be involved in anchoring the thin filaments into the Z-line.

D. Interaction of Myosin and Actin

There are a number of complex models that have been developed to explain the interaction of myosin, actin, and ATP (Sleep and Smith, 1981; Adelstein and Eisenberg, 1980). A simplified model derived from Lymn and Taylor (1971) is shown below; A is actin and

$$
\begin{array}{ccc}
& \text{ADP} + \text{P}_i & \\
\text{A} \cdot \text{M} \cdot \text{ADP} \cdot \text{P}_i \xrightarrow{\hspace{2cm}} & \text{A} \cdot \text{M} \\
\uparrow & & \downarrow \hspace{0.2cm} \text{ATP} \\
\text{A} + \text{M} \cdot \text{ADP} \cdot \text{P}_i \xleftarrow{\hspace{2cm}} & \text{A} + \text{M} \cdot \text{ATP}
\end{array}
$$

M is myosin. The initial step is the binding of ATP to the actin—myosin complex, which results in dissociation of actin from myosin. The second step is cleavage of the myosin bound ATP to form myosin ADP·Pi. Then the actin·myosin·ADP·P_i complex is formed and finally, the ADP and P_i are released. If there is no ATP to initiate the next cycle, the actin·myosin will

Table IV
Concentrations of High-Energy Phosphate
Compounds in Skeletal Muscle

Compound	μmol/g tissue
ATP	5–8.5
ADP	~1
AMP	0.2–0.3
Creatine phosphate	13–23

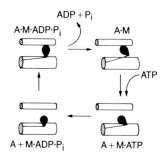

Fig. 11   Model of thick and thin filament interaction during a contraction cycle. A is actin, located in the smaller filament, and M is myosin, located in the thicker filament. Note the angle change of the myosin head during the cycle.

remain complexed.   After an animal is slaughtered and the muscle ATP is consumed, actin – myosin complex results in rigor. The physiological substrate of the myosin ATPase reaction is magnesium-ATP but the magnesium ion is not shown to simplify the nomenclature.

The energy for the force-generating mechanism in muscle is derived from cleavage of the high-energy phosphate bond of ATP. The concentrations of high-energy phosphate compounds in skeletal muscle are listed in Table IV. The ATPase is located in the myosin head and there is evidence that myosin heads can bind to the thin filament at two different angles (Goody and Holmes, 1983). The movement of the myosin heads during contraction are depicted in Fig. 11. In this model, the cleavage of ATP to $ADP \cdot P_i$ results in the cocking of the myosin head; once the cocked head binds actin, the energy is released, resulting in movement of the thin filaments. The end result is that the Z lines come closer together and the sarcomere is contracted.

As discussed previously, contraction in vertebrate skeletal muscle is regulated by the troponin – tropomyosin complex on the thin filament. When calcium is absent, tropomyosin blocks actin from binding with the cocked myosin head. When calcium is bound to troponin, this results in movement of tropomyosin such that actin can bind the myosin head and contraction occurs.

E. Contractile Protein Stoichiometry

Stoichiometry of the major contractile proteins from rabbit skeletal muscle myofibrils has been determined using SDS gel electrophoresis (Potter, 1976).

Table V
Molar Ratios of Major Contractile Proteins[a]

Myosin	1
Actin	7
Tropomyosin	1
Troponin	1

[a] Data from Potter (1976).

Table V shows the molar ratios of myosin, actin, troponin, and tropomyosin. The number of myosin molecules per thick filament was calculated to be 254.

The protein content of the myofibril has been determined (Table VI). The most abundant myofibrillar protein on a weight basis is myosin, followed by actin. The values given for the novel proteins titin and nebulin need further verification (Table VI).

F. Other Myofibrillar Proteins

A number of other proteins have been purified from skeletal muscle myofibrils (Greaser *et al.*, 1981). The most interesting are the large molecular weight proteins called titin or connectin (Wang *et al.*, 1979; Maruyama *et al.*, 1977b). Titin proteins have molecular weights (determined by SDS gel electrophoresis) of ~700,000 and 1,000,000. The proteins are thought to be located throughout the sarcomere and are present in high concentrations at the A–I junction. The proteins may form very thin filaments in muscle that have previously been described as gap filaments (Locker and Leet, 1976). These proteins are thought to be elastic in nature and could account for some of the elastic properties of skeletal muscle.

Another high molecular weight protein called nebulin has a molecular weight of ~600,000 (Wang and Williamson, 1980). This protein is located at the N_2 line and may constitute 5% of the total myofibrillar protein.

A number of other myofibril proteins have been isolated but are present in small amounts (Greaser *et al.*, 1981). The protein β-actinin is thought to bind the free end of F actin and could regulate the length of thin filaments. Filamin, a high molecular weight F-actin binding protein, has been identified in skeletal muscle; however, its function in this tissue is not understood (Bechtel, 1979).

Table VI
Myofibril Protein Content

	Percentage (W/W)	Reference
Thick filament		
Myosin	43	Yates and Greaser (1983)
C Protein	2	Offer *et al.* (1973)
M Protein	2	Trinick and Lowey (1977)
Thin filament		
Actin	22	Yates and Greaser (1983)
Troponin	5	Yates and Greaser (1983)
Tropomyosin	5	Bailey (1948)
Z Line		
α-Actinin	2	Suzuki *et al.* (1976)
Others		
Titin	10	Wang (1982)
Nebulin	5	Wang (1982)

Table VII
Major Proteins of the Sarcoplasm[a]

Protein	Concentration (mg protein/g tissue)
Glyceraldehyde phosphate dehydrogenase	12
Aldolase	6
Enolase	5
Creatine kinase	5
Lactate dehydrogenase	4
Pyruvate kinase	3
Phosphorylase	2.5
Myoglobin	0.1–20

[a] Data from Scopes (1970).

Skeletal muscle contains 10-nm intermediate filaments that are composed of desmin and other intermediate filament proteins. Desmin has a molecular weight of 55,000 and has been purified from skeletal muscle. Intermediate filaments are thought to be located around the periphery of Z lines and to provide a connecting link between Z lines of adjacent myofibrils (Robson *et al.*, 1981).

G. Sarcoplasmic Proteins

The concentration of sarcoplasmic proteins in skeletal muscle is ~ 55 mg/ml (Scopes, 1970). Some of the most abundant proteins found in the sarcoplasm are listed in Table VII. Most are enzymes involved in intermediate metabolism. Concentrations of these enzymes will vary with animal species and muscle fiber type.

The protein myoglobin is found only in skeletal, cardiac, and smooth muscles. The function of myoglobin is to store and transport oxygen in the muscle cell. The protein is soluble at low ionic strength and can contribute up to 95% of the red pigment of muscle. The content of myoglobin in skeletal muscle will vary depending on the metabolic profile of the muscle, animal species, and age of the animal. As a generality, the slow-twitch muscles contain more myoglobin than fast-twitch muscles. The pectoralis muscle from chicken contains myoglobin concentrations < 1 mg per gram of muscle, and muscles from older cattle can have myoglobin concentrations in the range 10–20 mg/g muscle. In many animal species, the concentration of myoglobin in skeletal muscle increases as the animal grows older.

The physical and chemical properties of myoglobin have been studied extensively. The sequence of its 153 amino acids is known and its three-dimensional structure has been determined. The protein has a molecular weight of

~17,500. Each myoglobin molecule contains a heme group with an iron atom. Depending on the status of the heme moiety, the protein will have different spectral properties and the effects of pH, salt concentration, and common myoglobin reaction products have been well documented.

H. Muscle Glycogen

High concentrations of glycogen are found in skeletal muscle and liver. The content of glycogen in skeletal muscle is often in the range 0.3 – 1% depending on muscle type, nutritional status, and muscle activity. Glycogen is a branched polysaccharide made from α-D-glucose units that are linked through α-1,6-glucosidic and α-1,4-glucosidic bonds (White *et al.*, 1978).

Glycogen polymers can have molecular weights in the millions. Glycogen is formed enzymatically and the glycogen content of muscle can increase after an animal has been well fed. The function of muscle glycogen is to provide glucose units for energy production. Because glycogen is stored inside the muscle cell, it can rapidly be broken down to provide substrate quickly for ATP production.

Glycogen is slowly synthesized in muscle and the key regulator enzyme is glycogen synthase, which forms the α-1,4-glucosidic linkage. However, glycogen can be broken down very rapidly and the key regulatory enzyme is phosphorylase, which cleaves the α-1,4-glucosidic linkage as shown below.

$$\text{Glycogen}_n + P_i \xrightarrow{\text{phosphorylase}} \text{glycogen}_{n-1} + \text{glucose-1-P}$$

Skeletal muscle phosphorylase has been extensively studied (Preiss and Walsh, 1981) and the complete amino acid sequence and the three-dimensional structure determined. The active form of the enzyme has a molecular weight of ~400,000 and is composed of four identical subunits, each having a molecular weight of ~96,500. Each subunit contains one bound pyridoxal phosphate (Vitamin B-6) molecule. The enzyme is highly regulated and can be found in an active (phosphorylase A) or inactive form (phosphorylase B). It is possible to convert phosphorylase A to phosphorylase B as shown below.

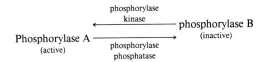

In response to stress, the hormone epinephrine is released. One effect of epinephrine on skeletal muscle is to initiate a cascade of enzymatic reactions that eventually activates phosphorylase and results in the breakdown of glycogen. Phosphorylase is also regulated by AMP, which activates phosphorylase

B, and glucose-1-phosphate, which exhibits end-product inhibition. High calcium concentrations activate phosphorylase kinase, which then activates phosphorylase. The same concentrations of calcium that initiate muscle contraction will activate phosphorylase kinase.

I. Collagen

Collagen is the most abundant protein in higher animals and can constitute up to one-third of the total body protein in mature animals. Tissues having large amounts of collagen include bone, cartilage, tendon, and skin. The collagen molecule has a length of ~300 nm and a diameter of 1.5 nm and is formed from three subunits that are coiled together (Ramachandran and Reddi, 1976). There are several distinct regions of the collagen molecule. The first 2–3% of the molecule from both the C- and N-termini does not contain the same helical structure as the rest of the molecule but these regions do contain cross-linking sites. The central region of the molecule (95%) contains the characteristic minor and major helical structures associated with collagen.

There are at least four major types of collagen; these can be distinguished by their subunit composition (White et al., 1978). Bone and tendon are predominately type I and muscle type III collagen. Collagen has a distinctive amino acid composition and contains approximately 33% glycine and 23% proline plus hydroxyproline. Collagen contains the unusual amino acids 3- and 4-hydroxyproline and 5-hydroxylysine.

Collagen molecules form collagen fibrils, which are cylindrical structures with diameters ranging from 5 to 200 nm. The collagen molecules in the fibril are formed by end-to-end and side-to-side joining and parallel collagen molecules are in a quarter stagger. This stagger results in the 64-nm repeat spacing seen with the electron microscope.

Collagen can be synthesized by a number of cell types including fibroblasts. Collagen is synthesized in a procollagen form that is much larger, having peptide extensions on both the C- and N-termini. The proline and lysine hydroxylation reactions occur after the procollagen subunits are synthesized but before helix formation. After helix formation, the protein is glycosylated and then secreted from the cell by Golgi vesicles. The procollagen molecule then undergoes both N- and C-terminus cleavage, forming collagen (White et al., 1978). During the aging process, collagen molecules in fibrils and fibers become covalently cross-linked. The cross-links are initiated by the enzyme lysyl oxidase, which converts lysine to allysine and hydroxyallysine. These products then form stable shift-base and aldol cross-links.

Collagen is degraded by collagenases. These enzymes cleave collagen into several large proteolytic fragments, which are then further broken down by other proteolytic enzymes. When heated, collagen will undergo a sharp transition over a small temperature range (Aberle and Mills, 1983). The transition is associated with the melting or destabilization of the triple chain structure. If the collagen is then cooled, some renaturation will take place.

REFERENCES

Aberle, E. D., and Doolittle, D. P. (1976). Skeletal muscle cellularity in mice selected for large body size and in controls. *Growth* **40**, 133.

Aberle, E. D., and Mills, E. W. (1983). Recent advances in collagen biochemistry. *Proc. Annu. Reciprocal Meat Conf.* **36**, 125.

Adams, R. D. (1975). "Diseases of Muscle; A Study in Pathology," 3rd ed. Harper & Row, Hagerstown, Maryland.

Adelstein, R. S., and Eisenberg, E. (1980). Regulation and kinetics of the actin–myosin–ATP interaction. *Annu. Rev. Biochem.* **49**, 921.

Allen, C. E., Beitz, D. C., Cramer, D. A., and Kauffman, R. G. (1976). "Biology of Fat in Meat Animals," North Cent. Reg. Res. Publ. No. 234. College of Agriculture and Life Sciences, University of Wisconsin, Madison.

Allen, R. E. (1978). Endocrinology of myogenic cells. *Proc. Annu. Reciprocal Meat Conf.* **32**, 99.

Arey, L. B. (1974). "Developmental Anatomy." Saunders, Philadelphia, Pennsylvania.

Arndt, I., and Pepe, F. A. (1975). Antigenic specificity of red and white muscle myosin. *J. Histochem. Cytochem.* **23**, 159.

Ashmore, C. R., and Doerr, L. (1971). Postnatal development of fiber types in normal and dystrophic skeletal muscle of the chick. *Exp. Neurol.* **30**, 431.

Ashmore, C. R., Addis, P. B., and Doerr, L. (1973). Development of muscle fibers in fetal pig. *J. Anim. Sci.* **36**, 1088.

Ashmore, C. R., Parker, W., Stokes, H., and Doerr, L. (1974). Comparative aspects of muscle fiber types in fetuses of the normal and "double-muscles" cattle. *Growth* **38**, 501.

Bailey, K. (1948). Tropomyosin: A new asymmetric protein component of the muscle fibril. *Biochem. J.* **43**, 271.

Bárány, M. (1967). ATPase activity of myosin correlated with speed of muscle shortening. *J. Gen. Physiol.* **50**, Suppl., 197.

Bárány, M., and Bárány, K. (1980). Phosphorylation of the myofibrillar proteins. *Annu. Rev. Physiol.* **42**, 275.

Barnett, J. G., Holley, R. G., and Ashmore, C. R. (1980). Stretch induced growth in chicken wing muscles: Biochemical and morphological characterization. *Am. J. Physiol.* **239**, C39.

Bechtel, P. J. (1979). Identification of a high molecular weight actin-binding protein in skeletal muscle. *J. Biol. Chem.* **254**, 1755.

Beerman, D. H., Hood, L. F., and Lipoff, M. (1983). Satellite cells and myonuclei populations in rat soleus and extensor digitorum longus muscles after maternal nutritional deprivation and realimentation. *J. Anim. Sci.* **57**, 1618.

Bird, J. W. C., Carter, J. H., Triemer, R. E., Brooks, R. M., and Spanier, A. M. (1980). Proteinases in cardiac and skeletal muscle. *Fed. Proc., Fed. Am. Soc. Exp. Biol.* **39**, 20.

Bloom, W., and Fawcett, D. W. (1975). "A Text of Histology," 10th ed. Saunders, Philadelphia, Pennsylvania.

Bourne, G. H., (1974). "The Structure and Function of Muscle," 2nd ed., Vol. 4. Academic Press, New York.

Callaway, J. E., and Bechtel, P. J. (1981). C-protein from rabbit soleus (red) muscle. *Biochem. J.* **195**, 463.

Carlson, B. M. (1979a). The regeneration of skeletal muscle — A review. *Am. J. Anat.* **137**, 119.

Carlson, B. M. (1979b). Relationship between tissue and epimorphic regeneration of skeletal muscle. *In* "Muscle Regeneration" (A. Mauro, ed.), p. 57. Raven Press, New York.

Cohen, C., and Holmes, K. C. (1963). X-ray diffraction evidence for α-helical coiled-coils in native muscle. *J. Mol. Biol.* **6**, 423.

Curtis, S. E. (1981). "Environmental Management in Animal Agriculture." Animal Environmental Services, Mahomet, Illinois.

Elzinga, M., Collins, J. H., Kuehl, W. M., and Adelstein, R. S. (1973). Complete amino-acid sequence of actin of rabbit skeletal muscle. *Proc. Natl. Acad. Sci. U.S.A.* **70**, 2687.

Enesco, M., and Puddy, D. (1964). Increase in the number of nuclei and weight in skeletal muscle of rats at various ages. *Am. J. Anat.* **114**, 235.

Fischmann, D. A. (1972). Development of striated muscle. *In* "The Structure and Function of Muscle" (G. H. Bourne, ed.), 2nd ed., Vol. 1, p. 75. Academic Press, New York.

Florini, J. R., Nicholson, M. L., and Dulak, N. C. (1977). Effects of peptide anabolic hormones of growth of myoblasts in culture. *Endocrinology (Baltimore)* **101**, 32.

Goldberg, A. L., Tischler, M., DeMartin, G., and Griffin, G. (1980). Hormonal regulation of protein degradation and synthesis in skeletal muscle. *Fed. Proc., Fed. Am. Soc. Exp. Biol.* **39**, 31.

Goldspink, G., and Williams, P. E. (1980). Development and growth of muscle. *Adv. Physiol. Sci.* **24**, 87.

Gollnick, P. D., Timson, B. F., Moore, R. L., and Riedy, M. (1981). Muscular enlargement and number of fibers in skeletal muscles of rats. *J. Appl. Physiol.* **50**, 936.

Goody, R., and Holmes, K. (1983). Cross-bridges and the mechanism of muscle contraction. *Biochim. Biophys. Acta* **726**, 13.

Gospodarowicz, D., Weseman, J., Moran, S., and Lindstrom, J. (1976). Effect of fibroblast growth factor on the division and fusion of bovine myoblasts. *J. Cell Biol.* **70**, 395.

Greaser, M. L., Yamaguchi, M., and Brekke, C. (1972). Troponin subunits and their interactions. *Cold Spring Harbor Symp. Quant. Biol.* **37**, 235.

Greaser, M. L., Wang, S., and Lemanski, L. F. (1981). New myofibrillar proteins. *Proc. Annu. Reciprocal Meat Conf.* **34**, 12.

Guerriero, V., and Florini, J. R. (1978). Stimulation by glucocorticoids of myoblast growth at low cell densities. *Cell Biol. Int. Rep.* **2**, 441.

Gunn, H. M. (1978). Differences in the histochemical properties of skeletal muscles of different breeds of horses and dogs. *J. Anat.* **127**, 615.

Guth, L., and Samaha, F. J. (1969). Qualitative differences between actomyosin ATPase of slow and fast mammalian muscles. *Exp. Neurol.* **25**, 138.

Hall-Craggs, E. C. B. (1970). The longitudinal division of fibres in overloaded rat skeletal muscle. *J. Anat.* **107**, 459.

Huszar, G. (1972). Developmental changes of the primary structure and histidine methylation in rabbit skeletal muscle myosin. *Nature (London), New Biol.* **240**, 260.

Huxley, H. E. (1973). Molecular basis of contraction in cross-striated muscles. *In* "The Structure and Function of Muscle" (G. H. Bourne, ed.), 2nd ed., Vol. 1, p. 302. Academic Press, New York.

Knappeis, G. G., and Carlsen, F. (1962). The ultrastructure of the Z-disc in skeletal muscle. *J. Cell Biol.* **13**, 323.

Knappeis, G. G., and Carlsen, F. (1968). The ultrastructure of the M-line in skeletal muscle. *J. Cell Biol.* **38**, 202.

Laurent, G. J., and Sparrow, M. P. (1977). Changes in RNA, DNA and protein content and the rates of protein synthesis and degradation during hypertrophy of the anterior latissimus dorsi muscle of the adult fowl (gallus domesticus). *Growth* **41**, 249.

Layman, D. K., Hegarty, P. V. J., and Swan, P. B. (1980). Comparison of morphological and biochemical parameters of growth in rat skeletal muscles. *J. Anat.* **130**, 159.

Layman, D. K., Swan, P. B., and Hegarty, P. V. J. (1981). The effect of acute dietary restriction on muscle fiber number in weanling rats. *Br. J. Nutr.* **45**, 475.

Lehrer, S. S. (1975). Intramolecular crosslinking of tropomyosin via disulfide bond formation: Evidence for chain register. *Proc. Natl. Acad. Sci. U.S.A.* **72**, 3377.

Locker, R. H., and Leet, N. G. (1975). Histology of highly-stretched beef muscle. I. The fine structure of grossly stretched single fibers. *J. Ultrastruct. Res.* **52**, 64.

Locker, R. H., and Leet, N.G. (1976). Histology of high-stretched beef muscle. II. Further evidence on the location and nature of gap filaments. *J. Ultrastruct. Res.* **55**, 157.

Lockhart, R. D. (1972). Anatomy of muscles and their relationship to movement and posture. *In* "The Structure and Function of Muscle" (G. H. Bourne, ed.), 2nd ed., Vol. 1, p. 1. Academic Press, New York.

Lowey, S., Slayter, H. S., Weeds, A. G., and Baker, H. (1969). Substructure of the myosin molecule. I. Subfragments of myosin by enzymic degradation. *J. Mol. Biol.* **42**, 1.

Lymn, R. W., and Taylor, E. W. (1971). Mechanism of adenosine triphosphate hydrolysis by actomyosin. *Biochemistry* **10**, 4617.

Mani, R. A., and Kay, C. M. (1978). Isolation and characterization of the 165,000 dalton component of the M-line of rabbit skeletal muscle and its interaction with creatine kinase. *Biochim. Biophys. Acta* **533**, 248.

Mannherz, H. G., and Goody, R. S. (1976). Proteins of contractile systems. *Annu. Rev. Biochem.* **45**, 427.

Maruyama, K., Kimura, S., Ishii, T., Kuroda, M., Ohashi, K., and Muramatsu, S. (1977a). β-actinin, a regulatory protein of muscle. Purification, characterization, and function. *J. Biochem. (Tokyo)* **81**, 215.

Maruyama, K., Matsubara, S., Natori, R., Nonomura, Y., Kimura, S., Ohashi, K., Murakami, F., Handa, S., and Eguchi, G. (1977b). Connectin an elastic protein of muscle: characterization and function. *J. Biochem. (Tokyo)* **82**, 317.

Mauro, A. (1961). Satellite cell of skeletal muscle fibers. *J. Biophys. Biochem. Cytol.* **9**, 493.

Mauro, A., ed. (1979). "Muscle Regeneration." Raven Press, New York.

Mercola, D., Bullard, B., and Priest, J. (1975). Crystallization of troponin-C. *Nature (London)* **254**, 634.

Millward, D. J., Garlick, P. J., Stewart, R. J., Nnanyelugo, D. O., and Waterlow, J. C. (1975). Skeletal muscle growth and protein turnover. *Biochem. J.* **150**, 235.

Mong, F. S. (1977). Histological and histochemical studies on the nervous influence on minced muscle regeneration of triceps surae of the rat. *J. Morphol.* **151**, 451.

Moos, C., Offer, G., Starr, R., and Bennett, P. (1975). Interaction of C-protein with myosin, myosin rod and light meromyosin. *J. Mol. Biol.* **97**, 1.

Moss, F. P. (1968). The relationship between the dimensions of fibers and the number of nuclei during normal growth of skeletal muscle in the domestic fowl. *Am. J. Anat.* **122**, 555.

Murray, J. M., and Weber, A. (1974). The cooperative action of muscle proteins. *Sci. Am.* **230**, 59.

Offer, G., Moos, C., and Starr, R. (1973). A new protein of the thick filaments of vertebrate skeletal myofibrils. *J. Mol. Biol.* **74**, 653.

Peachy, L. D., and Porter, K. R. (1959). Intracellular impulse conduction in muscle cells. *Science* **129**, 721.

Pearson, M. L., and Epstein, H. F. (1982). "Muscle Development Molecular and Cellular Control." Cold Spring Harbor Lab., Cold Spring Harbor, New York.

Peters, J. B., Barnard, R. J., Edgerton, V. R., Gillespie, C. A., and Stempel, K. E. (1972). Metabolic profiles of three fiber types of skeletal muscle in guinea pigs and rabbits. *Biochemistry* **11**, 2627.

Peterson, E. R., and Crain, S. M. (1972). Regeneration and innervation in cultures of adult mammalian skeletal muscle coupled with fetal rodent spinal cord. *Exp. Neurol.* **36**, 136.

Potter, J. D. (1976). The content of troponin, tropomyosin, actin, and myosin in rabbit skeletal muscle myofibrils. *Arch. Biochem. Biophys.* **162**, 436.

Preiss, J., and Walsh, D. A. (1981). The comparative biochemistry of glycogen and starch. *In* "Biology of Carbohydrates" (V. Ginsburg, ed.), Vol. 1, p. 200. Wiley, New York.

Ramachandran, G. N., and Reddi, A. H. (1976). "Biochemistry of Collagen." Plenum, New York.

Robson, R. M., Yamaguchi, M., Huiatt, T. W., Richardson, F. L., O'Shea, J. M., Hartzer, M. K., Rathbun, W. E., Schreiner, P. J., Kasang, L. E., Stromer, M. H., Pang, Y., Evans, R. R., and Ridpath, J. F. (1981). Biochemistry and molecular architecture of muscle cell 10-nm filaments and Z-line: Roles of desmin and α-actinin. *Proc. Annu. Reciprocal Meat Conf.* **34**, 5.

Rushbrook, J. I., and Stracher, A. (1979). Comparison of adult, embryonic, and dystrophic myosin heavy chains from chicken muscle by sodium dodecyl sulfate/polyacrylamide gel electrophoresis and peptide mapping. *Proc. Natl. Acad. Sci. U.S.A.* **76**, 4331.

Scopes, R. K. (1970). Characterization and study of sarcoplasmic proteins. *In* "The Physiology and Biochemistry of Muscle as a Food, 2" (E. J. Briskey, R. G. Cassens, and B. B. Marsh, eds.), p. 471. Univ. of Wisconsin Press, Madison.

Sleep, J. A., and Smith, S. J. (1981). Actomyosin ATPase and muscle contraction. *Curr. Top. Bioregl.* **2**, 239.

Snow, M. H. (1977). Myogenic cell formation in regenerating rat skeletal muscle injured by mincing. *Anat. Rec.* **188**, 201.

Squire, J. M. (1975). Muscle filament structure and muscle contraction. *Annu. Rev. Biophys. Bioeng.* **4**, 137.

Stockdale, F., and Holtzer, H. (1961). DNA synthesis and myogenesis. *Exp. Cell Res.* **24**, 508.

Stone, D., and Smillie, L. B. (1978). The amino acid sequence of rabbit skeletal α-tropomyosin. *J. Biol. Chem.* **253**, 1137.

Suzuki, A., and Cassens, R. G. (1980). A histochemical study of myofiber types in muscle of the growing pig. *J. Anim. Sci.* **51**, 1449.

Suzuki, A., Goll, D. E., Singh, I., Allen, R. E., Robson, R. M., and Stromer, M. H. (1976). Some properties of purified skeletal muscle α-actinin. *J. Biol. Chem.* **251**, 6860.

Trenkle, A. (1974). Hormonal and nutritional interrelationships and their effects on skeletal muscle. *J. Anim. Sci.* **38**, 1142.

Trinick, J., and Lowey, S. (1977). M-protein from chicken pectoralis muscle: isolation and characterization. *J. Mol. Biol.* **113**, 343.

Turner, D. C., Wallimann, T., and Eppenberger, H. M. (1973). A protein that binds specifically to the M-line of skeletal muscle is identified as the muscle form of creatine kinase. *Proc. Natl. Acad. Sci. U.S.A.* **70**, 702.

Vandekerckhove, J., and Weber, K. (1979). The complete amino acid sequence of actins from bovine aorta, bovine heart, bovine fast skeletal muscle, and rabbit slow skeletal muscle. *Differentiation* **14**, 123.

Wagner, P. D., and Ginger, E. (1981). Hydrolysis of ATP and reversible binding to F-actin by myosin heavy chains free of all light chains. *Nature (London)* **292**, 560.

Wang, K. (1982). Purification of titin and nebulin. *In* "Methods in Enzymology" (D. W. Frederiksen and L. W. Cunningham, eds.), Vol. 85, p. 264. Academic Press, New York.

Wang, K., and Williamson, C. L. (1980). Identification of an N_2-line protein of striated muscle. *Proc. Natl. Acad. Sci. U.S.A.* **77**, 3256.

Wang, K., McClure, J., and Tu, A. (1979). Titan a major myofibrillar component. *Proc. Natl. Acad. Sci. U.S.A.* **76**, 3698.

Waterlow, J. C., Garlick, P. J., and Millward, D. J. (1978). "Protein Turnover in Mammalian Tissue." Elsevier, Amsterdam.

Weeds, A. G. (1980). Myosin light chains, polymorphism and fiber types in skeletal muscles. *In* "Plasticity of Muscle" (D. Pette, ed.), p. 55. de Gruyter, New York.

Weeds, A. G., and Pope, B. (1977). Studies on the chymotryptic digestion of myosin. Effects of divalent cations on proteolytic susceptibility. *J. Mol. Biol.* **111**, 129.

Whalen, R. G. (1980). Contractile protein isozymes in muscle development: The embryonic phenotype. *In* "Plasticity of Muscle" (D. Pette, ed.), p. 177. de Gruyter, New York.

White, A., Handler, P., Smith, E. L., Hill, R. L., and Lehman, I. R. (1978). "Principles of Biochemistry," 6th ed. McGraw-Hill, New York.

Wilkinson, J. M., and Grand, R. J. (1978). Comparison of amino acid sequence of troponin I from different striated muscles. *Nature (London)* **271**, 31.

Yates, L. D., and Greaser, M. L. (1983). Quantitative determination of myosin and actin in rabbit skeletal muscle. *J. Mol. Biol.* **168**, 123.

Zechel, K., and Weber, K. (1978). Actins from mammals, bird, fish and slime mold characterized by isoelectric focusing in poly-acrylamide gels. *Eur. J. Biochem.* **89**, 105.

Zhenevskaya, R. P. (1962). Experimental histologic investigation of striated muscle tissue. *Rev. Can. Biol.* **21**, 457.

2

Conversion of Muscle to Meat

M. L. GREASER

Muscle Biology Laboratory
University of Wisconsin
Madison, Wisconsin

I. INTRODUCTION

Skeletal muscle is a tissue that serves the function of locomotion in higher animals. This tissue also serves as an important source of food for man and is referred to as meat in this context. The processes of change between muscle and meat are complex and involve metabolic, physical, and structural alterations. At the death of an animal, the blood flow to and from the muscle ceases and as a result, there are no new sources of energy for muscle function and the supply of oxygen is cut off. In addition, the products of metabolism can no longer be removed and thus accumulate in the tissue.

The purpose of this chapter is to detail the changes in muscle that occur in the first one to two days postmortem. The descriptions center primarily on processes occurring in beef, pig, and sheep muscle; however, studies with rabbit and rat muscle have been included when particular approaches have not been used with the large meat animals.

Copyright © 1986 by Academic Press, Inc.
All rights of reproduction in any form reserved.

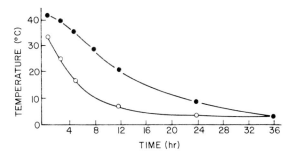

Fig. 1   Effect of distance from the surface on cooling rates in beef semimembranosus muscles. Carcasses were placed in a 2–3°C chiller at approximately 1 hr postmortem. ○, 2 cm; ●, 8 cm. From Follett *et al.* (1974), with permission.

An important concept that must be emphasized from the beginning is that all muscles and all parts of a muscle do not change in a uniform manner postmortem. First, muscle consists of two major fiber types, red and white, containing differing amounts of myoglobin, oxidative and glycolytic enzymes, and substrates (for review, see Cassens and Cooper, 1971). Different muscles contain differing ratios of white to red fibers, and some muscles have major regions of predominantly white and/or predominantly red fibers (e.g., the pig semitendinosus). As a result, one could expect a heterogeneous pattern of postmortem change even in adjacent cells. Swatland (1975b) has shown that adjacent fibers have different glycogen depletion patterns. Second, the effect of temperature on metabolic processes is profound and it must be realized that the rate of muscle cooling postmortem is not uniform, even within the same muscle. In most cases, carcasses are transferred to 0–5°C coolers 30–60 minutes postmortem. The postmortem cooling rates of bovine semimembranosus muscles at 2 and 8 cm from the surface are shown in Fig. 1. Clearly, a temperature gradient exists between the surface and the interior of the muscle. The center of the pig longissimus dorsi takes approximately 4 hr to cool to 10°C while it takes 8 hr for the center of the leg to reach the same temperature (Taylor and Dant, 1971). The rate of cooling is also affected markedly by the size of the carcass, the temperature of the cooler, the rate of air movement past the carcasses, and the thickness of the insulating fat layer over the surface. Figure 2 shows the effects of fat thickness and cooler conditions on the cooling rate in the center of beef longissimus dorsi muscles. The temperatures that the muscles are experiencing vary greatly and as a result one must define muscle location and cooling conditions to make meaningful comparisons. Thus, generalities about the rate of postmortem processes must be given with some caution; several examples will be presented in subsequent sections of the effect of temperature on muscles removed from the carcass to demonstrate temperature effects.

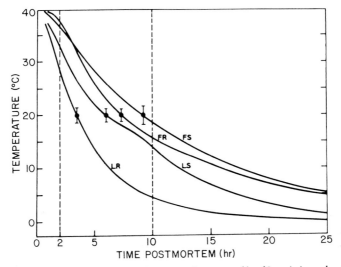

Fig. 2 Effect of fat cover and air velocity on cooling rates of beef Longissimus dorsi muscles. Animals were classified as fat (F) or lean (L) with mean fat thickness over the twelfth rib of 2.6 ± 0.7 cm or 0.5 ± 0.1 cm, respectively. One side of each animal was placed in either a rapid (R) (−2°C, air movement at 90 m/min) or slow (S) (9°C, no forced air movement) chill cooler at 40 min postmortem. The temperatures were measured at the middle of the longissimus dorsi opposite the twelfth thoracic vertebra. From Lochner *et al.* (1980), with permission.

The following review is divided into different sections in which separate aspects of postmortem changes are summarized. It should be realized that many of the metabolic processes are closely linked and that discussion of ATP changes apart from rigor mortis and pH changes apart from lactic acid formation are artificial divisions.

II. POSTMORTEM METABOLISM

A. High-Energy Compounds

1. ENZYME SYSTEMS REQUIRING ATP

Adenosine triphosphate is the major high-energy compound in cells and is used to drive a large variety of metabolic reactions including ion transport. Its presence is absolutely essential for cell survival. The postmortem muscle cell thus attempts to maintain its ATP level at all costs. In order to understand postmortem processes in muscle, the potential reactions and enzyme systems in which ATP is split must be considered. ATP is split at the rate of approximately 0.5 to 0.65 μmol/g muscle per minute at 38°C (Bendall, 1973a; Scopes, 1973). The enzyme possessing the highest potential ATPase activity in muscle is the

M. L. Greaser

Table I
Effect of Temperature on ATP Turnover Rate[a]

	Temperature (°C)							
	38	35	30	25	20	15	10	2
Rabbit psoas	100[b]	83.3	61.4	49.3	41.7	36.4	32.4	28.0
Beef sternomandibularis	100[b]	83.3	61.4	49.3	44.2	39.4	35.8	38.5
Myosin ATPase	100[b]	84.0	62.9	46.3	33.8	24.4	17.4	9.8

[a] From Bendall (1973a), with permission.
[b] Values normalized to 100% at 38°C.

myosin ATPase. In normal resting muscle, the myosin heads are not bound to actin and the milliomolar levels of magnesium present strongly inhibit the rate of ATP splitting (for review, see Taylor, 1979). However, there is a continual slow hydrolysis of ATP to ADP and phosphate. Because the concentration of myosin is so large (constituting about one-third of the total muscle protein), this resting ATPase activity is probably the major ATP depletion reaction postmortem (Greaser et al., 1969a; Bendall, 1973a,b). In fact, the effect of temperature on ATP turnover in postmortem muscle closely parallels the temperature dependence of myosin ATPase at temperatures between 38 and 25–30°C (Table I).

The myosin ATPase activity may be rapidly increased at least 300 to 500-fold if the cytoplasmic calcium concentration is increased from levels below 10^{-8} to 10^{-6} or 10^{-5} M (Bendall, 1973a, 1978). In living muscle, this calcium increase occurs during muscle contraction and causes myosin to combine with actin to form a complex whose ATPase activity is stimulated by magnesium. Thus, the cytoplasmic calcium level may have a dramatic effect on postmortem metabolism, especially when visible contraction occurs.

A second ATPase system that must contribute to postmortem ATP degradation is the sarcoplasmic reticulum Ca-ATPase. This protein, which is embedded in the membrane, functions to move calcium out of the cytosol during the rest cycle in an ATP-dependent transport process (for review, see Hasselbach, 1977). Since no membrane is absolutely impermeable to ions, the leakage of calcium from the sarcoplasmic reticulum means that this ATPase must continue to function at a finite rate to maintain the calcium gradient. The contribution this ATPase makes to the total ATP turnover is probably minor. However, since the sarcoplasmic reticulum controls the cytosolic calcium concentration, its function is clearly intertwined with the myosin ATPase in the myofibril described above.

Three other ATP splitting systems which must be considered are the mitochondrial ATPase, the plasmalemma Na^+,K^+-ATPase, and the plasmalemma Ca-ATPase. When mitochondria become anaerobic, they hydrolyze ATP instead of synthesizing it. The specific activity of this enzyme is very high at temperatures $> 30°C$ in isolated mitochondria and could account for a major portion of the ATP turnover (J. R. Mickelson and B. B. Marsh, personal communication). Haworth and co-workers (1981) have demonstrated that the mitochondrial ATPase of *heart* cells accounted for a significant drain on energy sources in the quiescent state. However, the temperature dependence of the mitochondrial ATPase is much greater than that of ATP turnover in intact muscle, so it too probably plays only a minor role. The Na^+,K^+-ATPase is found in the plasmalemma and functions to transport sodium out and potassium in, thus maintaining cell polarity (for reviews, see Robinson and Flashner, 1979; Wallick *et al.*, 1979). This enzyme should continue to operate postmortem and might accelerate if the plasmalemma became more leaky. The Ca^{2+}-ATPase of the sarcolemma (described by Caroni and Carafoli, 1980) and other uncharacterized transport ATPases might also contribute to the ATP turnover rate, although probably in a minor way.

Several cyclic enzyme pathways that use ATP have also been postulated to have a role in ATP turnover. These include the combinations phospho-fructokinase–fructose diphosphatase, phosphorylase–glycogen synthetase, and phosphorylase b kinase–phosphorylase a phosphatase (Scopes, 1974b). These systems probably are not significant since conditions that lead to activation of one enzyme of the pair suppress the activity of the other one (Scopes, 1974b).

It is difficult to project results with isolated ATPases to explain ATP turnover rates in intact muscle. However, it appears that the great majority of ATP splitting events can be accounted for by the activity of myosin ATPase in the resting state plus calcium activation of the same enzyme combined with actin in situations of more rapid ATP degradation. Since the observed range of ATP turnover rates from very slow to extremely rapid is only abut 10-fold, only a very small increase in the average cytosolic Ca^{2+} levels could account for the maximum ATP splitting rates observed.

2. CHANGES IN ATP AND CREATINE PHOSPHATE (CP) WITH TIME POSTMORTEM

Although the cell uses ATP as its primary source of energy to drive its biochemical reactions, another high-energy compound, creatine phosphate (CP), is present in significant concentrations and serves to buffer ATP levels by participating in the following reaction:

$$CP + MgADP \underset{\text{kinase}}{\overset{\text{creatine}}{\longleftrightarrow}} MgATP + Cr$$

Thus if the muscle cell is called upon to do work, even though the ATP splitting rate is rapid the cell ATP concentration remains stable. Similar processes occur

Table II
Resting pH and Metabolite Levels in Various Species[a]

Parameter	Rabbit psoas	Beef longissimus dorsi	Pig longissimus dorsi	Sheep pectoral
Initial pH	7.10	7.08	7.18–7.30	7.18
Total creatine[b]	42.0	42.0	44.0	34.0
Creatine phosphate[b]	23.0	19.0	18.0–19.0	13.1
ATP[b]	8.1	5.7	6.6–6.8	5.9
ADP[b]	1.1	0.9	1.0	0.9
Glycogen (glucose equivalents)[b]	~60	~50	~55–65	~45
Initial lactate[b]	13.0	16.0	6.0–11.2	9.4

[a] From Bendall (1973b), with permission.
[b] Values expressed as μmol/g of muscle.

in postmortem muscle, and it was first observed many years ago that the ATP levels remained more or less constant until ~70% of the CP had been degraded (Bendall, 1951). At this point the ATP levels begin to decline rapidly and ultimately fall to almost zero.

The concentration of CP in muscle removed from a live animal by biopsy or within 10 min after death is already much lower than that found in resting muscle. Cutting the muscle to remove the sample invariably causes a contractile response with a resulting depletion of a part of the CP present. There is also normally considerable contractile activity resulting from stunning or electrocution at the time of death. Resting levels have been obtained, however, by anesthetizing animals with $MgSO_4$ or curare to block nerve stimulation during sampling (Bendall, 1966, 1973b; Sair et al., 1970). The resting values of CP and ATP for beef, pig, sheep, and rabbit muscle are shown in Table II. In all species the total creatine is approximately $30-45 \mu$mol/g and of this $13-23 \mu$mol/g is in the phosphorylated form. In pig, muscles with low myoglobin (such as the longissimus dorsi and the white portion of the semimembranosus) have CP levels around 20μmol/g while muscles containing more myoglobin (such as the vastus intermedius) have ~10μmol/g (Bendall, 1975). The highest ratio of phosphorylated to total creatine has been ~65% even though the equilibrium of the creatine kinase reaction with the substrate concentrations present in muscle should be at 90% in the phosphorylated form (Scopes 1973).

Resting levels of ATP are also shown in Table II and vary between ~5.0 and 8.5μmol/g (Bendall, 1973a). There is some variation between muscles and in the pig the white muscles have $7-8 \mu$mol/g while the red ones contain $6-7 \mu$mol/g (Bendall, 1975).

The initial levels of CP in nonanesthetized muscles vary considerably with the species and the particular muscle examined. In muscles such as the beef longissimus dorsi and sternomandibularis the CP concentration is very similar to that of anesthetized muscles (Bodwell *et al.,* 1965a; Bendall, 1973b).   Beef psoas, however, has CP levels of 4 μmol/g or less within a few minutes after death (Bendall, 1973b).   Similar low levels are invariably found in all pig muscles even within 2 or 3 min of death (Kastenschmidt, 1970; Sair *et al.,* 1970; Bendall, 1973b).

The decline in CP levels in beef longissimus muscles left on the carcass until the time of sampling is reasonably slow with 1.5 μmol/g still remaining after 12 hr postmortem (Bodwell *et al.,* 1965a).   In isolated beef sternomandibularis muscles held at 1, 5, or 15°C the CP levels were totally depleted 10–12 hr postmortem (Newbold and Scopes, 1967).   CP is virtually depleted by 1 hr postmortem in beef psoas (Tarrant and Mothersill, 1977).   In normal pig longissimus dorsi, however, CP levels drop to zero within 2 to 3 hr after death (Kastenschmidt, 1970; Sair *et al.,* 1970).

ATP levels normally remain fairly constant until the CP drops below 4 μmol/g (Bendall, 1973b).   At 2 μmol/g CP the ATP level has usually dropped 50%.   The ATP concentrations in beef sternomandibularis held at 10 to 15°C are at ~ 5 μmol/g at 1.5 hr postmortem and decline slowly to 3.5 μmol/g until 8 or 9 hr postmortem, when a more rapid decrease begins (Newbold and Scopes, 1967).   If the same muscle is held at 38°C, the rapid decline begins at ~ 2.5 hr postmortem and ATP concentrations are below 0.5 μmol/g by 6 to 7 hr postmortem (Newbold and Scopes, 1967).

The ATP in the deeper portion of the semimembranosus may be virtually depleted by 5 hr postmortem (Follett *et al.,* 1974).   If this muscle is removed from the carcass, however, and held at − 5, 0, 5, 10 or 15°C the ATP depletion rate is drastically retarded with at least 12 hr required to lower ATP levels to < 20% of the initial values.

Pig muscle ATP levels are normally depressed significantly from the resting concentrations by the earliest times samples can be obtained (1–2 min postmortem).   Typical values are usually 3–4 μmol/g in the longissimus dorsi (Kastenschmidt, 1970; Sair *et al.,* 1970).   The ATP concentration declines rapidly to 1.0 μmol/g or less by 3 hr postmortem (Kastenschmidt, 1970).   With certain muscles and certain animals the ATP and CP levels decline much more rapidly; this condition is discussed in greater detail in Section IV,B.   In contrast, muscles from curare-immobilized pigs have ATP levels of 8 μmol/g at 25 min postmortem and they remain at this level for > 3 hr at 38°C before starting to decline (Bendall, 1966).

The effect of temperature on CP and ATP depletion rates is anomalous. Muscles held at temperatures between 37 and 15°C have decreasing rates of ATP depletion with decreasing temperature, but the rates at 15, 5, and 1°C were quite similar (Cassens and Newbold, 1966; Newbold and Scopes, 1967).   The 1° rate was actually faster than that found at 15°C.   A faster rate was also found at − 2 or − 3 than at 10°C (Behnke *et al.,* 1973).   The reason for this remains unclear, but it appears to be related to the cold-shortening phenomenon (Locker and Hagyard, 1963), which is discussed in Section III,A,1.

Even though there is wide variation in the rate of ATP depletion between muscles and species, the proportion of the original ATP concentrations at specific postmortem pHs is rather constant. Approximately one-eighth of the ATP has been broken down by the time the muscle pH has dropped to pH 6.3, one-half is depleted by pH 5.9–6.0, and 90% is gone when the pH is 5.6–5.7 (Bendall, 1973a; Bendall *et al.*, 1976).

3. NUCLEOTIDE BREAKDOWN PRODUCTS

ATP is split into ADP and inorganic phosphate by the ATPase discussed above (see Section II,A,1). The ADP is usually rephosphorylated to ATP with CP or by glycolysis. However, with increasing time postmortem the CP is depleted and glycolysis can not keep up with ATP resynthesis. As a result, the following reaction occurs in an attempt to restore the resting ATP concentrations:

$$2 \text{ ADP} \xrightleftharpoons[]{\text{myokinase}} \text{ATP} + \text{AMP}$$

The AMP is then deaminated to inosine monophosphate (IMP) by the following reaction:

$$\text{AMP} \xrightarrow{\substack{\text{AMP} \\ \text{deaminase}}} \text{IMP} + \text{NH}_3$$

It has been shown that a > 0.99 correlation exists between ammonia production in postmortem muscle and the disappearance of adenine nucleotides (Bendall and Davey, 1957).

Both inosine diphosphate (IDP) and inosine triphoshate (ITP) also appear in postmortem muscle, and it has been postulated (Bendall, 1973a) that they may be formed in the following manner:

$$2 \text{ ATP} + 2 \text{ IMP} \rightleftharpoons 2 \text{ ADP} + 2 \text{ IDP}$$
$$2 \text{ IDP} \rightleftharpoons \text{ITP} + \text{IMP}$$
$$\text{ADP} \rightleftharpoons \text{IDP} + \text{NH}_3$$

The IMP is further degraded by the pathway below (Lee and Newbold, 1963):

$$\text{IMP} \rightarrow \text{I} + \text{PO}_4 \xrightarrow{\substack{\text{nucleoside} \\ \text{phosphorylase}}} \text{hypoxanthine} + \text{ribose-1-phosphate}$$
$$\xrightarrow{\hspace{5cm}} \text{hypoxanthine} + \text{ribose}$$

The ADP levels in resting muscle are approximately 1.0 μmol/g (Bendall, 1973b). Much of this ADP is in a bound form, either complexed with actin (~0.5 μmol/g) or on the myosin heads (~0.24 μmol/g (Bendall, 1973b). The actin ADP is tightly bound and probably accounts for the major portion of the remaining muscle ADP content at 1 to 2 days postmortem. The myosin ADP continues to cycle and will no longer be present when the ATP is fully depleted. In pig longissimus dorsi the ADP levels begin to decline ~1 hr postmortem (Kastenschmidt, 1970) while in beef sternomandibularis the decline does not start until 5 to 6 hr after death (Newbold and Scopes, 1967).

The concentration of AMP in resting muscle is low ($\sim 0.2 - 0.3$ μmol/g) (Bendall, 1973b). This level may increase two to three times in beef sternomandibularis during the first 10 hr postmortem but declines to the original level by 24 hr after death (Newbold and Scopes, 1967). In pig longissimus dorsi the levels remain fairly constant with time postmortem (Kastenschmidt, 1970).

Inosine nucleotide levels in resting muscle are approximately 0.5 μmol/g (Bendall, 1973b). Most of this is IMP and the concentration of this compound increases postmortem in almost direct correspondence with the loss of ATP. In rabbit muscle the 10-min, 10-hr, and 24-hour values for ATP and IMP were 6.8, 1.7, and 0.7 μmol and 0.8, 6.6, and 8.3 μmol/g respectively (Bendall and Davey, 1957). The IMP levels increase to about 5 μmol/g between 12 and 24 hr postmortem in beef sternomandibularis and semitendinosus (Disney *et al.*, 1967; Newbold and Scopes, 1967) while similar levels are reached within 3 hr in pig longissimus dorsi (Tsai *et al.*, 1972). IDP concentrations are near zero in resting muscle, but reach concentrations of 0.4 to 0.8 μmol/g between 3 and 72 hr postmortem in pig longissimus dorsi (Tsai *et al.*, 1972). ITP has also been detected in rabbit (Bendall and Davey, 1957) and pig (Tsai *et al.*, 1972) but in lower levels than IDP.

The inosine and hypoxanthine levels are normally near zero in resting muscle (Rhodes, 1965). The appearance of these compounds depends on the IMP levels, and thus their production occurs primarily after glycolysis ceases (Bendall, 1973a). The production of hypoxanthine may reach 20–40% of the original ATP concentration by 24 hr postmortem in beef (Howard *et al.*, 1960; Disney *et al.*, 1967) and the level increases very slowly thereafter to a maximum after 60 days of storage at 2°C (Rhodes, 1965). Beef longissimus dorsi, however, was found to contain ~1.0 μmol/g each of inosine and hypoxanthine at 5 to 6 days postmortem (Hamm and van Hoof, 1974). The inosine plus hypoxanthine level of pig longissimus dorsi reaches >1.0 μmol/g by 3 hr postmortem and rises to about 2.0 μmol/g by 72 hr (Tsai *et al.*, 1972). In all cases the sum of the total adenine and inosine compounds remains constant with time postmortem (Bendall and Davey, 1957; Tsai *et al.*, 1972).

The concentrations of the nicotinamide adenine dinucleotides (NAD) also change postmortem. Resting muscles contain ~0.5–1.0 μmol/g (Bendall, 1973b). These levels gradually decline to ~35% of their initial value by 8 hr postmortem in the deep portions of semimembranosus muscles left on the carcass (Follett *et al.*, 1974) and remain at this level through 36 hr. The decline is much smaller (to 66 to 75% of initial values) in muscles removed from the carcass and held at 15°C or less. The NAD content of beef biceps femoris dropped from 0.8 to 0.5 μmol/g between 15 min and 24 hr postmortem (Hatton *et al.*, 1972). Pig muscle NAD concentrations decline by about one-third in the first 3 hr after death (Kastenschmidt, 1970). Lamb muscle NAD also declines to ~50% of initial level by 24 to 48 hr postmortem (Atkinson and Follett, 1973). The degradation of NAD is markedly accelerated in muscle minces (Newbold and Scopes, 1971) and homogenates (Severin *et al.*, 1963; Bernowsky and Pankow, 1973).

B. ATP Resynthesis Systems

1. GLYCOLYSIS

In life ATP is replenished by either oxidative metabolism in the mitochondria or by anaerobic glycolysis (Fig. 3). When the blood supply ceases, there is no longer a source of oxygen and the mitochondrial ATP production is halted. No new glucose can be transported to the cell and thus glycolysis of the glycogen stores is the only way that ATP can be resynthesized. The end product of these reactions is lactic acid, which increases for some time postmortem. As a result, the muscle pH drops from ~7.1 to 7.3 to the 5.5–5.7 range. In this section the changes in glycogen, glycolytic intermediates, lactic acid, and pH are described.

a. Glycogen

The resting levels of muscle glycogen range from about 45 μmol (glucose equivalents) per gram in beef sternomandibularis to 65 μmol/g in the longissimus dorsi of certain strains of pigs (Bendall, 1973b). Most beef muscles sampled immediately after death contain glycogen concentrations comparable to the resting state. In the beef longissimus dorsi the glycogen content at 0, 6, 12, 24, and 48 hr postmortem was 56.7, 41.6, 30.4, 10.1, and 10.0 μmol glucose equivalents per gram (Bodwell *et al.*, 1965a). Howard and Lawrie (1956) also found initial and 24-hr levels for glycogen of 51.9 and 8.3 μmol/g in beef longissimus dorsi. This contrasts with pig muscle, in which the glycogen concentrations may have dropped by 30 to 50% by 10 min postmortem (Kastenschmidt, 1970). There also appear to be significant breed differences in both resting (Bendall, 1973b) and 5–10-min postmortem glycogen levels (Sayre *et al.*, 1963c). Glycogen declines more rapidly postmortem in pig muscle, reaching final levels at 3 to 5 hr after death. The amount of glycogen remaining when glycolysis ceases may vary from 2 to 30% of resting levels (Sayre *et al.*, 1963c; Beecher *et al.*, 1965a; Bodwell *et al.*, 1965a; Newbold and Lee, 1965). The fact that not all the glycogen is depleted may be due to its compartmentalization

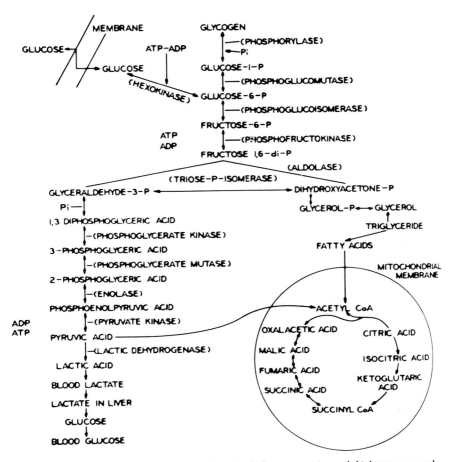

Fig. 3 Metabolic pathways in muscle. After death glucose entry (upper left), lactate removal (lower left), and mitochondrial ATP synthesis (lower right) all cease. From Kastenschmidt (1970), with permission.

(Ottaway and Mowbray, 1977) or to adenine nucleotide depletion (see Section II,B,1,e).

b. Glycolytic Intermediates

The postmortem changes in the various glycolytic intermediates in beef and pig muscle are shown in Table III. Levels of the hexose phosphates, α-glycerol phosphate, and lactate are much greater than those of the other intermediates at all times postmortem. Glucose-6-phosphate may double in concentration in the bovine semimembranosus between 0 and 8 hr postmortem (Follett *et al.*, 1974). The high hexose phosphate concentrations suggest that phosphofruc-

Table III

Postmortem Changes in Glycolytic Intermediates in Beef and Pig Muscle

	Beef[a]		Pig[b]	
	1.5 hr	24 hr	0 hr	3 hr
Glucose-1-phosphate ⎫			0.3	0.4
Glucose-6-phosphate ⎬	4.5	4.7	4.5	6.5
Fructose-6-phosphate ⎭			0.7	1.8
Fructose-diphosphate ⎫			5.8	0.2
Dihydroxyacetone-phosphate ⎪			0.2	~0
Glyceraldehyde-3-phosphate ⎪			0.1	~0
3-Phosphoglyceric acid ⎬	0.08	0.04	0.5	0.1
2-Phosphoglyceric acid ⎪			0.06	0.03
Phosphoenolpyruvate ⎭			0.11	0.03
α-Glycerol-phosphate	2.6	1.2	2.2	1.8
Lactate	32.0	98	40	90

[a] From Newbold and Scopes (1967). Sternomandibularis muscles held at 15°C. Values expressed as μmol/g of muscle.

[b] From Kastenschmidt (1970). Longissimus dorsi muscles (slow-glycolyzing) left on the carcass until time of sampling. Values expressed as μmol/g of muscle.

tokinase activity may be limiting the rate of glycolysis (Newbold and Scopes, 1967). However, in no case do the total concentrations of the intermediates between glycogen and lactic acid exceed 15% of the initial glycogen glucose equivalents.

c. Lactic Acid

The concentrations of lactic acid in resting muscle of beef, pig, sheep, and rabbits range from 6 to 16 μmol/g (Table II). Muscle samples taken from beef longissimus dorsi immediately after death fall in this range (Bodwell et al., 1965a), while pig muscles usually have at least 30–40 μmol/g (Bendall et al., 1963; Kastenschmidt et al., 1964; Kastenschmidt, 1970). The levels of lactate increase gradually in beef semimembranosus muscles left on the carcass with 1-, 3-, 6-, 13-, and 24-hr values of 26, 38, 56, 73, and 77 μmol/g respectively (Disney et al., 1967). Similar patterns occur in beef longissimus dorsi and sternomandibularis giving final concentrations of 80 to 100 μmol/g at 24 hr postmortem (Bodwell et al., 1965a; Newbold and Scopes, 1967). Pig muscle lactate levels increase more rapidly, with >80 μmol/g already present at 3 hr postmortem (Kastenschmidt, 1970). The ultimate levels may reach 130 μmol/g (Bendall et al., 1963).

d. pH

The muscle pH declines postmortem as a result of the accumulation of lactic acid. Thus the measurement of pH has been a reliable and sensitive indicator of the rate and extent of postmortem glycolysis. Measurements of muscle pH have been made either with probe electrodes (Bendall and Wismer-Pedersen, 1962; Sayre et al., 1963c) or by homogenizing in iodoacetate to arrest glycolysis (Bate-Smith and Bendall, 1949). Both methods have limitations (Bendall, 1973b). Homogenizations done with iodoacetate dissolved in water result in an alteration of the pK values of the muscle buffers because of the change in ionic strength. Most pH measurements done before 10 years ago were performed in this way. More recently 150 mM KCl has been included in the iodoacetate solutions. Temperature has an important effect on the muscle buffer pK values and as a result the measured pH will be 0.36 units higher if the solutions are at 0 than at 38°C (Bendall and Wismer-Pedersen, 1962). Homogenization of muscle causes the release of CO_2 and as a result the pH of the homogenate rises. Furthermore, although iodoacetate stops glycolysis by inactivating glyceraldehyde phosphate dehydrogenase, the remaining CP and ATP are degraded and as a result the homogenate pH increases. The true muscle pH at 38°C should equal the measured pH at 20°C − 0.2 units (iodoacetate effect) −0.055 (CO_2 effect) −0.15 (temperature effect) (Bendall, 1973b). It would seem that the probe electrode would be preferable, but anomalously high pH values for freshly excised beef muscles have been obtained, presumably because of interfering membrane potentials (Bendall, 1973b). Pig muscle does not display this problem and the probe electrode appears to be satisfactory for postmortem pH measurements in this species.

The initial pH values of resting muscles from the different species are shown in Table II; they range from 7.08 to 7.34. In nonanesthetized muscles sampled within 10 to 15 min postmortem the pH usually remains at 6.9 to 7.0 in beef (Bodwell et al., 1965a; Cassens and Newbold, 1966); pH is 6.8 – 7.0 in lamb at 40 to 80 min postmortem (Marsh and Thompson, 1958), and pH is 6.6 – 6.8 in pig at 10 to 15 min (Hallund and Bendall, 1965). These values for the longissimus dorsi decine normally to an ultimate level of 5.4 to 5.7 (Bate-Smith and Bendall, 1949; Marsh, 1954; Marsh and Thompson, 1958; Briskey and Wismer-Pedersen, 1961). In extreme cases the ultimate pH in certain pigs has reached <4.8 (Lawrie et al., 1958). Not all muscles behave similarly, and in particular several lamb muscles have ultimate pH values above 6.0 (B. B. Marsh and W. A. Carse, personal communication).

The rate of pH decline is markedly affected by temperature, as shown in Fig. 4. The pH fall is rapid at higher temperatures in the range from 37° to 7°C (Marsh, 1954). However, at temperatures <10°C the rates plateau or even increase slightly (Cassens and Newbold, 1967a; Newbold and Scopes, 1967). This effect of temperature is related to the variation of pH decline of muscles left on the carcass. The time for beef semimembranosus muscles to reach their ultimate pH was 24 – 48 hr at 1.5 cm from the surface, 12 hr at 5 cm, and 6 hr at 8 cm (Tarrant, 1977). In another study samples from beef psoas and semimem-

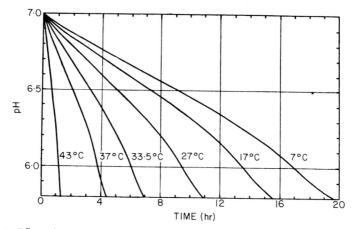

Fig. 4 Effect of temperature on postmortem pH decline in beef longissimus dorsi muscles. Muscles were removed from the carcass and held at the temperatures indicated. From Marsh (1954), with permission.

branosus at 1.5 cm from the surface reached pH 6.0 at 2.2 and 13.6 hr respectively (Tarrant and Mothersill, 1977).

The pH decline rate is species dependent with pig muscle changing at 0.64 units/hr at 37°C while beef, sheep, and rabbit muscles drop at ~0.27 to 0.40 units/hr (Marsh and Thompson, 1958; Hallund and Bendall, 1965). Treatment of pigs with curare (a neuromuscular blocking agent) decreases the pH decline rate to levels similar to those of other species (Hallund and Bendall, 1965). The rate of decline is pH dependent with lower rates near 6.7 and below 5.8 in the pig (Bendall, 1973b). Beef semimembranosus declines at 0.13 units/ hr between pH 7.0 and 6.7 and at 0.25 units/hr from 6.6 to 5.8 with the rates corrected to a constant temperature of 38°C (Bendall, 1978). The minimum at pH 6.7 does not occur in beef sternomandibularis, but the rate of decline falls off rapidly below pH 6.0 (Bendall, 1973b).

Considerable variability exists between different animals in the rate of glycolysis even with the same rate of cooling. Bendall (1978) has observed twofold differences in rates between 0.17 and 7 hr postmortem and threefold differences between 3 and 7 hr in beef biceps femoris. Marsh and co-workers (1980) found a 1.3-unit range in pH at 3 hr postmortem among the longissimus dorsi muscles of 40 beef sides. They also found differences as large as 0.5 units among adjacent sites in a single muscle. Even larger variations in glycolytic rate exist in pigs: the rapid glycolysis condition is discussed in greater detail in Section IV,B.

The pH values discussed so far are those far enough away from the muscle surface that they are measured in oxygen-deprived tissue. This means that samples homogenized in iodoacetate are removed from the interior of the muscle or probe electrodes are applied to freshly cut surfaces. A somewhat different pattern of pH decline occurs in the surface of muscles exposed to the air. Oxidative metabolism continues (see Section II,B,2) and as a result the lactic acid buildup is retarded. The muscle pH at 3 days postmortem ranged from 6.3 to 6.6 in 13 lamb muscle sites (Carse and Locker, 1974). Similar values were also found in beef.

e. Glycolysis Control

It is now clear that the rate of glycolysis is linked directly to the rate of ATP splitting in muscle. This has been demonstrated quite clearly in a reconstituted muscle glycolytic system (Scopes, 1974a,b) in which all the enzymes, substrates, and cofactors were mixed together at concentrations similar to those found in muscle. Glycolysis requires ADP and phosphate to proceed, and the system stops when these compounds are converted to ATP (Scopes, 1974b; Wu and Davis, 1981). The glycolytic rate increases in direct proportion to the amount of ATPase activity added to the system (Scopes, 1974a). The concentration and activity of the glycolytic enzymes is sufficient to synthesize ATP at the rate of 100 μmol/g per min (Scopes, 1970). This contrasts with typical ATP turnover rates of 0.5 to 0.65 μmol/g per min in normal muscle postmortem and maximal rates of about 10 μmol/g per min in extremely fast glycolyzing muscle (Bendall, 1973b) (see Section IV,B).

The extent of pH decline in this reconstituted system was similar to that found in muscle, namely, an ultimate pH of 5.3 to 5.5 (Scopes, 1974b). The factor that was most closely related to the final pH was the proportion of phosphorylase in the *a* form. Most phosphorylase in muscle is in the *b* form, which has less than 0.1% of phosphorylase a's activity (Scopes, 1973). Phosphorylase b is also strongly inhibited by ATP. The ultimate pH levels in reconstituted glycolytic systems were 5.74, 5.54, 5.32, and 5.3 when the phosphorylase a percentages were <0.1, 0.25, 2.0, and 40 respectively (Scopes, 1974b). The other factor that affected ultimate pH was the AMP deaminase activity. Higher amounts increase the ultimate pH by about 0.1 unit (Scopes, 1974b). The activity of this enzyme is maximal at pH 6.2, and therefore a competition exists for nucleotide, with the result that total adenine nucleotides are progressively and irreversibly converted to IMP. As a result glycolysis ceases because there is no longer any ADP available for rephosphorylation. It has been reported many times that glycolysis ceases because of enzyme inactivation while part of the glycogen remains. However, the most sensitive glycolytic enzyme to temperature and pH effects is phosphofructokinase, and it is still active at pH 5.35 and 37°C (Scopes, 1974b).

2. OXIDATIVE METABOLISM

Without blood flow to muscle the continuing supply of oxygen is stopped. The residual blood level in muscle is very low (<1%) and does not seem to vary depending on whether the animal is exsanguinated or not (Warriss, 1978). The oxygen associated with the blood hemoglobin and muscle myoglobin is rapidly used up, probably within 5 min or less (Bendall, 1973a). Thus the whole mitochondrial system (Fig. 3) stops ATP synthesis very soon postmortem.

Citrate and pyruvate levels decline somewhat postmortem (Beecher *et al.*, 1969; Kastenschmidt, 1970), but the levels of these and the other tricarboxylic acid cycle intermediates never limit oxygen consumption rates if the O_2 supply is reestablished (Bendall and Taylor, 1972). There appears to be no evidence for significant postmortem metabolism of fatty acids (Urbin and Wilson, 1961; Currie and Wolfe, 1977; Lazarus *et al.*, 1977).

This lack of oxidative metabolism is not found, however, in the surface layers of muscle exposed to air. The outer 0.7 mm of beef muscle has been found to remain in the prerigor state for up to 1 month (Leet and Locker, 1973). The lactic acid formed in the anaerobic deeper layers diffuses toward the surface and is oxidized, thus sparing the surface fibers' glycogen supply. The oxygen consumption rate is determined solely by the ATP turnover rate in thin muscle strips held between 38 and 15°C (Bendall, 1972). However, the respiration rate continues to decline with lower temperatures while the ATP turnover rate remains steady or increases slightly. As a result mitochondrial rephosphorylation cannot keep up with the ADP produced and muscle held at temperatures below 10°C must rely partially on anaerobic glycolysis (Bendall, 1972).

III. MACROMOLECULAR ALTERATIONS

A. Myofibrillar Proteins

1. RIGOR MORTIS AND SHORTENING

The most evident change in muscle postmortem is its transformation from being soft, pliable, and stretchable to a more rigid and inextensible state. This has been measured by alternately loading and unloading a muscle strip and then following the degree of stretching with time postmortem (Bate-Smith and Bendall, 1949; Briskey *et al.*, 1962). The muscle lengthening under load decreases 20- to 40-fold between the initial and final states (Bendall, 1973b). Immediately postmortem a load of 50 g/cm^2 stretches muscle to 120% of its slack length (Bendall, 1973a). After the completion of rigor a similar load stretches the muscle to 101% of its slack length. Increasing the load 10-fold results in a maximum lengthening to 102%.

The extensibility changes in postmortem rabbit muscle under various conditions are shown in Fig. 5. The rigor mortis patterns can be divided into three phases — delay, rapid, and postrigor (Bate-Smith and Bendall, 1949). The delay phase is that period of time in which there is no change in the muscle elasticity.

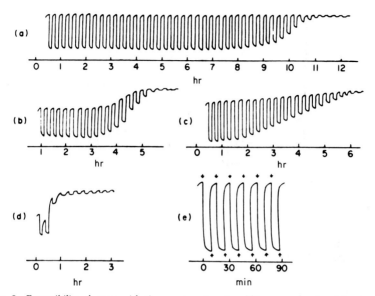

Fig. 5 Extensibility changes with time postmortem in rabbit psoas muscle. Muscles were alternately loaded and unloaded every 8 min until rigor was complete. (a) Well-fed animal, anesthetized with Myanesin, muscle held at 17°C. (b) Same as (a), except muscle held at 38°C. (c) Muscle from an animal that struggled at death, 17°C. (d) Muscle from an animal whose glycogen had been depleted prior to death by insulin injection, 17°C. (e) Enlarged tracing of the early stage of rigor. Upper arrows indicate that the load is on, lower arrows indicate that the load is off. From Bendall (1973b), with permission.

The rapid phase begins when the muscle extensibility begins to decline and ends when the extensibility reaches its minimum. The postrigor phase refers to the period after the rapid phase is completed.

It was recognized very early that the loss of extensibility was directly related to the loss of ATP concentrations postmortem (Bate-Smith and Bendall, 1947; Bendall, 1951). The rigidity occurs because the myosin heads firmly attach to actin and thus no longer allow free sliding of thick filaments past the thin ones. The ~50% decline in elasticity that occurs by the time that the ATP levels drop by half (Bendall, 1973b) requires explanation. It is known that MgATP levels above 0.1 to 0.2 mM are sufficient to prevent rigor in glycerinated muscle fibers (Izumi et $al.$, 1981). However, elasticity has begun to fall when the ATP levels are still at 2 to 4 mM. Since the ATP concentrations are averages for the whole muscle sample, there must be considerable variation between individual fibers, bundles, or motor units such that a proportion of the muscle fibers have ATP concentrations <0.1 mM and thus decrease the elasticity (Bendall, 1973a).

The length of time of the various rigor phases varies with species, muscle, holding temperature, and extent of struggling at the time of death. There is much more variation in the length of the delay phase than in the rapid phase.

The time to rigor completion (end of the rapid phase) is greater in beef and lamb muscles (Marsh, 1954; Marsh and Thompson, 1958) than in pig muscle (Briskey et al., 1962). Muscles with greater white fiber proportion have a longer delay phase than those that have predominantly red fibers (Briskey et al., 1962; Beecher et al., 1965b). Rigor mortis is completed more rapidly at higher temperatures in the range 37 – 15 °C. Locker and Daines (1975), using beef sternomandibularis, found times of 7 hr at 37 °C, 10 hr at 34 °C, 10 hr at 28 °C, 12 hr at 24 °C, and 24 hr at 15 °C for rigor completion. Struggling at death markedly shortens the rigor process in rabbits (Bate-Smith and Bendall, 1949). Curare, a neuromuscular blocking agent, also greatly prolongs the prerigor period in pig muscle (Bendall, 1966).

Muscle will shorten if unrestrained (or with light loads) as it goes into rigor or will develop tension if maintained at constant length. Temperature effects this with beef muscle shortening at $1 > 37 > 5 > 20, 15$ °C (Busch et al., 1967; Cassens and Newbold, 1967b). Tension development in beef muscle is maximum at 1 or 2 °C, smaller at 37 °C, and very little at 16 or 25 °C (Busch et al., 1972; Nuss and Wolfe, 1981). In pig or rabbit muscle greater tension occurs at 37 °C than at any of the lower temperatures (Busch et al., 1972). Tension declines after rigor mortis completion, and this may explain the softening that occurs in muscle with increasing time postrigor (Jungk et al., 1967).

At higher temperatures the initiation of shortening under light loads is pH dependent. Currie and Wolfe (1979) found that shortening began at pH 6.15 – 6.35 with loads of 5 g/cm² and pH 5.75 – 5.85 with loads greater than 25 g/cm². When the pH reaches 5.75, all muscles will contract. The extent of shortening at this pH depends on the remaining ATP concentration and the load on the muscle (Currie and Wolfe, 1979). This tendency to shorten during rigor is reflected in the decreased sarcomere lengths of postrigor muscle versus prerigor (Asghar and Yeates, 1978). The maximum force developed during rigor contraction is $< 5\%$ of that of living muscle (Bate-Smith and Bendall, 1947), presumably because only a small proportion of the fibers actually contract (Bendall, 1973b). Maximum isometric tension occurs at the same time as the completion of rigor (Schmidt et al., 1970a). This rigor shortening can be prevented by antemortem injection of Ca^{2+} chelators (Weiner and Pearson, 1969).

The pattern and cause of shortening that occurs at low temperatures in muscle postmortem is quite different from that at higher temperatures. Locker and Hagyard (1963) were first to show that muscle removed from a beef carcass immediately postmortem would shorten by up to 50 or 60% when placed at 2 °C. This shortening is called *cold shortening* and is distinct from rigor shortening because the former occurs more rapidly (in a few minutes after reaching low temperatures), at high pH (usually > 6.7), and with ATP levels of 5 to 6 mM while the latter develops slowly (many hours), at low pH (usually primarily

< 6.0), and at ATP levels < 1 mM (Busch *et al.*, 1967; Bendall, 1975). Muscles that have cold-shortened will also undergo a second contraction at the time of the rapid ATP decline, and it will be expressed as either tension development or shortening (Bendall, 1975). Cold shortening occurs only in muscles with a significant proportion of red muscle fibers and thus its extent is much less in most rabbit and pig muscles. The extent of shortening of pig longissimus dorsi and biceps femoris is < 20% at 2°C (Hendricks *et al.*, 1971). The beef sterno-mandibularis from young calves will only shorten ~ 20% compared to 60% in older animals (Davey and Gilbert, 1975). Cold shortening occurs in muscle with temperatures < 11°C (Locker and Hagyard, 1963; LaCourt, 1972) and more rapidly with lower temperatures in the range 0–11°C. It is also reversible, with three cycles of contracture and relaxation attainable in changing the muscle temperature between 15° and 2°C (Bendall, 1973a).

The cause of cold shortening is not fully understood. It has been found that fiber pieces will cold shorten while myofibrils will not (Davey and Gilbert, 1974). Shortening is believed to occur as a result of a rise in intracellular calcium levels in the region of the myofibrils. This calcium may be released by the sarcoplasmic reticulum because of its reduced efficiency in the cold (Kanda *et al.*, 1977) or from mitochondria (Buege and March, 1975).

2. CONTRACTILE AND ENZYMATIC PROPERTIES

Myofibrils from postmortem muscle usually are fully capable of contraction when supplied with ATP, Mg^{2+}, and Ca^{2+}. Partmann (1963) found that all fibers in muscle homogenates contracted upon addition of ATP. The ultimate pH is critical, however, in maintaining contractile function. With ultimate pH levels of 5.5 or greater, 100% of the myofibrils from 24-hr postmortem pig muscle would contract while none would shorten with ATP if the muscle pH was < 5.3 (Sung *et al.*, 1976).

The measured ATPase activity of myofibrils prepared at different times postmortem from muscles left on the carcass often increases 20–60% during the first 24 hr. This has been observed in pig (Galloway and Goll, 1967; Goll and Robson, 1967; Greaser *et al.*, 1969a; Cheng and Parrish, 1978a), beef (Parrish *et al.*, 1973), and rabbit myofibrils (Yang *et al.*, 1970; Ikeuchi *et al.*, 1978). This increase is probably an artifact and may be due to either the drastic morphological and/or extraction changes that occur when muscle is homogenized prerigor (Greaser *et al.*, 1969b), thus lowering prerigor myofibril activity, or to the increase in myofibril and myofibril bundle disruption with increasing time postmortem (see Section III,A,3). Increases in myofibril ATPase activity have been shown to occur with increased homogenization time (Yang *et al.*, 1970).

The temperature at which muscle is held postmortem has an effect on the myofibrillar ATPase activity. In rabbit muscle held at 37°C, the Ca^{2+}, Mg^{2+}-

ATPase activity declined by ~25% within 6 hr (Ikeuchi *et al.*, 1980b). Pig muscle myofibrils prepared from muscle held 8 hr at 37°C had lost ~75% of the ATPase activity compared to preparations made at death (Galloway and Goll, 1967). However, holding temperatures of 30°C and below resulted in no loss of pig myofibrillar ATPase activity during the first 6 hr postmortem (Penny, 1977) and 37°C treatment for 3 hr resulted in no decrease in activity (Penny, 1967). As a result of these studies it appears unlikely that any significant decline in myofibrillar ATPase will occur in muscles left on the carcass, even in the deepest regions, which cool slowly.

The ATPase activity of actomyosin from postmortem muscle increases markedly (Robson *et al.*, 1967; Strandberg *et al.*, 1973). This increase appears to be due to the increasing proportion of actin extracted with time postmortem (Fujimaki *et al.*, 1965c; Goll *et al.*, 1970).

Myofibrillar ATPase activities measured in the presence of a calcium chelator will give a measure of potential damage to the tropomyosin – troponin control system. In general the EGTA-ATPase increases with time postmortem, but the increases are quite small (Goll and Robson, 1967; Parrish *et al.*, 1973; Cheng and Parrish, 1978a; Ikeuchi *et al.*, 1980a). High-temperature incubation decreases the myofibrillar EGTA-ATPase activity as well as that activated by Ca and Mg (Galloway and Goll, 1967).

3. PROTEOLYTIC MODIFICATIONS

The action of proteolytic enzymes in postmortem muscle is remarkably limited. Sharp (1963) and Zender and co-workers (1958) observed a very slow increase in nonprotein nitrogen (NPN) in muscles held aseptically at 25° or 37°C for several weeks. Locker (1960) found no increase in either NPN or free amino acids between 0 and 2 days postmortem at 2 or 21°C. Parrish and co-workers (1969) found only minor increases in NPN and amino acids after 24 hr storage of beef muscle at 37°C. Davey and Gilbert (1966) followed changes in the ratio of NPN to total nitrogen over a 28-day period and concluded that <2.3% of the protein had been degraded. In fact part of the increase in NPN may have been from nucleic acid breakdown (Davey and Gilbert, 1966; Petropakis *et al.*, 1973). There also appear to be no significant increases in N-termini in beef muscle held 2 days at 2°C (Locker, 1960).

Subtle changes do occur in muscle proteins, as evidenced by the increasing degree of fragmentation that myofibrils undergo upon homogenization with time postmortem (Davey and Gilbert, 1967; Stromer and Goll, 1967a). This increase in fragmentation can be completely arrested by 5 mM EDTA (Hattori and Takahashi, 1979). The fragmentation that occurs upon homogenization is much greater in muscles left attached to the carcass before rigor mortis, suggesting that rigor tension may be physically altering the myofibrils (Hattori and

Takahashi, 1979). The yield of thick and thin filaments from purified myofibrils upon homogenization in ATP increases from $\sim 5\%$ at 0 to 2 days postmortem to $\sim 50\%$ after 6 days storage (Takahashi et al., 1981). The percentage of extractable myofibrillar protein in 0.1 M pyrophosphate increases from 59% immediately after death to 68% at 1 day postmortem (Penny, 1968). The actin content of myosin B increases from 23.1 to 28.1% between 0 and 2 days postmortem (Fujimaki et al., 1965c). Both of these latter observations point to a weakening of the attachment of actin filaments to the Z lines. The content of α-actinin, a Z-line protein, does not change in muscle postmortem (Arakawa et al., 1970; Penny, 1972; Yang et al., 1978). The yields of tropomyosin and troponin also do not change much in the first 24 hr after death (Arakawa et al., 1970; Yang et al., 1978).

Numerous studies have examined changes in the sodium dodecyl sulfate (SDS) gel electrophoretic patterns with time postmortem. The overall band patterns remain relatively constant. Troponin-T may be partially degraded, and new bands in the 25,000–30,000-dalton region appear in postmortem pig (Penny, 1976), beef (Olson et al., 1977), and rabbit muscle (Ikeuchi et al., 1980a; Takahashi et al., 1981). The appearance of the 30,000-dalton fragment is barely detectable in muscle held for 24 hr at 2°C (Olson et al., 1977) but greater amounts occur in muscle held at 37°C (Yates, 1977; Ikeuchi et al., 1980b). As much as three-fourths of the troponin-T may be degraded in electrically stimulated muscle held 1 or 2 days at 25 to 35°C (Penny and Dransfield, 1979). Beef muscle left on the carcass also shows significant amounts of the 30,000-dalton component at 24 hr postmortem (MacBride and Parrish, 1977; Young et al., 1980). There is, however, muscle–muscle variation with no evidence for troponin-T degradation in psoas muscle during the first 24 hr after death (Olson et al., 1977).

There is also evidence for myosin degradation postmortem. Yates (1977) found that myosin heavy chain was split at several different points in postmortem beef muscle, especially after holding at 37°C. The fragments formed included components in the 50,000–100,000-dalton region, one of which appeared to be subfragment-1. A 110,000-dalton band from the myofibrillar fraction, presumably from myosin, increased between 0 and 2 days postmortem in beef sternomandibularis held at 15°C (Young et al., 1980). Ikeuchi and co-workers (1980b) observed an increase in the number and amounts of bands in the 100,000–200,000-dalton region on SDS gels during the first 12 hr storage of rabbit muscle at 37°C. The amount of chromatographically separated myosin from beef semitendinosus muscle held at 0 to 4°C for 0, 5, and 7 days was 11.2, 8.3, and 0.24 mg, respectively (Gunther, 1974).

Changes have been observed in two myofibrillar proteins called connectin and desmin. The connectin content of the KI-insoluble residues of beef muscle decreased with time postmortem (Young et al., 1980). The amount of connectin isolated from rabbit muscle declined by ~ 50% in the first 3 days after death (Takahashi and Saito, 1979). Similarly, the content of the 55,000-dalton desmin band on SDS gels declined in beef muscle held for 2 days at 15°C (Young *et al.*, 1980). Robson and co-workers (1980) also found that desmin content declined postmortem at about the same rate as troponin-T was degraded. Since desmin is a Z-line protein, these observations may be related to the postmortem increase in fragility of myofibrils upon homogenization and the ultrastructural changes in the Z lines after death.

B. Connective Tissue

Postmortem changes in connective tissue appear to be very limited. Sharp (1963) found no increase in collagen fragments soluble in 0.1 *M* KCl after 6 months sterile storage of beef muscle at 37°C. The hydroxyproline content of the soluble fractions and the alkali-insoluble protein (primarily collagen) did not change between 3 and 15 days postmortem (Wierbicki *et al.*, 1954, 1955). The lack of increase in soluble hydroxyproline is not too surprising since few enzymes will attack native collagen in its helical regions (Sims and Bailey, 1981). No changes in the first 2 weeks postmortem were found in the percentage of neutral salt-soluble collagen, percentage of acid-soluble collagen, collagen melting temperature, SDS gel electrophoresis patterns of acid-soluble collagen, or total collagen content (Chizzolini *et al.*, 1977; Jeremiah and Martin, 1981).

In contrast Gunther (1974) found a twofold increase in the acid-soluble collagen proportion between 0 and 5 days postmortem in beef semitendinosus stored at 0 to 4°C. Field and co-workers (1970) found that pig epimysial collagen isolated at 24 hr postmortem had a lower melting point than that isolated at death. Postmortem pig and lamb collagen was less stable than 0-hr samples when measured by differential scanning calorimetry (Chizzolini *et al.*, 1975). McClain and co-workers (1970) found that the yield of intramuscular connective tissue dropped by ~ 50% in bovine longissimus dorsi muscles stored at 4°C for 10 hr. However, this change may be partially the result of cold shortening, which would affect the physical configuration of the connective tissue. It is known that the orientation of the collagen and reticular networks are drastically altered by muscle shortening (Rowe, 1974; Stanley and Swatland, 1976). McClain and co-workers also observed a 34% decrease in connective tissue yield in pig muscle between 0 and 72 hr postmortem.

Mucopolysaccharides appear to be relatively stable with time postmortem. McIntosh (1967) found an increase from 45 to 58% in the soluble hexosamines after 14 days aging of beef. In contrast no postmortem changes in soluble glucosamine (Shelef and Jay, 1969) or hexosamines (Stanley and Brown, 1973) were found in muscle kept under relatively sterile conditions.

C. Organelles and Membranes

1. SARCOPLASMIC RETICULUM

The first evidence for a postmortem change in the sarcoplasmic reticulum (SR) was the observation of the decay of a soluble factor that affected muscle fiber swelling in homogenates (Marsh, 1952). This factor was found to have reduced activity in postrigor muscle. The soluble factor was later identified as fragments of the SR.

Studies on SR fractions isolated at various times postmortem from pig longissimus dorsi showed that the ability to accumulate Ca^{2+} decreased markedly (Greaser et al., 1967, 1969a). Values for both normal and rapid glycolysis (PSE) muscles are shown in Table IV. The calcium-accumulating capacity declined by $\sim 40\%$ in the first 3 hr and by 80% at 24 hr postmortem. A study using highly purified reticulum fragments demonstrated an $\sim 95\%$ decline in pig muscle calcium uptake in the first 24 hr (Greaser et al., 1969c). The loss of calcium-accumulating ability paralleled the loss of muscle extensibility, although some calcium binding activity remained when muscles were fully in rigor (Schmidt et al., 1970b). Similar losses in calcium binding ability have been observed in pig muscle homogenates postmortem (Kryzwicki, 1971). Beef sarcoplasmic reticulum from the rectus abdominus had calcium uptake values that declined five- to 10-fold between 0 and 24 hr postmortem (LaCourt, 1971). The calcium binding ability of SR fragments from beef sternomandibularis held at $15°C$ drops to near zero by 24 hr after death (Kanda et al., 1977). Rabbit muscle held at $37°C$ shows a five- to sixfold decline in calcium uptake ability between death and the time of maximum isometric tension (roughly, at rigor completion) (Goll et al., 1973).

Even though the calcium uptake declines postmortem, the sarcoplasmic reticulum Ca^{2+}-ATPase remains constant or increases. In pig muscle preparations the ATPase activity appeared to increase almost twofold by 24 hr (Table IV). This increase was not seen in highly purified SR preparations in comparing 0- and 24-hr muscle (Greaser et al., 1969c). Beef muscle sarcoplasmic reticulum fractions have also been shown to have either constant or increasing ATPase activity with time postmortem (LaCourt, 1971).

2. MITOCHONDRIA

Mitochondria are fairly resistant to postmortem alteration. Storage of bovine sternomandibularis muscle for 48 to 96 hr at 1 to $4°C$ before mitochondria isolation resulted in preparations with succinoxidase, cytochrome oxidase, respiratory control index, and state 3 respiration rate values $> 90\%$ of those from 0-hr samples (Cheah and Cheah, 1971). Holding beef muscle at 10 or $20°C$ for 2 days also does not change the mitochondrial properties significantly

Table IV

Comparison of Macromolecular Changes Postmortem in Normal and Rapid Glycolysis (PSE) Pig Muscle

Parameter	Normal	PSE	Significance	Reference
Time for rigor mortis completion	Typical	Shorter	—	Bendall et al. (1963); Forrest et al. (1966)
Myofibrillar ATPase (μmol PO_4/mg·min)	0 hr 0.14	0.14	NS[a]	Greaser et al. (1969a)
	30 min 0.12	0.09	NS	Greaser et al. (1969a)
	1 hr 0.12	0.07	$p < .05$	Greaser et al. (1969a)
	3 hr 0.16	0.07	$p < .01$	Greaser et al. (1969a)
	24 hr 0.22	0.06	$p < .01$	Greaser et al. (1969a)
Myofibril contractility	Yes	No	—	Sung et al. (1977)
Myosin extractability	10× higher	—	—	Sung et al. (1977)
Myosin ATPase	24 hr 2.5	1.2	—	Sung et al. (1981)
Myosin phosphate burst	1.1–1.2	0	—	Sung et al. (1981)
Salt-soluble collagen	3.20%	4.29%	$p < .05$	McClain et al. (1967)
Heat-labile collagen	15%	23%	$p < .01$	McClain et al. (1967)
Sarcoplasmic reticulum ATPase	0 hr 0.11	0.14	NS	Greaser et al. (1969a)
	30 min 0.15	0.13	NS	Greaser et al. (1969a)
	1 hr 0.16	0.11	NS	Greaser et al. (1969a)
	3 hr 0.22	0.11	$p < .05$	Greaser et al. (1969a)
	24 hr 0.23	0.10	$p < .01$	Greaser et al. (1969a)
Sarcoplasmic reticulum calcium uptake	0 hr 2.20	1.85	NS	Greaser et al. (1969a)
	30 min 2.09	1.06	$p < .01$	Greaser et al. (1969a)
	1 hr 1.83	0.64	$p < .01$	Greaser et al. (1969a)
	3 hr 1.25	0.42	$p < .01$	Greaser et al. (1969a)
	24 hr 0.40	0.15	$p < .01$	Greaser et al. (1969a)
Mitochondrial Ca^{2+} efflux rate (μmol/min-mg)	121	216	$p < .05$	Cheah and Cheah (1979)
Response to electrical stimulation	Slow loss	Rapid loss	—	Forrest et al. (1966); Forrest and Briskey (1967)
AMP deaminase activity, 0 hr	1.04	1.83	—	Tsai et al. (1972)
3 hr	0.83	0	—	
Phosphorylase activity, 45–60 min	6.1	1.8	—	Fischer et al. (1979)

[a] NS, not significant.

(Cheah and Cheah, 1974). Leaving muscles on the carcass for 24 hr before mitochondrial isolation results in larger changes. Ashmore and co-workers (1972) found that beef muscle mitochondria at 24 hr postmortem (muscle pH of 5.7) had large decreases in respiratory activity. In muscles having a 24-hr pH of ~6.7 the changes were minor. The major change that occurred was a gradual leakage of cytochrome c, which can amount to as much as 40% after 8 days storage at 2°C (Cheah and Cheah, 1974). This loss of cytochrome c may be the major cause of lowered oxygen consumption in mitochondria from postmortem muscle (Cheah, 1973). In beef or pig muscle stored at 1–4°C, functional mitochondria could always be prepared from postmortem muscle as long as the muscle pH exceeded 5.5 (Cheah and Cheah, 1971; Cheah, 1973). Sheep muscle mitochondria prepared at 48 hr postmortem had twofold higher respiratory activity if the muscle pH was ~6.5 versus 5.8 (Ashmore et al., 1973).

Mitochondria from pig muscle stored at 1°C lost their functional properties more rapidly, with oxidation rates for pyruvate plus malate or succinate declining 30–40% by 24 hr postmortem (Cheah, 1973). The loss of cytochrome c from pig mitochondria is more rapid than from beef, particularly with muscles having a more rapid pH decline rate. In pig muscle left on the carcass before mitochondrial isolation, the state 3 and state 4 respiration, the respiratory control index, and the ADP/O ratio is reduced by the time the muscle pH reaches ~6.3–6.4 and no respiratory activity remains if the muscle pH is <5.9 (Campion et al., 1975).

The proportion of glutamic–oxaloacetic transaminase activity bound to mitochondria remains the same between 0 and 24 hr postmortem in both beef and pig muscle stored at 4°C (Hamm et al., 1969).

3. LYSOSOMES

Normal skeletal muscle cells contain almost no morphologically identifiable lysosomes. However, there may be some lysosomes present in other cells found in muscle tissue. Catheptic enzymes, which normally are found in lysosomes, have recently been localized inside muscle cells (Stauber and Ong, 1981). It appears that these enzymes are associated with specialized regions of the SR system.

Studies have been conducted on the relative proportion of catheptic enzymes that are bound in membrane particles with time postmortem. Lutalo-Bosa (1970) found that the cathepsin C in beef psoas was 50% free in 2-hr postmortem muscle and did not change with subsequent storage time. Cathepsin B was 25% free and increased to 40% free while Cathepsin D changed from 1% free to 20% free with postmortem storage. In contrast Suzuki and Fujimaki (1968) found less than a twofold increase in soluble cathepsin D between 0 and 3 days postmortem in rabbit muscle. Dutson and Lawrie (1974) found that the non-sedimentable β-glucuronidase increased twofold between 1 and 24 hr postmortem in beef longissimus stored at 2°C. Moeller and co-workers (1976) showed that holding muscle at 22°C for 4 hr, and then 12°C for 8 hr resulted in higher proportions of free cathepsin C and β-glucuronidase than when muscle was stored at 2°C for 12 hr. Incubation of beef muscle at 37°C resulted in a 50% increase in the free β-glucuronidase (Wu *et al.*, 1981). In contrast the percentage activity of β-glucuronidase and acid ribonuclease in the membrane-bound state did not change between 3 and 24 hr postmortem in beef muscle held at 3°C (Ono, 1971). Thus temperature appears to have an effect, and muscle from intact carcasses would be expected to show intermediate properties between the 0–4 versus 37°C incubation extremes.

4. PLASMALEMMA

The plasmalemma, or outer cell membrane, provides a barrier between the cell contents and the extracellular space. In living muscle the interior of the cell has a resting potential of -70 to -80 mV (Gallant *et al.*, 1979). Nerve impulses travel down the axons and cause the release of acetylcholine at the motor end plates. The acetylcholine diffuses to the plasmalemma and causes a local depolarization due to an increase in membrane permeability. This depolarization travels along the surface of the fiber and down the connected T tubules, initiating contraction. The membrane potential is then reestablished by the Na^+, K^+-ATPase.

The ability of the nerves to stimulate muscle contraction decays quite rapidly postmortem. Electrical stimulation (which affects primarily the nerve pathways) elicits no muscle contractions after ~ 1 hr in beef (Bendall *et al.*, 1976; Bendall, 1980), 30–40 min in lambs (Chrystall *et al.*, 1980), and 15 min in pigs (Swatland, 1975a). High voltage stimulation (600 V) causes only feeble contractions in beef muscle at 4 hr postmortem (Bendall, 1980). A response to electrical stimulation has been observed as late as 7 hr postmortem in beef (Partmann, 1963). Lambs treated with curare (a neuromuscular blocking agent) do not respond to 250-V electrical stimulation (Bendall, 1980). Thus nerve pathways appear to be the major mode of stimulation and direct muscle cell effects are of minor importance (Swatland, 1977; Bendall, 1980). Increasing voltages are required with time postmortem to cause contraction (Forrest *et al.*, 1966). A 12-V stimulus causes contraction in lamb muscle immediately after death but not at 30 min while a 200-V stimulus is effective at both times (Chrystall *et al.*, 1980). Electrical stimulation depletes white muscle glycogen levels more rapidly than those in red fibers (Swatland, 1975b).

Postmortem changes occur in the electromyographic (EMG) activity of muscle. Schmidt and co-workers (1972) found that EMG activity declined rapidly

Table V
ATP Turnover and pH Changes during Electrical
Stimulation[a]

Time (sec)	Cumulative ATP turnover[b]	pH
0	0	7.10
30	40	6.61
60	52	6.50
90	64	6.43
120	68	6.36
240	70	6.31
Full rigor	170	5.50

[a] From Bendall (1980), with permission. Beef carcasses were stimulated at 700 V (peak), 25 pulses per sec. Values were obtained from the triceps brachii.
[b] Expressed as μmol/g of muscle.

in pig muscle after death. Activity ceases at 4, 5, 6, 8, 10, and 38 min postmortem in pig muscles left on the carcass (Swatland, 1976b). Swatland (1976a) also showed that EMG activity lasted for 1 to 4 hr postmortem in beef sternomandibularis with an average of ~2 hr. The EMG activity ceased upon cooling and reappeared when the temperature was raised to 25 or 26°C. Placing a freshly excised muscle in a N_2 atmosphere stopped its twitching and EMG activity (Swatland, 1976a). Restoration of oxygen caused the EMG activity to resume. These experiments demonstrate the importance of the surface nerves to these activities.

The response to continuous electrical stimulation of muscle declines very rapidly (Bendall, 1980). Table V shows the effects of 700-V (peak) stimulation at 25 pulses/sec on beef triceps brachii. The initial rate of ATP turnover is extremely high, but declines to almost zero after 2 min. Most of the pH fall occurs in the first 30 sec, and the pH rarely goes lower than 6.3 in beef at the end of stimulation (Bendall, 1980). Pig muscle pH may decline as low as 6.0 after electrical stimulation.

Measurements of capacitance and electrical resistivity in muscle after death have been made. The capacitance of normal pig muscles falls from 7.8 nF at death to 3 at 6 hr postmortem and 2.5 after 24 hr (Swatland, 1980, 1981). Resistivity also declines by one-third to one-half in the first 5–6 hr after death (Swatland, 1980). The electrical resistance is much greater across the fibers than along them immediately postmortem (Callow, 1947). When the muscle pH drops to ~5.7, muscle fibers shrink and exude fluid (Bendall and Taylor, 1972). At this point the electrical resistance is now equal in both directions (Callow, 1947). The diffusion coefficient for NaCl was equal both parallel and at right angles to the fibers in 2-day postmortem pig muscle (Wood, 1966).

Fenichel and Horowitz (1965) found that the efflux of protein from muscle increased dramatically after rigor mortis development. Cerrella and Massaldi (1978) showed that the diffusion rate of soluble proteins out of 72-hr postmortem beef semitendinosus was equal from the surfaces parallel and perpendicular to the muscle fibers. The ability of the plasmalemma to exclude trypsin from the cell interior persisted for 8 hr in rabbit muscle strips bathed at 37°C (Busch et al., 1972).

Damage to the plasmalemma also is related to postmortem changes in the extracellular space. Heffron and Hegarty (1974) measured the inulin space of mouse muscle held at 18 to 20°C for various times postmortem. They found that the "extracellular" space increased from 17.1 ml/100 g to 40.3 ml/100 g between 0 and 4 hr postmortem. No changes occurred during the first 2 hr. By 24 hr after death the value was 74.6 ml/100 g. The conclusion was that the plasmalemma became partially permeable by 4 hr (the time to rigor mortis in these muscles) and fully permeable by 24 hr postmortem. Currie and Wolfe (1980) also found a large increase in the extracellular space at the time of rigor mortis onset. The change in membrane permeability results in exudation of liquid from the muscle surface. The extent of this drip loss varies with muscle holding temperature. Penny (1977) found that pig longissimus dorsi held for 6 hr postmortem at 10, 20, 25, 30, 34, and 37°C had drip losses of 1, 2, 4, 5.5, 8, and 11%, respectively. Taylor and Dant (1971) also found that drip loss was greater from muscles left on the carcass, which had a slow temperature decline. Thus it appears that the integrity of the plasmalemma declines dramatically at about the time of rigor mortis completion.

D. Enzyme Activation and Inactivation

The various enzymatic processes that occur in muscle undergo change during the postmortem time period. From the preceding discussions it is clear that all enzyme systems that use ATP become inactivated because of substrate depletion. Also, the mitochondrial ATP production is inactivated because of lack of O_2 availablility. Other processes may be altered by the postmortem drop in temperature and pH. In general lower temperatures decrease the rate of enzyme reactions while pH decline may either inactivate or activate certain enzymes. A distinction must be made between the altered rate of reaction of an enzyme versus a denaturation process. In the former the reaction rate may be restored by changing the reaction conditions while in the latter an irreversible change has occurred.

1. ORGANELLES

The myofibrillar ATPase normally resists denaturation postmortem (see Section III,A,7). Its activity, however, depends on the muscle pH. Bentler (1977)

Tsai and co-workers (1972) have followed the changes in several enzymes that affect nucleotides in pig muscle postmortem. They found that adenosine deaminase remained fully active at 24 hr after death and that 5' nucleotidase increased twofold at 3 and 24 hr compared to at death. In contrast the AMP deaminase activity had declined to less than half that in normal muscle at 24 hr postmortem.

One of the enzymes most sensitive to postmortem denaturation is creatine kinase. Scopes (1965) found that this enzyme was fairly stable down to pH 6.1 at 30°C but was rapidly inactivated at below pH 6.0. Tarrant and Mothersill (1977) showed that this enzyme had a 12% loss in activity at 48 hr postmortem in beef muscle at 1.5 cm from the surface but an 84% loss at 8 cm depth. Presumably, the slower temperature decline in the deep muscle regions led to a faster postmortem pH decline and as a result the high temperature plus low pH denatured the enzyme. This inactivation will have little consequence for postmortem metbolism, however, since the CP levels will have fallen to very low levels by the time pH 6.0 is reached.

Several enzyme activities have been followed histochemically. Bodwell and co-workers (1965b) found that alcohol dehydrogenase, glutamate dehydrogenase, glucose 6-phosphate dehydrogenase, and β-hydroxybutyrate dehydrogenase activities were either very weak or absent at 24 to 48 hr postmortem in beef muscle. Succinic dehydrogenase, isocitrate dehydrogenase, glutamate dehydrogenase, lactate dehydrogenase, and alcohol dehydrogenase activites all declined in pig muscle by 24 hr after death. In addition all uridine diphosphate glucose – glycogen transferase activity was gone in muscle held at 37°C for 3 hr (Bodwell et al., 1965b). Dutson and co-workers (1971) found that esterase activity in pig muscle was gone by 3 hr postmortem in pig muscle. However, acid phosphatase activity was fully retained.

3. PROTEASES

Although the amount of protein degradation in postmortem muscle appears to be very limited, there are several proteases present that may be altered in activity after death, particularly by changes in pH. The proteases can be divided into two major groups, namely, neutral proteases and cathepsins. The former group is not normally membrane-bounded and is optimally active at pH levels near 7.0. The calcium-activated factor (CAF) is the best-characterized member of this class (Dayton et al., 1976a,b). It requires millimolar levels of calcium for full activity and thus would be minimally active in the early postmortem period while there is still ATP-driven calcium binding by the sarcoplasmic reticulum. Free calcium levels rise to ~ 0.1 mM after rigor mortis (Nakamura, 1973). CAF activity is lower but still present at pH 5.5 – 5.8 (Penny and Ferguson-Pryce, 1979). The enzyme activity declines from 80 to 90 units at death to 60 to 70 units by 24 hr postmortem in beef longissimus dorsi and semitendinosus (Olson *et al.*, 1977). The decline is much greater in beef psoas, falling from 35 to 5 units in the first day. This enzyme has been implicated in the postmortem degradation of troponin-T (TN-T) and the postmortem break-down of the Z line (see Section III,E,1). Incubation of muscle pieces and freeze-dried muscle with calcium plus CAF results in structural and physical changes similar to those occurring in postmortem muscle (Penny *et al.*, 1974; Cheng and Parrish, 1977; Penny and Ferguson-Pryce, 1979). The crude enzyme has also been shown to degrade native collagen (Kang *et al.*, 1981). However, there is evidence that calcium activates a nonCAF protease that is involved in the increased myofibril susceptibility to fragmentation postmortem (Hattori and Takahashi, 1979).

Cathepsins are normally localized inside membrane compartments in living muscle. Most of these proteases require a low pH for optimal activity. They would also presumably have to be released from their compartments to act on sarcoplasmic or myofibrillar proteins. Cathepsin B degraded native myosin with an optimum pH of 5.2 but was only 50% as active at pH 5.7 and 20% as active at pH 6.2 (Schwartz and Bird, 1977). However, the purified enzyme did not degrade any proteins in the myofibril at the pHs normally encountered in postmortem muscle (Okitani *et al.*, 1980). Cathepsin C activity is partially destroyed by 37°C incubation of beef muscle for 12 hr versus 2°C incubation (Moeller *et al.*, 1977). Cathepsin D has optimum activity at pH 4.0 (Schwartz and Bird, 1977), about one-sixth of optimum activity at pH 5.3, and almost no activity above pH 6.3 (Suzuki and Fujimaki, 1968). Robbins and co-workers (1979) found that incubation of myofibrils with cathepsin D for 24 hr at 25 or 37°C and pH 5.2 – 5.3 resulted in Z-line degradation and partial destruction of myosin and troponin. In contrast there were no changes found in myofibrillar ATPase activity after cathepsin D incubation for 10 days at 3°C and pH 5.5 (Okitani *et al.*, 1972). Cathepsin L, a newly described muscle protease, was found to destroy troponin and/or tropomyosin in myofibrils with an optimum pH of 6.0 to 7.0 (Okitani *et al.*, 1980). It also degraded the myosin heavy chain in myofibrils at acid pH (optimum 4.1). Little myosin breakdown occurred in the pH range between 5.9 and 7.0. The enzyme has optimum stability in the pH 4.0 – 6.0 range (Okitani *et al.*, 1980) so the postmortem pH changes should not be deleterious. The same conclusions probably hold for all of the cathepsins.

Eino and Stanley (1973) showed that total water-soluble catheptic activity from beef muscle declined minimally between 0 and 24 hr postmortem. The activity of macrophage cathepsins B and D to degrade connective tissue matrices was found to be optimal at pH 5.5 (Werb *et al.*, 1980). Otsuka and co-workers (1976) purified an aminopeptidase from rabbit muscle that had maximum activity in the pH 6.0 – 7.0 range and retained 50% activity at pH

5.3. A catheptic protease with a pH optimum of 5.0 to 5.5 has been extracted from rabbit myofibrils and been found to degrade myosin heavy chains and α-actinin upon incubation with myofibrils (Arakawa *et al.*, 1976).

Okitani and Fujimaki (1972) have studied the effect of pH on protein degradation in muscle homogenates. They found that the greatest amounts of ninhydrin-positive materials were produced at pH levels >6.0 while more peptides were liberated at pH levels <5.5. Similar results were obtained using whole muscle in which the pH had been arrested at varying points by iodoacetate injection (Okitani *et al.*, 1973). Minimum proteolysis occurred at pH 6.3. These results lead to the conclusion that there are two separate proteolytic systems operating at neutral and acid pH.

E. Morphological Changes

1. MYOFIBRILS

The fact that myofibrils occupy 70 or 80% of the volume of the muscle cell has meant that these structures have received primary attention from investigators looking for changes postmortem. Many of these investigators have examined the effect of incubation temperature on myofibril structure. Maintaining beef sternomandibularis at 15°C for 24 hr yields myofibrils with normal A and I bands, H zones, and Z lines (Davey and Gilbert, 1967). Holding beef muscle at 2°C for 24 hr results in much shorter sarcomere lengths compared to muscle held at 16°C (Fig. 6) (Stromer and Goll, 1967a,b; Stromer *et al.*, 1967). Muscles with beginning sarcomere lengths averaging 2.7 μm shorten to 2.0 μm at 16°C and 1.2 μm at 2°C. Only about half of the fibers actually cold shorten (Voyle, 1969). Similar effects occur in lamb held at 0 or 5°C (Cook and Wright, 1966). Pig and rabbit muscle myofibrils shorten to a lesser extent. Shortening also occurs at 37°C in beef, pig, and rabbit muscle (Henderson *et al.*, 1970). The extent of high- and low-temperature shortening will be much less in intact carcasses because of the restraints due to muscle attachments. Myofibrils isolated from muscle immediately after death show more abnormal structure than those at 24 hr postmortem (Stromer and Goll, 1967b; Greaser *et al.*, 1969b). However, these differences result because of the structural damage that occurs when myofibrils are homogenized in the presence of ATP (Locker *et al.*, 1976).

Higher temperature incubation results in damage primarily in the Z-line region. Beef and lamb muscle held at 25 to 37°C for 24 hr show missing patches from the Z lines and occasional tearing of the attachments between the thin filaments and Z lines (Cassens *et al.*, 1963b; Cook and Wright, 1966; Henderson *et al.*, 1970; Gann and Merkel, 1978). Pig and rabbit muscle have

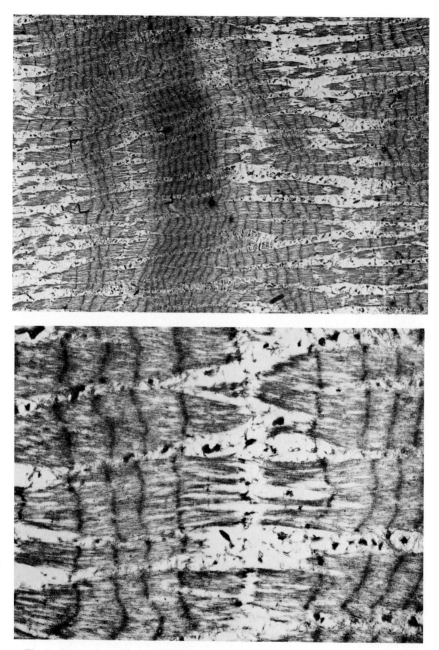

Fig. 6 Contraction nodes in cold-shortened beef muscle. Beef sternomandibularis muscle was removed from the carcass at 20 min after slaughter, held unrestrained at 15°C for 3 hr, and then placed in a cold room at 0°C until 48 hr postmortem. The muscle shown had shortened by 55%. Samples were fixed for electron microscopic examination. Note the alternating zones of contraction with regions of myofibril stretch and breakage. (top) ×2960; (bottom) ×12,600. From Marsh *et al.* (1974), with permission.

Fig. 7 Myofibrils from pig muscle at 24 hr postmortem. Pigs were slaughtered and the carcasses transferred to a 4°C cooler at 30 min postmortem. Samples were removed from the longissimus dorsi muscle and homogenized in 0.1 M KCl, 5 mM histidine (pH 7.2). Myofibrils were collected by centrifugation and fixed for electron microscopic examination. The filament structure was well preserved, but regions of Z-line degradation were evident (see arrow). ×18,000. From Greaser *et al.* (1969b). Copyright by Institute of Food Technologists.

more rapid Z-line degradation than beef (Henderson *et al.*, 1970) and pig myofibril Z lines may be almost totally absent after 8 hr incubation at 37°C (Galloway and Goll, 1967). Z-line damage is commonly observed in myofibrils from pig muscle left on the carcass for 24 hr (Fig. 7) (Greaser *et al.*, 1969b). This results in the appearance of lengthwise splits in the myofibrils (Abbott *et al.*, 1977). The Z lines appear virtually intact in 24-hr postmortem beef muscle (cf. Figs. 8 and 9), however. Z-line damage is much more extensive in the white or type II fibers compared to red or type I fibers in both pig and beef muscle (Dutson *et al.*, 1974; Abbott *et al.*, 1977; Gann and Merkel, 1978).

The M lines are destroyed in pig and rabbit muscle held at temperatures of 25°C and above for 24 hr but they remain unchanged in beef muscle incubated under similar conditions (Henderson *et al.*, 1970). The thin filaments of pig myofibrils may undergo side-to-side aggregation when the muscle is left on the carcass for 24 hr (Cassens *et al.*, 1963b) or when the muscle is held at 20 to 25°C until 9 hr postmortem (Hegarty *et al.*, 1973).

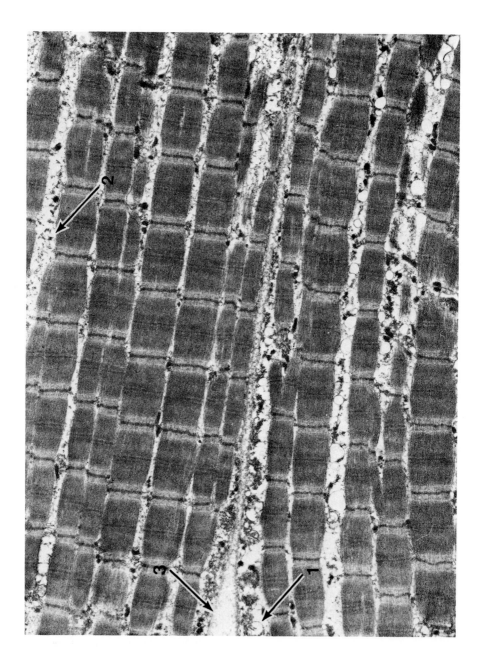

2. SARCOPLASMIC RETICULUM

The SR network becomes increasingly disrupted with time postmortem. Swelling of these membrane elements is often apparent within 1 hr postmortem (Wesemeier, 1974; Will et al., 1980) and extensive degeneration is evident at 24 hr postmortem (Stromer et al., 1967; Dutson et al., 1974; Abbott et al., 1977). Virtually no triad structures or T tubules are visible at 24 hr after death in pig muscle (Dutson et al., 1974). Ultrastructural studies on isolated SR vesicles obtained at death or at 24 hr postmortem, however, revealed no differences in membrane integrity or in the nature of the surface particles (Greaser et al., 1969b,c; Kanda et al., 1977).

3. MITOCHONDRIA

The mitochondria of muscle typically show some swelling and disruption with time postmortem (Fig. 9). This is usually evident by 6 to 24 hr in beef muscle (Will et al., 1980) and much earlier postmortem in pig muscle (Cassens et al., 1963b; Hegarty et al., 1973; Wesemeier, 1974). However, many mitochondria may appear relatively normal at 24 hr postmortem with distinct and closely packed cristae (Dutson et al., 1974; Abbott et al., 1977). The density of the material between the cristae declines and some cristae aggregation may occur (Dutson et al., 1974; Abbott et al., 1977). Occasionally, paracrystalline inclusions appear in mitochondria postmortem (Cheah et al., 1973; Ota et al., 1973). Remnants of mitochondria may appear as large, empty vesicles (Dutson et al., 1974; Will et al., 1980). Mitochondria near the sarcolemma show less structural alteration than those deeper inside the cell (Abbott et al., 1977). The variable stability of mitochondria postmortem probably depends on the pH and temperature conditions that the muscle has experienced. Mitochondria with normal morphology can be isolated from beef muscle stored 144 hr at 4°C (Cheah and Cheah, 1971, 1974) or pig muscle stored 24–48 hr at 1°C (Cheah, 1973). In contrast muscle left on the carcass before isolation typically yield mitochondria that show swelling (Greaser et al., 1967, 1969b). The rather steep inactivation of mitochondrial function in the pH 5.5 region (see Section III,D,1) may also be linked to possible morphological alterations.

Fig 8 Bovine longissimus dorsi muscle fixed at 24 hr postmortem. The muscle was left on the carcass and chilled in the conventional manner until sample removal. Remaining mitochondria (arrow 1) are swollen and have a lower matrix density. The spaces between the myofibrils are greater and the material appears more clumped (arrow 2). The plasmalemma has been damaged and the distance between adjacent fibers increases (arrow 3). The myofibril structure, however, is well preserved and the Z lines are intact. ×10,000. Unpublished photograph from L. E. Kasang and M. H. Stromer, Iowa State University, used by permission.

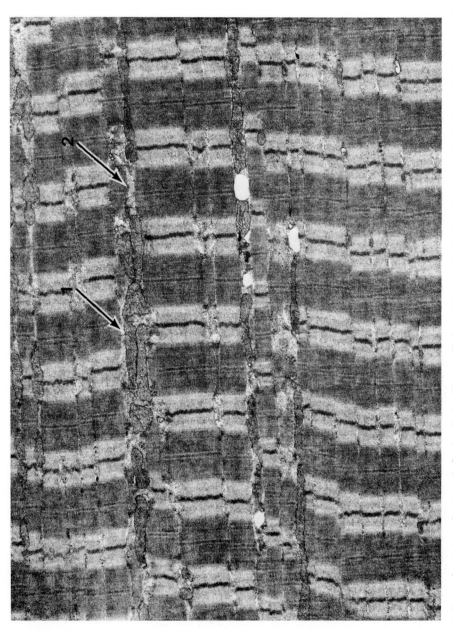

Fig. 9 Bovine longissimus dorsi muscle fixed at death. The myofibrils are tightly packed in the cell and mitochondria have closely apposed cristae (arrow 1). The spaces between the myofibrils mostly have a fine granular nature (arrow 2). ×10,000. Unpublished photograph from L. E. Kasang and M. H. Stromer, Iowa State University, used by permission.

4. NUCLEI

Muscle nuclei examined immediately postmortem are typically granular in appearance with dense regions of condensed chromatin. Nuclei remain intact postmortem and show little change in the first 24 hr after death (Abbott *et al.,* 1977). Their overall length gradually declines from 11 to 3 μm after 192 hr (Abbott *et al.,* 1977). An increase in density accompanies this decrease in size (Abbott *et al.,* 1977; Will *et al.,* 1980). The granularity also may become more irregular with time postmortem (Asghar and Yeates, 1978).

5. CYTOSOL

The appearance of the regions between the myofibrils changes with time postmortem. In muscle examined at death the myofibrils are tightly packed, with the spaces between them being filled with membrane elements and glycogen granules, and having a diffuse granularity (Fig. 8). However, by 24 hr postmortem (Fig. 9) the glycogen granules decline in number (Stromer *et al.,* 1967) and the distance between adjacent myofibrils increases (Abbott *et al.,* 1977; Will *et al.,* 1980). The space between the myofibrils may be quite clear, with the diffuse granularity replaced by coarser dense clumps of material (Henderson et al., 1970; Dutson *et al.,* 1974; Gann and Merkel, 1978).

6. PLASMALEMMA AND EXTRACELLULAR COMPONENTS

In living muscle the plasmalemma serves as a barrier between the inside and outside of the cell. During the postmortem period this barrier breaks down. Gaps are occasionally seen as early as 1 hr postmortem in beef muscle (Will *et al.,* 1980) and the membranes show considerable damage by 24 hr after death (Abbott *et al.,* 1977; Asghar and Yeates, 1978). The plasmalemma often separates from the basal lamina during this time as well (Will *et al.,* 1980). Structural damage is observed in the basal lamina of 24-hr postmortem beef muscle when observed by scanning election microscopy (Varriano-Marston *et al.,* 1976). The extracellular matrix appears more open and more space separates the individual fibers by 6 to 24 hr postmortem (Paul *et al.,* 1944; Paul, 1965). The collagen network remains fairly stable postmortem and only when these fibers alter their orientation as a result of shortening can structural changes be seen (Rowe, 1974).

F. Solubility

Protein solubilities may change with time postmortem. These changes reflect irreversible denaturation due to postmortem pH – temperature combinations, reversible precipitation due to lower pH, proteolysis, or altered protein – protein interactions as a result of changes in small molecule substrates (as, for example, myosin and actin with ATP). The amount of beef muscle sarcoplasmic protein extracted with water decreased between 1 and 24 hr postmortem (Fujimaki and Deatherage, 1964). No changes occurred, however, during 312 hr storage at 4°C in the sarcoplasmic protein extractable in a low ionic strength solution buffered near pH 7.0 (Goll *et al.*, 1964).

Deeper regions of beef muscle left on the carcass have 9 – 15% lower sarcoplasmic protein solubility by 24 hr postmortem (Disney *et al.*, 1967). No changes in solubility occurred between 1 and 9 hr postmortem in lamb muscle (van Eerd, 1972).

The changes in solubility of pig muscle sarcoplasmic protein with time postmortem are shown in Fig. 10. Little change occurs if the muscle pH is > 5.7 at the onset of rigor mortis (the beginning of loss of elasticity) and the muscle temperature is < 35°C (Sayre and Briskey, 1963). With lower pH and temperatures > 35°C, the sarcoplasmic protein solubility may drop almost 50%. Holding pig muscle at 37°C for 24 hr results in a decrease of ~ one-third in sarcoplasmic protein solubility (Scopes, 1964; Hapchuk *et al.*, 1976). Studies using

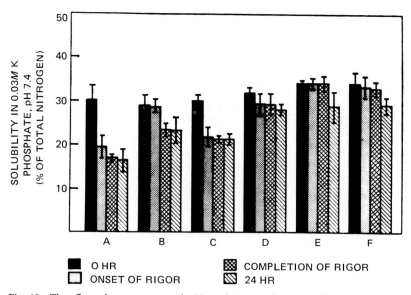

Fig. 10 The effect of temperature and pH conditions at the onset of rigor mortis on pig sarcoplasmic protein solubility. Conditions of rigor onset: A, pH 5.3 – 5.6; temperature > 35°C; B, pH 5.3 – 5.6, temperature < 35°C; C, pH 5.7 – 5.9, temperature > 35°C; D, pH 5.7 – 5.9, temperature < 35°C; E, pH 6.0+, temperature < 35°C; F, pH 6.0+, temperature < 35°C. From Sayre and Briskey (1963). Copyright by Institute of Food Technologists.

various temperature–pH combinations to treat isolated sarcoplasmic protein extracts also show that pH levels < 6.0 combined with temperatures > 35° lead to a significant decline in protein solubility (McLoughlin and Goldspink, 1963; Scopes, 1964; Charpenter, 1969; Goutefonga, 1971). The most sensitive protein to denaturation appears to be creatine kinase (Scopes and Lawrie, 1963; Scopes, 1964).

There is also an effect of pH on the solubility of myofibrillar proteins in high ionic strength solutions. Scopes (1964) found that myofibrils were 60–70% soluble in 1 M KCl in the pH range 7.0–5.8. However, at pH 5.5 only 40% of the protein was soluble and at pH 5.3 ~20% was soluble. Van Eerd (1972) found that the myofibrillar protein soluble in unbuffered 3% NaCl declined 15-fold between death and after rigor mortis completion. Changes in solubility of pig myofibrillar proteins in 1.1 M KI–0.1 M K phosphate, pH 7.4, with time postmortem are shown in Fig. 11. The solubilities remain virtually constant postmortem if the pH is > 6.0 at the time of rigor mortis onset. With lower pHs and muscle temperatures > 35°, the solubility at 24 hr postmortem may drop as much as fourfold. Myofibrillar protein solubility remains fairly constant postmortem in beef (Goll *et al.,* 1964; Aberle and Merkel, 1966), lamb (van Eerd, 1972), and rabbit (Ikeuchi *et al.,* 1980b) muscle either held at 2°C or left

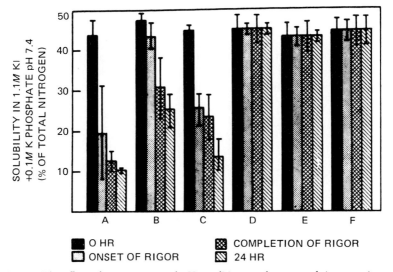

■ 0 HR   ▨ COMPLETION OF RIGOR
□ ONSET OF RIGOR   ▨ 24 HR

Fig. 11   The effect of temperature and pH conditions at the onset of rigor mortis on pig myofibrillar protein solubility. Conditions of rigor onset: A, pH 5.3–5.6; temperature > 35°C; B, pH 5.3–5.6, temperature < 35°C; C, pH 5.7–5.9, temperature > 35°C; D, pH 5.7–5.9, temperature < 35°C; E, pH 6.0+, temperature > 35°C; F, pH 6.0+, temperature < 35°C. From Sayre and Briskey (1963). Copyright by Institute of Food Technologists.

on the carcass before sample removal and extraction. Holding muscle at 37°C for varying periods may reduce solubility. Penny (1967) found that rabbit muscle held for 3 hr at 37°C had no change in myofibrillar protein solubility (muscle pH 5.95) while 4 hr holding (muscle pH 5.75) resulted in almost a 50% reduction. Beef and rabbit muscle held at 37°C until rigor mortis completion also had about a 50% reduction in the amount of soluble myofibrillar protein (Scopes, 1964; Ikeuchi et al., 1980b). A greater than twofold reduction in myofibrillar solubility occurred in pig muscle held for 24 hours at 37°C (Hapchuk et al., 1976). The ultimate muscle pH affects solubility, with much reduced values at pH levels below 5.5 (Sung et al., 1976).

Important differences in myofibrillar protein solubility with time postmortem are revealed when solvents other than KI are used. Potassium iodide is a highly disruptive solvent that causes F-actin to depolymerize. In contrast, solvents like 0.6–1.0 M KCl or 0.1 M pyrophosphate are less disruptive to actin. Chaudhry and co-workers (1969) found a 1.3- to twofold increase in myofibrillar protein soluble in 0.5 M KCl, 0.1 M K phosphate (pH 7.4) by 24 hr postmortem in beef muscle held at 2 to 25°C. Penny (1968) found that the protein extractable in 1 M KCl was the same between 0 and 1 day at 15 to 18°C, but the percentage of the myofibril soluble in 0.1 M pyrophosphate increased from 59 to 68% in the same period. He interpreted these results to mean that more actin and tropomyosin was being solubilized due to a weakening of the actin–Z line attachments. Fujimaki and co-workers (1965a,b) also found a higher proportion of actin in myosin B (actomyosin) extracts taken at 2 days postmortem versus those taken at death. The proportion of actin and α-actinin in Hasselbach–Schneider extracts of bovine muscle stored at 2°C increases between 0 and 2 days postmortem (Cheng and Parrish, 1978b). They also found increasing amounts of α-actinin, C protein, and myosin heavy chain in 1 mM Tris extracts of myofibrils at day 2 compared to immediately postmortem.

No differences were found in the yield of myosin from beef or rabbit muscle held 2 hr or 2 days at 2°C (Thomas and Frearson, 1971). However, myosin solubility declined markedly with ultimate pHs < 5.5 in pig muscle (Sung et al., 1976). It seems likely that changes in myosin solubility postmortem are the dominant determinants of myofibrillar protein solubility in any of the high ionic strength solvents mentioned above.

IV. QUALITY CONSIDERATIONS

The variety of postmortem changes described in the preceding sections may profoundly affect the suitability of muscle for food. In certain cases the appearance of the meat may be abnormal and thus present problems with consumer acceptance. The usefulness of the meat for processing may vary. Toughness of the meat may be affected, particularly in beef and lamb. Described below are two abnormal conditions that commonly occur in postmortem muscle and the effects of certain treatments on meat quality.

A. Abnormal Conditions

1. PALE, SOFT, EXUDATIVE MUSCLE

Ludvigsen (1954) first described a condition in pigs that he termed muscle degeneration. The condition was characterized by the postmortem development of a pale color, a soft texture, and considerable exudation from the muscle surface. This condition has come to be known as pale, soft, and exudative (PSE) (Briskey et al., 1959) and occurs in 5 to 20% of pig carcasses. It is directly related to porcine stress syndrome (animals may die as a result of mild stress) and malignant hyperthermia (both pigs and some humans may die after exposure to certain anesthetics) (for review, see Gronert, 1980). Considerable research has been conducted to describe the changes that occur and the cause of this condition. Briskey (1964) and Cassens et al. (1975) have reviewed this subject and should be consulted for more detailed discussions.

The PSE condition results from an extremely rapid postmortem glycolytic rate that drops the pH while muscle temperature is still high (Briskey and Wismer-Pedersen, 1961). The pH decline rate is about twice as fast in muscles that become PSE (1.04 units/hr versus 0.65 units/hr in normal muscle held at $37°C$) (Bendall et al., 1963). In certain cases the ultimate pH may be reached by 15 min postmortem.

A comparison of the changes in metabolites in normal verus PSE muscle is shown in Table VI. Even at the earliest times after death the glycogen level in PSE muscle has been severely depleted and the muscle lactate level is nearly double that in normal muscle. Lactate concentrations usually reach their maximum by 1 hr or less. CP and ATP concentrations are also lower at death and both are depleted by 1 hr postmortem. Hexose monophosphate levels usually remain higher than in normal muscle during the first 1 to 3 hr after death (Kastenschmidt, 1970; Fischer and Augustine, 1977).

PSE muscle also displays macromolecular alterations that differ from those found in normal muscle postmortem. The time before rigor mortis completion is much reduced and is in line with ATP depletion patterns (Bendall et al., 1963; Forrest et al., 1966). Muscle that becomes PSE shows declining myofibrillar ATPase activity postmortem, while this activity increases in normal muscle (Table IV). Myosin ATPase activity has been reduced twofold in 24-hr postmortem PSE muscle and the phosphate burst is absent (Sung et al., 1981). PSE myofibrils will not contract when mixed with MgATP (Sung et al., 1977).

Table VI

Comparison of Postmortem Levels of Metabolites in Normal and PSE Pig Muscle

Metabolite	Normal	PSE	Reference
Glycogen, 3 min (μmol glucose equiv/g)	35–100	23	Kastenschmidt et al. (1968)
Glycogen, 180 min (μmol glucose equiv/g)	20	0.8	Kastenschmidt et al. (1968)
Glucose, 3 min (μmol/g)	2.3	3.3	Kastenschmidt et al. (1968)
Glucose, 180 min (μmol/g)	4.3	6.8	Kastenschmidt et al. (1968)
Glucose 6-phosphate, 3 min (μmol/g)	4.5	8.5	Kastenschmidt (1970)
60 min (μmol/g)	5.0	7.0	Kastenschmidt (1970)
180 min (μmol/g)	6.5	7.5	Kastenschmidt (1970)
Lactate, 3 min (μmol/g)	30–40	~60	Kastenschmidt et al. (1968)
60 min (μmol/g)	40–60	105	Kastenschmidt et al. (1968)
180 min (μmol/g)	60–80	105	Kastenschmidt et al. (1968)
Creatine phosphate, 3 min (μmol/g)	6.0	3.0	Kastenschmidt et al. (1968)
60 min (μmol/g)	3.0	1.0	Kastenschmidt et al. (1968)
180 min (μmol/g)	2.0	1.0	Kastenschmidt et al. (1968)
ATP, 3 min (μmol/g)	5.5	3.5	Kastenschmidt et al. (1968)
60 min (μmol/g)	4.5	<0.5	Kastenschmidt et al. (1968)
180 min (μmol/g)	2.5	<0.5	Kastenschmidt et al. (1968)

Myosin B from PSE muscle does not superprecipitate as fast as do preparations from normal muscle (Park et al., 1975b) and there appears to be reduced troponin function in PSE muscle as well (Park et al., 1977). PSE myofibrils also contain more of a 165,000-dalton band on SDS–polyacrylamide gels; this is presumably a breakdown product of the myosin heavy chain (Park et al., 1975b). Both salt-soluble and heat-labile collagen are significantly higher in 48-hr postmortem PSE muscle (McClain et al., 1967).

The SR ATPase activity increases with time postmortem in normal muscle, while decreases occur in preparations from PSE muscle (Table IV) (Greaser et al., 1969a). Calcium uptake activity of the SR decreases much more rapidly in PSE muscle (Greaser et al., 1969a). However, there was no significant difference in calcium binding activity immediately after death (Greaser et al., 1969a,d). A reduced calcium binding activity by the sarcoplasmic reticulum might trigger the rapid ATP turnover in PSE muscle (Greaser et al., 1969a).

Mitochondria have also been postulated to trigger the PSE condition. Muscle mitochondria from stress-susceptible pigs have about twice the calcium efflux rate as those from normal animals (Cheah and Cheah, 1976, 1979). It has recently been suggested that this increased calcium activates phospholipase, which releases fatty acids from the mitochondrial membranes (Cheah and Cheah, 1981a,b). The fatty acids in turn inactivate the SR (Cheah, 1981), resulting in increasing cytosol calcium levels and a more rapid myofibrillar ATPase. It is not clear how the phospholipase would be controlled in vivo, and it appears that if this mechanism is operative, then something must be causing the mitochondrial calcium efflux to change.

Changes in response to electrical stimulation are much more rapid in PSE versus normal muscle. Muscles with rapid glycolytic rates had higher excitability thresholds and lower contraction strengths by 10 min postmortem (Forrest and Briskey, 1967). The resting membrane potential was 38–62 mV at 45 min postmortem in normal muscle and 21–27 mV in muscle with rapid glycolysis (Schmidt et al., 1972). Membrane capacitance and resistivity declined more rapidly postmortem in PSE muscle (Swatland, 1980, 1981). It has also been recently shown that halothane (an anesthetic that triggers malignant hyperthermia) changes the resting potential from -84.9 to -75.3 mV in muscle from stress-susceptible pigs while the change is insignificant in normal muscle (-83.6 to -82.8) (Gallant et al., 1979).

Morphological changes in PSE muscle postmortem are much more rapid than normal. A number of studies have shown that dense, irregular bands appear with greater frequency in PSE muscle (Bendall and Wismer-Pedersen, 1962; Cassens et al., 1963a,b; Lawrie et al., 1963; Elliott, 1965). Although these bands were originally postulated to consist of denatured sarcoplasmic protein, electron microscopic observations indicate that they are due to localized regions of myofibril contracture (Cassens et al., 1963a). Contractures are common even in biopsy samples, with a 30.1% frequency in fibers from malignant hyperthermia–susceptible animals and only a 7.6% incidence in normal muscle (Palmer et al., 1977). The increases in sarcoplasmic space between myofibrils developed more rapidly in PSE muscle, and I-band clumping and Z-line breaks were more common (Cassens et al., 1963b; Dutson et al., 1974). PSE muscle myofibers had a more granular appearance at 24 hr postmortem (Cassens et al., 1963b; Greaser et al., 1969b). This may be due to either the deposition of denatured sarcoplasmic proteins on the surface of the filaments or a denaturation of the myofibrillar proteins. Mitochondria and sarcoplasmic reticulum show signs of disruption as early as 15 min postmortem (Dutson et al., 1974).

PSE muscle protein solubility declines to a much greater extent than that of normal muscle. Muscles with rapid pH fall while the muscle temperature is high may decline ~50% in sarcoplasmic protein solubility and 75% in myofibrillar solubility (Wismer-Pedersen, 1959; Bendall and Wismer-Pedersen, 1962; Sayre and Briskey, 1963; Kastenschmidt et al., 1964; Penny, 1969; Park et al., 1975a) (Figs. 10 and 11). Creatine kinase appears to be the sarcoplasmic protein that is most easily denatured (Scopes and Lawrie, 1963). Phosphorylase content in the sarcoplasmic extracts declines by 45 to 60 min postmortem in PSE muscle and increases in the myofibrillar fraction (Fischer et al., 1977, 1979). The enzyme activity declines by threefold in this same time period (Table IV). There appears to be no change in myoglobin solubility in PSE muscle (Charpenter, 1969) and thus the pale color must result from the "white" precipitates of the other sarcoplasmic proteins, which mask the red color (Goldspink and McLoughlin, 1964).

The PSE condition can be artificially induced in most, but not all, cases by holding normal pig muscle at 37°C until completion of rigor (Bendall and Wismer-Pedersen, 1962; Bodwell *et al.*, 1966). A similar phenomenon occurs in some beef muscles held at 37°C (Locker and Daines, 1975). Beef psoas muscle typically has a more rapid pH decline postmortem and 64% of a group of 1395 animals had 30-min pH values <6.0 (Fischer and Hamm, 1980). The resulting muscle had a PSE appearance. Deep portions of beef rounds from double-muscled animals may also be somewhat pale and exudative (M. L. Greaser, personal observations).

2. DARK, FIRM, DRY MUSCLE (DFD)

Muscles that have a high ultimate pH (>6.0) appear much darker than normal. The dark appearance is due to a more active mitochondrial respiration rate, which reduces the depth of oxygen penetration and thus the proportion of visible oxymyoglobin. The higher pH also increases the water-holding capacity of the myofibrillar proteins because they are farther away from their isoelectric points.

DFD muscle occurs most often in beef animals, particularly with bulls (Hedrick *et al.*, 1959). It is often referred to as dark-cutting beef. As many as 7% of young bulls may have 24-hr muscle pH levels >6.0 (Fischer and Hamm, 1980). Pig muscle may also display this condition (Briskey *et al.*, 1959; Briskey, 1964). In both species the high pH is caused by substantial depletion of muscle glycogen before death, resulting in reduced lactic acid formation postmortem.

B. Effects of Antemortem Treatments

Several antemortem factors affect the incidence of PSE muscle. This subject has been reviewed by Cassens and co-workers (1975) and Asghar and Pearson (1980). Table VII lists a number of treatments that either increase or decrease the proportion of PSE muscle. PSE incidence is greater in animals that are stressed by excitement, exercise, and environmental temperature variations. Exposure to very warm temperatures causes the muscle temperature to rise and triggers rapid glycolysis. In contrast, warm air treatment followed by a cold water bath decreases muscle temperature from 38.6° to 33.4°C (Kastenschmidt *et al.*, 1964) and retards the glycolytic rate. Exhaustive exercise severely depletes muscle glycogen stores and gives a higher ultimate pH value. Thus, even if the glycolytic rate is high postmortem, pH values >5.8–6.0 will not result in protein denaturation and PSE muscle.

The dramatic effect of curare and $MgSO_4$ on PSE incidence and postmortem glycolytic rate deserves comment. These agents block nerve transmission at the cell membrane level and inhibit contractile activity. It therefore appears

Table VII
Antemortem Factors Affecting the Incidence of PSE Muscle

Treatment	Effect	References
Sucrose feeding	Increase	Briskey *et al.* (1959)
Excitement and mild exercise	Increase	Sayre *et al.* (1963b); Addis *et al.* (1974)
Exhaustive exercise	Decrease	Briskey *et al.* (1959)
Fluctuating air temperature	Increase	Howe *et al.* (1968)
Holding at 42–45°C for 30 min	Increase	Sayre *et al.* (1963a); Kastenschmidt *et al.* (1964); Forrest *et al.* (1965)
Injection of anterior pituitary extract	Increase	Kraeling *et al.* (1975); Kraeling and Rampacek (1977)
Thyroxine injection	Increase	Marple *et al.* (1975)
0.5°C Water bath, 30–40 min	Decrease	Sayre *et al.* (1961)
30–60 Min at 45°C followed by 1–3°C bath, 30 min	Decrease	Kastenschmidt *et al.* (1964)
30 Min at 45°C followed by 30 min at −29°C	Decrease	Kastenschmidt *et al.* (1965)
Magnesium injection	Decrease	Bendall (1966)
Curare injection	Decrease	Bendall (1966)
MgSO₄ injection	Decrease	Sair *et al.* (1970); Schmidt *et al.* (1970a)

that the rapid postmortem glycolysis condition is related to some defect in the motor nerves or muscle plasmalemma.

The dark-cutting condition in beef results from prolonged stress (24 hr or more), which causes a major reduction in muscle glycogen before death (Hedrick *et al.*, 1959). It may also be artificially induced by adrenaline injection several hours before slaughter (Hedrick *et al.*, 1959, 1964; Hatton *et al.*, 1972).

C. Effects of Postmortem Treatments

A number of postmortem factors may affect meat quality (for more extensive review, see Locker *et al.*, 1975). Some of these factors are listed in Table VIII. Rapid cooling of pig carcasses decreases the incidence of PSE muscle (Borchert and Briskey, 1964, 1965). Freezing lamb carcasses prerigor causes a marked toughening of the meat (Marsh *et al.*, 1968). Subjecting prerigor beef and lamb muscle to high pressures results in a significant increase in tenderness (MacFarlane, 1973). The pressurized muscle undergoes extremely rapid glycolysis (almost complete within 4 min), shortens 40–50%, and sustains considerable morphological damage, (contraction bands, loss of M lines, and breakage and aggregation of I bands) (MacFarlane, 1973; MacFarlane and Morton, 1978; Kennick *et al.*, 1980). The SR calcium uptake ability is also destroyed (Horgan,

Table VIII
Postmortem Factors Affecting Meat Quality

Factor	Effect	References
Liquid N_2 chilling of pig carcasses	Reduce PSE incidence	Borchert and Briskey (1964, 1965)
Chilling pig carcasses at $-29°C$	Reduce PSE incidence	Bodwell et al. (1966)
Prerigor freezing	Toughening	Marsh et al. (1968)
High pressure treatment — prerigor muscle	Tenderization	MacFarlane (1973)
Early postmortem excision, hold at $0-10°C$	Toughening	Locker and Hagyard (1963); Goll et al. (1964)
Early postmortem excision, hold at $37°C$	Tenderization	Locker and Daines (1975)
Altered carcass position (most muscles stretched)	Tenderization (most muscles)	Herring et al. (1965a); Buege and Stouffer (1974)
Rapid chilling of beef carcasses	Toughening	Lochner et al. (1980)
Rapid chilling of lamb carcasses	Toughening	Marsh et al. (1968)
Lamb carcasses, held at $45°C$, 4 hr	Toughening	Davey and Gilbert (1973)
Beef carcasses, held at $37°C$, 4 hr	Tenderization	Roschen et al. (1950); Lochner et al. (1980)
Electrical stimulation $(50-60$ Hz$)$	Tenderization	Harsham and Deatherage (1951)
Electrical stimulation (2 Hz)	Toughening	Marsh et al. (1980)

1979).   The tenderness increase is presumably due to structural damage in the myofibrils.

Low temperatures $(0-10°C)$ cause shortening in beef and lamb muscles excised prerigor.   This effect was first described by Locker and Hagyard (1963) and is referred to as cold shortening.   Muscles may shorten by up to 50 or 60% and shear force values may increase three- to fourfold (Marsh and Leet, 1966). No tenderness decrease was detectable with shortening of up to 20%, but tenderness declined to a minimum at ~35 to 40% shortening.   Paradoxically, muscles with 60% shortening were nearly as tender as ones which had not shortened.   The effect of muscle length on tenderness appears to be related to the degree of overlap of myofibril thick and thin filaments in the 0–40% shortening region (Marsh and Carse, 1974) and to contraction nodes and localized stretching plus breaking of sarcomeres (Fig. 6) in the 60%-shortened muscles (Marsh et al., 1974).   Muscles with fixed lengths can still cold shorten, decreasing sarcomere lengths in one region while stretching occurs in other areas (Marsh and Leet, 1966).   It is not clear, however, whether this occurs to a significant extent in muscles left on the carcass until rigor completion.   The cold shortening effect has important implications regarding tenderness of meat

Table IX
Effect of Temperature on Tenderness[a]

Temperature (°C)	15	24	28	34	37
Shortening (%)	11.8	15	17.7	24.5	31.5
Shear force	49.6	53.5	54.1	50.5	42.6

[a] From Locker and Daines (1975), with permission.

that is hot-boned prerigor. No problems were found if the meat was held for 24 hr at 15°C before final chilling (Schmidt and Gilbert, 1970).

Holding excised muscle at 37°C until rigor completion also causes significant shortening, but the meat becomes more tender than at temperatures between 15 and 34°C (Table IX). The reason for this tenderization is unknown but presumably involves an activation of some proteolysis system in the muscle. It is striking that tenderness actually increases at this temperature when the shortening (31.5%) would suggest near-maximum toughening. The effect of postmortem temperature was also evident in the results of Herring et al. (1965b). They left some muscles attached to the carcass until after rigor and compared them to stretched and restrained muscles held at 1°C. The sarcomere lengths of both groups were identical (2.6 μm) but the shear force values were 6.2 and 9.0 kg for the postrigor excised and prerigor excised–stretched muscle groups, respectively. Similar results were found in lambs with different outside fat thickness (Table X). Comparisons between the fat and intermediate groups showed no differences in sarcomere lengths but a significantly lower shear force in the fat animals (Smith et al., 1976). Tenderness of beef muscle left on the carcass appears to be more closely related to the muscle temperature at 2 hr postmortem than to temperatures at later times (Lochner et al., 1980). Carcasses with high 2-hr temperature generally were more tender than ones having more rapid

Table X
Effect of Fat Thickness on Cooling Rates, Sarcomere Lengths, and Shear Force Values in Lamb Longissimus Dorsi[a]

Group	Fat >7.5 mm	Intermediate 2.5–7.5 mm	Thin <2.5 mm
1-hr Temperature	19.0[b]	15.6[c]	11.5[d]
3-hr Temperature	10.7[b]	7.3[c]	3.8[d]
6-hr Temperature	3.6[b]	0.9[b]	0.7[b]
Sarcomere length (μm)	1.78[b]	1.78[b]	1.70[c]
Shear force (kg)	4.6[b]	6.1[c]	7.5[d]

[a] From Smith et al. (1976). Copyright by Institute of Food Technologists.

[b,c,d] Means in the same horizontal row with different superscripts are statistically different ($p < .05$).

cooling rates. Holding carcasses at 37°C for the first 2–4 hr postmortem improves tenderness (Roschen et al., 1950; Marsh et al., 1980).

The results of Herring et al. (1965b), Smith et al. (1976), and Lochner et al. (1980) all cast doubt on the importance of cold shortening in affecting tenderness of muscles left on the carcass. Instead, rapid chilling retards some tenderization process that is initiated very early postmortem.

The major postmortem treatment that is currently used to improve meat quality is electrical stimulation. This topic has been reviewed by Cross (1979) and Bendall (1980). Electrical stimulation has been used primarily in lamb and beef carcasses to improve tenderness and allow more rapid carcass chilling. Current is usually applied for 1 to 2 min at 200 to 600 V and frequencies of greater than 15/sec (Bendall, 1980). Stimulation of beef carcasses must be conducted by 1 hr postmortem to be effective (Bendall et al., 1976) and even earlier with lamb carcasses (Bendall, 1980). The muscles undergo violent contractions and the glycolytic rate increases more than 100-fold (Bendall, 1976; Bendall et al., 1976; Chrystall and Devine, 1978). After 2 min of stimulation the muscle no longer contracts and the muscle pH has dropped to about 6.3 (Chrystall and Devine, 1978; Bendall, 1980). Most of the pH decline occurs in the first 30 sec (Bendall et al., 1976). Because of the short-term acceleration in glycolytic rate the whole pH decline and rigor mortis time course is shifted much earlier. Lamb muscle that would typically reach pH 6.0 at 15 hr postmortem now drops below this level by 3 hr after death if stimulated (Carse, 1973). Beef carcasses that have been electrically stimulated are in full rigor by 4 hr compared to 15 to 20 hr normally (Bendall, 1980). As a result carcasses can be frozen or chilled much earlier postmortem without fear of thaw rigor or cold-shortening.

The mechanism of the tenderization process is not known. Stimulated muscles show more structural changes than controls with much higher incidences of contracture bands and stretched plus broken myofibrils (Savell et al., 1978; George et al., 1980; Will et al., 1980). It has also been postulated that the rapid attainment of low pH (<6.0) while the muscle temperature is still high would activate acid proteases to improve tenderness. An argument against this latter hypothesis has been provided by Marsh et al. (1980). They found that low-frequency stimulaton (2 Hz) caused rapid pH decline, minimal structural damage, and *tougher* meat. They thus suggested that electrical stimulation caused improved tenderness primarily by muscle fiber fracture.

V. CONCLUSIONS

Muscle is converted to meat by a complex series of chemical reactions involving breakdown of high-energy compounds and a locking together of myosin and actin in rigor mortis. The rates of these processes are much more rapid in pig muscle than beef or lamb. It appears that the motor nerve and/or the muscle plasmalemma exert primary control on the rate of postmortem glycolysis. The rates vary considerably between different muscles and at different distances from the surface because of the temperature gradients. Extremely rapid pH decline while muscle temperature is above 30°C leads to a denaturation of myosin, lowered protein solubility and decreased value of the meat for processing. The macromolecular changes that occur in myofibrils and lead to more tender meat in some animals remain to be defined. In particular the mechanisms that are operative early postmortem (2–4 hr) and markedly affect beef tenderness require intensified examination. Caution must be exercised in the use of antemortem and postmortem treatments since several factors may lead to decreased muscle quality.

ACKNOWLEDGMENT

This chapter is a contribution from the College of Agricultural and Life Sciences, University of Wisconsin, Madison.

REFERENCES

Abbott, M. T., Pearson, A. M., Price, J. F., and Hooper, G. R. (1977). Ultrastructural changes during autolysis of red and white porcine muscle. *J. Food Sci.* **42,** 1185–1188.

Aberle, E. D., and Merkel, R. A. (1966). Solubility and electrophoretic behavior of some proteins of post-mortem aged bovine muscle. *J. Food Sci.* **31,** 151–156.

Addis, P. B., Nelson, D. A., Ma, R. T.-I., and Burroughs, J. R. (1974). Blood enzymes in relation to procine muscle properties. *J. Anim. Sci.* **38,** 279–286.

Andrews, M. M., Guthneck, B. T., McBride, B. H., and Schweigert, B. S. (1952). Stability of certain respiratory and glycolytic enzyme systems in animal tissue. *J. Biol. Chem.* **194,** 715–719.

Arakawa, N., Goll, D. E., and Temple, J. (1970). Molecular properties of post-mortem muscle. 8. Effect of post-mortem storage on α-actinin and the tropomyosin–troponin complex. *J. Food Sci.* **35,** 703–711.

Arakawa, N., Fujiki, S., Inagaki, C., and Fujimaki, M. (1976). A catheptic protease active in ultimate pH of muscle. *Agric. Biol. Chem.* **40,** 1265–1267.

Asghar, A., and Pearson, A. M. (1980). Influence of ante- and post-mortem treatments upon muscle composition and meat quality. *Adv. Food Res.* **26,** 53–213.

Asghar, A., and Yeates, N. T. M. (1978). The mechanism for the promotion of tenderness in meat during post-mortem aging process. *CRC Crit. Rev. Food Sci. Nutr.* **10,** 115–145.

Ashmore, C. R., Parker, W., and Doerr, L. (1972). Respiration of mitochondria isolated from dark-cutting beef: Post-mortem changes. *J. Anim. Sci.* **34,** 46–48.

Ashmore, C. R., Carroll, F., Doerr, L., Tompkins, G., Stokes, H., and Parker, W. (1973). Experimental prevention of dark-cutting meat. *J. Anim. Sci.* **36,** 33–36.

Atkinson, J. L., and Follett, M. J. (1973). Biochemical studies on the discoloration of fresh meat. *J. Food Technol.* **8,** 51–58.

Bate-Smith, E. C., and Bendall, J. R. (1974). Rigor mortis and adenosine triphosphate. *J. Physiol. (London)* **106**, 177–185.

Bate-Smith, E. C., and Bendall, J. R. (1949). Factors determining the time course of rigor mortis. *J. Physiol. (London)* **110**, 47–65.

Beecher, G. R., Briskey, E. J., and Hoekstra, W. G. (1965a). A comparison of glycolysis and associated changes in light and dark portions of the porcine semitendinosus. *J. Food Sci.* **30**, 477–486.

Beecher, G. R., Cassens, R. G., Hoekstra, W. G., and Briskey, E. J. (1965b). Red and white fiber content and associated post-mortem properties of seven porcine muscles. *J. Food Sci.* **30**, 969–976.

Beecher, G. R., Kastenschmidt, L. L., Hoekstra, W. G., Cassens, R. G., and Briskey, E. J. (1969). Energy metabolites in red and white striated muscles of the pig. *J. Agric. Food Chem.* **17**, 29–33.

Behnke, J. R., Fennema, O., and Cassens, R. G. (1973). Rates of post-mortem metabolism in frozen animal tissues. *J. Agric. Food Chem.* **21**, 5–11.

Bendall, J. R. (1951). The shortening of rabbit muscle during rigor mortis: relation to the breakdown of adenosine triphosphate and creatine phosphate and to muscular contraction. *J. Physiol. (London)* **114**, 71–88.

Bendall, J. R. (1966). The effect of pre-treatment of pigs with curare on the post-mortem rate of pH fall and onset of rigor mortis in the musculature. *J. Sci. Food Agric.* **17**, 333–338.

Bendall, J. R. (1972). Consumption of oxygen by the muscles of beef animals and related species, and its effect on the colour of meat. I. Oxygen consumption in pre-rigor muscle. *J. Sci. Food Agric.* **23**, 61–72.

Bendall, J. R. (1973a). The biochemistry of rigor-mortis and cold contracture. *Proc. Eur. Meet. Meat Res. Workers* **19**, 1–27.

Bendall, J. R. (1973b). Post-mortem changes in muscle. In "The Structure and Function of Muscle" (G. H. Bourne, ed.), 2nd ed., Vol. 2, Part 2, pp. 243–309. Academic Press, London.

Bendall, J. R. (1975). Cold contracture and ATP-turnover in the red and white musculature of the pig, post-mortem. *J. Sci. Food Agric.* **26**, 55–71.

Bendall, J. R. (1976). Electrical stimulation of rabbit and lamb carcasses. *J. Sci. Food Agric.* **27**, 819–826.

Bendall, J. R. (1978). Variability in rates of pH fall and of lactate production in the muscles on cooling beef carcasses. *Meat Sci.* **2**, 91–104.

Bendall, J. R. (1980). The electrical stimulation of carcasses of meat animals. In "Developments in Meat Science" (R. Lawrie, ed.), Vol. 1, pp. 37–59. Applied Science Publishers, London.

Bendall, J. R., and Davey, C. L. (1957). Ammonia liberation during rigor mortis and its relation to changes in the adenine and inosine nucleotides of rabbit muscle. *Biochim. Biophys. Acta* **26**, 93–103.

Bendall, J. R., and Taylor, A. A. (1972). Consumption of oxygen by the muscles of beef animals and related species. II. Consumption of oxygen by post-rigor muscle. *J. Sci. Food Agric.* **23**, 707–719.

Bendall, J. R., and Wismer-Pedersen, J. (1962). Some properties of the fibrillar proteins of normal and watery pork muscle. *J. Food Sci.* **27**, 144–159.

Bendall, J. R., Hallund, O., and Wismer-Pedersen, J. (1963). Post-mortem changes in the muscles of Landrace pigs. *J. Food Sci.* **28**, 156–162.

Bendall, J. R., Ketteridge, C. C., and George, A. R. (1976). The electrical stimulation of beef carcasses. *J. Sci. Food Agric.* **27**, 1123–1131.

Bentler, W. (1977). Über postmortale Vorgänge im Skelettmuskel, vor allem bei Schlachtschweinen. VII. Vergleichende Untersuchungen über die Aktivität fibrillärer ATPasen der verschiedenen Schlachttiere bei unterschiedlichen pH-Werten und ihre Beziehung zu den Qualitätsmängeln im Schweinefleisch. *Fleischwirtschaft* **57**, 1671–1677.

Bernowsky, C., and Pankow, M. (1973). Protein binding of nicotinamide adenine dinucleotide and regulation of nicotinamide adenine dinucleotide glycohydrolase activity in homogenates of rabbit skeletal muscle. *Arch. Biochem. Biophys.* **156**, 143–153.

Blank, T. J. J., Gruener, R., Suffecool, S. L., and Thompson, M. (1981). Calcium uptake by isolated sarcoplasmic reticulum: Examination of halothane inhibition, pH dependence, and Ca^{2+} dependence of normal and malignant hyperthermic human muscle. *Anesth. Analg. (Cleveland)* **60**, 492–498.

Bodwell, C. E., Pearson, A. M., and Spooner, M. E. (1965a). Post-mortem changes in muscle. I. Chemical changes in beef. *J. Food Sci.* **30**, 766–772.

Bodwell, C. E., Pearson, A. M., and Fennell, R. A. (1965b). Post-mortem changes in muscle. III. Histochemical observations in beef and pork. *J. Food Sci.* **30**, 944–954.

Bodwell, C. E., Pearson, A. M., Wismer-Pedersen, J., and Bratzler, L. J. (1966). Post-mortem changes in muscle. II. Chemical and physical changes in pork. *J. Food Sci.* **31**, 1–12.

Borchert, L. L., and Briskey, E. J. (1964). Prevention of pale, soft, exudative porcine muscle through partial freezing with liquid nitrogen post-mortem. *J. Food Sci.* **29**, 203–209.

Borchert, L. L., and Briskey, E. J. (1965). Protein solubility and associated properties of porcine muscle as influenced by partial freezing with liquid nitrogen. *J. Food Sci.* **30**, 138–143.

Briskey, E. J. (1964). Etiological status and associated studies of pale, soft, exudative porcine musculature. *Adv. Food Res.* **13**, 89–178.

Briskey, E. J., and Wismer-Pedersen, J. (1961). Biochemistry of pork muscle structure. I. Rate of anaerobic glycolysis and temperature change versus the apparent structure of muscle tissue. *J. Food Sci.* **26**, 297–305.

Briskey, E. J., Bray, R. W., Hoekstra, W. G., Phillips, P. H., and Grummer, R. H. (1959). The effect of exhaustive exercise and high sucrose regimen on certain chemical and physical pork ham muscle characteristics. *J. Anim. Sci.* **18**, 173–177.

Briskey, E. J., Sayre, R. N., and Cassens, R. G. (1962). Development and application of an apparatus for continuous measurement of muscle extensibility and elasticity before and during rigor mortis. *J. Food Sci.* **27**, 560–566.

Brooks, J. (1929). Post-mortem formation of methaemoglobin in red muscle. *Biochem. J.* **23**, 1391–1400.

Buege, D. R., and Marsh, B. B. (1975). Mitochondrial calcium and postmortem muscle shortening. *Biochem. Biophys. Res. Commun.* **65**, 478–482.

Buege, D. R., and Stouffer, J. R. (1974). Effects of pre-rigor tension on tenderness of intact bovine and bovine muscle. *J. Food Sci.* **39**, 402–405.

Busch, W. A., Parrish, F. C., Jr., and Goll, D. E. (1967). Molecular properties of post-mortem muscle. 4. Effect of temperature on adenosine triphosphate degradation, isometric tension parameters, and shear resistance of bovine muscle. *J. Food Sci.* **32**, 390–394.

Busch, W. A., Goll, D. E., and Parrish, F. C., Jr. (1972). Molecular properties of post-mortem muscle. Isometric tension development and decline in bovine, porcine, and rabbit muscle. *J. Food Sci.* **37**, 289–299.

Callow, E. H. (1947). The actions of salts and other substances used in the curing of bacon and ham. *Br. J. Nutr.* **1**, 269–274.

Campion, D. R., Olson, J. C., Topel, D. G., Christian, L. L., and Kuhlers, D. L. (1975). Mitochondrial traits of muscle from stress–susceptible pigs. *J. Anim. Sci.* **42**, 1314–1317.

Caroni, P., and Carafoli, E. (1980). An ATP-dependent Ca^{+2}-pumping system in dog heart sarcolemma. *Nature (London)* **283**, 765–767.

Carse, W. A. (1973). Meat quality and the acceleration of post-mortem glycolysis by electrical stimulation. *J. Food Technol.* **8**, 163–166.

Carse, W. A., and Locker, R. H. (1974). A survey of pH values at the surface of beef and lamb carcasses, stored in a chiller. *J. Sci. Food Agric.* **25**, 1529–1535.

Cassens, R. G., and Cooper, C. C. (1971). Red and white muscle. *Adv. Food Res.* **19**, 1–74.

Cassens, R. G., and Newbold, R. P. (1966). Effects of temperature on post-mortem metabolism in beef muscle. *J. Sci. Food Agric.* **17**, 254–256.

Cassens, R. G., and Newbold, R. P. (1967a). Temperature dependence of pH changes in ox muscle post-mortem. *J. Food Sci.* **32**, 13–14.

Cassens, R. G., and Newbold, R. P. (1967b). Effect of temperature on the time course of rigor mortis in ox muscle. *J. Food Sci.* **32**, 269–272.

Cassens, R. G., Briskey, E. J., and Hoekstra, W. G. (1963a). Electron microscopic observations of dense, irregularly banded material occurring in some porcine muscle fibers. *Nature (London)* **198**, 1004–1005.

Cassens, R. G., Briskey, E. J., and Hoekstra, W. G. (1963b). Electron microscopy of post-mortem changes in porcine muscle. *J. Food Sci.* **28**, 680–684.

Cassens, R. G., Marple, D. N., and Eikelenboom, G. (1975). Animal physiology and meat quality. *Adv. Food Res.* **21**, 72–155.

Cerrella, E. G., and Massaldi, H. A. (1978.) Protein extraction from meat and the post-mortem permeability of muscle fibers. *J. Food Sci.* **43**, 1382–1385.

Charpenter, J. (1969). Influence de la température et du pH sur quelques caractéristiques physico-chimiques des protéines sarcoplasmiques du muscle de porc conséquences technologiques. *Ann. Biol. Anim., Biochim., Biophys.* **9**, 101–110.

Chaudhry, H. M., Parrish, F. C., Jr. and Goll, D. E. (1969). Molecular properties of post-mortem muscle. Effect of temperature on protein solubility of rabbit and bovine muscle. *J. Food Sci.* **34**, 183–191.

Cheah, A. M. (1981). Effect of long chain unsaturated fatty acids on the calcium transport of sarcoplasmic reticulum. *Biochim. Biophys. Acta* **648**, 113–119.

Cheah, K. S. (1973). Comparative studies of the mitochondrial properties of Longissimus dorsi muscles of Pietrain and Large White pigs. *J. Sci. Food Agric.* **24**, 51–61.

Cheah, K. S., and Cheah, A. M. (1971). Post-mortem changes in structure and function of ox mitochondria. Electron microscopic and polarographic investigation. *J. Bioenerg.* **2**, 85–92.

Cheah, K. S., and Cheah, A. M. (1974). Properties of mitochondria from ox neck muscle after storage in situ. *Int. J. Biochem.* **5**, 753–760.

Cheah, K. S., and Cheah, A. M. (1976). The trigger for PSE condition in stress-susceptible pigs. *J. Sci. Food Agric.* **27**, 1137–1144.

Cheah, K. S., and Cheah, A. M. (1979). Mitochondrial calcium efflux and porcine stress-susceptibility. *Experientia* **35**, 1001–1003.

Cheah, K. S., and Cheah, A. M. (1981a). Mitochondrial calcium transport and calcium-activated phospholipase in porcine malignant hyperthermia. *Biochim. Biophys. Acta* **634**, 70–84.

Cheah, K. S., and Cheah, A. M. (1981b). Skeletal muscle mitochondrial phospholipase A_2 and the interaction of mitochondria and sarcoplasmic reticulum in malignant hyperthermia. *Biochim. Biophys. Acta* **638**, 40–49.

Cheah, K. S., Cheah, A. M., and Voyle, C. A. (1973). Paracrystalline arrays in mitochondria following ageing of mitochondria in situ. *J. Bioenerg.* **4**, 383–389.

Cheng, C.-S., and Parrish, F. C., Jr. (1977). Effect of Ca^{+2} on changes in myofibrillar proteins of bovine skeletal muscle. *J. Food Sci.* **42**, 1621–1626.

Cheng, C.-S., and Parrish, F. C., Jr. (1978a). Effects of post-mortem storage conditions on myofibrillar ATPase activity of porcine red and white semitendinosus muscle. *J. Food Sci.* **43**, 17–21.

Cheng, C.-S., and Parrish, F. C., Jr. (1978b). Molecular changes in the salt-soluble myofibrillar proteins of bovine muscle. *J. Food Sci.* **43**, 461–463, 487.

Chizzolini, R., Ledward, D. A., and Lawrie, R. A. (1975). A differential scanning calorimetric study of intramuscular connective tissue collagen. *Proc. Eur. Meet. Meat Res. Workers* **21**, 29–31.

Chizzolini, R., Ledward, D. A., and Lawrie, R. A. (1977). Note on the effect of ageing on the neutral salt and acid soluble collagen from the intramuscular connective tissue of various species. *Meat Sci.* 1, 111–117.

Chrystall, B. B., and Devine, C. E. (1978). Electrical stimulation, muscle tension, and glycolysis in bovine sternomandibularis. *Meat Sci.* 2, 49–58.

Chrystall, B. B., Devine, C. E., and Davey, C. L. (1980). Studies in electrical stimulation: post-mortem decline in nervous response in lambs. *Meat Sci.* 4, 69–78.

Clark, F. M., Shaw, F. D., and Morton, D. J. (1980). Effect of electrical stimulation post-mortem of bovine muscle on the binding of glycolytic enzymes. Functional and structural implications. *Biochem. J.* 186, 105–109.

Cook, C. F., and Wright, R. G. (1966). Alterations in the contracture band patterns of unfrozen and prerigor frozen ovine muscle due to variations in post-mortem incubation temperatures. *J. Food Sci.* 31, 801–806.

Cross, H. R. (1979). Effects of electrical stimulation on meat tissue and muscle properties — a review. *J. Food Sci.* 44, 509–514, 523.

Currie, R. W., and Wolfe, F. H. (1977). Evidence for differences in post-mortem intramuscular phospholipase activity in several muscle types. *Meat Sci.* 1, 185–193.

Currie, R. W., and Wolfe, F. H. (1979). Relationship between pH fall and initiation of isotonic contraction in post-mortem beef muscle. *Can. J. Anim. Sci.* 59, 639–647.

Currie, R. W., and Wolfe, F. H. (1980). Rigor related changes in mechanical properties (tensile and adhesive) and extracellular space in beef muscle. *Meat Sci.* 4, 123–143.

Dalrymple, R. H., and Hamm, R. (1975). Postmortem glycolysis in prerigor ground bovine and rabbit muscle. *J. Food Sci.* 40, 850–853.

Davey, C. L., and Gilbert, K. V. (1966). Studies in meat tenderness. II. Proteolysis and the aging of beef. *J. Food Sci.* 31, 135–140.

Davey, C. L., and Gilbert, K. V. (1967). Structural changes in meat during ageing. *J. Food Technol.* 2, 57–59.

Davey, C. L., and Gilbert, K. V. (1973). The effect of carcass posture on cold, heat and thaw shortening in lamb. *J. Food Technol.* 8, 445–451.

Davey, C. L., and Gilbert, K. V. (1974). The mechanism of cold induced shortening in beef muscle. *J. Food Technol.* 9, 51–58.

Davey, C. L., and Gilbert, K. V. (1975). Cold shortening capacity and beef muscle growth. *J. Sci. Food Agric.* 26, 755–760.

Dayton, W. R., Goll, D. E., Zeece, M. G., Robson, R. M., and Reville, W. J. (1976a). A Ca^{2+}-activated protease possibly involved in myofibrillar protein turnover. Purification from porcine muscle. *Biochemistry* 15, 2150–2158.

Dayton, W. R., Reville, W. J., Goll, D. E., and Stromer, M. H. (1976b). A Ca^{+2}-activated protease possibly involved in myofibrillar protein turnover. Partial characterization of the purified enzyme. *Biochemistry* 15, 2159–2167.

Disney, J. G., Follett, M. J., and Ratcliff, P. W. (1967). Biochemical changes in beef muscle post-mortem and the effect of rapid cooling in ice. *J. Sci. Food Agric.* 18, 314–321.

Dutson, T. R., and Lawrie, R. A. (1974). Release of lysosomal enzymes during post-mortem conditioning and their relationship to tenderness. *J. Food Technol.* 9, 43–50.

Dutson, T. R., Pearson, A. M., Merkel, R. A., Koch, D. E., and Weatherspoon, J. B. (1971). Histochemical activity of some lysosomal enzymes in normal and pale, soft, and exudative pig muscle. *J. Anim. Sci.* 32, 233–238.

Dutson, T. R., Pearson, A. M., and Merkel, R. A. (1974). Ultrastructural post-mortem changes in normal and low quality porcine muscle fibers. *J. Food Sci.* 39, 32–37.

Eino, M. F., and Stanley, D. W. (1973). Catheptic activity, textural properties, and surface ultrastructure of post-mortem beef muscle. *J. Food Sci.* 38, 45–50.

Elliott, R. J. (1965). Postmortem pH values and microscopic appearance of pig muscle. *Nature (London)* **206**, 315-317.

Fenichel, I. R., and Horowitz, S. B. (1965). Nonelectrolyte transport in muscle during induced protein loss. *Science* **148**, 80-83.

Field, R. A., Pearson, A. M., Koch, D. E., and Merkel, R. A. (1970). Thermal behavior of porcine collagen as related to post-mortem time. *J. Food Sci.* **35**, 113-116.

Fischer, C., and Hamm, R. (1980). Biochemical studies on fast glycolysing bovine muscle. *Meat Sci.* **4**, 41-49.

Fischer, C., Hofmann, K., and Blüchel, (1977). Das Elektrophorese-Bild von Sarkoplasma und Myofibrillen bei wäBrigem, blassem Rindleisch. *Fleischwirtschaft* **57**, 2050-2051.

Fischer, C., Hamm, R., and Honikel, K. O. (1979). Changes in solubility and enzymatic activity of muscle glycogen phosphorylase in PSE-muscles. *Meat Sci.* **3**, 11-19.

Fischer, K., and Augustine, C. (1977). Stadien der postmortalen Glykogenolyse bei unterschiedlichen pH_1-Werten in Schweinefleisch. *Fleischwirtschaft* **57**, 1191-1194.

Follett, M. J., Norman, G. A., and Ratcliff, P. W. (1974). The anti-rigor excision and air cooling of beef semi-membranosus muscles at temperatures between $-5\,°C$ and $+15\,°C$. *J. Food Technol.* **9**, 509-523.

Forrest, J. C., and Briskey, E. J. (1967). Response of striated muscle to electrical stimulation. *J. Food Sci.* **32**, 483-488.

Forrest, J. C., Kastenschmidt, L. L., Beecher, G. R., Grummer, R. H., Hoekstra, W. G., and Briskey, E. J. (1965). Porcine muscle properties. B. Relation to naturally occurring and artificially induced variation in heart and respiration rates. *J. Food Sci.* **30**, 492-497.

Forrest, J. C., Judge, M. D., Sink, J. D., Hoekstra, W. G., and Briskey, E. J. (1966). Prediction of the time course of rigor mortis through response of muscle tissue to electrical stimulation. *J. Food Sci.* **31**, 13-21.

Fujimaki, M., and Deatherage, F. E. (1964). Chromatographic fractionation of sarcoplasmic proteins of beef skeletal muscle on ion-exchange cellulose. *J. Food Sci.* **29**, 316-326.

Fujimaki, M., Okitani, A., and Arakawa, N. (1965a). The changes of "myosin B" during the storage of rabbit muscle. I. Physicochemical studies on myosin B. *Agric. Biol. Chem.* **29**, 581-588.

Fujimaki, M., Arakawa, N., Okitani, A., and Takagi, O. (1965b). The dissociation of "myosin B" from stored rabbit muscle into myosin A and actin and its interaction with ATP. *Agric. Biol. Chem.* **29**, 700-701.

Fujimaki, M., Arakawa, N., Okitani, A., and Takagi, O. (1965c). The changes of "myosin B" ("Actomyosin") during storage of rabbit muscle. II. The dissociation of "myosin B" into myosin A and actin, and its interaction with ATP. *J. Food Sci.* **30**, 937-943.

Gallant, E. M., Godt, R. E., and Gronert, G. A. (1979). Role of plasma membrane defect of skeletal muscle in malignant hyperthermia. *Muscle Nerve* **2**, 491-494.

Galloway, D. E., and Goll, D. E. (1967). Effect of temperature on molecular properties of post-mortem porcine muscle. *J. Anim. Sci.* **26**, 1302-1308.

Gann, G. L., and Merkel, R. A. (1978). Ultrastructural changes in bovine Longissimus muscle during postmortem ageing. *Meat Sci.* **2**, 129-144.

George, A. R., Bendall, J. R., and Jones, R. C. D. (1980). The tenderizing effect of electrical stimulation of beef carcasses. *Meat Sci.* **4**, 51-68.

Goldspink, G., and McLoughlin, J. V. (1964). Studies on pig muscle. 3. The effect of temperature on the solubility of the sarcoplasmic proteins in relation to colour changes in post-rigor muscle. *Ir. J. Agric. Res.* **3**, 9-16.

Goll, D. E., and Robson, R. M. (1967). Molecular properties of post-mortem muscle. I. Myofibrillar nucleosidetriphosphatase activity of bovine muscle. *J. Food Sci.* **32**, 323-329.

Goll, D. E., Henderson, D. W., and Kline, E. A. (1964). Post-mortem changes in physical and chemical properties of bovine muscle. *J. Food Sci.* **29**, 590-596.

Goll, D. E., Arakawa, N., Stromer, M. H., Busch, W. A., and Robson, R. M. (1970). *In* "The Physiology and Biochemistry of Muscle as a Food, 2" (E. J. Briskey, R. G. Cassens, and B. B. Marsh, eds.). pp. 755–800. Univ. of Wisconsin Press, Madison.

Goll, D. E., Stromer, M. H., Robson, R. M., Temple, J., Eason, B. A., and Busch, W. A. (1971). Tryptic digestion of muscle components stimulates many of the changes caused by postmortem storage. *J. Anim. Sci.* **33**, 963–982.

Goutefonga, R. (1971). Influence du pH et de la température sur la solubilité des protéines musculaires du porc. *Ann. Biol. Anim., Biochim., Biophys.* **11**, 233–244.

Grant, N. H. (1955). The respiratory enzymes of meat. I. Identification of the active enzymes. *Food Res.* **20**, 250–253.

Greaser, M. L., Cassens, R. G., and Hoekstra, W. G. (1967). Changes in oxalate-stimulated calcium accumulation in particulate fractions from post-mortem muscle. *J. Agric. Food Chem.* **15**, 1112–1117.

Greaser, M. L., Cassens, R. G., Briskey, E. G., and Hoekstra, W. G. (1969a). Post-mortem changes in subcellular fractions from normal and pale, soft, exudative porcine muscle. I. Calcium accumulation and adenosine triphosphatase activities. *J. Food Sci.* **34**, 120–124.

Greaser, M. L., Cassens, R. G., Briskey, E. J., and Hoekstra, W. G. (1969b). Post-mortem changes in subcellular fractions from normal and pale, soft, exudative porcine muscle. 2. Electron microscopy. *J. Food Sci.* **34**, 125–132.

Greaser, M. L., Cassens, R. G., Hoekstra, W. G., and Briskey, E. J. (1969c). The effect of pH-temperature treatments on the calcium-accumulating ability of purified sarcoplasmic reticulum. *J. Food Sci.* **34**, 633–637.

Greaser, M. L., Cassens, R. G., Hoekstra, W. G., Briskey, E. J., Schmidt, G. R., Carr, S. D., and Galloway, D. E. (1969d). Calcium accumulating ability and compositional differences between sarcoplasmic reticulum fractions from normal and pale, soft, exudative porcine muscle. *J. Anim. Sci.* **28**, 589–592.

Gronert, G. A. (1980). Malignant hyperthermia. *Anesthesiology* **53**, 395–423.

Gunther, H. O. (1974). Probleme bei der Herstellung und Lagerung von Rindfeischkonserven. 1. Post-mortem-Vernanderungen in Muskel-und Bindegewebsproteinen. *Z. Lebensm.-Unters. -Forsch.* **156**, 211–223.

Hallund, O., and Bendall, J. R. (1965). The long-term effect of electrical stimulation on the post-mortem fall of pH in the muscles of Landrace pigs. *J. Food Sci.* **30**, 296–299.

Hamm, R. (1977). Postmortem breakdown of ATP and glycogen in ground muscle; a review. *Meat Sci.* **1**, 15–39.

Hamm, R., and van Hoof, J. (1974). Einfluss von Natriumchlorid auf Abbau und kolloidchemische Wirkung von zugesetztem Adenosintriphosphat in zerkleinertem Rindmuskel post rigor. *Z. Lebensm.-Unters.-Forsch.* **156**, 87–99.

Hamm, R., Kormendy, L., and Gantner, G. (1969). Transaminases of skeletal muscle. I. The activity of transaminases in post-mortem bovine and porcine muscles. *J. Food Sci.* **34**, 446–448.

Hapchuk, L. T., Pearson, A. M., and Price, J. F. (1976). Action of clostridium perfringens organisms on porcine muscle. *J. Food Sci.* **41**, 1042–1046.

Harsham, A., and Deatherage, F. E. (1951). Tenderization of meat. U.S. Patent 2,544,681.

Hasselbach, W. (1977). The sarcoplasmic calcium pump—a most efficient ion translocating system. *Biophys. Struct. Mech.* **3**, 43–54.

Hatton, M. W. C., Lawrie, R. A., Ratcliff, P. W., and Wayne, N. (1972). Effects of preslaughter adrenaline injection on muscle metabolites and meat quality of pigs. *J. Food Technol.* **7**, 443–453.

Hattori, A., and Takahashi, K. (1979). Studies on the post-mortem fragmentation of myofibrils. *J. Biochem. (Tokyo)* **85**, 47–56.

Haworth, R. A., Hunter, D. R., and Berkoff, H. A. (1981). Contracture in isolated adult rat heart cells. Role of Ca²⁺, ATP and compartmentation. *Circ. Res.* **49**, 1119–1128.

Hedrick, H. B., Boillet, J. B., Brady, D. E., and Naumann, H. D. (1959). Etiology of dark-cutting beef. *Res. Bull. — Mo., Agric. Exp. Stn.* **717**.

Hedrick, H. B., Parrish, F. C., Jr., and Bailey, M. E. (1964). Effect of adrenaline stress on pork quality. *J. Anim. Sci.* **23**, 225–229.

Heffron, J. J. A., and Hegarty, P. V. J. (1974). Evidence for a relationship between ATP hydrolysis and changes in extracellular space and fiber diameter during rigor development in skeletal muscle. *Comp. Biochem. Physiol. A* **49**, 43–56.

Hegarty, P. V. J., Dahlin, K. J., Benson, E. S., and Allen, C. E. (1973). Ultrastructural and light microscopic studies on rigor-extended sarcomeres in avian and porcine skeletal muscles. *J. Anat.* **115**, 203–219.

Henderson, D. W., Goll, D. E., and Stromer, M. H. (1970). A comparison of shortening and Z line degradation in post-mortem bovine, porcine, and rabbit muscle. *Am. J. Anat.* **128**, 117–136.

Hendricks, H. B., Lafferty, D. T., Eberle, E. D., Judge, M. D., and Forrest, J. C. (1971). Relation of porcine muscle fiber type and size to postmortem shortening. *J. Anim. Sci.* **32**, 57–61.

Herring, H. K., Cassens, R. G., and Briskey, E. J. (1965a). Further studies on bovine muscle tenderness as influenced by carcass position, sarcomere length, and fiber diameter. *J. Food Sci.* **30**, 1049–1054.

Herring, H. K., Cassens, R. G., and Briskey, E. J. (1965b). Sarcomere length of free and restrained bovine muscles at low temperature as related to tenderness. *J. Sci. Food Agric.* **16**, 379–384.

Horgan, D. J. (1979). ATPase activities of sarcoplasmic reticulum isolated from rabbit and bovine muscle subjected to pre-rigor pressure treatment. *J. Food Sci.* **44**, 492–500.

Howard, A., and Lawrie, R. A. (1956). Studies on beef quality. II. Physiological and biological effects of various pre-slaughter treatments. *Div. Food Preserv. Transp., Tech. Pap. (Aust., C.S.I.R.O.)* **2**, 18–51.

Howard, A., Lee, C. A., and Webster, H. L. (1960). Studies on beef quality. Part IX. Nucleotide breakdown in beef tissue: the extent of formation of hypoxanthine during storage as an indicator of ripening. *Div. Food Preserv. Tech. Pap. (Aust., C.S.I.R.O.)* **21**, 1–14.

Howe, J. M., Thomas, N. W., Addis, P. B., and Judge, M. D. (1968). Temperature acclimation and its effects on porcine muscle properties in two humidity environments. *J. Food Sci.* **33**, 235–238.

Ikeuchi, Y., Ito, T., and Fukazawa, T. (1978). Change in regulation of myofibrillar ATPase activity by calcium during postmortem storage of muscle. *J. Food Sci.* **43**, 1338–1339.

Ikeuchi, Y., Ito, T., and Fukazawa, T. (1980a). Change of regulatory activity of tropomyosin and troponin on acto-heavy-meromyosin ATPase during postmortem storage of muscle. *J. Food Sci.* **45**, 13–16, 20.

Ikeuchi, Y., Ito, T., and Fukazawa, T. (1980b). Changes in the properties of myofibrillar proteins during post-mortem storage of muscle at high temperature. *J. Agric. Food Chem.* **28**, 1197–1202.

Izumi, K., Ito, T., and Fukazawa, T. (1981). Effect of ATP concentration and pH on rigor tension development and dissociation of rigor complex in glycerinated rabbit psoas fiber. *Biochim. Biophys. Acta* **678**, 364–372.

Jeremiah, L. E., and Martin, A. H. (1981). Intramuscular collagen content and solubility. Their relationship to tenderness and alteration by postmortem aging. *Can. J. Anim. Sci.* **61**, 53–61.

Jungk, R. A., Snyder, H. E., Goll, D. E., and McConnell, K. G. (1967). Isometric tension changes and shortening in muscle strips during post-mortem aging. *J. Food Sci.* **32**, 158–161.

Kanda, T., Pearson, A. M., and Merkel, R. A. (1977). Influence of pH and temperature upon calcium accumulation and release by bovine sarcoplasmic reticulum. *Food Chem.* **2**, 253–266.

Kang, C. K., Donnelly, T. H., Jodlowski, R. F., and Warner, W. D. (1981). Partial purification and characterization of a neutral protease from bovine skeletal muscle. *J. Food Sci.* **46**, 702–707.

Kastenschmidt, L. L. (1970). The metabolism of muscle as a food. *In* "The Physiology and Biochemistry of Muscle as a Food, 2" (E. J. Briskey, R. G. Cassens, and B. B. Marsh, eds.), pp. 735–753. Univ. of Wisconsin Press, Madison.

Kastenschmidt, L. L., Briskey, E. J., and Hoekstra, W. G. (1964). Prevention of pale, soft, exudative porcine muscle through regulation of ante-mortem environmental temperature. *J. Food Sci.* **29**, 210–217.

Kastenschmidt, L. L., Beecher, G. R., Forrest, J. C., Hoekstra, W. G., and Briskey, E. J. (1965). Porcine muscle properties. A. Alteration of glycolysis by artificially induced changes in ambient temperature. *J. Food Sci.* **30**, 565–572.

Kastenschmidt, L. L., Hoekstra, W. G., and Briskey, E. J. (1968). Glycolytic intermediate and co-factors in "fast- and slow-glycolyzing" muscles of the pig. *J. Food Sci.* **33**, 151–158.

Kennick, W. H., Elgasim, E. A., Holmes, Z. A., and Meyer, P. F. (1980). The effect of pressurisation of pre-rigor muscle on post-rigor characteristics. *Meat Sci.* **4**, 33–40.

Kraeling, R. R., and Rampacek, G. B. (1977). Induction of a pale, soft, exudative-like myopathy and sudden death in pigs by injection of anterior pituitary extract. *J. Anim. Sci.* **45**, 71–80.

Kraeling, R. R., Ono, K., Davis, B. J., and Barb, C. R. (1975). Effect of pituitary gland activity on longissimus muscle postmortem glycolysis in the pig. *J. Anim. Sci.* **40**, 604–612.

Kronman, M. J., and Winterbottom, R. J. (1960). Post-mortem changes in the water-soluble proteins of bovine skeletal muscle during aging and freezing. *J. Agric. Food Chem.* **8**, 67–72.

Krzywicki, K. (1971). Relation of ATPase activity and calcium uptake to postmortem glycolysis. *J. Food Sci.* **36**, 791–794.

LaCourt, A. (1971). Action post mortem du pH et de la température sur la captage de calcium et l'activité ATPasique du reticulum sarcoplasmique fragmente du muscle de bovin. *Ann. Biol. Anim., Biochim., Biophys.* **11**, 681–694.

LaCourt, A. (1972). Contraction provoquée par le froid dans les muscles de veau. *Proc. Eur. Meet. Meat Res. Workers* **18**, 637–650.

Lawrie, R. A., Gatherum, D. P., and Hale, H. P. (1958). Abnormally low ultimate pH in pig muscle. *Nature (London)* **182**, 807–808.

Lawrie, R. A., Penny, I. F., Scopes, R. K., and Voyle, C. A. (1963). Sarcoplasmic proteins in pale, exudative pig muscles. *Nature (London)* **200**, 673.

Lazarus, C. R., Deng, J. C., and Watson, C. M. (1977). Changes in the concentration of fatty acids from the nonpolar phospho- and glycolipids during storage of intact lamb muscles. *J. Food Sci.* **42**, 102–107.

Lee, C. A., and Newbold, R. P. (1963). The pathway of degradation of inosinic acid in bovine skeletal muscle. *Biochim. Biophys. Acta* **72**, 349–352.

Leet, G. N., and Locker, R. H. (1973). A prolonged pre-rigor condition in aerobic surfaces of ox muscle. *J. Sci. Food Agric.* **24**, 1181–1191.

Lochner, J. V., Kauffman, R. G., and Marsh, B. B. (1980). Early postmortem cooling rate and beef tenderness. *Meat Sci.* **4**, 227–241.

Locker, R. H. (1960). Proteolysis in the storage of beef. *J. Sci. Food Agric.* **11**, 520–526.

Locker, R. H., and Daines, G. J. (1975). Rigor mortis in beef sternomandibularis muscle at 37°C. *J. Sci. Food Agric.* **26**, 1721–1733.

Locker, R. H., and Hagyard, C. J. (1963). A cold shortening effect in beef muscles. *J. Sci. Food Agric.* **14**, 787–793.

Locker, R. H., Davey, C. L., Nottingham, P. M., Haughey, D. P., and Law, N. H. (1975). New concepts in meat processing. *Adv. Food Res.* **21**, 157–222.

Locker, R. H., Daines, G. J., and Leet, N. G. (1976). Histology of highly-stretched beef muscle.

III. Abnormal contraction patterns in ox muscle produced by overstretching during prerigor blending. *J. Ultrastruct. Res.* **55**, 173–181.

Ludvigsen, J. (1954). Undersøgelser over den såkaldte "muskeldegeneration" hos svin. *Beret. Forsoegslab. (Copenhagen)* **1**, 272.

Lutalo-Bosa, A. J. (1970). Catheptic enzymes of bovine skeletal muscle. Ph.D. Thesis, McGill University, Montreal, Quebec.

MacBride, M. A., and Parrish, F. C., Jr. (1977). The 30,000-dalton component of tender bovine longissimus muscle. *J. Food Sci.* **42**, 1627–1629.

McClain, P. E., Pearson, A. M., Brunner, J. R., and Crevasse, G. A. (1967). Connective tissues from normal and PSE porcine muscle. 1. Chemical characterization. *J. Food Sci.* **34**, 115–119.

McClain, P. E., Creed, G. J., Wiley, E. R., and Hornstein, I. (1970). Effect of post-mortem aging on isolation of intramuscular connective tissue. *J. Food Sci.* **35**, 258–259.

Mac Farlane, J. J. (1973). Pre-rigor pressurization of muscle: Effects on pH, shear value, and taste panel assessment. *J. Food Sci.* **38**, 294–298.

Mac Farlane, J. J., and Morton, D. J. (1978). Effects of pressure treatment on the ultrastructure of striated muscle. *Meat Sci.* **2**, 281–288.

McIntosh, E. N. (1967). Effect of post-mortem aging and enzyme tenderizers on mucoprotein of bovine skeletal muscle. *J. Food Sci.* **32**, 210–213.

McLoughlin, J. V., and Goldspink, G. (1963). Post-mortem changes in the colour of pig longissimus dorsi muscle. *Nature (London)* **198**, 584–585.

Marple, D. N., Nachreiner, R. F., McGuire, J. A., and Squires, C. D. (1975). Thyroid function and muscle glycolysis in swine. *J. Anim. Sci.* **41**, 799–803.

Marsh, B. B. (1952). The effects of adenosine triphosphate on the fibre volume of a muscle homogenate. *Biochim. Biophys. Acta* **9**, 247–260.

Marsh, B. B. (1954). Rigor mortis in beef. *J. Sci. Food Agric.* **5**, 70–75.

Marsh, B. B., and Carse, W. A. (1974). Meat tenderness and the sliding-filament hypothesis. *J. Food Technol.* **9**, 129–139.

Marsh, B. B., and Leet, N. G. (1966). Studies in meat tenderness. III. The effects of cold shortening on tenderness. *J. Food Sci.* **31**, 450–459.

Marsh, B. B., and Thompson, J. F. (1958). Rigor mortis and thaw rigor in lamb. *J. Sci. Food Agric.* **9**, 417–424.

Marsh, B. B., Woodhams, P. R., and Leet, N. G. (1968). Studies in meat tenderness. 5. The effects on tenderness of carcass cooling and freezing before the completion of rigor mortis. *J. Food Sci.* **33**, 12–18.

Marsh, B. B., Leet, N. G., and Dickson, M. R. (1974). The ultrastructure and tenderness of highly cold-shortened muscle. *J. Food Technol.* **9**, 141–147.

Marsh, B. B., Lochner, J. V., Takahashi, G., and Kragness, D. D. (1980). Effects of early post-mortem pH and temperature on beef tenderness. *Meat Sci.* **5**, 479–483.

Mitchelson, K. R., and Hird, F. J. R. (1973). Effect of pH and halothane on muscle and liver mitochondria. *Am. J. Physiol.* **225**, 1393–1398.

Moeller, P. W., Fields, P. A., Dutson, T. R., Landmann, W. A., and Carpenter, Z. L. (1976). Effect of high temperature conditioning on subcellular distribution and levels of lysosomal enzymes. *J. Food Sci.* **41**, 216–217.

Moeller, P. W., Fields, P. A., Dutson, T. R., Landmann, W. A., and Carpenter, Z. L. (1977). High temperature effects on lysosomal enzyme distribution and fragmentation of bovine muscle. *J. Food Sci.* **42**, 510–512.

Morley, M. J. (1971). Measurement of oxygen penetration into meat using an oxygen micro-electrode. *J. Food Technol.* **6**, 371–387.

Nakamura, R. (1973). Estimation of water-extractable Ca in chicken breast muscle by atomic absorption. *Anal. Biochem.* **53**, 531–537.

Newbold, R. P., and Lee, C. A. (1965). Post-mortem glycolysis in skeletal muscle. The extent of glycolysis in diluted preparation of mammalian muscle. *Biochem. J.* **97**, 1–6.

Newbold, R. P., and Scopes, R. K. (1967). Post-mortem glycolysis in ox skeletal muscle. Effect of temperature on the concentrations of glycolytic intermediates and cofactors. *Biochem. J.* **105**, 127–136.

Newbold, R. P., and Scopes, R. K. (1971). Post-mortem glycolysis in ox skeletal muscle: Effect of adding nicotinamide-adenine dinucleotide to diluted mince preparations. *J. Food Sci.* **36**, 215–218.

Newbold, R. P., and Tume, R. K. (1981). Comparison of the effects of added orthophosphate on calcium uptake and release by bovine and rabbit muscle sarcoplasmic reticulum. *J. Food Sci.* **46**, 1327–1332.

Nuss, J. I., and Wolfe, F. H. (1981). Effect of post-mortem storage temperatures on isometric tension, pH, ATP, glycogen, and glucose-6-phosphate for selected bovine muscles. *Meat Sci.* **5**, 201–213.

Okitani, A., and Fujimaki, M. (1972). A neutral proteolytic system responsible for post-mortem proteolysis in rabbit skeletal muscle. *Agric. Biol. Chem.* **36**, 1265–1267.

Okitani, A., Suzuki, A., Yang, R., and Fujimaki, M. (1972). Effect of cathepsin D treatment on ATPase activity of rabbit myofibril. *Agric. Biol. Chem.* **36**, 2135–2141.

Okitani, A., Shinohara, K., Sugitani, M., and Fujimaki, M. (1973). A relation between the rate of increment in nonprotein nitrogenous compounds and muscle pH during post-mortem storage. *Agric. Biol. Chem.* **37**, 321–325.

Okitani, A., Matsukura, U., Kato, H., and Fujimaki, M. (1980). Purification and some properties of a myofibrillar protein-degrading protease, cathepsin L, from rabbit skeletal muscle. *J. Biochem. (Tokyo)* **87**, 1133–1143.

Olson, D. G., Parrish, F. C., Jr., Dayton, W. R., and Goll, D. E. (1977). Effect of post-mortem storage and calcium activated factor on the myofibrillar proteins of bovine skeletal muscle. *J. Food Sci.* **42**, 117–124.

Ono, K. (1971). Lysosomal enzyme activation and proteolysis of bovine muscle. *J. Food Sci.* **36**, 838–839.

Ono, K., and Woods, D. R. (1974). Adenosine $3',5'$-cyclic monophosphate in normal porcine muscles. *J. Food Sci.* **39**, 829–832.

Ota, S., Furuya, Y., and Shintaku, K. (1973). Studies on rigor mortis. *Forensic Sci.* **2**, 207–219.

Otsuka, Y., Okitani, A., Katakai, R., and Fujimaki, M. (1976). Purification and properties of an aminopeptidase from rabbit skeletal muscle. *Agric. Biol. Chem.* **40**, 2335–2342.

Ottaway, J. H., and Mowbray, J. (1977). The role of compartmentation in the control of glycolysis. *Curr. Top. Cell. Regul.* **12**, 107–208.

Palmer, E. G., Topel, D. G., and Christian, L. L. (1977). Microscopic observations of muscle from swine susceptible to malignant hyperthermia. *J. Anim. Sci.* **45**, 1032–1036.

Park, H. K., Ito, T., and Fukazawa, T. (1975a). Comparison between normal and PSE porcine muscle in the extractability of myosin B and in the rheological properties of those sausages. *Jpn. J. Zootech. Sci.* **46**, 360–366.

Park, H. K., Muguruma, M., Fukazawa, T., and Ito, I. (1975b). Relationship between superprecipitating activity and constituents of myosin B prepared from normal and PSE porcine muscle. *Agric. Biol. Chem.* **39**, 1363–1370.

Park, H. K., Ito, T., and Fukazawa, T. (1977). Calcium regulation of actomyosin prepared from PSE porcine muscle. *Jpn. J. Zootech. Sci.* **48**, 654–660.

Parrish, F. C., Jr., Goll, D. E., Newcomb, W. J., II, de Lumen, B. O., Chaudhry, H. M., and Kline, E. A. (1969). Molecular properties of post-mortem muscle. 7. Changes in nonprotein nitrogen and free amino acids of bovine muscle. *J. Food Sci.* **34**, 196–202.

Parrish, F. C., Jr., Young, R. B., Miner, B. E., and Andersen, L. D. (1973). Effect of post-mortem conditions on certain chemical, morphological, and organoleptic properties of bovine muscle. *J. Food Sci.* **38**, 690–695.

Partmann, W. (1963). Post-mortem changes in chilled and frozen muscle. *J. Food Sci.* **28**, 15–27.

Paul, P., Lowe, B., and McClurg, B. R. (1944). Changes in histological structure and palatability of beef during storage. *Food Res.* **9**, 221–233.

Paul, P. C. (1965). Storage- and heat-induced changes in the microscopic appearance of rabbit muscle. *J. Food Sci.* **30**, 960–968.

Penny, I. F. (1967). The effect of post-mortem conditions on the extractability and adenosine triphosphatase activity of myofibrillar proteins of rabbit muscle. *J. Food Technol.* **2**, 325–338.

Penny, I. F. (1968). Effect of aging on the properties of myofibrils of rabbit muscle. *J. Sci. Food Agric.* **19**, 518–523.

Penny, I. F. (1969). Protein denaturation and water-holding capacity in pork muscle. *J. Food Technol.* **4**, 269–273.

Penny, I. F. (1972). Conditioning of bovine muscle. III. The α-actinin of bovine muscle. *J. Sci. Food Agric.* **23**, 403–412.

Penny, I. F. (1976). The effect of conditioning on the myofibrillar proteins of pork muscle. *J. Sci. Food Agric.* **27**, 1147–1155.

Penny, I. F. (1977). The effect of temperature on the drip, denaturation, and extracellular space of pork *Longissimus dorsi* muscle. *J. Sci. Food Agric.* **28**, 329–338.

Penny, I. F., and Dransfield, E. (1979). Relationship between toughness and troponin T in conditioned beef. *Meat Sci.* **3**, 135–139.

Penny, I. F., and Ferguson-Pryce, R. (1979). Measurement of autolysis in beef muscle homogenates. *Meat Sci.* **3**, 121–134.

Penny, I. F., Voyle, C. A., and Dransfield, E. (1974). The tenderizing effect of a muscle proteinase on beef. *J. Sci. Food Agric.* **25**, 703–708.

Petropakis, H. J., Anglemier, A. F., and Montgomery, M. W. (1973). Changes in the low molecular weight nitrogenous compounds of excised bovine muscle. *J. Food Sci.* **38**, 59–62.

Rhodes, D. N. (1965). Nucleotide degradation during the extended storage of lamb and beef. *J. Sci. Food. Agric.* **16**, 447–451.

Robbins, F. M., Walker, J. E., Cohen, S. H., and Chatterjee, S. (1979). Action of proteolytic enzymes on bovine myofibrils. *J. Food Sci.* **44**, 1672–1680.

Robinson, J. D., and Flashner, M. S. (1979). The $(Na^+ + K^+)$-activated ATPase. Enzymatic and transport properties. *Biochim. Biophys. Acta* **549**, 145–176.

Robson, R. M., Goll, D. E, and Main, M. J. (1967). Molecular properties of post-mortem muscle. 5. Nucleoside triphosphatase activity of bovine myosin B. *J. Food Sci.* **32**, 544–549.

Robson, R. M., Stromer, M. H., Huiatt, T. W., O'Shea, J. M., Hartzer, M. K., Richardson, F. L., and Rathbun, W. E. (1980). Biochemistry and structure of desmin and the recently discovered muscle cell cytoskeleton. *Proc. Eur. Meet. Meat Res. Workers* **26** (Vol. 1), 22–25.

Roschen, H. L., Ortscheid, B. J., and Ramsbottom, J. M. (1950). Tenderizing meats. U.S. Patent 2,519,931.

Rowe, R. W. D. (1974). Collagen fibre arrangement in intramuscular connective tissue. Changes associated with muscle shortening and their possible relevance to raw meat toughness measurements. *J. Food Technol.* **9**, 501–508.

Sair, R. A., Lister, D., Moody, W. G., Cassens, R. G., Hoekstra, W. G., and Briskey, E. J. (1970). Action of curare and magnesium on striated muscle of stress-susceptible pigs. *Am. J. Physiol.* **218**, 108–114.

Savell, J. W., Dutson, T. R., Smith, G. C., and Carpenter, Z. L. (1978). Structural changes in electrically stimulated beef muscle. *J. Food Sci.* **43**, 1606–1609.

Sayre, R. N., and Briskey, E. J. (1963). Protein solubility as influenced by physiological conditions in the muscle. *J. Food Sci.* **28**, 675–679.

Sayre, R. N., Briskey, E. J., Hoekstra, W. G., and Bray, R. W. (1961). Effect of pre-slaughter change to a cold environment on characteristics of pork muscle. *J. Anim. Sci.* **20**, 487–492.

Sayre, R. N., Briskey, E. J., and Hoekstra, W. G. (1963a). Alteration of post-mortem changes in porcine muscle by preslaughter heat treatment and diet modification. *J. Food Sci.* **28**, 292–297.

Sayre, R. N., Briskey, E. J., and Hoekstra, W. G. (1963b). Effect of excitement, fasting, and sucrose feeding on porcine muscle phosphorylase and post-mortem glycolysis. *J. Food Sci.* **28**, 472–477.

Sayre, R. N., Briskey, E. J., and Hoekstra, W. G. (1963c). Comparison of muscle characteristics and post-mortem glycolysis in three breeds of swine. *J. Anim. Sci.* **22**, 1012–1020.

Schmidt, G. R., and Gilbert, K. V. (1970). The effect of muscle excision before the onset of rigor mortis on the palatability of beef. *J. Food Technol.* **5**, 331–338.

Schmidt, G. R., Cassens, R. G., and Briskey, E. J. (1970a). Changes in tension and certain metabolites during the development of rigor mortis in selected red and white skeletal muscle. *J. Food Sci.* **35**, 571–573.

Schmidt, G. R., Cassens, R. G., and Briskey, E. J. (1970b). Relationship of calcium uptake by the sarcoplasmic reticulum to tension development and rigor mortis in striated muscle. *J. Food Sci.* **35**, 574–576.

Schmidt, G. R., Goldspink, G., Roberts, T., Kastenschmidt, L. L., Cassens, R. G., and Briskey, E. J. (1972). Electromyography and resting membrane potential in *Longissimus* muscle of stress-susceptible and stress-resistant pigs. *J. Anim. Sci.* **34**, 379–383.

Schwartz, W. N., and Bird, J. W. C. (1977). Degradation of myofibrillar proteins by cathepsins B and D. *Biochem. J.* **167**, 811–820.

Scopes, R. K. (1964). The influence of post-mortem conditions on the solubilities of muscle proteins. *Biochem. J.* **91**, 201–207.

Scopes, R. K. (1965). Acid denaturation of creatine kinase. *Arch. Biochem. Biophys.* **110**, 320–324.

Scopes, R. K. (1970). Characterization and study of sarcoplasmic proteins. *In* "The Physiology and Biochemistry of Muscle as a Food, 2" (E. J. Briskey, R. G. Cassens, and B. B. Marsh, eds.), pp. 471–492. Univ. of Wisconsin Press, Madison.

Scopes, R. K. (1973). Studies with a reconstituted muscle glycolytic system. The rate and extent of creatine phosphorylation by anaerobic glycolysis. *Biochem. J.* **134**, 197–208.

Scopes, R. K. (1974a). Studies with a reconstituted muscle glycolytic system. The anaerobic glycolytic response to simulated tetanic contraction. *Biochem. J.* **138**, 119–123.

Scopes, R. K. (1974b). Studies with a reconstituted muscle glycolytic system. The rate and extent of glycolysis in simulated post-mortem conditions. *Biochem. J.* **142**, 79–86.

Scopes, R. K., and Lawrie, R. A. (1963). Post-mortem lability of skeletal muscle proteins. *Nature (London)* **197**, 1202–1203.

Severin, S. E., Tseitlin, L. A., and Druzhinina, T. N. (1963). Enzymic breakdown of diphosphopyridine nucleotide in cardiac and skeletal muscle. *Biochemistry (Engl. Transl.)* **28**, 112–117.

Sharp, J. G. (1963). Aseptic autolysis in rabbit and bovine muscle during storage. *J. Sci. Food Agric.* **14**, 468–479.

Shelef, L. A., and Jay, J. M. (1969). Relationship between amino sugars and meat microbial quality. *Appl. Microbiol.* **17**, 931–932.

Sims, T. J., and Bailey, A. J. (1981). Connective tissue. *In* "Developments in Meat Science" (R. Lawrie, ed.), Vol. 2, pp. 29–59. Applied Science Publishers, London.

Smith, G. C., Dutson, T. R., Hostetler, R. L., and Carpenter, Z. L. (1976). Fatness, rate of chilling and tenderness of lamb. *J. Food Sci.* **41**, 748–756.

Stanley, D. W., and Brown, R. G. (1973). The fate of intramuscular connective tissue in aged beef. *Proc. Eur. Meet. Meat Res. Workers* **19**, 231–248.

Stanley, D. W., and Swatland, H. J. (1976). The microstructure of muscle tissue — a basis for meat texture measurement. *J. Text. Stud.* **7**, 65–75.

Stauber, W. T., and Ong, S.-H. (1981). Fluroescence demonstration of cathepsin B activity in skeletal, cardiac, and vascular smooth muscle. *J. Histochem. Cytochem.* **29**, 866–869.

Strandberg, K., Parrish, F. C., Jr., Goll, D. E., and Josephson, S. A. (1973). Molecular properties of post-mortem muscle: Effect of sulfhydryl reagents and post-mortem storage on changes in myosin B. *J. Food Sci.* **38**, 69–74.

Stromer, M. H., and Goll, D. E. (1967a). Molecular properties of post-mortem muscle. II. Phase microscopy of myofibrils from bovine muscle. *J. Food Sci.* **32**, 329–331.

Stromer, M. H., and Goll, D. E. (1967b). Molecular properties of post-mortem muscle. 3. Electron microscopy of myofibrils. *J. Food Sci.* **32**, 386–389.

Stromer, M. H., Goll, D. E., and Roth, L. E. (1967). Morphology of rigor-shortened bovine muscle and the effect of trypsin on pre- and post-rigor myofibrils. *J. Cell Biol.* **34**, 431–445.

Sung, S. K., Ito, T., and Fukazawa, T. (1976). Relationship between contractility and some biochemical properties of myofibrils prepared from normal and PSE porcine muscle. *J. Food Sci.* **41**, 102–107.

Sung, S. K., Izumi, K., Ito, T., and Fukazawa, T. (1977). Contractility of porcine "ghost" myofibril irrigated with myosin from normal and PSE porcine muscle. *Agric. Biol. Chem.* **41**, 1087–1089.

Sung, S. K., Ito, T., and Izumi, K. (1981). Myosin ATPase and acto-heavy meromyosin ATPase in normal and pale, soft, and exudative (PSE) porcine muscle. *Agric. Biol. Chem.* **45**, 953–957.

Suzuki, A., and Fujimaki, M. (1968). Studies on proteolysis in stored muscle. Part II. Purification and properties of a proteolytic enzyme, cathepsin D, from rabbit muscle. *Agric. Biol. Chem.* **32**, 975–982.

Swatland, H. J. (1975a). Survival time of intact neuromuscular pathways in slaughtered pigs. *Can. Inst. Food Sci. Technol. J.* **8**, 122–123.

Swatland, H. J. (1975b). Relationships between mitochondrial content of muscle fibers and patterns of glycogen depletion postmortem. *Histochem. J.* **7**, 367–374.

Swatland, H. J. (1976a). Motor unit activity in excised prerigor beef muscle. *Can. Inst. Food Sci. Technol. J.* **9**, 177–181.

Swatland, H. J. (1976b). Electromyography of electrically stunned pigs and excised bovine muscle. *J. Anim. Sci.* **42**, 838–844.

Swatland, H. J. (1977). Sensitivity of prerigor beef muscle to electrical stimulation. *Can. Inst. Food Sci. Technol. J.* **10**, 280–283.

Swatland, H. J. (1980). Postmortem changes in electrical capacitance and resistivity of pork. *J. Anim. Sci.* **51**, 1108–1112.

Swatland, H. J. (1981). Electrical capacitance measurements on intact carcasses. *J. Anim. Sci.* **53**, 666–669.

Takahashi, K., and Saito, H. (1979). Post-mortem changes in skeletal muscle connectin. *J. Biochem. (Tokyo)* **85**, 1539–1542.

Takahashi, K., Nakamura, F., and Inoue, A. (1981). Postmortem changes in the actin-myosin interaction of rabbit skeletal muscle. *J. Biochem. (Tokyo)* **89**, 321–324.

Tappel, A. L., and Martin, R. (1958). Succinoxidase activity of some animal tissues. *Food Res.* **23**, 280–282.

Tarrant, P. V. (1977). The effect of hot-boning on glycolysis in beef muscle. *J. Sci. Food Agric.* **28,** 927–930.

Tarrant, P. V., and Mothersill, C. (1977). Glycolysis and associated changes in beef carcasses. *J. Sci. Food Agric.* **28,** 739–749.

Taylor, A. A., and Dant, S. J. (1971). Influence of carcass cooling rate on drip loss in pigmeat. *J. Food Technol.* **6,** 131–139.

Taylor, E. W. (1979). Mechanism of actomyosin ATPase and the problem of muscle contraction. *CRC Crit. Rev. Biochem.* **6,** 103–164.

Thomas, J., and Frearson, N. (1971). Comparison of myosins prepared from rabbit and bovine muscle immediately after death and after storage at 2°C. *J. Food Sci.* **36,** 1110–1113.

Tsai, R., Cassens, R. G., Briskey, E. J., and Greaser, M. L. (1972). Studies on nucleotide metabolism in porcine Longissimus muscle postmortem. *J. Food Sci.* **37,** 612–616.

Urbin, M. C., and Wilson, G. D. (1961). The post-mortem oxygen requirements of bovine tissue. *J. Food Sci.* **26,** 314–317.

Van Eerd, J. P. (1972). Emulsion stability and protein extractability of ovine muscle as a function of time postmortem. *J. Food Sci.* **37,** 473–475.

Varriano-Marston, E., Davis, E. A., Hutchinson, T. E., and Gordon, J. (1976). Scanning electron microscopy of aged free and restrained bovine muscle. *J. Food Sci.* **41,** 601–605.

Voyle, C. A. (1969). Some observations on the histology of cold-shortened muscle. *J. Food Technol.* **4,** 275–281.

Wallick, E. T., Lane, L. K., and Schwartz, A. (1979). Biochemical mechanism of the sodium pump. *Annu. Rev. Physiol.* **41,** 397–411.

Warriss, P. D. (1978). Factors affecting the residual blood content of meat. *Meat Sci.* **2,** 155–159.

Weiner, P. D., and Pearson, A. M. (1969). Calcium chelators influence some physical and chemical properties of rabbit and pig muscle. *J. Food Sci.* **34,** 592–596.

Werb, Z., Banda, M. J., and Jones, P. A. (1980). Degradation of connective tissue matrices by macrophages. I. Proteolysis of elastin, glycoproteins, and collagen by proteases isolated from macrophages. *J. Exp. Med.* **152,** 1340–1357.

Wesemeier, H. (1974). Elektronenmikroskopische Untersuchungen an der Skelettmuskulatur von unbelasteten sowie experimentall belasteten Fleischschweinen. *Arch. Exp. Vet.* **28,** 329–383.

Whiting, R. C. (1980). Calcium uptake by bovine muscle mitochondria and sarcoplasmic reticulum. *J. Food Sci.* **45,** 288–292.

Wierbicki, E., Kunkle, L. E., Cahill, V. R., and Deatherage, F. E. (1954). The relation of tenderness to protein alterations during post-mortem aging. *Food Technol. (Chicago)* **8,** 506–511.

Wierbicki, E., Cahill, V. R., Kunkle, L. E., Klosterman, E. W., and Deatherage, F. E. (1955). Effect of castration on biochemistry and quality of beef. *J. Agric. Food Chem.* **3,** 244–249.

Will, P. A., Ownby, C. L., and Henrickson, R. L. (1980). Ultrastructural postmortem changes in electrically stimulated bovine muscle. *J. Food Sci.* **45,** 21–25, 34.

Wismer-Pedersen, J. (1959). Quality of pork in relation to rate of pH change post-mortem. *Food Res.* **24,** 711–727.

Wood, F. W. (1966). The diffusion of salt in pork muscle and fat tissue. *J. Sci. Food Agric.* **17,** 138–140.

Wu, J. J., Dutson, T. R., and Carpenter, Z. L. (1981). Effect of postmortem time and temperature on the release of lysosomal enzymes and their possible effect on bovine connective tissue components of muscle. *J. Food Sci.* **46,** 1132–1135.

Wu, T.-F. L., and Davis, E. J. (1981). Regulation of glycolytic flux in an energetically controlled cell-free system: The effects of adenine nucleotide ratios, inorganic phosphate, pH and citrate. *Arch. Biochem. Biophys.* **209,** 85–99.

Yang, R., Okitani, A., and Fujimaki, M. (1970). Studies on myofibrils from stored muscle. Part 1.

Post-mortem changes in adenosine triphosphatase activity of myofibrils from rabbit muscle. *Agric. Biol. Chem.* **34**, 1765–1772.

Yang, R., Okitani, A., and Fujimaki, M. (1978). Postmortem changes in regulatory proteins of rabbit muscle. *Agric. Biol. Chem.* **42**, 555–563.

Yasui, T., Sumita, T., and Tsunogae, S. (1975). Stability of myofibrillar EDTA-ATPase in rabbit psoas fiber bundles. *J. Agric. Food Chem.* **23**, 1163–1168.

Yates, L. D. (1977). Molecular and ultrastructural changes in bovine muscle. M.S. Thesis, Texas A&M University, College Station.

Young, O. A., Graafhuis, A. E., and Davey, C. L. (1980). Post-mortem changes in cytoskeletal proteins of muscle. *Meat Sci.* **5**, 41–55.

Zender, R., Lataste-Dorolle, C., Collet, R. A., Rowinski, P., and Mouton, R. F. (1958). Aseptic autolysis of muscle: biochemical and microscopic modifications occurring in rabbit and lamb muscle during aseptic and anaerobic storage. *Food Res.* **23**, 305–326.

3

Physical and Biochemical Changes Occurring in Muscle during Storage and Preservation

A. M. PEARSON

Department of Food Science and Human Nutrition
Michigan State University
East Lansing, Michigan

I. INTRODUCTION

Storage and preservation of meat are a consequence of the inability of man to utilize fully the fresh muscle and organs of animals slaughtered for human consumption. Industrialization and changing from an essentially rural to a mainly urban society have also made it necessary to transport meat for considerable distances, thus making storage essential. In fact, some writers have claimed that development of the refrigerator car was a major contributing factor to industrial development in the United States (Vaughan, 1945). The use of refrigeration has permitted slaughtering of livestock near the point of pro-

Copyright © 1986 by Academic Press, Inc.
All rights of reproduction in any form reserved.

duction with shipment of meat to centers of consumption. The end result has been that virtually all meat is consumed after the onset of rigor mortis, followed by either storage and/or processing.

Storage and processing of meat may result in quite different products than that from fresh prerigor muscle. Marked physical and biochemical changes are known to occur during storage and processing. Such changes can influence flavor, tenderness, and other palatability characteristics of meat. Thus, the physical and biochemical events that occur during conversion of prerigor muscle to fresh meat and during processing and/or cooking may be both desirable and undesirable. These changes and their significance to the properties of meat are covered in this chapter.

II. SOME FACTORS INFLUENCING POSTMORTEM CHANGES

Physical and biochemical changes begin in muscle almost immediately following death and continue at varying rates until the tissues are completely degraded (Moulton and Lewis, 1940; Bendall, 1961). The extent and speed is dependent upon a number of factors, including the temperature, the relative humidity, the level and activity of the indigenous proteolytic enzymes, and the microflora present. Cold temperatures, near or below the freezing point, are generally beneficial in that they retard bacterial degradation and also slow up the action of the indigenous muscle enzymes. However, cold temperatures are not universally beneficial, as they can cause excessive muscle shortening (cold shortening) and toughness in prerigor meat (Locker and Hagyard, 1963). This phenomenon is discussed in Section IV,B.

Relative humidity also has a marked influence upon microbial growth. High relative humidities provide conditions favorable to microbial proliferation, whereas low humidities inhibit microbial growth. High relative humidity at low temperatures, however, accelerates the rate of heat transfer from the carcass and speeds up chilling. High humidity also retards surface evaporation and helps to reduce shrinkage during the chilling process. Thus, humidity can be regulated to achieve the fastest chilling of meat with minimum shrinkage or to give the best conditions for aging to control microbial growth.

There are a number of proteolytic enzymes present in muscle (Bird *et al.*, 1980). The most important ones from the standpoint of degrading myofibrillar proteins appear to be the Ca^{2+}-activated proteinase (CAF) and cathepsins B and D. Little is known about the levels of these proteolytic enzymes and the factors in muscle that control their activity, either antemortem or postmortem. However, these enzymes are capable of breaking down the myofibrillar proteins and probably play important roles in changes occurring in postmortem muscle

(Abbott, 1976; Abbott *et al.*, 1977). Some conditions that are favorable and unfavorable to the activity of these indigenous muscle enzymes and their probable impact upon the important physical properties of meat are discussed later.

The type and extent of microbial contamination can also influence degradation of meat (Hasegawa *et al.*, 1970a,b; Rampton *et al.*, 1970; Tarrant *et al.*, 1971). The spoilage bacteria have been shown to degrade meat by elaborating proteolytic enzymes that attack the muscle proteins (Porzio and Pearson, 1975; Hapchuk and Pearson, 1978; Hapchuk *et al.*, 1979). Thus, many of the changes occurring during storage of meat may be the result of microbial contamination and growth.

Normally, these factors are controlled to extend the storage life of meat. This is usually accomplished either by refrigeration near the freezing point or by freezing itself. Processing, cooking, and other treatments may also be used to produce conditions that inhibit degradation and extend the storage life of meat.

Early postmortem changes occur very quickly in meat as it passes into rigor mortis. Some of these changes are desirable, whereas others are unfavorable to the organoleptic properties of the meat. Thus, the present discussion centers on both the changes occurring as the meat goes into rigor mortis and those occurring during subsequent holding after passing into rigor.

Meat is commonly thought of only as the musculature of the carcass, but it also includes both lipid components and supporting connective tissues. Both of these tissue components also change during storage of meat and influence the physical, chemical, and organoleptic properties of meat, so they are also considered in discussion of the effects of storage and preservation on the properties of meat.

III. PHYSICAL CHANGES OCCURRING IN PRE- AND POSTRIGOR MEAT

A. Gross Physical Changes

Prerigor muscle is soft, pliable, and highly extensible for several hours following slaughter. Rigor mortis, however, sets in within about 24 hr after death and is accompanied by stiffening and hardening of the muscles (Moulton and Lewis, 1940). Careful observations show that the muscles tend to shorten and lose their elasticity. At the onset of rigor, the tissue no longer contracts on stimulation by electric current. Although the physical changes occur gradually, the lack of response to electrical stimulation can be taken as the criterion that rigor mortis has actually occurred. As the meat passes through rigor mortis, it loses some of the rigidity, but does not become soft and extensible again until autolysis reaches an advanced stage (Bendall, 1961).

B. Microscopic and Ultrastructural Changes

Birkner and Auerbach (1960) noted that muscle fibers from a freshly killed animal seen under the light microscope are poorly differentiated, being straight to slightly wavy. They also exhibit distinct longitudinal striations and show no transverse breaks, which become evident only after aging. After the onset of rigor mortis, the fibers become more distinctly outlined and exhibit regular cross-striations, which are not readily apparent in the prerigor state. The presence of fiber differentiation immediately postmortem or in fresh biopsies of muscle observed under the electron microscope suggests that the lack of fiber differentiation under ordinary light microscopy is an artifact. Apparently, the freshly prepared fibers are refractive to fixation and staining by normal procedures.

Kinkiness and contracture nodes, which are also called rigor nodes, are indicative of development of rigor mortis (Paul *et al.*, 1944, 1952). Figure 1(A) shows a characteristic rigor node, which is believed to be produced by active contraction. Passive contraction also seems to occur almost simultaneously and is thought to be responsible for the bending and kinking of the fibers shown in Figure 1(B). Evidence suggests that the kinkiness is caused by tension developed by the collagen as it changes from the flaccid prerigor state to a more rigid postrigor structure (Voyle, 1969).

Aging for periods as short as 2 days under refrigeration results in disappearance of some of the cross-striations in a few areas of muscle seen under the light

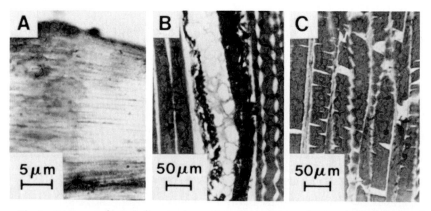

Fig. 1   Responses of muscle fibers to rigor and postrigor changes.   (A) Longitudinal section of uncooked beef biceps femoris muscle showing rigor node (Paul *et al.*, 1944).   (B) Longitudinal section of fresh raw semitendinosus beef muscle showing straight (left) and kinked fibers (right) on either side of strand of connective tissue (Birkner and Auerbach, 1960).   (C) Longitudinal section of aged (14 days) longissimus dorsi beef muscle showing breaks in relation to fiber kinks.   From Birkner and Auerbach (1960).

microscope (Birkner and Auerbach, 1960). Transverse breaks also become apparent at this time. Aging for longer periods of time (4–9 days) results in an increase in the frequency and extent of disintegration of the cross-striations, and also in the disappearance of the kinks and waviness of the fibers. The transverse breaks become more frequent and are believed to be related to a simultaneous increase in meat tenderness. The breaks in the fibers are shown in Figure 1(C) and would appear to release some of the stress imposed by the kinking of the fibers on undergoing rigor mortis.

At this time, it is not clear whether fiber kinkiness is the result of (1) simple folding of passive fibers due to the actively contracting adjacent or straight fibers (Bendall, 1961), (2) myofibrillar supercontraction producing fiber kinkiness (Johnson and Bowers, 1976), or (3) normal myofibrillar contraction of <40%. However, the fact that uncooked muscle, either pre- or postrigor, shows fiber kinkiness but has no supercontraction bands, would suggest that it is a normal phenomenon in raw muscle in full rigor.

Studies with the electron microscope have shown that the weak point in the structure of myofibrils is the Z line. Degradation of the myofibrils occurs first on each side of the Z disc. This viewpoint is supported by comparing fresh untreated samples (Fig. 2A) with enzymatically treated myofibrils (Fig. 2B) and with aseptic samples showing natural autolysis (Fig. 3C). There is evidence that breakage of the myofibrils at the Z line is correlated with meat tenderness; in fact, myofibrillar fragmentation has been shown to be useful for prediction of meat tenderness (Olson et al., 1976; Olson and Parrish, 1977; Parrish et al., 1979).

C. Effects of Cooking

The first change that is evident in cooking of fresh meat is in the color, which gradually begins to change from bright red to gray. As heating proceeds further, the color becomes brown, with the amount of browning being dependent on the temperature and the amount of reducing sugars being present (Sharp, 1957; Pearson et al., 1962, 1966). Part of the color change observed during heating is the result of denaturation of myoglobin and residual hemoglobin (Kramlich et al., 1973).

Coincident with the color changes that take place in meat during cooking, coagulation and denaturation of the sarcoplasmic and myofibrillar proteins occur. The extent of alteration is dependent upon the temperature and length of heating. The changes first become evident when the temperature reaches 30–50°C and are accompanied by alterations in the water-holding capacity, in the rigidity and separation of the muscle fibers, and in several other properties of the proteins (Hamm, 1966). The latter include changes in pH, the isoelectric

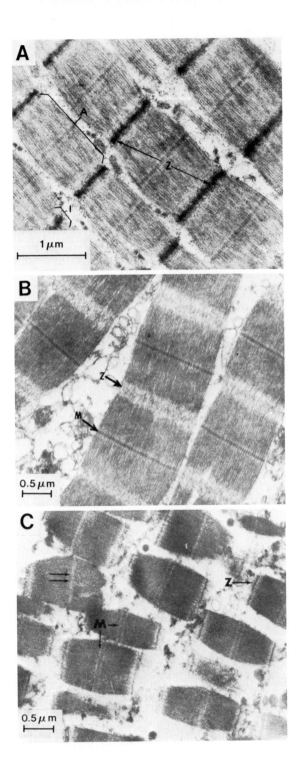

point (IP), solubility, the number of sulfhydryl (SH) groups, dye-binding capacity, and the ability of the muscle proteins to bind Ca^{2+} and Mg^{2+} ions. These biochemical characteristics and their significance are covered in greater detail in Section IV.

Ramsbottom and Strandine (1949) and Paul *et al.* (1952) have demonstrated that prerigor muscle is initially quite tender if cooked rapidly. At the onset of rigor, however, it reaches maximum toughness. This is followed by gradual improvement during aging, until it reaches approximately the original degree of tenderness again. Contraction nodes are evident in the rapidly cooked prerigor muscle but are absent in postrigor muscle cooked under the same conditions (Paul *et al.*, 1952). Cia and Marsh (1976) have theorized that the tenderizing effect observed following rapid cooking of prerigor muscle is due to shattering of the myofibrils caused by supercontraction during cooking. If this is true, there may be advantages to microwave cooking, in which the rate of heat penetration would be much faster and more uniform than that occurring during conventional methods of cookery (Cia and Marsh, 1976).

Figure 3(A) presents a transmission electron micrograph showing the heavy contraction bands formed in prerigor muscle on cooking by boiling; the less extensive fragmentation on cooking of prerigor meat by microwaves is shown in Fig. 3(B). Figure 3(C) presents a scanning electron micrograph of prerigor muscle cooked by boiling and illustrates the distortion and damage to the fibers caused by cooking.

Histological studies have shown that there is some shortening of the sarcomeres during development of rigor mortis (Johnson and Bowers, 1976). However, the amount of shortening that is found in cooked muscle in full rigor does not appear to be adequate to account for its increased toughness in comparison to cooked prerigor muscle. It is also difficult to understand why some adjacent fibers in the same muscle undergo supercontraction, whereas others contract much less. Voyle (1969) has shown that those fibers undergoing supercontraction in prerigor muscle upon cooking are straight and have shorter sarcomeres than adjacent passively contracting or kinked fibers. Figure 4 shows both types of fibers in the same sections in both transmission and scanning electron micrographs (Johnson and Bowers, 1976).

Locker (1960) first observed a relationship between muscle shortening and

---

Fig. 2 Comparison of degradation of pig myofibrils due to enzymatic action. (A) Characteristic myofibrils at 15 min postmortem. Z, Z lines; A, A band; and I, I band. From Dutson *et al.* (1974). (B) Myofibrils incubated with an enzyme isolated from *Pseudomonas fragi* for 72 hr at 10°C. Z, Z line and M, M line. Note absence of Z-line structure. From Tarrant *et al.* (1973). (C) Aseptic sample showing natural autolysis on storage at 30°C for 96 hr. M, M lines; Z, Z-line material; double arrows show absence of M line. Fragmentation at Z line is characteristic of aged and enzymatically treated muscle (Abbott, 1976).

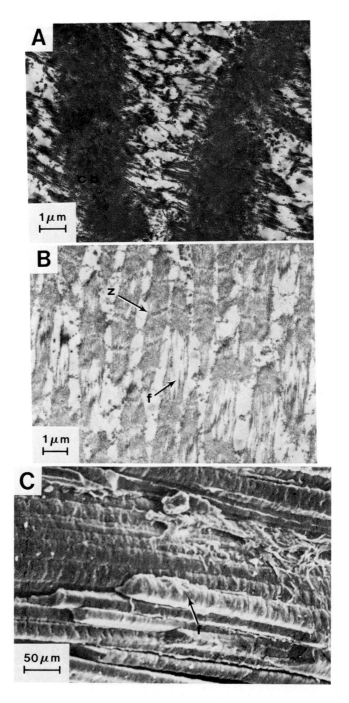

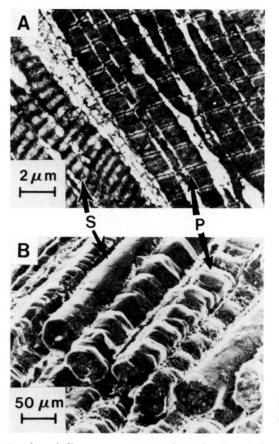

Fig. 4 Shortening of muscle fibers as a consequence of rigor showing actively contracted (S) and passively contracted or kinked fibers (P) in adjacent areas. (A) Transmission electron micrograph. (B) Scanning electron micrograph of same area. Note supercontracted and passively contracted fibers in adjacent areas of both micrographs. Both taken from Johnson and Bowers (1976).

Fig. 3 Effects of cooking on muscle. (A) Transmission electron micrograph of prerigor beef sternomandibularis muscle cooked by boiling showing contraction band (CB) caused by heating. (B) Transmission electron micrograph of prerigor beef sternomandibularis muscle cooked by microwaves showing fragmented areas (f) and Z line (Z). Fragmentation is more evenly distributed and less extensive than shown in (A). (C) Scanning electron micrograph of prerigor beef sternomandibularis muscle cooked by boiling. Note distortion and damage to surface. All taken from Hsieh et al. (1980).

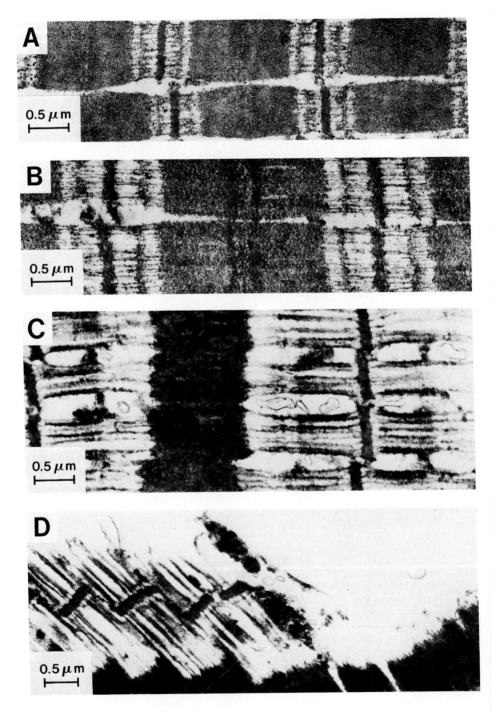

tenderness in which he demonstrated shortened muscle to be tougher. Later Locker and Hagyard (1963) demonstrated that exposure of prerigor muscle to cold caused massive shortening, which caused toughness in the cooked meat. These authors coined the term *cold shortening* to describe the phenomenon of shortening of prerigor meat on exposure to cold and the associated cold-induced toughness. Cold shortening is discussed in greater detail in Section IV,B. It is, however, difficult to explain why prerigor meat may shorten during cooking yet remain tender, while cold-shortened prerigor meat is tough after cooking (Locker, 1977). It may be related to tearing and distortion during cooking of prerigor meat, whereas cold-shortened meat does not tear or contract during cooking (Hsieh *et al.,* 1980).

Another factor in tenderization during cooking is the hydrolysis of the myofibrillar and connective tissue proteins. Locker (1977) has discussed meat tenderness and proposed that gap filaments limit the tensile strength of both raw and heat-denatured myofibrils. Gap filaments form a core in each thick filament and emerge at one end to pass between the thin filaments through the Z line then between the thin filaments in the adjacent sarcomere and into the next thick filament where they terminate. They are highly elastic and appear to maintain the thick and thin filaments in register. Recent evidence obtained by R. H. Locker (personal communication, 1981) suggests that gap filaments remain strong and elastic, even after extreme cooking, and are the only myofibrillar survivors in cooked meat. Thus, gap filaments appear to be responsible for much of the tensile strength or lack of tenderness in cooked meat.

Figure 5(A) shows the gap filaments in relation to the thick and thin filaments in cooked muscle. The thick filaments have been fused by cooking but are still distinguishable. Figure 5(B) shows the same sample stretched by 77% after cooking, illustrating the elasticity of both the A and I bands. The gap filaments are much clearer and easier to see on both sides of the Z line. Figure 5(C) shows muscle that is set in rigor at twice excised length and then restrained and cooked. The gap filaments are clearly evident between the coalesced A and I bands, which have both shortened during cooking. Figure 5(D) presents an electron micrograph of broken gap filaments, which have yielded on both sides of the Z line. These micrographs illustrate the structural strength and elasticity of the gap filaments in cooked muscle.

Both myosin and actin undergo denaturation and shrinkage early in cooking

---

Fig. 5 Gap filaments in cooked meat showing their relationship to thick and thin filaments under different amounts of restraint. (A) Sample allowed to go into rigor at 150°C at excised length and then cooked at 80°C under restraint to prevent further shortening. (B) Same sample stretched by 77% after cooking. (C) Muscle allowed to set in rigor at twice excised length and then cooked. (D) Same sample as C after stretching beyond breaking point. Note broken ends of gap filaments attached to Z lines. From Locker (1977), with permission.

(Hamm, 1966). The first change observed is that of coagulation, which begins at 30 to 40°C and is almost complete at 55°C. Coagulation of isolated myosin begins at 35°C (Greenstein and Edsall, 1940) and is largely complete at 53°C (Locker, 1956). Simultaneous with coagulation, there is a decrease in the water-holding capacity of meat that begins at about 35°C but occurs primarily between 40 and 50°C (Hamm, 1966). Coagulation of the proteins and release of juice is not complete even at 60°C but continues to a lesser extent even above this temperature. Apparently, the gap filaments are the last myofibrillar proteins that are broken down by cooking (R. H. Locker, personal communication, 1981).

Collagen undergoes thermal shrinkage upon heating, with the temperature being dependent upon the source. Locker *et al.* (1975) suggested that collagen on undergoing thermal shrinkage imposes stress on the muscle fibers and is responsible for the cooking crimp seen under the microscope. As the temperature increases further, collagen becomes degraded and finally is gelatinized. The degradation of collagen begins at ~ 70°C but complete gelatinization does not take place even at 100°C unless heating is continued for a prolonged period of time (Lawrie, 1966). Complete gelatinization of collagen occurs quickly on pressure cooking at 115 to 125°C (Bendall, 1946). The tenderizing effect of cooking at low temperatures for long periods of time or of raising the final internal temperature can be explained by the effects of heat upon collagen. Cooking at high temperatures in dry heat, however, will harden the meat and cause excessive shrinkage, so it is not commonly recommended.

D. Effects of Processing

There are several manufacturing procedures that play important functions in processing of meat. These include curing, smoking, and production of sectioned and formed meat products. These procedures are not discussed here because they are described by others (Kramlich *et al.*, 1973; Schmidt, 1978). However, changes in the physical and biochemical properties as a result of processing are covered in this chapter.

The most obvious change occurring during curing of meat is related to color. The myoglobin reacts with nitrite to produce a stable pink pigment known as nitric oxide myoglobin (Kramlich *et al.*, 1973). If nitrite is not added to meat during curing, the myoglobin is either oxidized to metmyoglobin, which is brown in color, or in deep tissues may be reduced to form reduced myoglobin, which is deep red or purple. Heat and light will accelerate the rate of the reactions. Physically, there are few other changes in meat during curing. The addition of salt does, however, alter the water-binding capacity of the tissues (Hamm, 1960; Hamm and Deatherage, 1960). Lowering of the water activity (a_w) by curing is largely responsible for the stable nature of cured meat, since low a_w values inhibit microbial growth and meat spoilage.

The physical changes that occur in processing of sectioned and formed meats are more dramatic, with the visible alterations taking place much faster than in normal curing as a result of tumbling or massaging. In addition, the individual pieces of meat or separate muscles accumulate a tacky coating of muscle proteins, which upon stuffing in a casing followed by heating and cooling fuse into a single piece (Schmidt, 1978). Examination of the final product shows that the pieces are bonded together by the myofibrillar proteins that are extracted onto the surface by the action of the added salt in combination with tumbling or massaging. More details on this process and the products produced are given in Chapter 7.

Grinding or chopping causes easily recognized physical changes in meat that are essentially the same in pre- and postrigor meat. The rapid alteration in surface color is readily apparent, as is the change in texture. Grinding and chopping also accelerate a number of biochemical processes, including oxidation of the lipids and glycolysis in prerigor muscle. Grinding also increases the surface area, improves the water-binding capacity, and decreases the drip losses in frozen meat. Some of these changes are desirable, whereas others are deleterious to meat quality. Addition of curing salts during chopping speeds up the action and results in rapid color changes. Thus, some of the processing procedures can be combined and improve the stability of the final products.

IV. BIOCHEMICAL CHANGES OCCURRING IN PRE- AND POSTRIGOR MEAT

A. Glycolysis

The breakdown of glycogen with formation of lactic acid is an important change that greatly influences the properties of meat and has been discussed in considerable detail by Bate-Smith (1948). Normally, muscle contains appreciable amounts of high-energy compounds at the time of slaughter. Bodwell *et al.* (1965) showed that creatine phosphate, total reducing sugars, and ATP in beef muscle declined rapidly during the first 12 hr after slaughter, with an approximately stoichiometric increase in lactic acid. This was accompanied by a drop in pH from 6.99 immediately after death to 5.46 by 48 hr. Most of the decline in pH occurs during the first 12–15 hr, and is virtually complete at 24 hr after death.

Although development of rigor mortis can often be followed by the decline in pH, rigor will occur in the absence of the pH drop in the case of muscle that contains little or no energy reserve at the time of slaughter. This condition was first recognized by Claude Bernard (1877), the noted French physiologist, who called it *alkaline rigor*. Various manifestations of this condition are not uncommon in meat animals, occurring to varying degrees in dark-cutting beef and in dark, firm, and dry (DFD) pig muscle. However, low-quality muscle in the pig, a condition closely related to pale, soft, and exudative (PSE) muscle, exhibits only minor differences from normal muscle at 24 hr postmortem when observed under the electron microscope, except for a greater amount of disruption (Dutson *et al.*, 1974). Some of the factors involved in development of these conditions are discussed in greater detail in Chapter 4.

Glycolysis is also greatly accelerated by grinding or mincing of prerigor muscle, with full rigor resulting in ~ 6 hr following grinding compared to ~ 16 hr for intact sheep muscle (Pearson *et al.*, 1973a). Addition of a combination of $CaCl_2$ and epinephrine to ground muscle has also been shown to further accelerate glycolysis, reducing the time required to go into rigor mortis by an additional $1-2$ hr. The mechanism of this action is not known. It is not unlikely that addition of NaCl also may speed up glycolysis in meat, but the author has no knowledge of any investigations in this area.

Injection of Ca^{2+} ions has been demonstrated to cause excessive shortening in prerigor intact muscle (Weiner and Pearson, 1969). The shortening increases toughness, the phenomenon appearing similar to that occurring in cold shortening. Calcium chelators have also been shown to prevent shortening and produce more tender meat (Weiner and Pearson, 1966, 1969; Pearson *et al.*, 1973b). Pyrophosphate has also been demonstrated to prevent the interaction of myosin and actin *in vitro* (Weiner *et al.*, 1969) and may explain the beneficial effects of phosphates in improving the water-binding capacity and tenderness of cured meats.

B. Cold Shortening

Cold shortening was first observed by Locker and Hagyard (1963), who noted that prerigor, excised, unrestrained beef muscle shortened rapidly at 0°C, but suffered minimum shortening at 14 to 19°C. The relationship between shortening and temperature is shown in Fig. 6. The data clearly illustrate that maximum shortening occurs at low temperatures (< 5°C) and then increases again after reaching ~ 35°C (Locker and Hagyard, 1963).

Figure 7 demonstrates that there is a clearcut relationship between percentage shortening and shear force (Marsh and Leet, 1966). Maximum toughness was shown to develop at between 25 and 50% of muscle shortening. With either more or less shortening the meat was more tender, which suggests that extreme shortening (50–60%) also caused tenderization.

Marsh *et al.* (1968) observed that holding prerigor meat at ~ 16°C until the

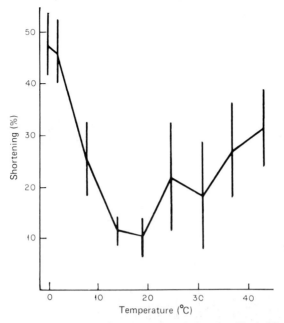

Fig. 6 Effects of temperature upon shortening of prerigor meat. Vertical lines are standard deviations. From Locker and Hagyard (1963).

onset of rigor mortis will prevent cold shortening and the accompanying toughening. This resulted in development of the process known as conditioning and aging, which was later used on all New Zealand lamb carcasses intended for export to North American markets. In the process, the lamb carcasses were held in a conditioning and aging room in which the temperature was maintained at ~16°C until they passed into rigor mortis. Normally, under these conditions ~16–18 hr are required for development of full rigor. Although this method was widely used in New Zealand, it has been largely replaced by electrical stimulation in the past few years.

The basic underlying cause of cold shortening appears to be related to the inability of the sarcoplasmic reticulum (SR) to sequester and bind excess Ca^{2+} ions released from the SR and mitochondria under the influence of cold temperatures and declining pH values in prerigor muscle (Kanda *et al.*, 1977). It is well known that rabbit muscle does not suffer from cold shortening, and pig muscle does not appear to be greatly affected (Marsh *et al.*, 1972). Muscles from these two species contain a greater proportion of white to red fibers than beef and sheep meat (Cassens, 1971), both of which are susceptible to cold shortening. White fibers contain fewer mitochondria and have a better developed SR

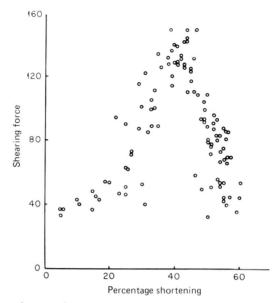

Fig. 7   Diagram showing relationship between percentage shortening and shear force (tenderness) of muscle. Shortening was achieved by placing samples at 2°C at different time intervals during onset of rigor and is expressed as percentage of initial excised length. From Marsh and Leet (1966). Copyright by Institute of Food Technologists.

system (Cassens, 1971), both of which appear to contribute to the resistance of white muscle to cold shortening.

Buege and Marsh (1975) proposed that the SR in red muscle releases large quantities of Ca^{2+} under cold temperatures. Although this may be a major factor in initiation of cold shortening in red muscle, it does not explain the reversibility of cold shortening if the temperature is raised. Cornforth *et al.* (1980) have proposed a role for both the SR and mitochondria that explains the reversibility of cold shortening upon rewarming of prerigor muscle. In this theory, the initiator of cold shortening is exposure of prerigor red muscle to cold temperatures ($<10°C$), which cause the mitochondria and the already overloaded SR to spill excess Ca^{2+} ions into the intracellular spaces. Since the SR is already saturated, the Ca^{2+} ions trigger the interaction of myosin and actin to form actomyosin and cause the muscles to contract and shorten. The drop in pH also contributes to the release of Ca^{2+} ions since both the SR and mitochondria lose more Ca^{2+} at low pH values. However, raising the temperature will again prevent Ca^{2+} ions from being released and will also increase the capacity of the SR to take up the excess Ca^{2+}, so the free Ca^{2+} is sequestered and the muscles relax,

C. Prevention of Cold Shortening

As already mentioned, conditioning and aging of meat so that it passes into rigor at a relatively high temperature will prevent cold shortening. This method has been successful in producing tender meat and was used in New Zealand until recently (Marsh and Leet, 1966; Marsh *et al.*, 1968). Even though it was successful, it has been criticized by the industry for using temperatures around 15 to 16°C, which would be more favorable for microbial growth than lower temperatures. Although this method was successful in producing tender meat without any problems from pathogenic microorganisms under carefully controlled conditions (Locker *et al.*, 1975), public health officials have considered it to pose spoilage and food poisoning possibilities. Furthermore, New Zealand freezing works have considered space requirements for the process to be expensive.

Electrical stimulation has largely replaced conditioning and aging. It has been used extensively by New Zealand freezing works to produce lamb carcasses for the export markets and has also been widely used in United States on beef carcasses. Although the original process was first used and patented by Harsham and Deatherage (1951), its modern use traces to the basic work of Carse (1973). Since that time, numerous researchers (Chrystall and Hagyard, 1975, 1976; Bendall, 1976; Bendall *et al.*, 1976; Davey *et al.*, 1976; Gilbert *et al.*, 1977; McCollum and Henrickson, 1977; Savell *et al.*, 1977, 1978a,b,c; Cross, 1979; Will *et al.*, 1979a, b; Dutson *et al.*, 1980a,b; Pearson and Dutson, 1985) have studied electrical stimulation in relation to meat tenderization.

Electrical stimulation has been shown to not only increase meat tenderness but also to improve lean meat color, to enhance ribeye marbling scores, to prevent development of heat ring (a two-toned appearance with an outer dark ring in the ribeye), to reduce the time required for aging, to enhance flavor scores, and to extend retail caselife slightly, according to a recent review by Seideman and Cross (1982). The action appears to be due, at least in part, to the depletion of ATP and hastening of the onset of rigor mortis. However, not all the improvement in tenderness comes from prevention of cold shortening; some may be due to stimulation of proteolysis by activation of the lysosomal enzymes (Dutson *et al.*, 1980a,b). Another mode of action by which tenderness may be improved by electrical stimulation is by disruption of the tissues occurring as a consequence of the violent contraction during stimulation (Dutson *et al.*, 1980b).

The other procedure for preventing cold shortening that has been successfully employed on a commercial basis is the tender-stretch method, in which the carcass is hung by the obturator foramen (aitch bone) instead of by conventional hanging by the Achilles tendon (Hostetler *et al.*, 1975). Hanging by this method places several of the major muscles under tension and apparently prevents shortening, so the postrigor meat is more tender than similar meat hung by the conventional process (Hostetler *et al.*, 1970, 1972, 1973, 1975). However, tenderness of some muscles is not improved by the tender-stretch method, either because they are already restrained so that they do not shorten or else are not under sufficient tension to prevent shortening.

Although other procedures for preventing cold shortening and toughening have been proposed, they have not been used commercially. These include altered posture (Herring *et al.*, 1965, 1967; Davey and Gilbert, 1973), a restraining device to place the longissimus dorsi muscle under stress (Buege and Stouffer, 1974), and the use of carbon dioxide anesthesia to accelerate glycolysis (Pearson *et al.*, 1973b). Other research has shown that glycolysis is not accelerated in muscles of sheep that are injected with epinephrine, since the stress of slaughtering is apparently sufficient to produce the maximum amount of stimulation via this pathway (Pearson *et al.*, 1973b).

D. Thaw Rigor

Thaw rigor or thaw contracture are terms used to describe the shortening that occurs upon thawing of meat frozen in the prerigor state. Usually it is characterized by massive shortening, as is illustrated in Fig. 8 (Forrest *et al.*,

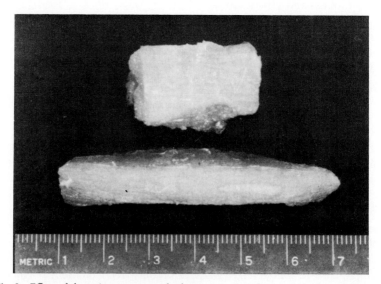

Fig. 8 Effect of thaw rigor upon muscle shortening. Sample at top was frozen immediately postmortem and then thawed rapidly. Sample at bottom was handled the same except that it was frozen postrigor. Top sample is only 42% of its original length. From *Principles of Meat Science* by John C. Forrest, Elton D. Aberle, Harold B. Hedrick, Max D. Judge, and Robert A. Merkel. W. H. Freeman and Company. Copyright 1975.

1975). Although thaw rigor has been recognized for many years, its practical significance was first recognized during freezing and thawing of whale meat for human consumption (Sharp and Marsh, 1953).

Thaw rigor appears to be due to the occurrence of an extensive salt flux on thawing, which results in release of excessive amounts of Ca^{2+} ions such that the SR is saturated (Bendall, 1961). The excess Ca^{2+} ions move into the intracellular spaces and cause extensive contraction.

The onset of thaw rigor occurs when the amount of ATP is relatively high, that is, $\sim 40\%$ (Newbold, 1966). In this respect it closely resembles cold shortening. Thaw rigor, like cold shortening, also results in appreciable toughness in meat as compared to control samples frozen after passing into rigor. Thaw contracture also causes excessive losses of drip from the tissues upon thawing. The relationship between drip and thaw rigor is shown in Fig. 9 (Marsh and Thompson, 1958). Thus, thaw rigor is of great practical importance to the meat industry and has led to the past recommendation that meat not be frozen until it is in full rigor. Obviously, this means that the meat must be chilled and allowed to go into rigor mortis before being frozen.

Marsh et al. (1968) have demonstrated that thaw rigor is even more damaging to meat tenderness if meat that is frozen prerigor is subjected to cooking from the frozen state. Meat that is thawed very rapidly has been shown to suffer less from thaw rigor than similarly handled meat that is thawed less rapidly. Appar-

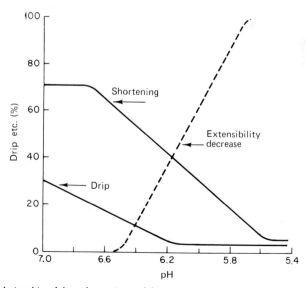

Fig. 9 Relationship of thaw shortening and drip expressed as percentage of initial values. The values for pH and the decrease in extensibility are indicative of the various pre- and postrigor states. From Marsh and Thompson (1958).

ently, fast thawing minimizes movement of the salt flux into the intercellular spaces, and thus, causes less shortening.

E. Prevention of Thaw Contracture

The obvious way to prevent thaw rigor is to freeze meat only after it has reached full rigor. Although this practice will prevent both cold shortening and thaw rigor, it would often be advantageous to freeze meat in the prerigor state. Thus, recent research showing that electrical stimulation can be used to prevent thaw contracture has allowed freezing of prerigor meat without any problems from thaw rigor (Davey and Gilbert, 1973). Freezing of electrically stimulated prerigor meat has been especially useful for meat intended for use in the sausage industry. Furthermore, it can also be utilized for retail meat cuts and makes more efficient use of the energy used in cooling and freezing.

Locker *et al.* (1975) have pointed out that the effects of thaw rigor can also be minimized by thawing very slowly as compared to cooking from the frozen state. They have also demonstrated that glycolysis continues in the frozen state and that a sufficient period of frozen storage will avert thaw rigor. Thus, the problems of meat quality associated with thaw rigor can be avoided by a number of procedures. However, electrical stimulation appears to be the most generally acceptable method for solving the problem of thaw rigor.

F. Effects of Aging

Aging or ripening of meat has long been recognized as resulting in improvement of meat tenderness. During this process, the meat is held at refrigerated temperatures ($1-5°C$) for extended periods of time. It has generally been accepted that natural proteolysis of connective tissues and muscle fibers occurs during aging due to the presence of indigenous muscle proteinases (Whitaker, 1959; Balls, 1960). The physical breaks at the Z line in aged muscle are obvious on examining the muscle fibers by transmission electron microscopy, as discussed in Section III,B. However, it has been difficult to prove that there are any major chemical changes during meat aging (Bate-Smith, 1948; Whitaker, 1959; Bodwell and Pearson, 1964a,b).

The catheptic enzymes are thought to play an important role in the breakdown of the muscle and connective tissue proteins, but the extent of proteolysis has been less than would be expected in view of the degree of improvement in tenderness (Hoagland *et al.*, 1917; Whitaker, 1959). Although added exogenous enzymes have been shown to cause extensive fragmentation of isolated myofibrillar proteins and treated muscle (Gergely, 1950; Szent-Györgi, 1960; Koszalka and Miller, 1960; Landmann, 1963), the effects of endogenous muscle enzymes upon the muscle proteins are more subtle. Whitaker (1959) concluded that the changes occurring during aging of meat are not what would have been expected in view of the improvement in tenderness but pointed out that even minor changes in the structure of proteins can cause major alterations in their physical properties.

Bird *et al.* (1980) have pointed out that of all the indigenous muscle enzymes CAF and cathepsins D and B are most likely to be involved in degradation of the myofibrillar proteins. There is good evidence that CAF can degrade Z lines in a way similar to that seen after aging of meat (Dayton *et al.*, 1976). CAF has also been shown to remove the 400-Å periodicity associated with troponin in the I band and to degrade the M lines partially, according to Dayton *et al.* (1976). Such changes could well account for the improvement in tenderness as a consequence of meat aging.

Further evidence for the role of aging in meat tenderness is found in the recent work of Locker and Wild (1982a), who have developed a simple device for measuring the yield point in strips of raw muscle by registering maximum tension generated at a low rate of extension. The machine, which they called the yieldmeter, is shown in Fig. 10. Traces obtained by using the yieldmeter for measuring tension development against time are presented for muscle strips from the same carcass in Fig. 11. The plots show that less tension is developed in aged muscle, even after relatively short periods of time. Results suggest that this instrument is complimentary to the various shear devices and is particularly well suited to following changes occurring during aging of meat.

Locker and Wild (1982b) have shown that the yield point of raw muscle decreases markedly upon aging. This is demonstrated in Fig. 12, which presents plots of yield points and shear force values for similar muscle strips aged at 2 and 15°C (Locker and Wild, 1982a). The plots clearly show that the yield points obtained by use of the yieldmeter give a more sensitive measurement of changes taking place during meat aging than do those for the shear device. This is true since, unlike shear measurements, which must be carried out on cooked meat, the yield point data are not complicated by heat denaturation or the stress imposed by heating of the collagen net. Electron micrographs suggest that aging weakens both actin and gap filaments so that the I band is readily fractured, thereby improving tenderness.

There also probably are alterations in the connective tissue proteins, collagen and elastin, that result in greater meat tenderness following aging. Bodwell and McClain (1971) have discussed the structure and function of collagen in meat and some possible modes of breakdown that occur during meat aging. They suggested that the amount of inter- and intramolecular cross-linking can alter the properties of muscle, especially tenderness. Certain enzymes are also known to hydrolyze collagen and could play a part in altering the properties of the connective tissues during meat aging.

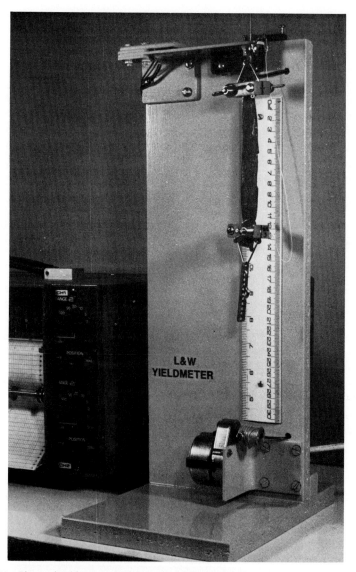

Fig. 10 Photo of yieldmeter, which provides an objective method for measuring the yield point of raw muscle strips during aging.   From Locker and Wild (1982a).

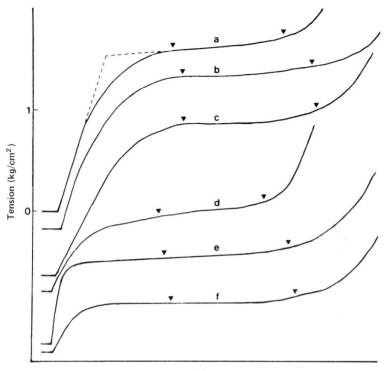

Time

Fig. 11   Machine traces for muscle strips taken from a single steer carcass. Since this is a montage, the origins on the ordinate differ in position.   Traces a, b, and c show rigor strips run at 25.0, 10.8, and 5.4% per min at chart speeds of 30, 15, and 7.5 mm per min, respectively.   Traces d, e, and f show strips aged for 1 day at 15°C run under the same conditions as given for a, b, and c above in the same order.   Arrowheads on each curve denote 50 and 100% extension in order of their appearance.   From Locker and Wild (1982a).

Locker (1977) has suggested that both cooking and aging of meat may influence the breakdown of the connective tissues and aid in meat tenderization. Although the exact mechanism for tenderization during meat aging is not completely clear, it appears to be related to partial hydrolysis of both myofibrillar and connective tissue proteins by indigenous enzymes.   Microbial proliferation may also contribute exogenous enzymes that will hydrolyze different proteins in meat.   Regardless of the exact mode of action, evidence suggests enzymatic breakdown that occurs in meat on storage is mainly responsible for tenderization during aging.

Another factor that unquestionably is involved in postrigor changes in meat is the alteration in water binding that occurs coincident with aging, which has been discussed in considerable detail by Hamm (1960).   Such changes also

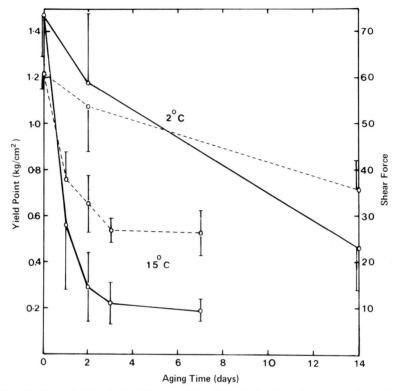

Fig. 12   Plots of yield point (solid lines) and shear force (broken lines) for muscle strips aged at 2 and 15°C. From Locker and Wild (1982a).

occur on adding of phosphates and salt during meat processing and have a marked effect upon the physical properties of meat, including tenderness. Wismer-Pedersen (1971) has discussed the factors influencing water-binding capacity and their interactions as well as methods for measuring the water-binding capacity in meat. (See also Chapter 4.)

V. FACTORS INFLUENCING THE STABILITY OF LIPIDS

Lipids are found in meat in considerable quantities and can have a marked influence upon its acceptability. Love and Pearson (1971) concluded that the lipids can be classified as depot and tissue lipids, depending upon their location. Depot lipids are generally stored as relatively large deposits in specialized connective tissues, whereas tissue lipids are integrated into the muscle tissues and are distributed in small quantities throughout the lean tissues. The intra-cellular or tissue lipids are found in close association with the proteins and contain a large percentage of phospholipids in relation to the triglycerides, whereas the reverse is true for depot lipids (Watts, 1962).

Although lipids play an important role in meat by contributing the character-istic species flavor, they are also subject to lipid oxidation (Pearson *et al.*, 1977). Thus, the lipids may impart either desirable or undesirable flavors to meat. They also contribute to both juiciness and tenderness in meat, although their contribution is relatively low (Blumer, 1963). It seems likely that these attributes are primarily due to the tissue lipids, even though a role for the triglycerides can not be ruled out (Wasserman and Talley, 1968).

There are two main types of oxidation in meat, that is, normal oxidation that occurs during long-term storage under refrigeration either above or below freezing and the extremely rapid oxidation that occurs following cooking (Watts, 1962). Early work on meat recognized only the normal oxidative changes that occur in breakdown of the triglycerides. Increased proportions of polyunsaturated fatty acids make the meat more susceptible to normal oxida-tion, but even then development of oxidation generally requires weeks or even months. In contrast, the rapid oxidative changes that occur after cooking and grinding result from breakdown of the phospholipids and develop in a matter of a few hours. Since their development is most frequently encountered after cooking and appears to speed up on reheating, this type of oxidation was called *warmed-over flavor* (WOF) by Tims and Watts (1958).

A. Normal Oxidation

Oxidation of meat during frozen storage is largely due to changes in the triglyceride fraction (Igene *et al.*, 1979a). The phospholipid content undergoes little change during holding for periods of over a year at $-18°C$. There are differences among the meats from different species in their susceptibility to lipid oxidation as well as among different tissues in the same species. For example, beef lipids are more stable than those of chicken white meat, which in turn are more stable than those in chicken dark meat (Igene and Pearson, 1979). The diets fed to animals before slaughter can also alter the susceptibility of the tissues to oxidation (Pearson *et al.*, 1977; Shorland *et al.*, 1981).

Temperature and length of holding can also alter the rate of oxidation of lipids in meat. Autoxidation proceeds faster during storage of meat at high temperatures (Lundberg, 1962). Thus, there are distinct advantages in holding meat at low temperatures in order to retard autoxidation during storage. Lipid oxidation occurs more rapidly above freezing and may cause adverse changes not only in flavor, but also in color (Greene, 1969).

B. Warmed-Over Flavor

WOF has become a more serious problem with an increasing number of meals being precooked by food vendors and then being reheated before being served. In contrast to normal oxidation, in which the triglycerides are primarily involved, the phospholipids are primarily responsible for development of WOF (Wilson *et al.*, 1976; Igene and Pearson, 1979; Igene *et al.*, 1979a). Disruption of the membranes during cooking or by grinding apparently exposes the phospholipids to oxygen so that autoxidation proceeds rapidly. Although WOF is not normally a serious problem in raw intact meat, it can occur in ground meat that is exposed to atmospheric oxygen when spread out in thin layers (Sato and Hegarty, 1971; Sato *et al.*, 1973).

Cured meats do not develop WOF, since phosphates, nitrites, and ascorbates, which are commonly used in curing, inhibit the development of WOF (Pearson *et al.*, 1977). Although the exact mechanism by which nitrite prevents WOF is not known, its inhibitory action is well recognized (Sato and Hegarty, 1971; Bailey and Swain, 1973; Fooladi *et al.*, 1979). It seems probable, however, that nitrite may chelate free ferrous iron (Fe^{2+}), which has been demonstrated to be the catalyst for WOF development (Love and Pearson, 1974). Igene *et al.* (1979b) have demonstrated that myoglobin does not catalyze WOF but is a source of nonheme iron, which is released from the heme pigments upon cooking. Since ascorbates and phosphates may also tie up free metal ions, their action in preventing WOF may be through similar chelating mechanisms.

Antioxidants are also known to inhibit WOF. Plant extracts have been shown to inhibit WOF development (Pratt and Watts, 1964; Pratt, 1972) and are sometimes used in sauces to prevent WOF in precooked meat items. Some products of the Maillard reaction have antioxidant activity such that WOF does not occur in heat-sterilized meat products (Sato *et al.*, 1973). Commercial antioxidants are also good inhibitors of WOF. The mechanism of antioxidative activity for this group of compounds seems to be by preventing formation of hydroperoxides through stabilization of the phospholipids and triglycerides (Lundberg, 1962).

VI. SUMMARY

Changes occurring during storage and preservation of meat can be either physical or biochemical. It is probable that both types of changes are involved. Physical changes include color alteration and coagulation of the proteins that occur during cooking. Other physical changes observed during cooking include development of contraction nodes on heating, which appear to be related to the greater tenderness of meat cooked before the onset of rigor mortis. Processing also causes changes in the color of meat as the myoglobin reacts to produce the characteristic pink pigment observed in cured meats. Sectioned and formed meats are prepared by tumbling or massaging, which accelerate curing while simultaneously extracting the myofibrillar proteins so that the small pieces of meat are bound together after heating followed by cooling.

Biochemical changes also occur during the aging of meat and appear to be related to the action of indigenous enzymes, including CAF and cathepsins B and D. Although only minor changes are observed in the myofibrillar proteins during aging, small changes may be responsible for the increase in tenderness and water binding. Cold shortening and thaw rigor are two physical changes that appear to be caused by the inability of prerigor muscle in a cold environment to hold Ca^{2+} ions bound by the sarcoplasmic reticulum, with the result that they flood the intracellular spaces and cause excessive shortening and toughening. Electrical stimulation, conditioning and aging, and the tender-stretch method for preventing toughness and shortening are discussed.

Changes in lipids during storage of meat were also reviewed. Lipid oxidation during normal refrigerated holding appears to be associated with hydroperoxide formation from the triglycerides while the phospholipids are quite stable. In cooked meat, the phospholipids are responsible for autoxidation, whereas the triglycerides are relatively more stable. Methods for preventing oxidation are discussed along with the mechanisms by which they cause autoxidation in meat.

REFERENCES

Abbott, M. T. (1976). Influence of bacterial growth on porcine muscle ultrastructure and headspace volatiles. Ph.D. Thesis, Michigan State University, East Lansing.

Abbott, M. T., Pearson, A. M., Price, J. F., and Hooper, G. R. (1977). Ultrastructural changes during autolysis of red and white porcine muscle. *J. Food Sci.* **42**, 1185.

Bailey, M. E., and Swain, J. W. (1973). Influence of nitrite on meat flavor. *Proc. Meat Ind. Res. Conf.* p. 29.

Balls, A. K. (1960). Catheptic enzymes in muscle. *Proc. Res. Conf. Am. Meat Inst. Found. Univ. Chicago* **12**, 73.

Bate-Smith, E. C. (1948). The physiology and chemistry of rigor mortis, with special references to the aging of beef. *Adv. Food Res.* **1**, 1.

Bendall, J. R. (1946). The effect of cooking on the creatine–creatinine, phosphorus, nitrogen and pH values of raw lean beef. *J. Soc. Chem. Ind. London* **65**, 226.

Bendall, J. R. (1961). Post-mortem changes in muscle. *In* "The Structure and Function of Muscle" (G. H. Bourne, ed.), Vol. 3, p. 227. Academic Press, New York.

Bendall, J. R. (1976). Electrical stimulation of rabbit and lamb carcasses. *J. Sci. Food Agric.* **27**, 819.

Bendall, J. R., Ketteridge, C. C., and George, A. R. (1976). The electrical stimulation of beef carcasses. *J. Sci. Food Agric.* **27**, 1123.

Bernard, C. (1877). "Leçons sur la diabète et la glycogénèse animals." Baillière et Fils, Paris.

Bird, J. W. C., Carter, J. H., Triemer, R. E., Brooks, R. M., and Spanier, A. M. (1980). Proteinases in cardiac and skeletal muscle. *Fed. Proc. Fed. Am. Soc. Exp. Biol.* **39**, 20.

Birkner, M. L., and Auerbach, E. (1960). Microscopic structure of animal tissues. *In* "The Science of Meat and Meat Products" (Am. Meat Inst. Found., eds.), p. 10. Freeman, San Francisco, California.

Blumer, T. N. (1963). Relationship of marbling to the palatability of beef. *J. Anim. Sci.* **22**, 771.

Bodwell, C. E., and McClain, P. E. (1971). Chemistry of animal tissues. Proteins. *In* "The Science of Meat and Meat Products" (J. F. Price and B. S. Schweigert, eds.), 2nd ed., p. 78. Freeman, San Francisco.

Bodwell, C. E., and Pearson, A. M. (1964a). Some properties of catheptic enzymes present in beef muscle. *Proc. Int. Congr. Food Sci. Technol., 1st, 1962* p. 71.

Bodwell, C. E., and Pearson, A. M. (1964b). The activity of partially purified catheptic enzymes on various natural and synthetic substrates. *J. Food Sci.* **29**, 602.

Bodwell, C. E., Pearson, A. M., and Spooner, M. E. (1965). Post-mortem changes in muscle. I. Chemical changes in beef. *J. Food Sci.* **30**, 766.

Buege, D. R., and Marsh, B. B. (1975). Mitochondrial calcium and postmortem muscle shortening. *Biochem. Biophys. Res. Commun.* **65**, 478.

Buege, D. R., and Stouffer, J. R. (1974). Effects of pre-rigor tension on tenderness of intact bovine and ovine muscle. *J. Food Sci.* **39**, 396.

Carse, W. A. (1973). Meat quality and the acceleration of postmortem glycolysis by electrical stimulation. *J. Food Technol.* **8**, 163.

Cassens, R. G. (1971). Microscopic structure of animal tissues. *In* "The Science of Meat and Meat Products" (J. F. Price and B. S. Schweigert, eds.), 2nd ed., p. 11. Freeman, San Francisco.

Chrystall, B. B., and Hagyard, C. J. (1975). Accelerated conditioning of lamb. Meat Ind. Res. Inst. N.Z. Publ. 470.

Chrystall, B. B., and Hagyard, C. J. (1976). Electrical stimulation and lamb tenderness. *N. Z. J. Agric. Res.* **19**, 7.

Cia, G., and Marsh, B. B. (1976). Properties of beef cooked before the onset of rigor. *J. Food Sci.* **41**, 1259.

Cornforth, D. P., Pearson, A. M., and Merkel, R. A. (1980). Relationship of mitochondria and sarcoplasmic reticulum to cold-shortening. *Meat Sci.* **4**, 103.

Cross, H. R. (1979). Effects of electrical stimulation on meat tissue and muscle properties — a review. *J. Food Sci.* **44**, 509.

Davey, C. L., and Gilbert, K. V. (1973). The effect of carcass posture on cold, heat and thaw shortening in lamb. *J. Food Technol.* **8**, 445.

Davey, C. L., Gilbert, K. V., and Carse, W. A. (1976). Carcass electrical stimulation to prevent cold-shortening toughness in beef. *N. Z. J. Agric. Res.* **19**, 13.

Dayton, W. R., Reville, W. J., Goll, D. E., and Stromer, M. H. (1976). A Ca^{2+}-activated protease possibly involved in myofibrillar protein turnover. Partial characterization of the purified protein. *Biochemistry* **15**, 2159.

Dutson, T. R., Pearson, A. M., and Merkel, R. A. (1974). Ultrastructural postmortem changes in normal and low quality porcine muscle fibers. *J. Food Sci.* **39**, 32.

Dutson, T. R., Smith, G. C., and Carpenter, Z. L. (1980a). Lysosomal enzyme distribution in electrically stimulated ovine muscle. *J. Food Sci.* **45**, 1097.

Dutson, T. R., Smith, G. C., Savell, J. W., and Carpenter, Z. L. (1980b). Possible mechanisms by which electrical stimulation improves meat tenderness. *Proc. Eur. Meet. Meat Res. Workers* **26** (Vol. 1), 84.

Fooladi, M. H., Pearson, A. M., Coleman, T. H., and Merkel, R. A. (1979). The role of nitrite in preventing development of warmed-over flavour. *Food Chem.* **3**, 283.

Forrest, J. C., Aberle, E. D., Hedrick, H. B., Judge, M. D., and Merkel, R. A. (1975). "Principles of Meat Science." Freeman, San Francisco.

Gergely, J. (1950). Relation of ATPase and myosin. *Fed. Proc., Fed. Am. Soc. Exp. Biol.* **9**, 176.

Gilbert, K. V., Davey, C. L., and Newton, K. G. (1977). Electrical stimulation and the hot boning of beef. *N. Z. J. Agric. Res.* **20**, 139.

Greene, B. E. (1969). Lipid oxidation and pigment changes in raw beef. *J. Food Sci.* **34**, 110.

Greenstein, J. P., and Edsall, J. (1940). The effect of denaturing agents on myosin. I. Sulfhydryl groups as estimated by porphyrindin titration. *J. Biol. Chem.* **133**, 397.

Hamm, R. (1960). Biochemistry of meat hydration. *Adv. Food Res.* **10**, 355.

Hamm, R. (1966). Heating of muscle systems. *In* "The Physiology and Biochemistry of Muscle as a Food, 1" (E. J. Briskey, R. G. Cassens, and J. C. Trautman, eds.), p. 363. Univ. of Wisconsin Press, Madison.

Hamm, R., and Deatherage, F. E. (1960). Changes in hydration, solubility and charges of muscle proteins during heating of meat. *Food Res.* **25**, 587.

Hapchuk, L. T., and Pearson, A. M. (1978). Isolation and partial purification of a proteolytic enzyme preparation from *Clostridium perfringens*. *Food Chem.* **3**, 115.

Hapchuk, L. T., Pearson, A. M., and Price, J. F. (1979). Effect of a proteolytic enzyme produced by *Clostridium perfringens* upon porcine muscle. *Food Chem.* **3**, 213.

Harsham, A., and Deatherage, F. E. (1951). Tenderization of meat. U.S. Patent 2,544,681.

Hasegawa, T., Pearson, A. M., Price, J. F., and Lechowich, R. V. (1970a). Action of bacterial growth on the sarcoplasmic and urea soluble proteins from muscle. 1. Effects of *Clostridium perfringens, Salmonella enteritidis, Achromobacter liquefaciens, Streptoccocus faecalis* and *Kurthia zopfii*. *Appl. Microbiol.* **20**, 117.

Hasegawa, T., Pearson, A. M., Price, J. F., Rampton, J. H., and Lechowich, R. V. (1970b). Effect of microbial growth upon sarcoplasmic and urea-soluble proteins from muscle. *J. Food Sci.* **35**, 720.

Herring, H. K., Cassens, R. G., and Briskey, E. J. (1965). Further studies on bovine tenderness as influenced by carcass position, sarcomere length, and fiber diameter. *J. Food Sci.* **30**, 1049.

Herring, H. K., Cassens, R. G., Suess, G. G., Brungardt, V. H. and Briskey, E. J. (1967). Tenderness and associated characteristics of stretched and contracted bovine muscles. *J. Food Sci.* **32**, 317.

Hoagland, R., McBryde, C. N., and Powick, W. C. (1917). Changes in fresh beef during cold storage above freezing. *U.S., Dep. Agric., Bull.* **433**.

Hostetler, R. L., Landmann, W. A., Link, B. A., and Fitzhugh, H. A., Jr. (1970). Influence of carcass position during rigor mortis on tenderness of beef muscles: Comparison of two treatments. *J. Anim. Sci.* **31**, 47.

Hostetler, R. L., Landmann, W. A., Link, B. A., and Fitzhugh, H. A., Jr. (1972). Effect of carcass suspension on sarcomere length and shear force of some major bovine muscles. *J. Food Sci.* **37**, 132.

Hostetler, R. L., Landmann, W. A., Link, B. A., and Fitzhugh, H. A., Jr. (1973). Effect of carcass suspension method on sensory panel scores for some major bovine muscle. *J. Food Sci.* **38**, 264.

Hostetler, R. L., Carpenter, Z. L., Smith, G. C., and Dutson, T. R. (1975). Comparison of postmortem treatments for improving tenderness of beef. *J. Food Sci.* **40**, 223.

Hsieh, Y. P. C., Cornforth, D. P., and Pearson, A. M. (1980). Ultrastructural changes in pre- and post-rigor beef muscle caused by conventional and microwave cookery. *Meat Sci.* **4**, 299.

Igene, J. O., and Pearson, A. M. (1979). Role of phospholipids and triglycerides in warmed-over flavor in meat model systems. *J. Food Sci.* **44**, 1285.

Igene, J. O., Pearson, A. M., Merkel, R. A., and Coleman, T. H. (1979a). Effect of frozen storage time, cooking and holding temperature upon extractable lipids and TBA values of beef and chicken. *J. Anim. Sci.* **49,** 701.

Igene, J. O., King, J. A., Pearson, A. M., and Gray, J. I. (1979b). Influence of heme pigments, nitrite and non-heme iron on development of warmed-over flavor (WOF) in cooked meat. *J. Agric. Food Chem.* **27,** 838.

Johnson, P. G., and Bowers, J. A. (1976). Influence of aging on the electrophoretic and structural characteristics of turkey breast muscle. *J. Food Sci.* **41,** 255.

Kanda, T., Pearson, A. M., and Merkel, R. A. (1977). Influence of pH and temperature upon calcium accumulation and release by bovine sarcoplasmic reticulum. *Food Chem.* **2,** 253.

Koszalka, T. R., and Miller, L. L. (1960). Proteolytic activity of rat skeletal muscle. *J. Biol. Chem.* **235,** 665.

Kramlich, W. E., Pearson, A. M., and Tauber, F. W. (1973). "Processed Meats." Avi Publ. Co., Westport, Connecticut.

Landmann, W. A. (1963). Enzymes and their influence on meat tenderness. *In* "Symposium on Meat Tenderness," p. 87. Campbell Soup Co., Camden, New Jersey.

Lawrie, R. A. (1966). "Meat Science," 1st ed. Pergamon, Oxford.

Locker, R. H. (1956). The dissociation of myosin by heat. *Biochim. Biophys. Acta* **20,** 515.

Locker, R. H. (1960). Degree of muscular contraction as a factor in the tenderness of beef. *Food Res.* **25,** 304.

Locker R. H. (1977). Meat tenderness and gap filaments. *Meat Sci.* **1,** 87.

Locker, R. H., and Hagyard, C. J. (1963). A cold shortening effect in beef muscle. *J. Sci. Food Agric.* **14,** 787.

Locker, R. H., and Wild, J. C. (1982a). A machine for measuring yield point in raw beef. *J. Text. Stud.* **13,** 71.

Locker, R. H., and Wild, J. C. (1982b). Yield point in raw beef muscle. The effects of ageing, rigor temperature and stretch. *Meat Sci.* **7,** 93.

Locker, R. H., Davey, C. L., Nottingham, P. M., Haughey, D. P., and Law, N. H. (1975). New concepts in meat processing. *Adv. Food Res.* **21,** 158.

Love, J. D., and Pearson, A. M. (1971). Lipid oxidation in meat and meat products. A review. *J. Am. Oil Chem. Soc.* **48,** 547.

Love, J. D., and Pearson, A. M. (1974). Metmyoglobin and nonheme iron as prooxidants in cooked meat. *J. Agric. Food Chem.* **22,** 1032.

Lundberg, W. O. (1962). Mechanisms. *In* "Symposium on Foods: Lipids and Their Oxidation" (H. W. Schultz, E. A. Day, and R. O. Sinnhuber, eds.), p. 31. Avi Publ. Co., Westport, Connecticut.

McCollum, P. D., and Henrickson, R. L. (1977). The effect of electrical stimulation on the rate of postmortem glycolysis in some bovine muscles. *J. Food Qual.* **1,** 15.

Marsh, B. B., and Leet, N. G. (1966). Studies on meat tenderness. III. Effect of cold shortening on tenderness. *J. Food Sci.* **31,** 451.

Marsh, B. B., and Thompson, J. F. (1958). Rigor mortis and thaw rigor in lamb. *J. Sci. Food Agric.* **9,** 417.

Marsh, B. B., Woodhams, P. R., and Leet, N. G. (1968). Studies on meat tenderness. V. Effects on tenderness of carcass cooling and freezing before completion of rigor mortis. *J. Food Sci.* **33,** 12.

Marsh, B. B., Cassens, R. G., Kauffman, R. G., and Briskey, E. J. (1972). Hot boning and pork tenderness. *J. Food Sci.* **37,** 179.

Moulton, C. R., and Lewis, W. L. (1940). "Meat Through the Microscope" (2nd ed.). Institute of Meat Packing, University of Chicago, Chicago.

Newbold, R. P. (1966). Changes associated with rigor mortis. *In* "The Physiology and Biochemistry of Muscle as a Food, 1" (E. J. Briskey, R. G. Cassens, and J. C. Trautman, eds.), p. 213. Univ. of Wisconsin Press, Madison.

Olson, D. G., and Parrish, F. C., Jr. (1977). Relationship of myofibril fragmentation index to measures of beef steak tenderness. *J. Food Sci.* **42**, 516.

Olson, D. G., Parrish, F. C., Jr., and Stromer, M. H. (1976). Myofibril fragmentation and shear resistance of three bovine muscles during postmortem storage. *J. Food Sci.* **41**, 1136.

Parrish, F. C., Jr., Vandell, C. J., and Culler, R. D. (1979). Effect of maturity and marbling on the myofibril fragmentation index of bovine longissimus muscle. *J. Food Sci.* **44**, 1668.

Paul, P., Lowe, B., and McClurg, B. R. (1944). Changes in histological structure and palatability of beef during storage. *Food Res.* **9**, 221.

Paul, P., Bratzler, L. J., Farwell, E. D., and Knight, K. (1952). Studies on tenderness of meat. I. Rate of heat penetration. *Food Res.* **17**, 514.

Pearson, A. M., and Dutson, T. R. (1985). Scientific basis for electrical stimulation. *Adv. Meat Res.* **1**, 185.

Pearson, A. M., Harrington, G., West, R. G., and Spooner, M. E. (1962). The browning produced on heating fresh pork. I. The relation of intensity of browning to chemical constituents and pH. *J. Food Sci.* **27**, 177.

Pearson, A. M., Tarladgis, B. G., Spooner, M. E., and Quinn, J. R. (1966). The browning produced on heating fresh pork. II. Nature of the reaction. *J. Food Sci.* **31**, 184.

Pearson, A. M., Carse, W. A., Davey, C. L., Locker, R. H., and Hagyard, C. J. (1973a). Influences of epinephrine and calcium upon glycolysis, tenderness and shortening of sheep muscle. *J. Food Sci.* **38**, 1124.

Pearson, A. M., Carse, W. A., Wenham, L. M., Fairbairn, S. J., Locker, R. H., and Jury, K. E. (1973b). Influence of various adrenergic accelerators and blocking agents upon glycolysis and some related properties of sheep muscle. *J. Anim. Sci.* **36**, 511.

Pearson, A. M., Love, J. D., and Shorland, F. B. (1977). Warmed-over flavor in meat, poultry and fish. *Adv. Food Res.* **23**, 1.

Porzio, M. A., and Pearson, A. M. (1975). Isolation of an extracellular neutral proteinase from *Pseudomonas fragi*. *Biochim. Biophys. Acta* **384**, 235.

Pratt, D. E. (1972). Water soluble antioxidant activity in soybeans. *J. Food Sci.* **37**, 322.

Pratt, D. E., and Watts, B. M. (1964). Antioxidant activity of vegetable extracts. I. Flavone aglycones. *J. Food Sci.* **29**, 27.

Rampton, J. H., Pearson, A. M., Price, J. F., Hasegawa, T., and Lechowich, R. V. (1970). Effect of microbial growth upon myofibrillar proteins. *J. Food Sci.* **35**, 511.

Ramsbottom, J. M., and Strandine, E. J. (1949). Initial physical and chemical changes in beef as related to tenderness. *J. Anim. Sci.* **8**, 398.

Sato, K., and Hegarty, G. R. (1971). Warmed-over flavor in cooked meats. *J. Food Sci.* **36**, 1198.

Sato, K., Hegarty, G. R., and Herring, H. K. (1973). The inhibition of warmed-over flavor in cooked meats. *J. Food Sci.* **38**, 398.

Savell, J. W., Smith, G. C., Dutson, T. R., Carpenter, Z. L., and Suter, D. A. (1977). Effect of electrical stimulation on palatability of beef, lamb and goat meat. *J. Food Sci.* **42**, 712.

Savell, J. W., Dutson, T. R., Smith, G. C., and Carpenter, Z. L. (1978a). Structural changes in electrically stimulated beef muscle. *J. Food Sci.* **43**, 1616.

Savell, J. W., Smith, G. C., and Carpenter, Z. L. (1978b). Effect of electrical stimulation on quality and palatability of light-weight beef carcasses. *J. Anim. Sci.* **46**, 1221.

Savell, J. W., Smith, G. C., and Carpenter, Z. L. (1978c). Beef quality and palatability as affected by electrical stimulation and cooler aging. *J. Food Sci.* **43**, 1666.

Schmidt, G. L. (1978). Sectioned and formed meat. *Proc. Annu. Reciprocal Meat Conf.* **31**, 18.

Seideman, S. C., and Cross, H. R. (1982). Utilization of electrical stimulation to improve meat quality: A review. *J. Food Qual.* **5**, 247.

Sharp, J. G. (1957). Deterioration of dehydrated meat during storage. II. Effect of pH and temperature on browning changes in dehydrated aqueous extracts. *J. Sci. Food Agric.* **8**, 21.

Sharp, J. G., and Marsh, B. B. (1953). Whale meat, production and preservation. *G. B., Dep. Sci. Ind. Res., Food Invest. Board, Spec. Rep.* **58**.

Shorland, F. B., Igene, J. O., Pearson, A. M., Thomas, J. W., McGuffey, R. K., and Aldridge, A. E. (1981). Effects of dietary fat and vitamin E on the lipid composition and stability of veal during frozen storage. *J. Agric. Food Chem.* **29**, 863.

Szent-Györgi, A. (1960). Proteins in the myofibril. *In* "The Structure and Function of Muscle" (G. H. Bourne, ed.), Vol. 2, p. 1. Academic Press, New York.

Tarrant, P. J. V., Pearson, A. M., Price, J. F., and Lechowich, R. V. (1971). Action of *Pseudomonas fragi* on the proteins of pig muscle. *Appl. Microbiol.* **22**, 224.

Tarrant, P. J. V., Jenkins, N., Pearson, A. M., and Dutson, T. R. (1973). Proteolytic enzyme preparation from *Pseudomonas fragi*: Its action on pig muscle. *Appl. Microbiol.* **25**, 996.

Tims, M. J., and Watts, B. M. (1958). Protection of cooked meats with phosphates. *Food Technol.* **12**, 241.

Vaughan, H. W. (1945). "Types and Market Classes of Live Stock," 23rd ed. College Book Co., Columbus, Ohio.

Voyle, C. A. (1969). Some observations on the histology of cold-shortened muscle. *J. Food Technol.* **4**, 245.

Wasserman, A. E., and Talley, F. (1968). Organoleptic identification of roasted beef, veal, lamb and pork as affected by fat. *J. Food Sci.* **33**, 219.

Watts, B. M. (1962). Meat products. *In* "Symposium on Foods: Lipids and Their Oxidation" (H. W. Schultz, E. A. Day, and R. O. Sinnhuber, eds.), p. 212. Avi Publ. Co., Westport, Connecticut.

Weiner, P. D., and Pearson, A. M. (1966). Inhibition of rigor mortis by ethylenediamine tetraacetic acid. *Proc. Soc. Exp. Biol. Med.* **123**, 185.

Weiner, P. D., and Pearson, A. M. (1969). Calcium chelators influence some physical and chemical properties of rabbit and pig muscle. *J. Food Sci.* **34**, 592.

Weiner, P. D., Pearson, A. M., and Schweigert, B. S. (1969). Turbidity, viscosity and ATPase activity of fibrillar protein extracts of rabbit muscle. *J. Food Sci.* **34**, 313.

Whitaker, J. R. (1959). Chemical changes associated with aging of meat with emphasis on the proteins. *Adv. Food Res.* **9**, 1.

Will, P. A., Ownby, C. L., and Henrickson, R. L. (1979a). Ultrastructural postmortem changes in electrically stimulated bovine muscle. *J. Food Sci.* **45**, 21.

Will, P. A., Henrickson, R. L., Morrison, R. D., and Odell, G. V. (1979b). Effect of electrical stimulation on ATP depletion and sarcomere length in delay-chilled bovine muscle. *J. Food Sci.* **45**, 1646.

Wilson, B. R., Pearson, A. M., and Shorland, F. B. (1976). Effect of total lipids and phospholipids on warmed-over flavor in red and white muscle from several species as measured by thiobarbituric acid analysis. *J. Agric. Food Chem.* **24**, 7.

Wismer-Pedersen, J. (1971). Water. *In* "The Science of Meat and Meat Products" (J. F. Price and B. S. Schweigert, eds.), 2nd ed., p. 177. Freeman, San Francisco.

4

Functional Properties of the Myofibrillar System and Their Measurements

R. HAMM

Bundesanstalt für Fleischforschung
Kulmbach, Federal Republic of Germany

I. THE IMPORTANCE OF THE MYOFIBRILLAR SYSTEM FOR THE WATER-HOLDING CAPACITY OF MEAT AND MEAT PRODUCTS

A. Introduction

Cross-striated muscle contains ~75% water. The state of water in the muscle cell and changes of water structure and water movements during muscular contraction are in close relation with the secret of life. Our knowledge in this field is very limited. Much more is known about the factors that influence the immobilization of water in the intact muscle postmortem or in comminuted tissue. The power with which tissue water and added water are bound by muscle proteins is of great importance for the quality of meat and meat products. The myofibrillar system plays a dominant role in water binding. Almost

Copyright © 1986 by Academic Press, Inc.
All rights of reproduction in any form reserved.

all procedures for the storage and processing of meat are influenced by the water-holding capacity (WHC) of the tissue; vice versa, these procedures can change the WHC of muscle. Such procedures include transportation, storage, aging, freezing and thawing, drying, mincing, salting, curing, smoking, canning, and cooking. It is well known that WHC is of particular importance for the quality of sausage of the emulsion type and canned ham. The great economic problem of weight losses during storage, freezing and thawing, or cooking of meat is related to the binding of water within the muscle. Thus, investigation of the WHC of meat is of considerable economic interest. Moreover, it gives information not only on the WHC of meat itself but also on alterations in meat proteins. Changes in WHC are a very sensitive indicator of changes in the charges and structure of myofibrillar proteins (Hamm, 1960, 1972, 1975a).

Only selected aspects of WHC are discussed because of the limited length of this chapter. Some important factors related to WHC that are not considered here are the influence of animal factors (species, sex, age, breeding, muscle type, conditions, and treatment of animals before slaughter) (for reviews, see Hamm, 1960, 1972) and the effect of WHC on other parameters of meat quality (tenderness, juiciness, color) (for reviews, see Hamm, 1960, 1972, 1977c). The influence of the collageneous system and of added nonmeat proteins on WHC of meat and meat products are not discussed because the effect of nonmyofibrillar proteins is not the subject of this chapter.

B. Determinants of the Water-Holding Capacity of Meat

1. THE STATE OF WATER IN THE MUSCLE CELL

There is no doubt that myofibrillar proteins are primarily responsible for the binding of water in muscle. It is also obvious that different types of water binding exist in the tissue. A very small part of the muscle water is present as *constitutional water* (~ 0.3 g H_2O per 100 g protein, i.e., $<0.1\%$ of the total tissue water), which is located within the protein molecules, protein–water and water–water binding energies being much greater than those existing in normal water (Fennema, 1977). It is generally accepted that a further part of tissue water (5–15% of the total water) shows a relatively restricted mobility (see Hamm, 1972, 1975a). According to the classification proposed by Fennema (1977), this fraction of tissue water might be defined as *interfacial water*. This water is located at the surface of the proteins, probably in multilayers and in small crevices. Either water–solute or water–water interactions are involved, and the binding energies are generally greater than those in normal water. All these result in decreased mobility of water molecules (intermediates between constitutional water and bulk water), in reduced vapor pressure and melting point. A part of this water remains liquid at $-40°C$.

The question arises as to whether the remaining bulk of cellular water is "free" in the physical–chemical sense of this term. The interpretation of nuclear magnetic resonance (NMR) studies for investigation of the state of water in muscle has aroused controversy. Numerous authors believe that essentially all of the water in skeletal muscle exists in a physical state different from that of pure water or diluted electrolyte solutions (see Blanshard and Derbyshire, 1975; Ling and Walton, 1976; Ernst and Hazlewood, 1978; Peschel and Belouschek, 1979). The alternative model, again citing NMR evidence, is that most of the muscle water behaves as bulk water in a diluted salt solution but that a small fraction of the water, that adjacent to the protein molecules, has a modified molecular structure. This fraction of tissue water corresponds to the interfacial water mentioned above. The authors representing this school conclude from their results that in the bulk of intracellular water no dramatic changes in the characteristics of molecular motion have been imposed by the constituents of the muscle cell (see Hansen, 1971; Cooke and Kuntz, 1974; Blanshard and Derbyshire, 1975; Fung et al., 1975; Fung, 1977; Garlid, 1979).

In addition to the fibers there exists an extracellular space in muscle. Hazlewood et al. (1974) found that ~ 10% of the total water in the living muscle must be associated with the extracellular space. The mobility of the extracellular water is supposed to be somewhere between that of pure water and that of cellular water. The dimension of the extracellular space depends on the state of swelling. Swelling of muscle fibers decreases the extracellular space (Tasker et al., 1959) whereas shrinkage of fibers as a result of development of rigor mortis, pH fall, or addition of divalent cations is associated with an increase in the extracellular space (Rome, 1967, 1968; April et al., 1972; Heffron and Hegarty, 1974). Therefore, swelling and shrinkage of muscle fibers in the intact tissue cause movements of water between intracellular and extracellular spaces.

According to Offer and Trinick (1983) it seems reasonable to suppose that water is held in muscle by capillarity, mostly in the interfilamental spaces within the myofibrils but a substantial part in the spaces between the myofibrils and in the extracellular space. However, methods for an exact or at least tentative differentiation between capillary forces and other forces restricting the mobility of water in animal tissues do not exist and, therefore, it is not clear whether and to which extent water is ordered in capillary biosystems.

2. WATER-HOLDING AND SWELLING CAPACITY OF MEAT, AND IMMOBILIZED WATER

The water in muscle that is not bound as interfacial water or hydration water, is more or less expressible during application of force, but it is not yet known if the differences in this WHC of muscle are due to differences in the structures of the bulk water. Therefore, the contradictory results of studies on the state of water in muscle mentioned above are of very limited value for an understanding of differences or changes in the water-binding properties of meat.

The remarkable changes in WHC occurring during storage and processing of meat are determined by the extent to which bulk phase water is immobilized within the microstructure of the intact or comminuted tissue (Hamm, 1972). Fennema (1977) called this immobilized bulk water "entrapped water" because it is physically entrapped in a fashion similar to that found in gels. In this chapter the entrapped water is called *immobilized water* because this term has been commonly used since the first review on the WHC of meat (Hamm, 1960). It should be mentioned that most of the changes or differences in WHC of meat that are of practical importance, are apparently not related to the fractions of constitutional water or interfacial water (Hamm, 1960, 1962a, 1972).

There seems to be a more or less continuous transition from water strongly immobilized within the tissue, which can be expressed with difficulty, to the "loose water" that can be squeezed out by very low pressure. It is not possible, therefore, to develop absolute figures for the immobilized part of bulk phase water because the amount of immobilized water measured depends on the method used, but by using a standardized method, one can measure relative differences in WHC quite accurately.

For the different phenomena of WHC and their measurements I propose the following terminology:

Drip loss (DL)	Formation of exudate from meat or meat systems (except thawing loss) without application of external forces.
Thawing loss (TL)	Formation of exudate from meat or meat systems after freezing and thawing without application of external forces.
Cooking loss (CL)	Release of fluid after heating of meat or meat systems either without or with application of external forces (e.g., centrifugation or pressing).
Expressible juice (EJ)	Release of juice from unheated meat or meat systems (also after freezing and thawing) during application of external forces such as pressing (e.g., filter-paper press method), centrifugation methods, or suction methods (e.g., capillary volumeter method).

According to the different methods used, WHC is defined as the ability of meat to hold fast its own or added water during application of any force. Swelling capacity of meat is defined as the spontaneous uptake of water from any surrounding fluid, resulting in an increase of weight and volume of muscle fibers. Swelling capacity often shows a close correlation with WHC (Hamm, 1960, 1972).

3. PRINCIPAL FACTORS AFFECTING THE IMMOBILIZATION OF WATER IN MEAT

The bulk phase water in muscle tissue is located within the filaments, in the interfilamental spaces, in the interfibrillar spaces (sarcoplasmic space), and in the extracellular space. The question arises as to whether these different compartments for the tissue water are related to WHC of meat.

It is to be expected that the extracellular water is more easily expressed than the cellular water. Fischer *et al.* (1976) measured the influence of load (pressure) on the volume of meat juice that was obsorbed by a gypsum diaphragm. It is possible that the observed plateau of the load–volume curve between a 300 and a 1000 g load indicates the border between extracellular and cellular water. There is little doubt that the extent of drip formation during storage of the intact muscle is greater with more of the tissue water present in the extracellular space.

Since myofibrils occupy ~70% of the volume of lean muscle, most of the tissue water must be located in the myofibrils. Therefore, changes in WHC of meat will be, in great part, due to variations in the immobilization of bulk phase water in the interfilamental spaces and the filaments themselves. I believe that the water-imbibing power of the myosin in the thick filaments plays an important role in WHC of meat, particularly after mincing; it is well known that myosin, which makes up about half of the myofibrillar protein, has an enormous capacity to imbibe water (Szent-Györgyi, 1973), so it is not surprising that in muscle with a high WHC and increased swelling, for example, meat after addition of NaCl (Offer and Trinick, 1983) or in the prerigor state (Hamm, 1985), the thick filaments are obviously greatly swollen.

As mentioned above, we do not know exactly which forces restrict the mobility of water in the tissue but we know much more about the factors that influence the immobilization of water. The immobilization of water in the tissue is apparently determined by the spatial molecular arrangement of the myofibrillar proteins (mainly myosin) or filaments (Hamm, 1972) (Fig. 1). If the attraction between adjacent molecules or filaments is decreased, as is caused by increasing electrostatic repulsion between similarly charged protein groups or by weakening of hydrogen bonds or hydrophobic bonds, the protein network is enlarged, the swelling increases, and more water can be immobilized within the larger meshes, that is, there is an increase in WHC (e.g., in terms of expressible juice or cooking loss) (Fig. 1A,B).

Examples of effects of this type are the increase in tissue pH above the isoelectric point IP of myosin (pH 5) (increase in negative net charge), addition of NaCl at pH > 5 (screening of positive protein charges by preferential binding of Cl^- ions), dissociation of linkages between myosin heads and thin filaments

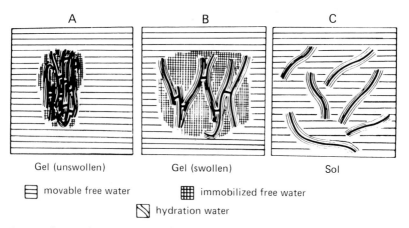

Fig. 1   Influence of cross-linking of proteins or filaments on the water-holding capacity (WHC) or swelling of meat.   Explanation is in the text.

by ATP (before onset of rigor mortis) or pyrophosphate (Hamm, 1960, 1972, 1975a) (see also Sections I,D,1, 3,a, b, 5, and 7), and weakening of the interaction between hydrophobic bonds (e.g., by interaction between hydrophobic protein groups and the hydrophobic part of smaller molecules such as lecithin) (Honikel and Hamm, 1983).   It is also possible that cleavage of linkages between Z-line material and thin filaments and between Z lines (by desmin?) during aging of meat increases WHC.

If, however, the attraction between adjacent molecules is increased, as happens when the electrostatic attraction between oppositely charged groups increases or by the effect of interlinking bonds, less space is available for the retention of immobilized water in the protein network.   Thus, during tightening of the network, shrinkage occurs, and a part of the immobilized water becomes free to move and flows out at low pressure (Fig. 1B,A) (Hamm, 1960, 1972, 1975a).

Examples of effects of the latter type are lowering the tissue pH to the IP of myosin (increase in the attraction between oppositely charged protein groups), addition of NaCl at pH < IP (lowering the repulsion between positively charged protein groups by the screening effect of Cl^- ions), association of thick and thin filaments during development of rigor mortis (binding of actin reduces the water-imbibing power of nondissolved myosin), and heat coagulation of proteins (interaction between hydrophobic protein groups) (see Sections I,D,1, 3,a, 5, and 6).   It should be mentioned that at temperatures $\geq 60°C$ the effect of shrinkage of coagulating myofibrillar proteins on the release of cell water is masked by the effect of shrinkage of the different types of muscle collagen (Bendall and Restall, 1983).   After heating, a greatly swollen system (Fig. 1B) entraps more water in the large meshes of the network of coagulated protein than a less swollen system (Fig. 1A).

From the greater swelling of thick filaments in meat with high WHC (see above) and from the shrinkage of thick filaments during development of rigor mortis (Hamm, 1985), we may conclude that changes in the WHC of meat are determined not only by alterations in the interaction between thick and thin filaments, but also by alterations in the interaction between the protein molecules of the myosin system.

As to the theory of swelling of muscle fibers, it was suggested that Donnan osmotic pressure phenomena are more important than the electrostatic repulsion between adjacent protein molecules or filaments (Hamm, 1972; Ledward, 1983). However, there exist convincing facts supporting the electrostatic theory of swelling (Hamm, 1972; Winger and Pope, 1981a,b; Offer, 1984).

Currie and Wolfe (1983) explained variation in the WHC of meat from different animals by differences in the permeability of the cell wall for water, measured by the uptake of labelled inulin by the intact muscle. This theory needs further proof because the experimental evidence is not very convincing as yet (Honikel *et al.*, 1985).

After comminution, the cell wall (sarcolemma) of the muscle fibers is, to a great extent, destroyed. Therefore, differences between extracellular and intracellular water are eliminated and the water-imbibing power of the thick filaments or myosin primarily determines the WHC of meat (Fukazawa, 1961a,b; Nakayama and Sato, 1971a,b). The rod moiety of the myosin molecule rather than the globular heads is responsible for the strong thermogelling properties of myosin (Samejima *et al.*, 1981a).

The relative effects of different factors such as pH change, salting, and rigor mortis on WHC and swelling are similar for comminuted meat and the intact muscle but after comminution the swelling capacity of the myofibrillar system is much less limited. It is not clear if the thin filaments also participate in the immobilization of water in comminuted meat.

The suggested concept for WHC of meat allows an explanation of the fact that the changes in drip loss are not necessarily the result of changes in WHC of myosin or actomyosin. Contraction of muscle at low temperature (cold shortening) or at elevated temperatures (rigor shortening) causes an increase in drip loss during storage of meat (Hamm, 1982; Honikel *et al.*, 1985) (see also Section I,D,3,a). However, neither cold shortening nor rigor contraction have a significant influence on the amount of water released upon heating of the muscle or of sausage batters prepared from such material (Hamm, 1982). Apparently, the thermal shrinkage of meat overshadows or masks the effect of shortening on fiber shrinkage. Drip loss can be increased by a shift of interfilamental and/or interfibrillar water into the extracellular space, perhaps mainly as a result of osmotic phenomena (Heffron and Hegarty, 1974; Honikel *et al.*, 1985) without changes in the water-imbibing power of the thick filaments, whereas cooking loss of tissue or WHC of comminuted meat is primarily determined by the ability of myosin to immobilize water in its protein network. There exists, however, ample experimental evidence that an increase in water retention by the myosin system lowers drip loss as well as cooking loss and mixture stability (Hamm, 1972) (see scheme in Fig. 2).

Another example of differences in drip loss without differences in other WHC phenomena is the influence of rates of freezing and thawing on the amount of thaw drip. Except for the effects of long-term freezer storage, the extent of thaw drip is determined by size and location (intra- or extracellular) of ice crystals rather than by the WHC of the myofibrillar proteins (Hamm et al., 1982) (see also Section I,D,4,a).

In conclusion I may suggest that the water-imbibing power of the thick filaments and not only interfilamental spacing or membrane permeability play a role for changes or differences in the WHC of meat (particularly after mincing) so the interactions between myosin molecules are of particular importance. In certain cases, however, the extent of drip loss can be altered by shifts of water between intracellular and extracellular space without changes in the water-imbibing power of the thick filaments.

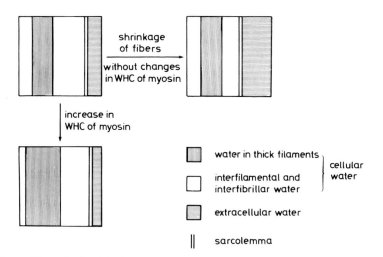

Fig. 2   Scheme for the shift of water between the muscle cell (to the left of the sarcolemma) and the extracellular space (to the right of the sarcolemma). An example of the shrinkage of muscle fibers without changes in the WHC of myosin (above) is cold shortening and its effect on drip loss (Fig. 10A). An example of an increase in WHC of myosin (below) is the effect of NaCl on meat (Hamm, 1985).

C. Methods for Measurement of Water-Holding Capacity

1. GENERAL REMARKS

From the preceding section it can be concluded that methods for the determination of the rather tightly bound interfacial water in muscle tissue are not appropriate for studies on water-holding capacity of meat, that is, for the determination of relative differences in the immobilization of bulk water; thus, data on NMR, a_w, osmotic pressure, relative vapor pressure, adsorption isotherms, solvent versus nonsolvent water, differential thermoanalysis and similar techniques do not give information on WHC of meat as defined in Section I,B,2. For the determination of WHC, a force such as pressure (pressing, centrifugation) or suction has to be applied to the meat sample and the amount of released water (loosely bound water) has to be determined. Pressure is also exerted by shrinking during storage (drip loss) or heating (cooking loss). A combination of different types of forces can be applied, but the vast number of methodological variations in the measurement of WHC are not discussed in this article; only a few examples of the different types of methods are presented.

When the methods of pressing and centrifugation are employed, the product undergoes deformation. It is not known whether water loss is entirely due to the force equilibration or if it originates in part from the alterations in the internal structure of the material. Therefore, one must be careful in interpreting WHC data (Labuza and Lewicki, 1978).

When the different measures of WHC are compared with one another, one can often obtain statistically significant correlations among values (e.g., see Tsai and Ockerman, 1981), but the correlations are often not precise enough to permit a single test to be used as a reliable predictor (Ranken, 1976). WHC measurements with raw, unground muscle do not necessarily reflect the water retention during cooking (Toscano and Autino, 1969; Scheper, 1975; Grosse, 1979) or other types of processing. Rigor mortis has little effect on the WHC of unsalted beef but causes a strong decrease of the WHC of salted muscle homogenates or sausage emulsions (see Section I,D,3,a, b). Therefore, for selection of an adequate technique, one must consider the purpose for which WHC data will be used (e.g., for information on weight loss during thawing or on water retention during cooking or on water binding of emulsion-type sausages). If necessary, correlations between a particular method for measuring WHC and the effect of certain technological parameters have to be established.

2. DETERMINATION OF DRIP LOSS

For measurement of drip loss, weighed samples (e.g., 30 g of 3-cm meat cubes) are sealed in plastic pouches under atmospheric pressure and stored at a specific temperature in such a way that contact between the meat sample and the released drip is prevented. The drip loss is calculated from the weight loss of the meat sample. For purposes of comparison, shape and weight of the meat cubes must be almost identical (Honikel *et al.*, 1980).

Measurement of drip loss can be carried out in such a manner that after the release of drip the remaining material can be used for the determination of cooking loss. In many cases, the measurement of cooking loss is meaningless if the extent of the preceeding drip loss (during storage, freezing and thawing, etc.) is not known.

3. FILTER-PAPER PRESS METHOD

The filter-paper press method (FPM) is widely used because of all the methods for measuring WHC it is one of the simplest, and it seems to be as reliable as the other procedures. In the original FPM, developed by Grau and Hamm (1953, 1957), 300 mg of the intact or ground muscle tissue is placed on a filter paper with a defined moisture content and pressed between two plexiglass plates to a thin film; the water squeezed out is absorbed by the filter paper. The area of the brown ring of expressed juice (*RZ*, ring zone) (Fig. 3), which is obtained by subtracting the area *M* of the meat film from the total area (*T*), is proportional to the amount of water in the *RZ*. The pressure produced by tightening the plates by hand is so great that individual differences of pressure do not influence the area of *RZ*. This simple technique can be carried out in a slaughterhouse. The accuracy of the method is satisfactory provided that the sample is precisely weighed.

The WHC can be expressed as the area *RZ* (cm²) or as the amount of loose water in *RZ* (calculated from the area *RZ*) related either to the weight of sample, to the total water content, or to the protein content of the sample. With samples of raw meat showing a great variation in WHC, a highly significant correlation ($r = .99$) between *M* and the amount of loosely bound water related

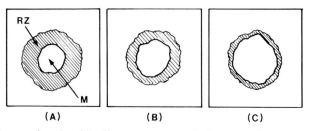

Fig. 3   Diagram of results of the filter-paper press method obtained from meat with low (A), medium (B), and high water-holding capacity (C). The area of the ring of released juice (ring zone, RZ) can be calculated by subtracting the area *M* of the film of pressed meat from the total area *T* ($T = RZ + M$).

to total water was found (Scheper, 1974). With regard to measurement of WHC with samples of raw, unground tissue picked from the muscles, the coefficients of repeatability reported in the literature vary from 0.68 to 0.98 (Janicki and Walczak, 1954; Fewson and Kirsammer, 1960; Engelke, 1961; Gravert, 1962; Fewson et al., 1964; Sonn, 1964). The error variance found by Fewson et al. (1964) was <5%. A number of modifications of this technique have been suggested. In most of them the application of defined pressure is recommended (e.g., Wierbicki and Deatherage, 1959) or the amount of released water is determined by weighing the meat sample or the filter paper before and after pressing (for reviews, see Hamm, 1960, 1972).

The assumption that almost all of the water released during pressing of the meat sample is located in the area RZ (Grau and Hamm, 1957; Wierbicki and Deatherage, 1959) cannot be correct because the amount of expressible water, determined by weight loss, is higher than the amount of water in RZ calculated from the area of RZ (Fig. 4A) (Hofmann, 1977; Hofmann et al., 1982). Therefore, the fluid that is squeezed out by pressing is adsorbed by the filter paper not only around but also below the meat area M. As Fig. 4A shows, with increasing pressure the amount of released water, determined by weight loss of the meat sample, increases until a load of about 40 kg is imposed; further increase in load does not result in an additional release of juice. It is interesting that the amount of bound water in the tissue, which cannot be squeezed out, is in the range of 10 to 15% of the total water (before pressing). This nonexpressible water corre-

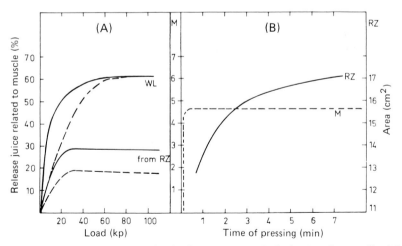

Fig. 4 Parameters of importance for the filter-paper press method. (A) Influence of load (kg) on the amount of released juice (water related to meat), measured by weight loss (WL) or calculated from the area of RZ (see Fig. 3). —— , Meat with lower WHC; – – –, meat with higher WHC. (B) Influence of time of pressing on the area of M and RZ (see Fig. 3).

sponds to the amount of interfacial water in muscle determined by other, more sophisticated methods (Section I,B,1).

At loads of ~40 kg and more, the amount of expressible water is independent of the WHC of meat (Fig. 4A). The area of RZ, however, depends strongly on the WHC also at loads >40 kg; pressing with 40 kg corresponds roughly to the pressure exerted by using the technique of Grau and Hamm (1957; Hofmann et al., 1982). The area T is much less dependent on WHC than RZ. From this observation it must be concluded that RZ is primarily determined by the area M of the meat film rather than by the amount of expressible water (Hofmann et al., 1982). The size of M depends on the deformability or plasticity of the meat sample, which is mainly a function of the protein–protein interactions in the myofibrillar system and, therefore, correlated with WHC. In meat with a weak protein–protein interaction (e.g., prerigor, high pH, added salts) a higher proportion of the total water is immobilized than in meat with a strong protein–protein interaction (Fig. 1). Consequently, under a certain pressure, meat with high WHC is more easily deformed (larger M) than meat with a low WHC. This explains the highly significant correlations between FPM and other methods for the determination of WHC (see below). It must be considered, however, that under certain circumstances differences in the deformability of the meat sample might not be related to differences in WHC but to other factors. This certainly plays a role if large amounts of water or fat are added to minced meat.

With the FPM the deformation of the meat sample to a thin film occurs very quickly; after a few seconds the meat film has reached its maximum size (Fig. 4B). The loosely bound water is squeezed out at the same rate but the adsorption of the released fluid by the filter paper takes a longer time. It takes several minutes for the areas RZ and T to reach their maxima (Fig. 4B) (Grau and Hamm, 1957; Hofmann et al., 1982).

From Fig. 4A it can be seen that the area of RZ indicates differences in WHC at relatively low pressures. According to Autino and Toscano (1970), who used loads between 100 and 1000 g, the lower loads give higher repeatability. With the application of low pressure (e.g., Brendl and Klein, 1971), the RZ might be determined not only by the area M of meat but also by the amount of expressible fluid.

In the usual application of the FPM (Grau and Hamm), the RZ is used as a measure for WHC. However, it would be sufficient and less time consuming to measure just M because the area of RZ is primarily a function of M, as shown above: both areas correlate with each other (Scheper, 1974). Following a proposal of K. O. Honikel (personal communication, 1977), George et al. (1980) expressed WHC in terms of the ratio M/RZ. Hofmann et al. (1982) proposed using the ratio M/T of the areas because within a certain weight range (200–400 mg) this ratio is independent of the weight of sample and determined by the WHC only.

Advantages of the FPM: the procedure is simple and quick; in comparison with other methods it is a sensitive test for recognition of small differences in WHC (Tsai and Ockermann, 1981); it is applicable to unground and ground tissue, with and without added water; heat-denatured meat can also be used; only 0.3 g tissue or homogenate is necessary; the technique requires only a few minutes and the result is fixed on the filter paper, permitting evaluation at any time. Disadvantages of the FPM: the technique is not applicable for samples containing large amounts of fat and/or water, such as sausage emulsions (Grau and Hamm, 1957; Tsai and Ockerman, 1981).

The FPM carried out with freshly slaughtered pork allows the prediction of weight loss during chilling (Scheper, 1975). In studies on the influence of different factors on WHC of meat, Tsai and Ockerman (1981) found highly significant correlations ($r = .88 – .99$) between the FPM (Wierbicki and Death-erage, 1959) and a centrifugation method (Miller et al., 1968). According to Ubertalle and Mazzocco (1979), the FPM (Grau and Hamm, 1957) is quicker and more easily reproduced than the low-speed centrifugation method of Wierbicki et al. (1957) or the ultracentrifugation technique of Bouton et al. (1971). Toscano and Autino (1969), who compared the FPM (Grau and Hamm) with a cooking loss method and a dehydration procedure, preferred the FPM.

4. CENTRIFUGATION METHODS

Methods of centrifugation (CM) can be used for measuring the WHC of intact (unground) tissue. The technique of Bouton et al. (1971, 1972) represents a reasonable centrifugation procedure, avoiding mincing and heating. In this method a weighed muscle sample (3–4 g) is centrifuged at 100,000 g for 1 hr in stainless steel tubes. Fischer et al. (1976) found that with centrifugation at 60,000 g for 30 min good results are obtained. The juice released from the meat is decanted off as quickly as possible (in order to avoid readsorption of water). The meat sample is removed from the tube with forceps, dried with tissue paper, and then reweighed to determine liquid loss. If the residue is dried in the tube at 105°C, the total water content of the sample can be determined, and WHC can be expressed as released or bound water as a percentage of total water. The coefficient of variation is $< 5\%$, and although the conditions are arbitrary, they are easily reproducible. The need for a high-speed centrifuge makes it almost impossible to use this type of method in a slaughterhouse. Recognizing this fact as well as a poor correlation with other methods for measuring WHC (FPM, low-speed centrifugation), Ubertalle and Mazzocco (1979) do not recommend this technique.

At relatively low speed and/or with minced meats the amount of juice, if any, released by CMs is very small (Wierbicki and Deatherage, 1959). Therefore, in a number of CMs the samples are heated in the centrifuge tube at temperatures between 40 and 100°C before centrifugation (for reviews, see Hamm, 1960, 1972).

In some techniques readsorption of water, released by centrifugation, is avoided by separation of the released juice from the sample on a fritted glass disk and separating the released juice from tissue by centrifugation into a smaller graduated section of the tube (e.g., Wierbicki et al., 1957). A similar tube (butyrometer) with a layer of glass beads instead of the fritted glass disk was used by Ender and Pfeiffer (1975). In the method of Penny (1975), the meat is placed on perforated Teflon disks, covered with pieces of fine nylon mesh and placed at the bottom of the centrifuge tube, which is partially filled with glass beads (Gumpen and Martens, 1977). The juice released by centrifugation can also be adsorbed by molecular sieves (Dagbjartsson and Solberg, 1972), filter paper (Jauregui et al., 1981), or gypsum (Hofmann et al., 1980), placed on the bottom of the centrifuge tube.

Different CMs with varying speeds (1200–4000 r.p.m.) and duration (8–20 min) using meat and tissue homogenates were found to be accurate and reproducible (Grosse et al., 1979). Highly significant correlations between the results of different CMs and the FPM were found (Janicki and Walczak, 1954; Engelke, 1961; Sonn, 1964; Holtz, 1966; Brendl and Klein, 1971). Goutefongea (1963, 1966) reported that both his modification of the FPM (using 5-g samples) and a CM showed the same variability and that FPM recorded smaller differences in WHC than the CM. Other authors, using smaller samples for the FPM (0.3–0.5 g) came to the conclusion that the CM is more exact and more reproducible (Janicki and Walczak, 1954; Gravert, 1962; Steinhauf et al., 1965). This may be due mainly to the larger size of samples used in the CM. Penny (1975) found a highly significant correlation ($r = .88$) between "spun drip" of frozen, thawed meat measured with his CM and the amount of "free drip" (released without centrifugation).

5. SUCTION METHOD

A newer type of method for the evaluation of WHC of meat is based on the application of capillary forces to the muscle tissue (Hofmann, 1975; Fischer and Hofmann, 1978). A gypsum plate is placed on the surface of the intact tissue with a relatively low pressure (e.g., load of 800 g) for a definite time (30–120 sec). The loosely bound water is sucked up into the porous material by the effect of capillary forces. The air displaced from the capillaries by the meat juice goes into an U-shaped calibrated glass tube containing a colored fluid (Fig. 5). The volume of the displaced air read from the shift of this fluid is equal to the volume of loosely bound water and is inversely proportional to the WHC of the tissue. It is not necessary to adjust the gypsum to a certain humidity. For each measurement a new disk of gypsum has to be used. The thickness of meat

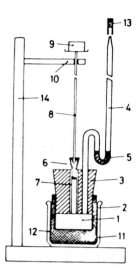

Fig. 5 Diagram of the capillary volumeter. 1, Porous gypsum disk; 2, plastic ring; 3, silicon stopper; 4, calibrated glass tube; 5, colored fluid; 6, stopper; 7, glass tube; 8, rod; 9, weight; 10, holding device; 11, meat sample; 12, weighing jar; 13, closing cap; 14, stand. After Hofmann (1975).

has no influence on the result. However, the volume of loosely bound water sucked up by the porous material depends on the pressure used. This volume increases with an increase in load to ~300 g, then a plateau follows; at loads >1000 g the volume increases again with rising load. In the low-pressure range the capillary forces of the gypsum are probably dominating; between 300 and 1000 g the capillary forces seem to be equal to the WHC and at loads >1000 g meat juice is squeezed out by the high pressure (Fischer *et al.*, 1976). It could be that the plateau of the volume–pressure curve indicates the border between extracellular and intracellular water, as mentioned in Section I,B,1. Therefore, measurements with this technique, using loads >1000 g, might indicate differences in the amounts of extracellular water. Surprisingly, with this so-called capillary volumeter method (CVM), the same values of WHC in terms of volume of loosely bound water are obtained if the tissue is cut either parallel or perpendicular to the fiber direction (Hofmann, 1975).

The CVM is highly reproducible; the coefficient of variation was found to be ~4%. The CVM could be successfully used for the evaluation of pork quality. Fischer *et al.* (1976) used simple equipment based on the same principle. The pressure was applied by a spring, measuring WHC directly on the carcass without removing a sample. A significant correlation between the results of the CVM and those of the CM of Bouton *et al.* (1972) was found. The correlation between the CVM and the FPM (Grau and Hamm, 1957) was smaller but still significant (Fischer *et al.*, 1976).

The CVM is advantageous since no weighing of the sample is required, the data obtained are directly related to WHC, and, also, kinetic measurements can be carried out. It is a disadvantage that the data cannot be related to total water or tissue weight, and that the CVM cannot be applied to studies on minced meat or homogenates (e.g., sausage emulsions). It is sometimes not easy to provide a tight fit of the gypsum disk on the meat surface in order to prevent the uptake of air.

6. DETERMINATION OF COOKING LOSS

Cooking loss is often used as a measure of WHC of meat. The amount of fluid released by cooking of ground or unground muscle can be determined without centrifugation (Ranken, 1973; Lee *et al.*, 1978; Honikel *et al.*, 1981a). About 5 g of the sample (or homogenate) is weighed into a preweighed centrifuge tube. After the tube is covered with a glass marbel, it is placed in a boiling water bath for 20 min. The tube with contents is then allowed to cool and the juice released by cooking is drained off. The cooked meat sample is blotted between two filter papers, then placed back into the tube, which is reweighed to determine the percentage moisture loss during cooking. Tsai and Ockerman (1981) described a quick technique for measuring cooking loss based on microwave heating and compared it with a modified cooking centrifugal technique (Miller *et al.*, 1968) and a FPM with raw meat (Wierbicki and Deatherage, 1959); in general, the results of the three methods were highly correlated. The microwave test does not work well in products containing high levels of salt and phosphate or fats.

In a number of techniques (reviewed by Hamm, 1960, 1972), the juice released by cooking is separated from the meat sample by centrifugation (see also Section I,C,3). However, when these methods are used, usually similar values for cooking loss are obtained as with techniques not employing centrifugation (e.g., with the method mentioned in the previous paragraph). Thus, the time-consuming centrifugation of the cooked sample is not necessary (Hofmann *et al.*, 1980). Tsai and Ockerman (1981) observed that the centrifugal technique of Wierbicki *et al.* (1957), modified by Miller *et al.* (1968) (cooking at 90°C) may be applicable to lean meat; however, when the fat content of the sample increases, some fat will be rendered and lost in the juice. Thus, the total juice loss includes water as well as fat.

Determination of cooking loss (amount of separated jelly) is the usual technique for evaluation of the binding quality of sausage emulsions. For this purpose, the raw batter can be heated in sausage casings (Puolanne and Ruusunen, 1978) or in cans (e.g., 200 g at 110°C). The separated fluid is poured off into a measuring cylinder (Honikel *et al.*, 1980). If no fat (forming a layer on the top of the water phase) or small amounts of juice are released, it is more accurate to drain off the fluid and weigh the residue. Methods for the determination of heat stability of sausage emulsions by heating and centrifugation were proposed by Pohja (1974) and by Nyfeler and Prabucki (1976) (see also Section II,B,2).

D. Factors Affecting Water-Holding Capacity of Meat

1. PH LEVELS

A good example of the importance of protein–protein interactions for the WHC and swelling of muscle tissue according to the scheme of Fig. 1 is the influence of pH on WHC and swelling of meat, either whole or ground (Fig. 6) (Grau *et al.*, 1953; Hamm, 1962b). A loosening of the microstructure and, consequently, an increase of immobilized water is caused by raising the protein net charge by the addition of acid or base (Fig. 7), which results in an increase of interfilament spacing (April *et al.*, 1972). The pH at which the WHC (or swelling) is at a minimum (pH 5.0) corresponds to the isoelectric point (IP) of myosin or actomyosin, which make up the bulk of myofibrillar proteins. At the IP the net charge of a protein is at a minimum; at this pH we should expect a maximum of intermolecular salt linkages between positively and negatively charged groups (Fig. 7); this explains the minimum of WHC at the IP. By measurement of the pH dependence of the amount of negatively and positively charged dyes bound to myofibrils it could be demonstrated that at pH 5 there does indeed exist a maximum interaction between the charged groups of the myofibrillar proteins; in the isoelectric pH range almost no dye ions are bound (Hamm, 1972), although at the IP almost all ionizable groups of the myofibrillar system are charged. In the range of pH 5.0–6.5, which is of particular practical

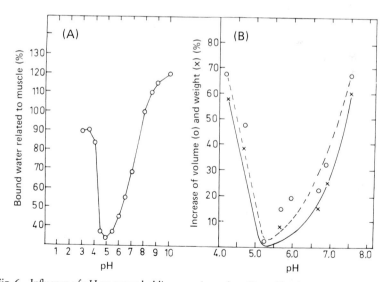

Fig. 6 Influence of pH on water-holding capacity and swelling of beef (post rigor). (A) WHC of a muscle homogenate, measured by the filter-paper press method (Hamm, 1962b). (B) Swelling of muscle cubes (3 mm) (Grau *et al.*, 1953).

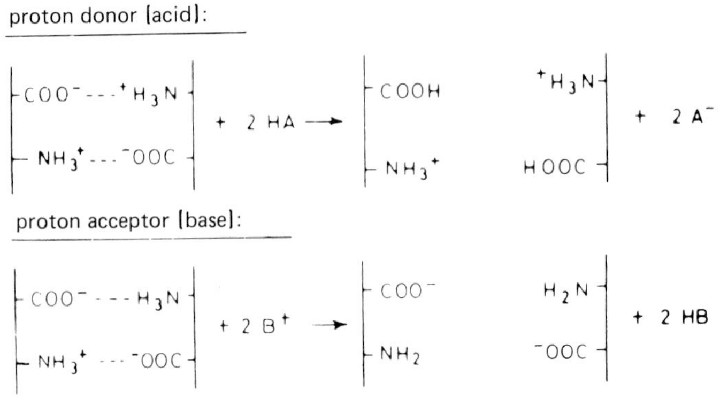

proton donor (acid):

$$\left[\begin{array}{l} COO^- \cdots {}^+H_3N \\ \\ NH_3^+ \cdots {}^-OOC \end{array}\right] + 2\,HA \longrightarrow \left[\begin{array}{ll} COOH & {}^+H_3N \\ \\ NH_3^+ & HOOC \end{array}\right] + 2\,A^-$$

proton acceptor (base):

$$\left[\begin{array}{l} COO^- \cdots H_3N \\ \\ NH_3^+ \cdots {}^-OOC \end{array}\right] + 2\,B^+ \longrightarrow \left[\begin{array}{ll} COO^- & H_2N \\ \\ NH_2 & {}^-OOC \end{array}\right] + 2\,HB$$

Fig. 7   Influence of acid (HA) and base (B$^+$) on the interaction between protein charges.   Left: Isoelectric protein (low WHC) (Hamm, 1975a).

interest, a change of pH has a considerable influence on the WHC that must be due to changes in the state of ionization of histidine and to a lesser extent of glutamic acid residues (Hamm, 1972).   The change in WHC due to changes in pH in this range are completely reversible whereas in the pH range >10 and <4.5 irreversible changes take place (Hamm, 1962c).

A pH dependence of muscle swelling was observed as early as 1925 (Weber). The results presented in Fig. 6 (Grau and Hamm, 1953; Hamm, 1962b) were confirmed for muscle tissue of different species by many authors (reviewed by Hamm, 1960, 1972).   The influence of pH as is shown in Fig. 6 is of practical importance because storage and processing of meat is usually coupled to changes in pH.   Differences in WHC of meat between animals of the same species could be related to pH differences; the WHC increases with rising pH provided that the pH variation is larger than the range 5.5 – 5.8 (Hamm, 1972, 1975a).   An improvement of the water binding of sausage emulsions with increasing pH of the raw material (beef) has been demonstrated (Puolanne and Matikkala, 1980).

However, the WHC of meat from different animals depends not only on muscle pH but also on other factors which are not yet sufficiently identified (Hamm, 1962g; Ranken, 1976); more or less pronounced cold shortening, rate of glycolysis, and perhaps also the content and properties of connective tissue present in muscle might influence WHC.

2. DIVALENT CATIONS

Divalent cations, the most important of which are magnesium and calcium, occur in different fractions of muscle tissue, for example, in the structured material (myofibrillar proteins, sarcoplasmic reticulum, mitochondria, etc.), in

Table I

Magnesium, Calcium, Zinc, and Iron in Postrigor Bovine Muscle[a]

| Total | Bound to or tightly associated with | | | Ions, free or very loosely associated |
	Myofibrillar system	Sarcoplasmic proteins	Sarcoplasmic nonproteins	
1000 Mg	110	20	0	870
100 Ca	60	3	4	33
50 Zn	32	~12	<0.5	6
60 Fe	17	39	4	0

[a] Expressed in μmol per 100 g wet tissue. From Hamm (1959b).

the fraction of sarcoplasmic proteins, bound to nonprotein compounds such as nucleotides, and as free ions that might be attracted by negatively charged protein groups but be easily extractable by water (Table I) (Hamm, 1959b). It was suggested that divalent cations are bound to myofibrils by at least three different types of binding: (1) very tight bonds that are stable at the IP of myosin (pH 5); (2) pH-dependent bonds that are relatively strong at pH 7 but are loosened at falling pH (perhaps complexes of a chelate type); and (3) loose electrostatic binding to negatively charged protein groups; ions bound in this way might be easily exchangeable by other cations and extractable by water (Hamm, 1959b, 1962f). The possibility of cross-linking of adjacent protein molecules by divalent cations cannot be excluded. According to Hamm (1962f) the release of Ca^{2+} and Mg^{2+} ions from the muscle structure within the first 48 hr postmortem, which was also observed by Nakamura (1973) for Ca^{2+}, is caused primarily by the postmortem fall of pH, whereas the amount of divalent cations bound in a pH-independent manner to myofibrils are not changed. The possibility of participation of protein-bound divalent cations in protein–protein interaction after depletion of ATP (e.g., actin–myosin interaction in the state of rigor) has been discussed (Hamm, 1962f). The effects of Ca^{2+} discussed here have to be distinguished from the effects of Ca^{2+} released from the sarcoplasmic reticulum on muscular contraction postmortem and rigor mortis (Hamm, 1979a).

Divalent cations such as Ca^{2+} and Mg^{2+} are supposed to lower the WHC of meat. The binding of cations reduces the electrostatic repulsion between the negatively charged groups by screening effects; therefore, the protein structure is tightened and shrinking occurs. Addition of different acetates to homogenized myofibrils shows this effect is much stronger with the magnesium salt than with the sodium salt (at the same ionic strength) (Hamm, 1962e) (Fig. 8) because Mg^{2+} ions are bound more strongly by the myofibrillar proteins than Na^+ ions (Hamm, 1957a). With increasing concentrations of added acetates

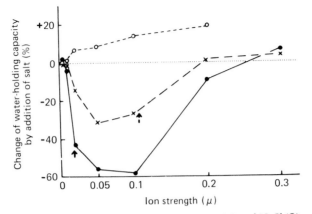

Fig. 8 Influence of magnesium acetate (●), sodium acetate (✕), and NaCl (○) at different ion strengths (μ) on the water-holding capacity (filter-paper press method) of homogenized myofibrils (60% added water, adjusted to pH 6.5). The arrows indicate the Mg²⁺ (left arrow) and the alkali ion concentration (right arrow) in the sarcoplasma (Hamm, 1962d).

the WHC-lowering (shrinking) effect decreases because the swelling effect of the anion (acetate) becomes important. With NaCl (or MgCl₂) no shrinking occurs even at low salt concentration because the WHC-increasing effect of the Cl⁻ ions superimposes the effect of cations on WHC. The opposite effect, that is, an increase of WHC, is observed if divalent cations, which are naturally present in muscle homogenates, are removed either by partial exchange against Na⁺ or K⁺ using cation-exchanging resins (Hamm, 1958b) or by certain chelating agents (Bozler, 1955; Hamm, 1958c). The WHC-lowering effect of divalent cations is at a minimum at the IP of myosin (pH 5) and increases with rising pH (Hamm, 1958b, c) because the strength of cation binding to myofibrillar proteins is increasing (Hamm, 1957a, 1962f). These results suggest that divalent cations present in muscle reduce its WHC and that sequestering of these ions or exchange against monovalent ions increase WHC.

3. POSTMORTEM CHANGES

a. Influence of Early Postmortem Changes on WHC of Unsalted Meat

It has been known for a long time (Hamm, 1958a, 1959a), and also demonstrated more recently (Hamm and Rede, 1972; Goussault, 1978; Currie and Wolfe, 1980; Tsai and Ockerman, 1981), that WHC of beef decreases within the first 48 hr postmortem; in muscles of other species similar postmortem changes of WHC occur but with different rates according to the different rates of rigor development (for references, see Hamm, 1960, 1972, 1975a; Regenstein and Rank Stamm, 1979). However, detailed studies on the dependence of WHC of normal muscles on postmortem metabolism and rigor mortis have not been published until recently (Honikel *et al.*, 1981a,b; Hamm, 1982; Hamm *et al.*, 1984; Kim *et al.*, 1985a,b). It has been shown that the development of rigor mortis in the intact bovine and porcine muscle, which occurs in muscle with normal glycogen content at pH 5.9 and at an ATP level of ~ 1 μmol/g tissue at 20°C, has no significant effect on the WHC of muscle pieces or unsalted muscle homogenates (Honikel *et al.*, 1981a; Kim *et al.*, 1985a). The cooking loss (Fig. 9), as well as the amount of fluid expressible from the unheated tissue (Jolley *et al.*, 1980–1981), increases slightly and continuously with the postmortem fall of pH. The question arises as to why the development of rigor mortis has no significant effect on the WHC of the unsalted beef. During the postmortem drop of pH from 7 to 5.9, at which the rigor occurs, the myofibrillar protein (mainly myosin) is approaching the IP; this leads to a loss of WHC (Szent-Györgyi, 1960) due to a tightening of the myofibrillar system by the formation of "salt" cross-linkages between proteins as explained in Section I,D,1. This pH-dependent type of intermolecular cross-linkage is so strong that additional cross-linking between myosin and actin filaments caused by rigor mortis cannot

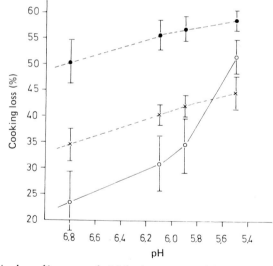

Fig. 9 Cooking loss of intact muscle (×) (bovine sternomandibularis) and of unsalted (●) and salted (○) muscle homogenates in relation to the postmortem pH fall in the intact muscle at temperatures between 0 and 30°C. The homogenates were prepared after the postmortem times indicated (abscissa). The bars indicate the standard deviation of the mean for all incubation temperatures (Honikel *et al.*, 1981b).

exert an additional significant effect on the WHC of meat at the normal pH of meat in absence of salt (Honikel *et al.*, 1981a; Kim *et al.*, 1985a). The cross-linking between thick and thin filaments during development of rigor mortis might reduce the water-imbibing power of muscle by hindering the swelling ability of the thick filaments rather than by the observed decrease in the space between filaments (Goldman *et al.*, 1979; Millman, 1981), the latter being probably more important for drip formation than for other WHC phenomena (see Section I,B,3). This theory, however, needs further investigation.

The influence of conditioning temperature on the postmortem changes in WHC is of practical importance. The results of earlier work on the effect of cold shortening on WHC of beef have been contradictory (Davey and Gilbert, 1974a, 1975; Locker and Daines, 1975, 1976; Powell, 1978), but more recent studies have clarified this situation (Honikel *et al.*, 1980, 1981b; Hamm, 1982; Kim *et al.*, 1985b). The WHC in terms of cooking loss or of the amount of juice expressible from the raw tissue, measured 24 hr postmortem in the intact muscle as well as in the unsalted muscle homogenates, is not significantly influenced by the temperature of conditioning and, therefore, by the rate of postmortem metabolism. Neither cold shortening (prerigor contracture) nor shortening at elevated temperatures (e.g., 30°C) (rigor contracture) exert an effect on WHC of beef 24 hr postmortem (Honikel *et al.*, 1981b; Kim *et al.*, 1985b). Of course, the WHC decreases faster postmortem if the rate of pH fall increases but the relationship between postmortem pH and WHC is independent of the rate of pH decrease.

Prerigor contracture (cold shortening) has no significant effect and rigor contracture at elevated temperatures only a small influence, on the drip loss within the first 24 hr postmortem. However, after a longer period of storage (at 0 to 4°C) higher amounts of drip are released from shortened muscles than from unshortened muscles (Honikel *et al.*, 1980; Kim *et al.*, 1985b) (Fig. 10A). A comparison of Fig. 10A with Fig. 10B shows that the amount of drip loss after 7 days storage increases with the extent of shortening (decrease in sarcomere length). Apparently, muscle contracture does not result in a significant change of muscle volume within 24 to 48 hr postmortem, but then shrinkage occurs that is higher in the more contracted muscles. As is shown below, the increased drip is not due to a lower WHC of the myofibrillar proteins, because cold shortening as well as the storage of cold-shortened muscle for 7 days have no detrimental influence on the binding properties of sausages prepared from such material (Honikel *et al.*, 1980). An explanation of the influence of muscular contracture postmortem on drip formation in terms of the migration of water between intracellular and extracellular spaces and of osmotic effects was given by Honikel *et al.* (1985).

It is interesting that the cooking loss (including the drip loss before heating) of cold-shortened beef measured after 7 days storage at 0°C is about the same as at 24 hr postmortem, and, therefore, also the same as that of nonshortened muscle

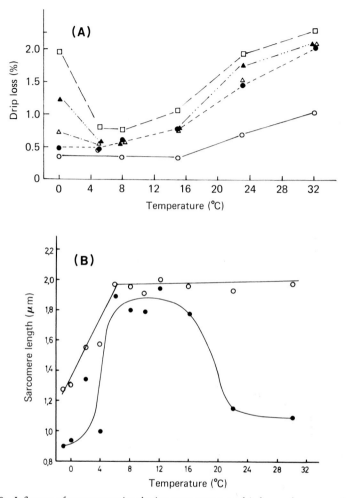

Fig. 10   Influence of postmortem incubation temperature on drip loss and sarcomere length of bovine neck muscles. (A) Drip loss of muscle samples stored at the temperatures indicated for 24 hr postmortem; from the second to the seventh day postmortem the samples were stored at 4°C (Honikel *et al.*, 1980). Postmortem age at testing: 1 day (O), 2 days (●), 3 days (△), 4 days (▲), or 7 days (□). (B) Sarcomere length of muscles incubated for 3 (O) and 24 (●) hr postmortem at different incubation temperatures (Honikel *et al.*, 1981b).

(Honikel *et al.*, 1982). Apparently the effect of pressure, caused by shrinkage of tissue during heating, on the myofibrillar system is not altered during aging and must be much higher than the effect of shrinkage of the raw tissue during storage.

If pieces of prerigor meat are cooked, a more or less extensive contracture, depending on cooking conditions, takes place. The prerigor muscle shortens considerably more during cooking than muscle in which rigor is established (Cia and Marsh, 1976; Ray *et al.*, 1980a) but its cooking loss is less (Paul *et al.*, 1952; Cia and Marsh, 1976; Cross and Tennent, 1980; Ray *et al.*, 1980a,b). Similar results have been obtained with ground beef (Cross *et al.*, 1979). Presumably, both the higher pH and the higher content of ATP of the early postmortem tissue contribute to fluid retention, the two influences being adequate to counteract the pressure that is undoubtedly generated in the tissue by drastic shortening (Cia and Marsh, 1976).

b. Influence of Early Postmortem Changes on WHC of Salted Comminuted Meat

Meat in the prerigor state has a higher WHC and better fat-emulsifying properties than in the rigor or postrigor state. It thus produces sausages of the frankfurter or bologna type with reduced release of moisture and less rendering out of fat when cooked (Hamm, 1960, 1975a, 1978). These superior processing properties are lost during the development of rigor mortis. Unlike the effect of rigor mortis on unsalted beef or pork, development of rigor mortis in the intact muscles exerts a substantial effect on WHC (e.g., cooking loss) of salted muscle homogenates prepared from this material. At the onset of rigor mortis (in beef with normal initial glycogen concentration at pH ~ 5.9) a considerable increase in cooking loss (decrease of WHC) (Honikel *et al.*, 1981a,b; Kim *et al.*, 1985a,b) (Fig. 9) or of the fluid that is expressible from the unheated homogenates (Hamm and Grabowska, 1979; Jolley *et al.*, 1980–1981; Kim *et al.*, 1985a) can be observed. Sausages manufactured from prerigor beef show a significantly lower release of moisture and fat during cooking than those prepared from postrigor beef (Honikel *et al.*, 1980).

Addition of salt to the tissue homogenate (or sausage emulsion) causes an increase in WHC (decrease in cooking loss) that is much more pronounced in homogenates of prerigor muscle than in those of rigor or postrigor muscle (Honikel and Hamm, 1978; Honikel *et al.*, 1980, 1981a,b; Kim *et al.*, 1985a,b) (Fig. 9). It can be demonstrated that the rather similar decrease of WHC of salted and unsalted muscle homogenates (prepared from the intact muscle after different times postmortem) during the prerigor phase (at pH > 5.9) (Fig. 9) is caused by the fall of pH only (ATP breakdown is of minor influence on WHC in this prerigor phase) (Hamm *et al.*, 1980; Honikel *et al.*, 1981a; Jolley *et al.*, 1981) and that at least two-thirds of the substantial loss of WHC of the salted muscle homogenates between pH 6.8–5.5 postmortem must be due to the development of rigor mortis (Hamm, 1956; Honikel *et al.*, 1981a).

What is the reason for this effect of rigor on salted beef? Addition of salt at pH values higher than the IP of myosin causes an increase in WHC and swelling that is related to a shift of the IP to lower pH values. The binding of salt ions increases the electrostatic repulsion between adjacent protein molecules. The resulting loosening of the protein network causes an increase of WHC (decrease of cooking loss) (see Section I,D,6). The formation of interfilamental cross-linkages during rigor will hinder this swelling effect of NaCl. Therefore, the effect of NaCl in increasing WHC of muscle homogenates or sausage emulsions is diminished with progressive development of rigor mortis (Hamm et al., 1980; Honikel et al., 1981a,b; Kim et al., 1985a,b) (Fig. 9).

Rigor mortis exerts an analogous effect on the influence of pH on WHC. The increase in electrostatic repulsion between adjacent protein molecules caused by raising the pH above the IP (see Section I,D,1) is hindered by the formation of stable interfilamental cross-linkages during rigor mortis; therefore, in homogenates prepared from prerigor beef the WHC increases with rising pH more strongly than in those made from rigor or postrigor muscle (Hamm, 1958a, 1959a).

In earlier work it was suggested that the high WHC of prerigor meat is caused by sequestering of divalent cations by ATP and that the decrease in WHC postmortem is due to the shrinkage of tissue by divalent cations (see Section I,D,2) released by the postmortem hydrolysis of ATP (Hamm, 1960, 1962f, 1972). However, it is not the postmortem breakdown of ATP itself but the development of rigor mortis initiated by the ATP depletion that causes the greater part of the postmortem loss of WHC of salted muscle homogenates or sausage emulsions prepared at different times postmortem. This does not exclude the participation of divalent cations in the formation of linkages between actin and myosin filaments at the state of rigor.

As was the case with unsalted meat, the WHC (cooking loss) of salted muscle homogenates is not significantly influenced by cold shortening or by contracture at elevated temperature (30°C) (Honikel et al., 1981b; Jolley et al., 1981; Kim et al., 1985b), but the development of rigor mortis in the intact muscle results in a remarkable decrease of WHC of salted muscle homogenates regardless of the temperature at which the muscle has gone into the state of rigor (Honikel et al., 1981b; Kim et al., 1985b) (Fig. 9).

It can be concluded that longitudinal alterations in muscle fibers as they occur during cold shortening or rigor contracture have much less influence on the WHC of comminuted salted meat than transversal changes, which are caused by interactions between protein groups (e.g., effect of pH or salt), and particularly by the formation of cross-linkages between the myofilaments during development of rigor mortis (Honikel et al., 1981b). It is interesting that the extent of shortening of sarcomeres, that is, the extent of overlapping of the thick and thin filaments, does not influence the WHC of salted muscle homogenates (Honikel et al., 1981b Kim et al., 1985b) or of emulsion-type sausages (Honikel et al., 1980). Apparently the formation of relatively few cross-linkages between actin and myosin filaments suffices to decrease the WHC of comminuted salted tissue after development of rigor mortis; the number of cross-linkages seems not to be of importance. Unlike WHC, the tenderness of meat is strongly influenced by the degree of overlapping of myofilaments.

c. Influence of Salting Meat in the Prerigor State

The high WHC of "hot" (prerigor) meat can be maintained for several days by coarsely grinding the lean prerigor muscle, salting it with 2 to 4% NaCl (or nitrite curing salt), and storing it under refrigeration. From such material, sausages of the frankfurters or bologna type of the same excellent quality can be obtained as from fresh hot meat (Hamm, 1960, 1972, 1973, 1978; Fischer *et al.*, 1982; Hamm and Rede, 1972; Reagan *et al.*, 1981) (Fig. 11). It is interesting that salt addition to the prerigor tissue causes an irreversible increase of WHC although it accelerates the breakdown of ATP (Hamm, 1977a). Similar effects can be obtained by salting poultry meat before onset of rigor mortis (Kijowski *et al.*, 1981a,b).

In order to obtain the beneficial presalting effect, the salt has to penetrate the tissue before the ATP concentration has fallen to a level at which the onset of rigor mortis takes place. Therefore, it is important to salt the hot meat before or immediately after grinding and before cooling the ground material. For the

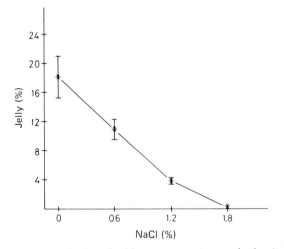

Fig. 11 Release of water (jelly) from frankfurter sausage mixtures after heating in cans at 80°C for 30 min. The lean beef used for the preparation of the sausage emulsions was ground and salted before onset of rigor mortis with different amounts of NaCl and then stored for 24 hr at 4°C. The final NaCl concentration in the mixtures was always the same (1.6%). The bars indicate standard deviation (*n* = 4) (Fischer *et al.*, 1982a).

same reason, a beneficial effect of curing hot-boned ham on the yield is to be expected if the brine penetrates the muscle before onset of rigor mortis.

The presalting effect, that is, the prevention of postmortem loss of WHC by salting prerigor, increases with increasing amounts of added salt up to ~1.8%; higher concentrations apparently do not cause a further significant improvement of emulsion stability (Fischer et al., 1982) (Fig. 11).

The irreversibility of the effect of NaCl on the WHC of prerigor meat is caused by a prevention of rigor mortis in the fibers, fiber fragments or myofilaments. This inhibition of rigor is probably due to a strong repulsion between adjacent protein molecules caused by the combined effects of ATP (see Section I,D,3,b), high tissue pH (see Section I,D,1), and high ionic strength (see Section I,D,6); the result is irreversible changes in the conformation of the myofibrillar protein (Hamm, 1972, 1975b, 1977a, 1981; Fischer and Honikel, 1980; Fischer et al., 1982). Hamm and van Hoof (1974) simulated exactly the same effect by the addition of ATP to comminuted postrigor beef, adjusted to pH 6.9 in the presence of NaCl. In this case the combined effect of ATP, high pH, and NaCl produced a considerable increase of WHC that remained at its high level although ATP was quickly hydrolyzed.

The idea that an irreversible solubilization of myofibrillar proteins causes the presalting effect is probably not correct, because the solubility of myofibrillar proteins in prerigor salted comminuted beef decreases postmortem to a similar extent as in unsalted beef, although the high WHC remains unchanged (Hamm, 1958a; Grabowska and Hamm, 1979). The principle of preventing the postmortem decrease of WHC by salting the comminuted meat in the prerigor state can be applied also to freezing (see Section I,D,4,b) and freeze-drying (Hamm, 1978).

d. Abnormal Postmortem Glycolysis

In the muscles of stress-susceptible pigs an acceleration of postmortem ATP breakdown and glycolysis can occur that results in undesirable pale, soft, and exudative (PSE) pork (Briskey, 1964; Hamm, 1972; Scheper et al., 1979a,b). Because of its low WHC PSE pork should not be used for the production of canned ham or of those emulsion-type sausages that consist mainly of pork (Wirth, 1972).

It is well accepted that the low WHC of PSE porcine muscle is due to the extremely fast drop of pH to ~5.8 or lower within the first hour after death while the temperature in the tissue is still $> 36°C$. High temperature and a low pH cause a partial denaturation of muscle proteins (Ludvigsen, 1954; Wismer-Pedersen and Briskey, 1961; Bendall and Lawrie, 1964; Charpentier, 1969; Goutefongea, 1971). This leads to the exudative and pale appearance (Hamm and Potthast, 1972; Scheper, 1972). It was postulated that denatured sarcoplasmic proteins are precipitated on the surface of the nondenatured myofibrils, which causes a decrease of WHC of the myofibrillar system (Bendall and Wismer-Pedersen, 1962). This view was supported by studies on the postmortem changes in solubility of sarcoplasmic enzymes in normal and PSE muscle (Fischer *et al.*, 1979). Protein denaturation and, therefore, low WHC can be prevented to some extent by high cooling rates (Honikel and Kim, 1985). Honikel and Kim (1985) suggested that changes in the cell membranes rather than protein denaturation are responsible for the high drip loss from PSE muscle.

Watery, pale beef occurring as a result of accelerated postmortem glycolysis in bovine muscles was also observed but to a much smaller extent than PSE pork (Fischer *et al.*, 1977; Hunt and Hedrick, 1977; Fischer and Hamm, 1980). As in the case of PSE pork, the low WHC could be attributed to an abnormally fast postmortem breakdown of ATP and glycogen (Fischer *et al.*, 1977; Fischer and Hamm, 1980), which results in a reduced solubility of sarcoplasmic proteins (Hunt and Hedrick, 1977). The quality deviation of fast-glycolyzing bovine muscle is much less severe than that of PSE pork and, therefore, might not present a serious economic problem (Fischer and Hamm, 1980). Martin and Fredeen (1974) could not find a significant correlation between the rate of postmortem pH fall and WHC of beef but their samples did not include beef with such high rates of postmortem pH drop as were observed by the above authors.

In muscles of stress-susceptible pigs a quality deviation can sometimes be observed that represents just the opposite of PSE muscle, namely, dark, firm, and dry (DFD) pork (Briskey, 1964). Such meat is characterized by high WHC and pH values 24 hours postmortem (pH_{24}) that are significantly higher than the pH_{24} measured in muscles from normal animals (Scheper, 1976). As shown in Section I,D,1, it is to be expected that an elevated pH results in a higher WHC. The use of such DFD pork is advantageous in the production of emulsion-type sausages or of cooked cured products; however, problems arise if it is used for raw cured products or fermented sausages (Wirth, 1976). Serious difficulties are caused by the reduced shelf life because the high pH supports the growth of microorganisms (Bem *et al.*, 1976). The high final pH of DFD pork is due to the fact that the glycogen (and ATP) concentration in such muscles at the time of death is abnormally low (Hamm and Potthast, 1972; Potthast and Hamm, 1976; Scheper *et al.*, 1979a,b). Potthast and Hamm (1976) suggested that both PSE and DFD conditions in the muscles of stress-susceptible pigs are caused by accelerated anaerobic glycolysis; however, in DFD muscles glycogen is metabolized just before death so that the metabolites leave the muscle tissue via the blood stream before interruption of blood circulation (causing low levels of glycogen, lactate, and hydrogen ions immediately after slaughter) whereas in PSE muscles accelerated glycolysis occurs after death (high levels of glycogen immediately after slaughter).

A quality deviation similar to DFD pork is found in beef and called dark-cutting beef; this is also characterized by a high final pH (pH 48 hr after slaughter 5.9 – 7.4) and high WHC (Lawrie, 1958; Hedrick *et al.*, 1959; Hunt and Hedrick, 1977). The main problem in the use of dark-cutting beef is the low microbial stability in comparison with beef normal final pH, whereas the high WHC can be of advantage for the production of emulsion-type sausages. DFD beef was found especially in young bulls (particularly housed untethered), heifers and cows being less often affected (Augustini and Fischer, 1979). As in DFD pork, the level of glycogen in dark-cutting beef soon after death is significantly lower than in normal muscles; therefore, only low amounts of lactic acid can be formed by postmortem glycolysis (Fischer and Hamm, 1980).

e. Electrical Stimulation

Electrical stimulation (ES) of lamb or beef carcasses soon after slaughter is used in order to prevent the detrimental effect of fast cooling on the tenderness of meat caused by cold shortening; if cold shortening is prevented by slow chilling, ES is supposed to improve tenderness and color to some extent (Bendall, 1980; Lawrie, 1981). In about 30 papers not cited here, different minor effects of ES on WHC of meat in terms of expressible juice, cooking loss, drip loss, weight loss, and thaw loss are reported. In many cases ES causes a small but mostly insignificant decrease of WHC, regardless of which form of ES had been applied. Therefore, it is to be expected that ES has also no significant detrimental effect on the functional properties of the myofibrillar system in emulsion-type sausages.

f. Hot-Boning

Hot-boning of carcasses, which was quite common before the era of refrigeration, offers a number of advantages such as facilitation of centralized processing, reduction of cooling space, energy input and chilling time, reduced shrinkage, and improved sanitation and shelf life (Cuthbertson, 1980). Furthermore, hot meat is particularly suitable for the production of emulsion-type sausages (Hamm, 1981). The effect of conditioning temperature and of cold shortening on the WHC of hot-boned meat in the absence or presence of added salt as well as the presalting of meat immediately after hot-boning were mentioned above.

The question arises whether the WHC of hot-boned meat cuts that are cooled under conditions avoiding cold shortening (slow chilling) is different from the WHC of meat from conventionally chilled carcasses. Most authors found that cooking or drip losses from hot-boned beef are the same or even smaller than those of meat from cold-boned beef if the hot-boned cuts were subjected to delayed chilling procedures (Kastner *et al.*, 1973; Follett *et al.*, 1974; Dransfield *et al.*, 1976; Cuthbertson, 1980; Taylor *et al.*, 1980 – 1981). Smulders *et al.* (1981) observed that hot-boning resulted in higher drip losses of beef cuts. Nevertheless, it is generally to be expected that the functional properties of the myofibrillar system of hot-boned and then chilled beef in emulsion-type sausages are about the same as those from conventionally chilled meat.

g. Influence of Aging after Development of Rigor Mortis

Earlier observations, which have shown an increase of WHC (decrease of expressible juice or of cooking loss) during storage of meat after development of rigor mortis (for references, see Hamm, 1972), were confirmed for beef (Brendl and Klein, 1970; El-Badawi et al., 1971; Dumont and Valin, 1973; Goussault, 1978; Anon and Calvelo, 1980) and for poultry (Wardlaw et al., 1973). In several cases no increase of WHC during postrigor storage could be observed (Parrish et al., 1969; Cagle and Henrickson, 1970; Hamm, 1972; Valin et al., 1975). An increase of WHC during aging can be partially explained by an increase of the pH of muscle tissue but might be due also to other effects such as the disintegration of the Z disks by tissue proteases (Davey and Winger, 1979; Ashgar and Pearson, 1980). It is possible that the increased fragility of muscle fibers contributes to a higher WHC of comminuted aged meat. Also, the enzymatic disintegration of certain proteins of the cytoskeleton such as connectin and desmin (Lawrie, 1983) has to be taken into consideration.

The change of drip loss from the intact tissue during postrigor aging follows a different pattern because it increases during storage of beef or pork (e.g., see Joseph and Connolly, 1977; Penny, 1977), indicating a continuous shrinkage of muscle whereby water migrates from the intracellular into the extracellular space (Penny, 1977). Apparently there exist at least two different types of changes during postrigor aging: (1) shrinkage of the muscle tissue resulting in an increase of the amount of extracellular fluid; and (2) an increase of pH and/or proteolytic disintegration of the sarcomeres resulting in an increase in water retention of the intact tissue or improvement of the WHC of comminuted products. However, the small significance of the latter effect would not justify a preference for aged meat for the production of emulsion-type sausages.

Varying temperatures and times of aging (24 or 48 hr at 7°C, then 2°C; 24 or 48 hr at 15°C, then 2°C; 24 hr at 21°C, then 2°C; continuously at 2°C) did not significantly influence the amount of expressible juice and cooking loss of bovine muscles (Parrish et al., 1969).

4. FREEZING AND THAWING

a. Freezing Postrigor

i. Freezing of Water in Muscle. Water in muscle tissue starts to freeze at about $-1°C$; at $-5°C$ ~80% of the freezable water is frozen and at $-30°C$

90% of the tissue water is present in the frozen state (Riedel, 1961; Love, 1966). Freezing under commercial conditions usually begins in extracellular space, increasing the concentration of solids in the extracellular fluid. This draws water osmotically from within the still unfrozen cell, which adds to the growing ice crystals. Slow freezing of tissue results in large ice crystals, which are located entirely in the extracellular areas; there is extensive translocation of water, and fibers in this slowly frozen meat have a shrunken appearance. Rapid freezing, on the other hand, results in numerous small ice crystals located uniformly throughout the specimen. The faster the transition from 0 to $-5\,^{\circ}\mathrm{C}$, the less is the translocation of water during freezing (Fennema, 1966; Love, 1966; Lawrie, 1968; Partmann, 1968, 1973; Goma et al., 1970; Powrie, 1973; Voyle, 1974; Bevilacqua et al., 1979; Nusbaum, 1979; Bevilacqua and Zaritsky, 1980).

ii. Influence of Freezing and Thawing on WHC: General Remarks. Meat should be frozen in a way that losses of water during thawing are avoided as far as possible. The drip loss during thawing (thawing drip) results in an economic disadvantage because of the weight loss and an unpleasant appearance of meat, and because the wet surface promotes bacterial spoilage. There is also a loss of valuable nutrients that are dissolved in the exudate, such as proteins, vitamins, and minerals, as well as a loss of flavor components.

The WHC of meat in terms of drip loss, expressible juice, or cooking loss has often been found to decrease following the process of freezing and thawing (without freezer storage) (Gunar et al., 1973; Locker and Daines, 1973; Anon and Calvelo, 1980; Petrovic et al., 1980; Tsai and Ockerman, 1981; for further references, see Hamm, 1960, 1972). Some authors, however, could not find a significant influence of freezing and thawing (Hamm, 1972; Skenderovic et al., 1975).

Some variation in the effect of freezing and thawing on the WHC of meat can be explained by differences in the WHC of the unfrozen material; the higher the WHC of meat is before freezing (e.g., due to higher pH or longer time of aging) the higher it is usually after defrosting (Hamm, 1960, 1972). The reabsorption of water after thawing is apparently influenced by the WHC of the tissue. Other influences such as the rate of freezing, the rate of thawing, and the effects of storage are discussed below.

iii. Influence of the Rate of Freezing. Differences in the effect of freezing and thawing on the WHC of meat can be due to variation in the velocity of freezing. The rate of freezing has a strong influence on the amount of defrosting drip. It is known that slow freezing results in more thawing drip and less expressible juice than quick freezing (Skenderovic et al., 1975; Rankow and Skenderovic, 1976; Khan and Lentz, 1977; Znamiecki et al., 1977; Nusbaum, 1979; for further references, see Hamm, 1960, 1972). Apparently, the extracellular water formed by melting of the large ice crystals in the extracellular space of slowly frozen meat is not as well reabsorbed by the muscle cell as the water formed by melting of intracellular ice crystals in the fast frozen tissue.

However, this opinion seems not to be in agreement with the fact that some authors could not find a significant influence of freezing velocity on the WHC of thawed meat in terms of drip loss or expressible juice (Brendl and Klein, 1971; Charpentier, 1971; Pap, 1972; Cutting, 1977; Nesterov *et al.*, 1977; for further references, see Hamm, 1960, 1972). An explanation for these conflicting results can be derived from the work of Bevilacqua *et al.* (1979) and Anon and Calvelo (1980). These authors found a rather complicated relationship between the characteristic freezing time (t_c = time necessary for a point to pass from -1 to $-7°C$) and the amount of thaw drip. This relationship is governed by a sequence of changes in location and size of ice crystals with decreasing t_c: intra- and extracellular ice formation (little thaw drip), growth of intracellular ice crystals causing fiber damage (increase in thaw drip), decrease of intracellular ice and increase of extracellular ice formation (some decrease in thaw drip), and exclusive formation of extracellular ice crystals of a constant size (no change in thaw drip). Thus the relationship between t_c and thaw drip shows a maximum of drip formation. Consequently, if measurements are carried out in different ranges of the t_c-thaw drip curve, different results concerning the influence of changes in freezing rate on the amount of thaw drip are obtained (Hamm *et al.*, 1982). These facts explain the contradictory results, which showed different effects of increasing freezing rates on drip loss upon thawing, and it emphasizes the urgent necessity to indicate in publications the exact conditions of freezing in terms of the characteristic freezing time.

It was postulated by many authors that the formation of larger amounts of exudate from slowly frozen meat would be due to a decrease in WHC of meat caused by a denaturation of myofibrillar proteins. However, there seems to exist (at least for red meats) little evidence for an irreversible denaturation of proteins (in terms of changes in solubility, ATPase activity, contractility of the myofibrillar system, free sulfhydryl groups, etc.) as a result of freezing and thawing (without freezer storage) (Sec and Modravy, 1970; Hofmann and Hamm, 1978; Anon and Calvelo, 1980; Kijowski and Niewiarowicz, 1980). The WHC of salted muscle homogenates and the stability of meat emulsions are sensitive indicators for denaturation of the myofibrillar system. Apparently, the freeze–thawing procedure itself has no significant influence on the stability of meat batters, as shown below. Therefore, freezing and thawing seem not to cause an irreversible protein denaturation.

Freezing damage to tissue is often attributed to an increased ion concentration in the unfrozen part of the cell water (Fennema, 1966; Love, 1966; Meryman, 1971). However, this factor may not play a dominant role in the loss of WHC due to freezing and thawing because salting of meat before freezing has no detrimental effect on WHC (Empey and Howard, 1954; Wierbicki *et al.*, 1957; Howard, 1960). Thus, differences in the amount of exudate released during thawing might be due to differences in the size and location (intracellular or extracellular) of ice crystals and to differences in the mechanical damage of muscle fibers (caused by different size and location of ice crystals) rather than to irreversible changes of myofibrillar or other proteins.

iv. Influence of the Rate of Thawing. Thawing must be regarded as a greater potential source of damage than freezing. During thawing the temperature rises rapidly to near the melting point and remains there throughout the long course of thawing, thus allowing considerable opportunity for chemical reactions and recrystallization (Fennema, 1973).

There is no doubt that the rate of thawing of frozen meat may have an important influence on the amount of drip loss but the information in the literature on the relationship between thawing velocity and WHC of the thawed meat does not present a uniform pattern (Singh and Essary, 1971; Ciobanu *et al.*, 1972; Woltersdorf and Schraml, 1974; for further references, see Hamm, 1972). Increase or decrease or no changes in WHC with rising thawing velocity have been found. For an explanation of these contradictory results it is important to consider the possible influence of freezing rate, that is, of the size and location of ice crystals on the effect of thawing rate in the light of the results obtained by Bevilacqua *et al.* (1979) and Anon and Calvelo (1980), although they have not studied the effect of the velocity of thawing.

A relationship between the rate of freezing and the rate of thawing on WHC of meat is to be expected. If fast-frozen meat with its small intracellular ice crystals is quickly thawed, the relatively small droplets of water formed within the cells will be readily reabsorbed; during slow thawing, however, the intracellular crystals will have the opportunity to grow by recrystallization before melting and, if the size of the intracellular columns is large enough, to cause damage to cell membranes, which would result in an increased drip loss. On the other hand, the reabsorption of water formed by thawing of extracellular ice present in slowly frozen meat will decrease with increasing thawing rate: if such meat is quickly defrosted, the water formed will be released as drip loss before it can migrate back into the muscle cells, whereas during slow thawing the gradually formed water will have the opportunity to migrate into the intracellular spaces following the gradient of ion concentration. With meat frozen at a medium rate both effects could compensate each other in a way that the rate of defrosting will not significantly affect the amount of drip loss (Hamm *et al.*, 1982).

These expectations correspond exactly to results obtained by Brendl and Klein (1971). Charpentier (1971) obtained similar results. However, more studies are necessary in order to prove this theory of the effect of thawing rate on the WHC of meat. Perhaps the conditions during thawing by microwaves are different from those discussed here. No evidence for protein denaturation during thawing of meat has been presented as yet.

v. Influence of Freezer Storage. Although meat quality can be lowered at a significant rate during thawing, loss of WHC is sometimes much greater during freezer storage (Fennema, 1973). Many researchers agree that the quantity of thaw exudate (sometimes also of cooking loss) increases with time of storage of meat in the frozen state (Charpentier, 1971; Valin *et al.,* 1971; Rahelic and Pribis, 1974; Skenderovic *et al.,* 1975; Rankow and Skenderovic, 1976; Khan and Lentz, 1977; Miller *et al.,* 1980; Petrovic *et al.,* 1980; for further references, see Hamm, 1972; Powrie, 1973). Other authors, however, did not observe a significant influence of freezer storage on the WHC of meat (Law *et al.,* 1967; Nilsson, 1969; Goma *et al.,* 1970; Hofmann *et al.,* 1971; Dhillon and Maurer, 1975; Nestorov *et al.,* 1977; Orr and Wogar, 1979; Park *et al.,* 1980). Two different reasons for the detrimental influence of storage of meat on thawing drip and other WHC phenomena are hypothesized: recrystallization of ice and protein denaturation. Storage of muscle in the frozen state can be combined with more or less pronounced growth of ice crystals in the extracellular spaces and a concomitant compression of the muscle fibers (Lover, 1966; Partmann, 1973). These changes are reflected by increased amounts of thawing drip. Growth of ice crystals in the extracellular space occurs if the temperature of the frozen tissue is allowed to rise above its eutectic temperature (Voyle, 1974), which is probably somewhere in the range between -40 and $-60°C$. Thus at the temperatures of storage commercially used growth of extracellular ice crystals during storage can occur. It is conceivable that the growth of small intracellular ice crystals, present in quickly frozen meat, during storage exerts a detrimental effect on WHC because of an increasing mechanical damage of cell membranes, whereas the large extracellular ice crystals present in very slowly frozen meat cannot grow further during storage and, therefore, recrystallization effects during storage of such material cannot be of importance for the WHC.

In numerous experiments a denaturation of sarcoplasmic proteins and sometimes also of myofibrillar proteins during freezer storage in terms of changes in solubility, enzyme activities, free sulfhydryl groups, electrophoretic mobility, contractility of the fibers, etc. has been demonstrated (Love, 1966; Partmann, 1968, 1971; Golovkin and Melusova, 1969; Hofmann *et al.,* 1971; Fennema, 1973; Skenderovic *et al.,* 1975; Khan and Lentz, 1977; Hofmann and Hamm, 1978; Kijowski and Nieviarowicz, 1980; Miller *et al.,* 1980; Samejima *et al.,* 1981b). These protein changes, which are accelerated by raising the temperature of storage between -15 and $-3°C$ are attributed to the effect of high ion concentration in the nonfrozen part of tissue water, but also to the effect of free radicals formed by lipid peroxidation and reacting with sulfhydryl groups and other reactive protein groups. Such protein changes during storage, which are supposed to be due to an aggregation rather than to an unfolding of the peptide chains, are much more pronounced in fish muscle than in red meats, changes in poultry meat being intermediate (Partmann, 1968, 1971). On the other hand, freezer storage of meat must not necessarily result in significant changes of individual muscle proteins or of the contractile myofibrillar system (Partmann, 1971; Valin *et al.*, 1971; Rahelic and Pribis, 1974).

Both recrystallization of ice and protein changes might contribute to an increase of thawing drip during storage of meat. However, in the case of red meats the importance of protein denaturation for the detrimental effect of storage on the WHC of meat seems usually to be overestimated. The location and size of ice crystals at the moment of thawing and the rate of thawing seems to be more important for the extent of thawing drip than changes of muscle proteins during storage.

vi. Frozen Meat for Comminuted Meat Products. If recrystallization of ice in the muscle tissue during storage is more important for the WHC of frozen and thawed meat than protein changes, it is to be expected that after the frozen and thawed meat is minced the WHC of the system would be about the same as that before freezing (and storage). This seems to be supported by the results of Skenderovic *et al.* (1975). During storage ($-18°C$) of intact beef muscles for 1 year the thermostability (WHC) of frankfurters manufactured from this meat after different periods of storage did not change although some decrease of the salt-soluble protein fraction was observed. The cooking losses of sausage emulsions prepared from fresh or frozen poultry muscles were about the same (Orr and Wogar, 1979). Miller *et al.* (1980) found a loss of total extractable protein and an increase of thaw exudate during storage ($-18°C$) of beef and pork for 37 weeks; however, the processing loss (difference in the weight of raw and heated and smoked sausage) as well as the cooking loss of the product (consumer cooking loss) of frankfurters produced from the same meat after different periods of storage did not show significant differences. So, the emulsion stability did not change. The authors speculate that an eventual failure of emulsion stability might occur through further storage of the raw material. This is supported by results obtained by Rankow and Skenderovic (1976) which showed that storage of pork at $-20°C$ for 2 years resulted in an increase of the amount of jelly and fat released during cooking of emulsions prepared from this material. Apparently protein changes in frozen meat exert a significant influence on WHC after thawing if the meat is stored longer than 1 year.

b. Freezing Prerigor

Under certain conditions of fast-freezing small carcasses (e.g., lambs) or hot-boned meat cuts it can happen that the muscle is frozen before onset of rigor mortis.   If such prerigor frozen meat is thawed, a severe contracture of muscle fibers occurs.   This thaw rigor or thaw contracture is accompanied by an accelerated breakdown of ATP and glycogen (Fig. 12) (Marsh and Thompson, 1958; Fischer and Honikel, 1980; Honikel and Fischer, 1980).   Thaw contracture takes place in both red and white muscles; it is completely prevented only if the ATP level is as low as 0.1 μmol/g (Davey and Gilbert; 1971a).

Thaw contracture causes not only a considerable toughening of meat but also a strong decrease of WHC (increase of thaw drip, cooking loss, or expressible juice) (Marsh and Thompson, 1958; Fischer and Honikel, 1980; for further references, see Hamm, 1960, 1972).   The drop of WHC during thaw contracture is determined by shortening of sarcomeres rather than by the development of rigor mortis (Fig. 12) (Fischer and Honikel, 1980).   As mentioned above (Section I,D,3,a and b), cold shortening does affect the drip formation during longer storage of meat but not the WHC of myofibrillar proteins; thaw contracture, however, results in loss of WHC of the myofibrillar proteins because it lowers considerably the WHC of salted muscle homogenates and the stability of sausage emulsions (Fischer and Honikel, 1980; Honikel and Fischer, 1980).

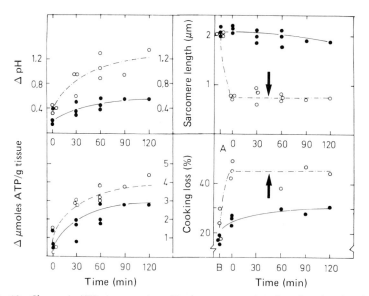

Fig. 12   Changes in ATP concentration, pH value, sarcomere length, and cooking loss (homogenates) after fast thawing ($\leq$ 30 min) of comminuted bovine neck muscle, frozen before onset of rigor mortis. (O), unsalted; (●), salted with 2% NaCl before freezing.   Points at letter A (abscissa) show sarcomere length in the frozen state; points at letter B (abscissa) show cooking loss of homogenate prepared from the frozen tissue without preceeding thawing.   At different times after thawing the homogenates were prepared with 50% added water; they contained always 2% NaCl. The arrows indicate onset of rigor mortis.   After Fischer and Honikel (1980).

Thus the damage to the myofibrillar protein system seems to be much more severe after thaw contracture than after cold shortening.

In order to prevent the detrimental effect of thaw contracture on WHC of meat, the ATP level in the tissue at the beginning of thawing should not be $>0.1 \ \mu mol/g$ (see above). Of course, the level of ATP in the frozen tissue depends on the rate of cooling and freezing and is influenced by the size of carcass or cut. With decreasing rate of freezing the ATP turnover and the rate of ATP depletion increase, and in comminuted meat ATP breakdown and glycolysis during freezing occur faster than in intact muscle (Fischer et al., 1980a). Freezing does stop the postmortem metabolism but only at about $-18°C$ and lower temperatures. Above $-18°C$ increasing temperatures of storage cause an increasing rate of ATP breakdown and glycolysis that is higher in the comminuted meat than in the intact tissue (Fischer et al., 1980b). If the ATP concentration in the frozen tissue falls below $\sim 1 \ \mu mol/g$ no contraction or rigor can occur because they are prevented by the rigid matrix of ice.

The drip formation after fast thawing of prerigor frozen beef is higher than that obtained after very slow thawing. The amount of drip increases during storage of slowly thawed meat and is not significantly higher than the drip obtained after storage of cold-shortened beef (Honikel and Fischer, 1980). During slow thawing of prerigor frozen meat (e.g., from -20 to $-1°C$ within 10 to 12 hr) a slow breakdown of ATP and glycogen takes place. If the critical ATP level of $\sim 1 \ \mu mol/g$ is reached while the meat is still frozen, the muscle fibers cannot contract and rigor cannot occur because these events are prevented by the solid mass of ice. During the subsequent thawing the lack of ATP allows a rigor development without contracture and severe thawing loss (Honikel and Fischer, 1980). For similar reasons it is possible to prevent thaw contracture by storage of the prerigor frozen meat for several days at $-3°C$ (Bendall, 1974) or for at least 20 days at $-12°C$ (Davey and Gilbert, 1976).

During cooling of prerigor beef prior to freezing in the temperature range between $10°C$ and the freezing point an accelerated breakdown of ATP and glycogen occurs because of the cold-shortening phenomenon (see Section I,D,3,a). The longer the meat remains in this temperature range the stronger is the ATP depletion before freezing (Fischer et al., 1980a). This explains the fact that some authors did not observe thaw contracture phenomena (high drip loss) during thawing of prerigor frozen beef and pork cuts (Rahelic and Pribis, 1974; Gorska and Duda, 1981) or broilers (Ristic et al., 1980).

The high WHC of prerigor muscle can be maintained for months by rapid freezing of the comminuted meat in a thin layer before the breakdown of ATP has started. If this material is processed (chopped with added salt and water) prior to thawing, the onset of rigor is prevented by the combined effects of ATP, high pH, and salt (see Section I,D,3,c); therefore, the WHC remains high and sausages of high emulsion stability are obtained (Dimitrijevic et al., 1970; Hamm, 1972, 1978; Honikel and Fischer, 1980).

As mentioned in Section I,D,3,c, the postmortem decrease of WHC of meat can be prevented by salting the comminuted beef in the prerigor state. This principle can be applied also to freezing. Breakdown of ATP postmortem during freezing and storage as well as during thawing of prerigor salted and frozen beef is not accompanied by contracture and rigor development in the fiber fragments (Fig. 12); consequently, no or only little decrease of WHC in the thawed material takes place (Fig. 12), and sausages manufactured from such material show an emulsion stability similar to that of sausages made from fresh hot beef (Hamm, 1972, 1978, 1981; Fischer and Honikel, 1980; Fischer *et al.*, 1980a; Honikel and Fischer, 1980).

5. HEATING

The most drastic changes in meat during heating, such as shrinkage and hardening of tissue and the release of juice, are caused by changes in the muscle proteins. The decrease of WHC (sum of the cooking loss plus the amount of water expressible from the remaining tissue) upon heating does not proceed continuously but shows a characteristic step between 50 and 55°C (Fig. 13) (Hamm and Deatherage, 1960; Hamm, 1966). As studies on muscle proteins, either in the isolated state or *in situ*, showed, the most drastic changes of the myofibrillar proteins occur between 30 and 50°C, almost reaching completion at 60°C (Hamm 1979b; Hamm and Grabowska, 1978). These changes are

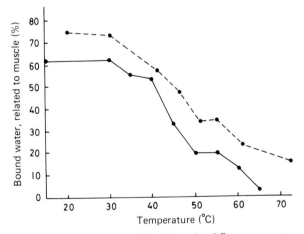

Fig. 13 Water-holding capacity of ground beef heated at different temperatures. After heating, the sample including cooking drip was homogenized with 60% added water and the WHC was measured using the filter-paper press method. Meat heated at its normal pH of 5.5 and then adjusted to pH 5.5 (——); meat heated at its normal pH of 5.5 but not adjusted after heating (– – –) (Hamm and Deatherage, 1960).

characterized by an unfolding of protein molecules, accompanied by an association of molecules and resulting in coagulation (Hamm, 1977b). From these results it can be concluded that in the first phase of decrease in WHC, that is, between 30 and 50°C, changes are due to heat coagulation of the actomyosin system. In the temperature range between 50 and 55°C neglegible changes in WHC occur. The second phase of WHC loss, between 60 and 90°C, seems to be due to denaturation of the collageneous system (shrinking and solubilization of collagen) (Davey and Gilbert, 1974b) and/or to the formation of new stable cross-linkages within the coagulated actomyosin system (Hamm, 1977b). It is not yet clear to what extent actomyosin changes and collagen denaturation contribute to the decrease of WHC in the second phase (temperatures > 55°C).

The increase in the amount of cooking loss from meat of different species with rising temperature was observed by many authors (for references, see Hamm, 1972). The cooking loss rises with increasing time of heating (Hamm and Iwata, 1962) (Fig. 14). During dry heating of beef, a longer time of heating at a lower temperature (250–450 min at 65°C) resulted in a higher cooking loss than shorter heating at a higher temperature (90–120 min at 93°C) (Bramblett and Vail, 1964). Hearne et al. (1978) found that heating rates influenced evaporation and total cooking losses (drip cooking loss plus evaporation cooking loss) of beef; cores heated at the slow rate had greater losses than cores

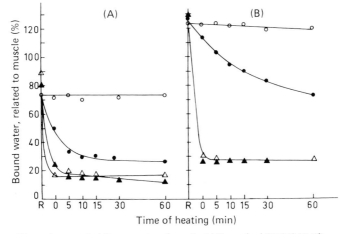

Fig. 14 Change in water-holding capacity of unsalted (A) or salted (B) (2% NaCl) comminuted beef during heating at different temperatures. R, raw tissue (20°C); 0 time at which the temperature indicated was just reached; temperatures: 30°C (○), 50°C (●), 70°C (▲), and 90°C (△). After heating for the period indicated, the samples including the cooking drip were homogenized with 30% added water and the WHC was measured by the filter-paper press method (Hamm and Iwata, 1962).

heated at the fast rate. According to the results obtained by Godsalve *et al.* (1977), who studied dry-cooking of bovine muscle, the decrease of WHC of meat during heating seems to determine the water loss rates. The moisture loss rate possessed two constant rate periods followed by two distinct falling rate periods. The first constant rate period was caused by water vaporizing from a boiling front moving inwardly in the vicinity of the sample surface. The second constant rate period began when the proteins at the interior of the sample started to be denatured by heat. The water released by denaturation flowed to the surface of the sample in the temperature range of $57-67°C$, washed out the effect of the traveling boiling front, and reestablished surface evaporation.

Generally it can be said that weight loss of meat during cooking (frying, etc.) depends on two types of factors: (1) the factors that modify the binding of water in meat; and (2) the factors that determine the migration of water from inside to the outside surface. This migration depends mainly on the shape and geometric characteristics of the sample. Moreover, the influence of pH, state of contraction, and degree of aging have to be considered (Laroche, 1978).

As to the factors that modify meat WHC, relative differences in WHC of raw meat are often reflected in corresponding differences in the amount of cooking loss or in the WHC of meat after mincing and heating. This has been demonstrated for the effect of pH, degree of aging, freezing and thawing, thaw contracture, cooking conditions, salting (see Fig. 14), addition of polyphosphates, etc. (for references, see Hamm, 1972; Laroche, 1978). Generally, it can be postulated that an increase of swelling of the myofibrillar system counteracts to some extent the shrinkage of fibers or fiber fragments during heating because of its tendency to impede heat coagulation of the proteins. From the results mentioned above it can be concluded that heating of the intact or comminuted muscle tissue (without added fat) above $30°C$ before or during chopping will exert a detrimental effect on the stability of sausage batters prepared from such material.

In sausage batters, meat loaves, and patties, in products utilizing chunks of meat, etc., not only the changes in the insoluble myofibrillar proteins upon heating but also the heat-induced gelation of solubilized myofibrillar proteins might be of importance for the WHC of heated products. It has been found that only myosin gives an invariant influence on heat gelation in a model system and that actin exhibits a favorable effect on binding properties only with myosin. As Yasui *et al.* (1979) showed, gelation of myosin occurs during elevation of temperature between 30 and $60°C$. Myosin gel formed by heating has a micronetwork structure. The heat-induced gelation of myosin may be the result of the development of a three-dimensional network structure that holds water in a less immobilized state (Yasui *et al.*, 1979). However, immobilization of water in a heated sausage batter might be due to the heat coagulation of a swollen actomyosin network rather than to the heat gelation of solubilized myosin (Section II,A). Myofibrillar protein was found to gel much better than the sarcoplasmic protein (Hamm and Grabowska, 1978).

6. SALTING

It is well known that the addition of sodium chloride to meat systems causes swelling and an increase of WHC (for references, see Hamm, 1960, 1972). Beside the effect of pH (Section I,D,1), the influence of salt is a typical example of the importance of protein charges and their changes for the WHC of meat. The effect of NaCl on WHC or swelling depends on the pH of the tissue: NaCl increases the WHC at pH > IP and decreases it at pH < IP; around the IP of the myofibrillar system (pH 5), NaCl has no significant influence on WHC (Fig. 15) (Hamm, 1962d). The effect of NaCl is predominantly due to an association of Cl⁻ ions with positively charged groups of myosin or actomyosin. This adsorption of Cl⁻ ions, which results in a shift of the IP to lower pH (Fig. 15), causes a weakening of the interaction between oppositely charged groups at pH > IP (Fig. 16) and, therefore, an increase of swelling and WHC (see Fig. 1A, B) and it causes a weakening of intermolecular repulsive forces at pH < IP (Fig. 16) that results in a shrinkage, that is, loss of WHC (see Fig. 1B, A) (Hamm, 1957a).

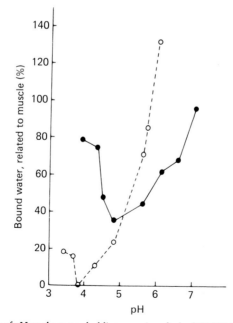

Fig. 15 Influence of pH on the water-holding capacity of salted (O) (2% NaCl) and unsalted (●) comminuted beef (50% added water; filter-paper press method) (Hamm, 1962g).

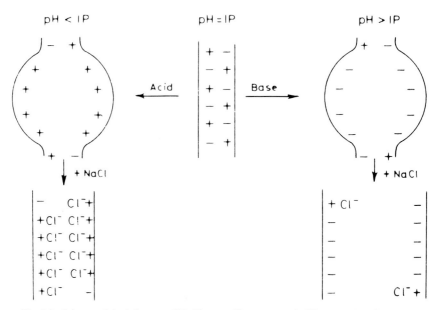

Fig. 16 Scheme of the influence of NaCl on swelling or water-holding capacity of meat at pH values above or below the isoelectric point (Hamm, 1975a).

This pH dependence of the effect of NaCl on WHC governs also the influence of pH on WHC of comminuted meat in the prerigor state (Fig. 9). The strong effect of NaCl on the myofibrillar system at pH $>$ IP is considerably restricted by interfilamental cross-linking taking place during development of rigor mortis (Fig. 9) (Section I,D,3,b). The curves of Fig. 15 explain the fact that the same amount of added NaCl improves water binding and stability of sausage batters more the higher the pH of the system is, and that a decrease of pH in the sausage mixture below the IP during storage of fermented sausages facilitates the release of water.

Ohashi *et al.* (1972) suggested that the addition of NaCl may lead to liberation of bivalent cations (Mg^{2+}, Ca^{2+}) from muscle proteins, and thus also a loosening of the microstructure of the tissue (see Section I,D,2) may take place. It has also been suggested that swelling of the myofilamental system caused by addition of NaCl is due not only to an increase of electrostatic repulsion by Cl^- binding but also to removal of structural constraints (presumably transversal elements such as Z and M lines) (Offer and Trinick, 1983; Offer, 1984).

The opinion that the effect of NaCl and other neutral salts on swelling and WHC of muscle tissue is due to an adsorption of ions to myofibrillar proteins rather than to osmotic effects (Hamm, 1972) is supported by the results of Winger and Pope (1981a,b), who found that muscle samples soaked in NaCl or

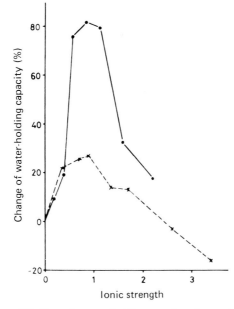

Fig. 17 Influence of NaCl on the water-holding capacity of comminuted beef (filter-paper press method). ×, No added water; ●, 60% added water (Hamm, 1957a).

KCl solutions showed a much stronger swelling than in mannitol solutions of the same osmolality.

For the practice of salting and curing it is important that the maximum WHC is achieved at a NaCl addition corresponding to an ionic strength of 0.8 to 1.0 (Fig. 17), which means ~ 5% NaCl, calculated on the meat; at a water addition of 60% the maximum WHC is reached at a NaCl concentration of 8% on the meat (Hamm, 1957a; Shults *et al.*, 1972; Ranken, 1973; Wierbicki *et al.*, 1976; Grabowska and Hamm, 1979). This is also valid for emulsion-type sausages (Puolanne and Ruusunen, 1980).

A decrease of WHC at high salt concentration might be due to a reduced repelling activity of carboxyl groups of myosin by the effect of cations (Shut, 1976) and/or to the salting-out phenomenon, that is, to the interaction between nonpolar groups that come to the surface of the protein following an salt-induced unfolding of the peptide chains (Eagland, 1975; von Hippel, 1976).

Presalting of meat after onset of rigor mortis, that is, addition of salt to ground rigor or postrigor muscle and storage under refrigeration for about 24 hr, usually does not exert a beneficial influence on WHC of the material and on the stability of sausage batters (Niinivaara and Rynnänen, 1953; Hamm, 1957b, 1958d; Bartels and Gerigk, 1966). Ranken (1973) and Puolanne and Ruusunen (1980) observed a positive effect of presalting on WHC. These authors, however, compared the WHC of the presalted meat with that of meat that was stored in the ground unsalted state for the same time (either without or with added water); storage of unsalted ground meat is not common practice.

The influence of presalting meat in the prerigor state on WHC is described in Section I,D,3,c, whereas freezing of prerigor salted meat is mentioned in Section I,D,4,b. The importance of solubilization of myofibrillar proteins by added salt for WHC and sausage batter stability is mentioned in Section II,A.

7. ADDITION OF POLYPHOSPHATES AND CITRATE

Polyphosphates are widely used to improve the WHC of meat products (Michels, 1968; Mahon et al., 1971; Iles, 1973; Cassidy, 1977; Halliday, 1978; Hamm, 1980). About 400 publications exist on the effect and application of phosphates for meat and meat products. Here only the mechanism of the effect of pyrophosphate (diphosphate, DP) or tripolyphosphate (TP) in combination with NaCl on the WHC of meat and on the stability of sausage emulsions (Brühwurst) are discussed.

Of the different polyphosphates used in food processing, DP and TP increase the WHC of meat most; a plateau of increase in WHC is reached with the addition of $\sim 0.5\%$ polyphosphate (Shults et al., 1972; Shults and Wierbicki, 1973; Wierbicki et al., 1976; Puolanne and Ruusuunen, 1980). The addition of DP to sausage mixtures of the frankfurter type reduces the release of water (jelly) (Fig. 18); this effect is the more pronounced the higher the temperature during chopping or the higher the temperature of sausage cooking (Fig. 18). Both $Na_4P_2O_7$ and $Na_3HP_2O_7$ improve WHC and emulsion stability but the former to a higher extent because it simultaneously causes an increase of pH (see below). $Na_2H_2P_2O_7$ lowers the pH of the meat and, therefore, shows little or no effect on WHC.

In addition to general effects due to the increase of pH and ionic strength, DP affects WHC of meat and solubility of myofibrillar proteins by its interaction with actomyosin (Hamm, 1971, 1972; van den Oord and Wesdrop, 1978). In the absence of DP, in the rigor or postrigor muscle actin and myosin are tightly connected; addition of DP causes a cleavage of bonds between the filaments that results in a loosening of the protein network and, consequently, in an increase of WHC and swelling (Fig. 1A,B). In the presence of NaCl, at the interfaces between protein gel and surrounding fluid the filaments fall apart and protein is solubilized (Fig. 1,B,C) (Hamm, 1971, 1972). Apparently, DP exerts a swelling effect on the myofibrillar system similar to the effect of ATP in prerigor muscle tissue. Proof of the dissociating effect of DP on actomyosin in comminuted salted meat was furnished by rheological measurements (Hamm and Rede, 1972; Hamm, 1975b). Salted homogenates of prerigor muscle showed low viscosity and yield values due to the dissociation of actin and myosin by

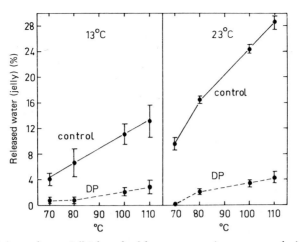

Fig. 18 Release of water (jelly) from frankfurter sausage mixtures prepared with addition of 0.3% diphosphate (DP) (related to meat plus fat) or without DP (control) at 13 or 23°C upon heating at different temperatures (Hamm, 1980).

ATP and the rheological data were not significantly affected by addition of DP; also, the WHC of such homogenates was not substantically changed by DP. After storage of muscle for 24 hr postmortem the salted homogenates prepared from this material showed considerably higher viscosity and yield values than the prerigor muscle homogenates because of the association of actin and myosin filaments after depletion of ATP (rigor mortis). Addition of DP to such homogenates lowered the rheological values to a level similar to the prerigor state and increased the WHC significantly. These results also show that it is not useful to add DP to prerigor meat.

Increasing the pH and/or the NaCl concentration within the limits valid for meat products increases the effect of DP on WHC of meat remarkably. This fact can be explained based on the electrostatical theory of swelling of protein system (Hamm, 1971, 1972). An increase of pH at the basic side of the IP (pH > 5) causes an increase of the negative net charge and, therefore, a repulsion of equally (negatively) charged groups (see Fig. 7). A similar effect is caused by the adsorption of salt ions (Fig. 16). However, this effect of elevated pH and/or addition of NaCl on WHC or swelling of rigor or postrigor muscle tissue is restrained by cross-linkages between actin and myosin filaments (see also Section I,D,3,b). If these linkages are split by the addition of DP, the increase of WHC or swelling by pH elevation and/or NaCl is much stronger than in the absence of DP (Hamm, 1971, 1980).

It has to be realized, however, that cleavage of linkages between actin and myosin by DP exerts little or no effect on WHC or swelling if the electrostatic repulsion between adjacent protein molecules or filaments is low; this is the case at the IP of myosin (pH 5) and in the absence of salt ions. Thus, at pH values close to 5 and in absence of added NaCl DP exerts no influence on WHC, but with increasing pH the effect of DP on WHC increases (Hamm, 1962d). Regenstein and Rank Stamm (1979) found an increase of WHC upon addition of DP with muscle tissue but not with gels of natural actomyosin or with glycerinated muscle fibers. This effect can be explained simply by the fact that the salt ions, which are in the sarcoplasma originally, cause a considerable increase of WHC of the postrigor muscle tissue (Hamm, 1962e) that is intensified by cleavage of cross-linkages in the actomyosin system by DP (see above). The sarcoplasmic salts, however, are not present in natural actomyosin or glycerinated muscle fibers. For this reason, the swelling capacity of the system is low and cleavage of cross-linkages has little or no influence on WHC.

The question arises as to whether the swelling of the insoluble myofibrillar system or a solubilization of the proteins is decisive for the increase of WHC and sausage emulsion stability upon addition of DP. It has long been known that DP causes a strong swelling of ground salted muscle tissue (Bendall, 1954). This seems to be in disagreement with the findings of other authors who observed no swelling upon addition of DP, sometimes even a shrinkage, which was accompanied by a solubilization of myofibrillar proteins (e.g., with 0.3% DP up to 35% of the myofibrillar protein) (see Kotter, 1960; Grabowska and Hamm, 1979). An explanation for this contradiction can be derived from the following results. Addition of DP to salted homogenates of postrigor bovine muscle caused instantaneously strong swelling of meat with very little solubilization of myofibrillar proteins; within a few hours, however, the protein solubilization increased considerably, whereas the swelling decreased to a level below that of the control without DP (Hamm and Neraal, 1977). In an investigation of Hamm and Egginger (1979), 0.3 or 0.5% DP was added to salted (2% NaCl) homogenates of postrigor bovine muscle and the amount of cooking loss was determined. Immediately after addition of DP, the cooking loss was strongly reduced by DP, whereas little myofibrillar protein was solubilized. Sixty min after addition of DP the amount of solubilized protein had increased considerably, but the cooking loss had reached values similar to those of the control (without DP). From these results it can be concluded that for the favorable effect of DP on WHC the increased swelling of the insoluble myofibrillar system rather than the solubilization of protein is decisive (Hamm, 1980).

The release of water (and fat) upon cooking of emulsion-type sausages (Brühwurst) is reduced by tripolyphosphate (TP) to a similar extent as by DP but the conclusion that the mechanism of the function of TP and DP is the same is not correct. An increase in WHC of comminuted salted muscle after addition of TP occurs only to the extent to which TP is split to DP by tissue tripolyphosphatase (Fig. 19) (Hamm and Neraal, 1977). In contrast to DP, TP apparently does not cause a dissociation of the actomyosin system. This observation is in

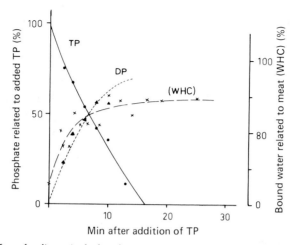

Fig. 19 Effect of sodium tripolyphosphate (TP) on the water-holding capacity of ground beef (postrigor) at different times after addition of TP. The amount of TP decreases and the diphosphate (DP) concentration increases because of the enzymatic breakdown of TP (Hamm and Neraal, 1977).

agreement with results of Yasui *et al.* (1964) obtained with actomyosin solutions. The enzymatic breakdown of added TP to DP in ground meat is usually so fast that after completion of the preparation of the sausage emulsion most of the TP has been transformed to DP, which then affects the WHC of meat as would added DP (Hamm and Neraal, 1977). The enzymatic breakdown of DP to inorganic phosphate (P_i) in the ground meat by tissue diphosphatases takes place much more slowly than the dephosphorylation of TP (Neraal and Hamm, 1977a,b). It is interesting that this breakdown of DP in ground salted meat does not result in a decrease of the high WHC evoked by the addition of DP, although P_i has only little effect on WHC (Hamm and Neraal, 1977). This phenomenon corresponds to the effect of presalting prerigor meat, in which breakdown of ATP does not result in a decrease of WHC (see Section I,D,3,c). The irreversibility of the effect of DP on salted meat might be for the same reason as the presalting effect.

Sodium citrate also improves the WHC of meat systems but to a lesser extent than DP (Hamm, 1972). The mechanism of the function of citrate must be completely different from the DP effect, because citrate is not able to split linkages between actin and myosin filaments. Hamm (1958c) explained the effect of citrate on WHC of meat by a sequestering of bivalent cations. As mentioned in Section I,D,2, Ca^{2+} and Mg^{2+} ions lower the WHC of tissue; thus their binding by citrate anions should result in an increase of WHC. This seems to be still the most reasonable explanation. It should be mentioned, however, that not all compounds that have a sequestering effect on bivalent cations improve the WHC of meat; probably, special configuration of the anion is necessary in order to reach the cations associated with the myofibrillar proteins (Hamm, 1972). It is interesting that citrate improves the WHC of meat in a fat-containing sausage emulsion much more than in a muscle homogenate with a similar content of NaCl. For an explanation of this fact more investigation will be necessary.

II. THE IMPORTANCE OF THE MYOFIBRILLAR SYSTEM FOR THE RETENTION OF FAT IN MEAT PRODUCTS OF THE EMULSION TYPE

A. The Importance of Water-Holding Capacity, Protein Solubilization, and Fat Emulsification for the Stability of Sausage Batters

When meat is minced for the production of emulsion type sausages the muscle cell wall (the sarcolemma), which sheathes the myofibrils, is destroyed. The constituent myofilaments of the myofibril are then no longer restrained and the myofibrillar system is thus transformed from one of limited swelling to one of unlimited swelling accompanied by increasing WHC (Section I,B,3). At the surface of contact between the swollen particles and the surrounding fluid (sarcoplasma plus added water) varying amounts of the myofibrillar proteins are dissolved. The extent of solubilization depends on the ionic strength (usually determined by the amount of added NaCl), the pH, the presence of ATP or added polyphosphates, and other factors (Grabowska and Hamm, 1979; Hamm and Egginger, 1979; Hamm and Grabowska, 1979).

Swelling of a protein network means increase of distances between molecules or protein aggregates. The larger the distance, the lower the interactions (e.g., attractive and repulsive forces between molecules). Finally these forces approach zero, that is, the proteins are solubilized (see Fig. 1B,C). Thus swelling and dissolution of a protein system are different states of the same event. There is a continuous transition between both states.

During heating, dissolved myofibrillar proteins form stable gels at much lower concentration than do the sarcoplasmic proteins. The heat gelation of different myofibrillar proteins has been mentioned (Section I,D,5). Most of the factors that increase WHC of meat cause also a better solubility of the myofibrillar proteins and result in an improvement of the sausage quality, particularly in a reduction of the amount of water and fat released during smoking and/or cooking (Hamm and Grabowska, 1980a). For this reason, some authors emphasized the importance of the solubilization of the myosin–actin system in sausage batters for production of sausages with a high capacity to bind water and fat. Dissolved myofibrillar proteins are believed to be solely capable of forming a sufficiently strong network during heating, which immobilizes the water present (Kotter and Prändl, 1956; Kotter and Palitzsch, 1970; Kramlich, 1971; Kotter and Fischer, 1975). Moreover it is suggested that only the dissolved myofibrillar proteins show a high fat-emulsifying capacity by surrounding the fat globules with a thin protein layer; from this fact it was concluded that meat batters are emulsions of the oil-in-water type (meat emulsions) (Saffle, 1968; Kramlich, 1971; Cunningham and Froning, 1972; Sulzbacher, 1973; Schmidt et al., 1981).

The idea that a sausage batter represents a fat-in-water emulsion in which the solubilized myofibrillar proteins act as an efficient emulsifier (e.g., see Hansen, 1960; Saffle, 1968; Evans, 1971; Sulzbacher, 1973) and that solubilized myofibrillar proteins are responsible for water retention after heat coagulation became something like a dogma. This can be derived from the fact that many authors studied the effect of solubilized salt-soluble muscle proteins on the emulsification of vegetable oil in water and on the stability of such emulsions in order to obtain by this way information on the factors affecting the stability of meat batters. However, such model systems are completely different from a sausage batter, that is, a mixture of comminuted meat and fat tissue (see below). A vast quantity of data on the emulsifying capacity and the emulsion stability of such model systems exists but it is questionable whether these data are related to the stability of sausage batters.

Actually, a number of facts speak against the importance of solubilized myofibrillar proteins for the stability of sausage batters and against the concept that meat batters represent oil-in-water emulsions (Brown, 1972; Hamm, 1973, 1981; van den Oord and Visser, 1973; Johnson, 1976; Randall and Voisey, 1977; Theno and Schmidt, 1978; Hamm and Grabowska, 1979, 1980a,b; Pyrcz et al., 1981; Honikel et al., 1982). Hamm (1972, 1973, 1981) suggested that the swelling of the nondissolved myofibrillar protein system rather than the solubilized myofibrillar protein fraction is the decisive factor that determines the sausage quality and that the formation of an emulsion cannot be of great importance. Most of the factors that increase WHC and swelling of comminuted meat also improve the binding of fat in the final product. Under conventional conditions of sausage production, solid or semisolid particles of added fat tissue are mixed with a system of fiber fragments, myofilaments, sarcoplasm, and added water or ice. During heating of the sausage mixture (smoking, cooking) the coagulating network of proteins or filaments surrounds the melting fat particles and the fat cells in such a way that the melted fat cannot coalesce because of mechanical fixation (immobilization) within the meshes of the coagulated protein matrix. The larger these cavities are, that is, the stronger the swelling of the protein system, the more fat can be immobilized, the less coalescence will occur, and the more water can be immobilized (Section I,B,3). Such an effect might be more important for a desirable distribution of fat in the sausage than an actual emulsification of the fat caused through coating fat droplets by layers of dissolved muscle proteins (or added proteins). The WHC or swelling of the insoluble protein matrix rather than the amount of dissolved myofibrillar proteins is decisive for the immobilization ("emulsification") of fat in the sausage (see also Section II,D). It was suggested that the strength of the gelled batter is important in the fat-binding properties of comminuted meat systems (Acton et al., 1983; Ziegler and Acton, 1984).

Girard (1981b) and Girard et al. (1983) showed that during prolonged chopping of the batter a fat particle distribution comparable to that of an emulsion occurs; this agrees with the theory of fat emulsion within a protein sol. However, for shorter periods of chopping, as used in practice, Hamm's theory of fat distribution within a protein gel would be more appropriate.

During the conventional preparation of batters in the chopper the fat is not in the fluid state (Townsend et al., 1968; van den Oord and Visser, 1973; Schut et al., 1978); it is even, at least partially, localized within fat cells (van den Oord and Visser, 1973; Evans and Ranken, 1975; Paardekooper and Tinbergen, 1978; Tinbergen and Olsman, 1979). Therefore, conditions for the formation of a true emulsion are not given. It is to be expected, however, that during thermal treatment of the sausage (hot smoke, cooking, heat preservation) the fat is melted before the protein matrix coagulates. Coalescence of the liquid fat within the period between melting of fat and protein coagulation might be prevented by the relatively high viscosity of the batter. Nevertheless, in this phase of processing it seems to be possible that the fat droplets are surrounded by a layer of dissolved proteins in a similar way as in a true emulsion. Microscopical studies of sausages revealed that indeed fat droplets surrounded by protein layers are present (Hansen, 1960; Helmer and Saffle, 1963; Borchert et al., 1967; Cassens et al., 1977; Theno and Schmidt, 1978; Jones and Mandigo, 1982), but it is still questionable whether this phenomenon is of any importance for the heat stability of sausage batters. It should be mentioned that coalescence inhibitors can be built up also by the concentration of solid particles, perhaps myofilaments, at the interface (Friberg, 1976; Rydhag, 1977).

B. Methods for Measurement of the Emulsion Capacity of Muscle Proteins and of the Emulsion Stability of Meat Products

1. MEASUREMENT OF EMULSION CAPACITY OF MUSCLE PROTEINS

The outlines of all the methods now in use are based upon the system first reported by Swift et al. (1961). A fixed amount of meat homogenate, or of a protein solution of known concentration, is stirred vigorously with a high-speed mixer together with a fixed amount of melted lard or vegetable oil. During mixing, more melted lard or oil is added at a specific rate from a separating funnel. The oil – water emulsion formed becomes increasingly more viscous with increasing amounts of fat and then suddenly changes and the viscosity decreases. The addition of fat is interrupted when the emulsion turns and the added amount of melted lard or oil expressed per unit of protein is called the emulsifying capacity (EC) of the particular protein. Ever since, many authors have reported their work, based on such a model system either with minor changes or more important modifications (for references, see Saffle, 1968; Cunningham and Froning, 1972; Sulzbacher, 1973; Webb, 1974; Johnson, 1976; Schut, 1976, 1978). An important improvement of the procedure was the detection of the endpoint (emulsion collapse) by measurement of the electrical resistance (e.g., Webb et al., 1970; Haq et al., 1972; Porteous, 1979).

Although some knowledge concerning meat emulsions has been derived from such model experiments, it has to be realized that data obtained by this way are of very limited value if they should be transferred to actual sausage production (Brown, 1972; Schut, 1976). The model system uses relatively simple materials, mostly meat protein fractions and melted fat or oil. The various methods used for extraction of the meat proteins may cause great differences in results; moreover, they cannot compare with the complex meat that is used in batters. The emulsion viscosity, which turned out to have a great effect on both emulsion capacity and stability, is only a small part in the model system emulsions compared to sausage batters. Last but not least, the shear force applied in model systems is considerably greater than is achieved in practical equipment. The great shear force combined with the use of melted fat (or oil) and thin protein solutions may result in emulsion systems completely different from those used in actual practice (Schut, 1976). When increasing amounts of melted fat or oil are emulsified in a protein solution in a model system, the total interfacial surface gradually increases. If it exceeds the critical surface area of a particular protein, the protein film no longer exhibits any resistance against the stress imposed and the emulsion will collapse. In the model system of Swift this occurs at a meat protein/fat ratio of 1 : 1000 (Swift and Sulzbacher, 1963). It is obvious that in actual sausage batter formulations this ratio never will be reached.

2. MEASUREMENT OF EMULSION STABILITY

The terminology usually used for model systems is that emulsion capacity (EC) is the amount of oil or melted fat added continuously to the system until the emulsion suddenly breaks down. Stability of an emulsion is usually defined as the time required for a stable emulsion (a specific amount of fat emulsified by a specific amount of protein) to break. From a sausage maker's standpoint stability is important only from the time a batter is made until it is cooked, which is usually only a very few hours (Saffle, 1968). Swift (1965) has stated the importance of estimating both capacity and stability in applying emulsification measurements for the evaluation of meats. He also stated that it could appear advisable to go even further and to determine routinely the stability of heated emulsions as a means of obtaining the most realistic guidance for practical sausage making.

As an example of the methods for determination of emulsion stability (ES) of model system a heat stability test used by Porteous (1979) should be mentioned that is a modification of the method of Carpenter and Saffle (1964). Emulsions were made from solutions of salt-soluble muscle proteins using 80% of the volume of soybean oil required to cause emulsion collapse. A weighed portion of this emulsion was cooked at 78°C in a constant temperature bath for 30 min and then centrifuged at 1000 r.p.m. (210 g). The separated oil was decanted and the weight of the remaining emulsion determined. The emulsion stability factor was expressed as fraction percent. A similar procedure was described by Haq *et al.* (1972).

The conditions existing in such a model system can be hardly used for predicting the stability of sausage batters, as explained in the previous section. It is obvious that the EC of muscle proteins can be evaluated in such model systems only but heat stability tests can be carried out also with meat batters. There is no doubt that such a test is more useful than measurements of the ES of model systems. Sausage batters do not or only to some extent represent actual emulsions (Section II,A). Therefore, the retention of water and fat upon heating of batters should not be called emulsion stability but heat stability or cooking stability. For such studies it is recommended to prepare the experimental batters in a laboratory chopper of ~5 kg content (no use of Waring blendor systems or similar mixers or homogenizers) in which relatively small amounts of batters (~1 kg) can be prepared under conditions that are with regard to effects of temperature, time, and degree of combination similar to commercially used procedures. For the stability test, 200 g of the batter is heated in cans (e.g., 99 × 36 mm) at the desired temperature (e.g., 80 and 110°C) for 35 min. After storage at ~4°C for 24 hr the cans are rewarmed to ~45°C and opened. The separated liquid is poured off into a calibrated cylinder and the volumes of the watery phase and the lipid phase are measured, or both phases are weighed and related to the weight of the batter sample before heating (Pyrcz *et al.*, 1981). Variations of this type of stability test have been described; the samples were heated in test tubes with or without subsequent centrifugation (Meyer *et al.*, 1964; Rongey, 1965; Saffle *et al.*, 1967; Townsend *et al.*, 1968; Pohja, 1974; Johnson *et al.*, 1977; Hermansson, 1980) (see also Section I,C,5).

C. Emulsifying Capacity of Meat Proteins

Meat proteins, like most other proteins, are excellent emulsifiers. As they lower the interfacial tension and are adsorbed in the fat–water interfaces by the orientation of the polar groups towards the water and of nonpolar groups towards the fat, their steric configuration is distorted and may cause unfolding of the molecule to some extent. The result of the strong activity at the interface is that the fat particles are surrounded by a film of protein, giving them a mechanical protection against coalescence. The stability of the oil-in-water emulsions is therefore believed to be influenced primarily by the viscoelasticity of protein film (for references, see Schut, 1976; Lee *et al.*, 1981).

Most of the factors that increase the EC of myofibrillar proteins also improve the heat stability of sausage batters, but they also increase WHC and swelling of the lean meat in the batter. Therefore, it is uncertain if the EC of solubilized myofibrillar proteins plays any role in the stability of batters and in sausage quality. As explained in Section II,A, the state of swelling of the nondissolved myofibrillar system and the viscosity of the batter seem to be more important for the immobilization of fat than the EC of the solubilized myofibrillar protein fraction. Thus the value of EC data of extracted muscle proteins for predicting and interpretation of the stability of sausage batters is still questionable. The same is true for added nonmeat proteins; such proteins (e.g., soy protein) might act as stabilizer by supporting the formation of a heat-coagulating protein matrix, which entraps fat droplets and water, rather than as real emulsifiers. For these reasons the vast literature on the EC of isolated meat proteins is not discussed here.

D. Factors Affecting Fat Retention in Sausage Batters

1. IMPORTANCE OF THE WATER-HOLDING CAPACITY OF LEAN MEAT

With regard to sausage quality, the measurement of heat stability of experimental sausage batters (Section II,B,2) prepared similarly to commercial sausage production provides more reliable information than the EC and ES data obtained with the extracted proteins. It is interesting that almost all factors that increase WHC and swelling of the lean meat also improve the retention or immobilization of fat in sausage batters upon heating. Such factors include increasing pH (Randall, 1977), increasing salt concentration (Honikel *et al.*, 1982), processing of prerigor meat (Froning and Neelakantan, 1971; Radetic *et al.*, 1973; Hamm, 1979b; for further references, see Hamm, 1972; Cunningham and Froning, 1972), and addition of DP (Hamm, 1980; Ambrosiadis and Klettner, 1981). Detrimental influences not only on WHC of meat but also on

fat retention include freezer storage of lean meat (Orr and Wogar, 1979; Miller
et al., 1980) and processing of PSE pork (Gonzales, 1975; Kijowski and Nieviarowicz, 1978). Sausage stability in terms of fat immobilization is closely related
to the water-binding properties of the product (Brown, 1972; Schut, 1976;
Hamm and Grabowska, 1980a).

2. EFFECT OF TEMPERATURE AND TIME OF CHOPPING; INFLUENCE OF LIQUID LIPIDS ON THE COAGULATION OF MUSCLE PROTEINS

In numerous studies particular attention was paid to the effect of comminution and temperature on the stability of meat batters. If the time of chopping is
too short and/or the consistency of the fat is too firm, the fat is not evenly
dispersed; thus the fat particles or the clusters of fat cells are relatively large and
do not fit into the meshes of the myofibrillar protein network of the minced
muscle tissue. Consequently, during thermal processing of the batter, large
drops of fat are formed that might disturb the formation of a coagulated protein
matrix and coalesce. Therefore, a certain degree of fat dispersion seems to be
favorable for cooking stability of the batter and overall quality of the sausage.
The importance of a sufficient fat dispersion during chopping was emphasized
also by Schut (1976) and Cassens and Schmidt (1979).

However, increasing the intensity of comminution does not necessarily cause
an improvement in cooking stability of the batter. It was suggested that rupture of the cells of adipose tissue results in a reduced retention of fat during
heating. This effect seems to depend on the type of fatty tissue (Schut, 1976).
Within the period between melting of fat and protein coagulation during thermal processing the fat probably coalesces more easily if it is released from the
cells by rupture of the cell wall. It could be that the degree of disintegration of
the fat cells depends on chopping temperature, but this problem has not yet
been sufficiently investigated.

These results explain the fact that there exists an optimum time of chopping
with regard to sausage stability (Brown and Toledo, 1975; Paardekooper and
Tinbergen, 1978; Ambrosiadis and Klettner, 1981); this optimum time depends
on different factors such as formulation of the batter, type of fat tissue, temperatures of lean meat, fat, and added water, and preblending of the lean.

Studies on the effect of comminuting have to be carried out at constant
temperature (temperature control, e.g., by addition of liquid nitrogen). During
the process of commercial sausage manufacture, the temperature of batter rises
with increasing time of chopping. The more adipose tissue is added and/or the
harder the fat is, the more the temperature increases after a given period of
chopping.

A rise in temperature up to ~10°C might be an advantage because it facilitates the dispersion of fat particles in the batter. However, at higher temperatures the fat starts to melt; the liquefied fat might disturb the formation of a coherent myofibrillar protein network in which the fat droplets can be immobilized during heat coagulation of the protein matrix. A detrimental effect of liquid fat on protein network formation, which seems to be enhanced by addition of lecithins (Pyrcz *et al.*, 1981; Honikel *et al.*, 1982), could be due to an interaction between the lipids and the nonpolar, hydrophobic groups of the myofibrillar proteins, because the hydrophobic protein groups are involved in the process of heat coagulation of proteins (Hamm, 1977b). Therefore, an association of these groups with lipid molecules could impede the formation of a coherent network of coagulated protein. This assumption finds support in the observation that meat emulsions made with liquid oil appear after heat treatment as an incoherent agglomerate of loose coagulated protein parts (Schut, 1976). The surprising fact that in such systems the oil does not generally separate from the emulsion during thermal processing could be caused by an interaction of the small, finely distributed oil droplets, which are characterized by a tremendous interfacial area, with the hydrophobic side chains of the muscle proteins.

This theory of a disturbance of the protein network formation by liquid fats is supported also by the observation that batters were destabilized not only by increasing the chopping temperature beyond the softening point of the fat but also by incorporating soft fat (e.g., mixtures of fully hydrogenated soy fat with soy oil), which caused a disruption of the protein network and a coalescence of the fat upon cooking. When a very hard fat (e.g., hydrogenated soy fat or beef kidney fat) was added to the batter of 26°C, the batter remained stable, although that fat was not distributed as uniformly as in the batter with a medium hard fat at 26°C, which was instable upon heating (Lee *et al.*, 1981). Therefore, the high instability of batters prepared with the normally used porcine adipose tissue at a chopping temperature of 26°C and the incoherent protein network formed under these conditions cannot be caused by denaturation of the myofibrillar proteins but must be due to the effect of melted fat on protein coagulation.

The optimal range of chopping temperature for normal batter formulations (with pork fat), which should not be exceeded, is ~10–15°C (Radetic *et al.*, 1973; Hermansson, 1980; Girard, 1981a; Ambrosiadis and Klettner, 1981; Lee *et al.*, 1981; for further references, see Schut, 1976) (see also Fig. 18). In order to obtain an optimal degree of comminution without exceeding the optimum temperature, frozen lean meat and frozen adipose tissue can be used, ice can be added instead of water, or the batter can be cooled by other ways, for example, by addition of dry ice. As mentioned above, the adipose tissue should not be comminuted too intensively; on the other hand, the lean meat should be sufficiently minced to develop optimum binding properties. Therefore, it is recommended to chop the lean meat or to put it through a very fine mincer before the fat is added (Ambrosiadis and Klettner, 1981).

3. INFLUENCE OF BATTER FORMULATION

A further factor affecting the cooking stability of the batter is the formulation, that is, its content of lean meat, fatty tissue, added water, and salt. The lean meat, which exists in the batter in various degrees of disintegration ranging from relatively big pieces of muscle fibers via fragments of myofibrils of different sizes to small myofilaments, participates in the immobilization of fat by creating a protein matrix of network character. The highly swollen matrix of filamental material seems to play a particularly important role in determining the consistency of the final product since this material solidifies to a firm gel at heating binding all components (including fat and water) together to form a coherent product (Schut and Brouwer, 1971). This means that a certain ratio of protein matrix volume to fat volume must exist at which the batter stability is optimal. Up to a certain fat/lean ratio (e.g., 1), the lean meat content of the batter exerts little effect on the batter stability and sausage quality (Honikel *et al.*, 1982) but if the fat/lean ratio rises above this limit, water (jelly) and fat are increasingly released with increasing fat content. Therefore, the protein matrix of the comminuted lean meat is able to immobilize relatively large amounts of fat. For the same reason, at a constant concentration of lean meat (e.g., 45%) a relatively large amount of the added water (e.g., 65%) can be replaced by fat before the batter becomes unstable; this is also the case if the amount of salt, related to lean meat plus added water, remains constant (Honikel *et al.*, 1982).

If in a batter with constant contents of lean meat (e.g., 45%) and salt, related to the lean meat (e.g., 4%), the amount of added water is replaced by increasing amounts of fatty tissue, the cooking stability of the batter increases with increasing fat content. Honikel *et al.* (1982) could demonstrate that this phenomenon is mainly due to an increasing salt concentration in the watery phase (lean meat plus added water) because only a little salt might be taken up by the lipid phase. The increasing salt concentration in the lean meat moiety of the batter causes an increase of WHC of the myofibrillar system (Section I,D,6) and, therefore, an increase in the cooking stability of the batter. If the salt concentration in the batter related to lean meat plus added water remains constant, then increasing replacement of water by fat does not result in an increase of batter stability (Honikel *et al.*, 1982).

With regard to normal formulation of sausage batters, the water/fat ratio is apparently more important for cooking stability than the fat/lean meat ratio (Pizza *et al.*, 1978; Honikel *et al.*, 1982). A maximum batter stability exists at a added water/fat ratio of ~ 1 (Morrison *et al.*, 1971; Radetic *et al.*, 1973; Beloussow *et al.*, 1981; Honikel *et al.*, 1982). The formation of a maximum of batter stability may be primarily a function of the WHC of the lean meat because with increasing amounts of water added to the lean meat a maximum of WHC was observed (Schut, 1969; Grabowska and Hamm, 1979). At low ice water concentrations, the lean meat cannot reach its optimal swelling; on the other hand, the addition of ice water causes dilution of the sarcoplasmic ions and, in normal formulations of sausage batters, dilution of the added salt; an increased dilution of the myofibrillar systems also lowers the degree of fiber disintegration during chopping (Hamm, 1972). These factors result in a decrease of WHC and, therefore, in a loss of batter stability.

4. GENERAL CONCLUSIONS

The idea of the importance of solubilized myofibrillar proteins for the stability of sausage batter, and the concept that sausage batters are oil-in-water emulsions, initiated a large number of studies on the solubilization of muscle proteins, on the EC of the dissolved proteins, and on the stability of such emulsions. However, the results obtained by such studies are only of very limited value for an understanding of the conditions that exist in sausage batters, because the meat batters are not real meat emulsions. It is interesting that Schut, an outstanding expert on "meat emulsions" who carried out many experimental studies on muscle protein emulsions as well as on batter systems, finally came to the conclusion that "many factors affect emulsion stability of which the most important are: (1) water-holding capacity of the meat selected which depends on the type and quality of meat, the pH, and the condition of the meat (e.g., time post mortem, fresh or frozen state, etc.); (2) the formulation which involves the ratio meat to added ice water, the ratio of matrix volume (meat + ice water) to fat volume, the ionic strength, and the use of additives; (3) mechanical treatment, which includes the order of addition of the ingredients, the chopping time, disintegration of meat fibers and myofibrils, the chopping temperature, and the degree of disintegration of the fat (interfacial surface); and (4) heat processing, which involves smoking, pasteurization and/or sterilization. . . . it appears that all aspects of meat emulsions production and stability cannot be understood when the WHC and the fat emulsification are treated as two separate items as has frequently been done in the past. . . . since it has been clearly demonstrated (Hamm, 1972) that a relationship exists between the WHC of the meat before and after heat treatment, the general condition, and especially the WHC of the meat is decisive for both water holding and fat holding of the meat emulsion during thermal processing, assuming that the emulsion formulation and the production process are adequate" (Schut, 1976).

After all, it seems to be justified to replace the term "meat emulsion" by the term "meat batter," as was already suggested by Brown (1972).

REFERENCES

Acton, J. L., Ziegler, G. P., and Burg, D. L. (1983). *CRC Crit. Rev. Food Sci. Nutr.* **18**, 99.
Ambrosiadis, J., and Klettner, P.-G. (1981). *Fleischwirtschaft* **61**, 1621.
Anon, M. C., and Calvelo, A. (1980). *Meat Sci.* **4**, 1.
April, E. C., Brandt, P. W., and Elliott, G. F. (1972). *J. Cell. Biol.* **53**, 53.
Ashgar, A., and Pearson, A. M. (1980). *Adv. Food Res.* **26**, 53.
Augustini, C., and Fischer, K. (1979). *Fleischwirtschaft* **59**, 1871.
Autino, C., and Toscano, G. (1970). *Atti Soc. Ital. Sci. Vet.* **24**, 407.
Bartels, H., and Gerigk, K. (1966). *Fleischwirtschaft* **46**, 611.
Belloussow, A. A., Wojakin, M. P., Gorschko, G. P., Kusnezowa, T. W., Plotnikow, W. A., and
    Salawatulin, R. M. (1981). *Proc. Eur. Meet. Meat Res. Workers* **27** (Vol. 1), C 20.
Bem, I., Hechelmann, H., and Leistner, L. (1976). *Fleischwirtschaft* **56**, 985.
Bendall, J. R. (1954). *J. Sci. Food Agric.* **10**, 468.
Bendall, J. R. (1974). *Proced. Meat Ind. Res. Symp.* No. 3, p. 7.1.
Bendall, J. R. (1980). *Dev. Meat Sci.* **1**, 37.
Bendall, J. R., and Lawrie, R. A. (1964). *Fleischwirtschaft* **44**, 416.
Bendall, J. R., and Restall, D. J. (1983). *Meat Sci.* **8**, 93.
Bendall, J. R., and Wismer-Pedersen, J. (1962). *J. Food Sci.* **27**, 144.
Bevilacqua, A. E., and Zaritzky, N. E. (1980). *J. Food Technol.* **15**, 589.
Bevilacqua, A. E., and Zaritzky, N. E., and Calvelo, A. (1979). *J. Food Technol.* **14**, 237.
Blanshard, J. M. V., and Derbyshire, W. (1975). *In* "Water Relations of Food" (R. B. Duckworth,
    ed.), p. 559. Academic Press, New York.
Borchert, L. L., Greaser, M. L., Bard, J. C., Cassens, R. G., and Briskey, E. J. (1967). *J. Food Sci.*
    32, 419.
Bouton, P. E., Harris, P. V., and Shorthose, W. R. (1971). *J. Food Sci.* **36**, 435.
Bouton, P. E., Harris, P. V., and Shorthose, W. R. (1972). *J. Food Sci.* **37**, 351.
Bozler, E. (1953). *J. Gen. Physiol.* **38**, 735.
Bramblett, V. D., and Vail, E. (1964). *Food Technol. (Chicago)* **18**(2), 123.
Brendl, J., and Klein, S. (1970). *Sb. Vys. Sk. Chem.-Technol. Praze, Potraviny* **E28**, 117.
Brendl, J., and Klein, S. (1971). *Dtsch. Lebensm.-Rundsch.* **67**, 353.
Briskey, E. J. (1964). *Adv. Food Res.* **13**, 89.
Brown, D. D. (1972). Ph.D. Thesis, University of Georgia, Athens.
Brown, D. D., and Toledo, R. T. (1975). *J. Food Sci.* **40**, 1061.
Cagle, E. D., and Henrickson, R. L. (1970). *J. Food Sci.* **35**, 260.
Carpenter, J. A., and Saffle, R. L. (1964). *J. Food Sci.* **29**, 774.
Cassens, R. G., and Schmidt, R. (1979). *J. Food Sci.* **44**, 1256.
Cassens, R. G., Schmidt, R., Terrell, R., and Borchert, L. L. (1977). *Res. Rep. — Univ. Wis., Coll.*
    Agric. Life Sci., Res. Div., R2878.
Cassidy, J. P. (1977). *Food Prod. Dev.* **11**, 74.
Charpentier, J. (1969). *Ann. Biol. Anim., Biochim., Biophys.* **9**, 101.
Charpentier, J. (1971). *Rev. Gen. Froid* **62**, 913.
Cia, G., and Marsh, B. B. (1976). *J. Food Sci.* **41**, 1259.
Ciobanu, A., Berescu, V., Christea, S., Nacea, M., Nicolescu, L., Savu, A., and Vasilescu, S. (1972).
    Lucr. Cercet. — Inst. Cercet. Project. Aliment. **9**, 63; *Food Sci. Technol. Abstr.* **6**, 1 S 9 (1974).
Cooke, R., and Kuntz, I. D. (1974). *Annu. Rev. Biophys. Bioeng.* **3**, 95.
Cross, H. R., and Tennent, I. (1980). *J. Food Sci.* **45**, 765.
Cross, H. R., Berry, B. W., and Muse, D. (1979). *J. Food Sci.* **44**, 1432.
Cunningham, F. E., and Froning, G. W. (1972). *Poult. Sci.* **51**, 1715.
Currie, R. W., and Wolfe, F. H. (1980). *Meat Sci.* **4**, 123.

Currie, R. W., and Wolfe, F. H. (1983). *Meat Sci.* **8**, 147.

Cuthbertson, A. (1980). *Dev. Meat Sci.* **1**, 61.

Cutting, C. (1977). *Aust. Refrig. Air Cond. Heat.* **31**(2), 25.

Dagbjartsson, B., and Solberg, M. (1972). *J. Food Sci.* **37**, 499.

Davey, C. L., and Gilbert, K. V. (1974a). *J. Food Technol.* **9**, 51.

Davey, C. L., and Gilbert, K. V. (1974b). *J. Sci. Food Agric.* **25**, 931.

Davey, C. L., and Gilbert, K. V. (1975). *J. Sci. Food Agric.* **26**, 1721.

Davey, C. L., and Gilbert, K. V. (1976). *J. Sci. Food Agric.* **27**, 1085.

Davey, C. L., and Winger, R. J. (1979). *In* "Fibrous Proteins: Scientific, Industrial and Medical Aspects" (D. A. D. Perry and L. K. Creamer, eds.), Vol. 1, p. 97. Academic Press, New York.

Dhillon, A. S., and Maurer, A. J. (1975). *Poult. Sci.* **54**, 1407.

Dimitrijevic, M., Panin, J., and Miljevic, M. (1970). *Tehnol. Mesa* **11**, 341.

Dransfield, E., Brown, A. J., and Rhodes, D. M. (1976). *J. Food Technol.* **11**, 901.

Dumont, B., and Valin, C. (1973). *19th Proc. Eur. Meet. Meat Res. Workers* **19** (Vol. 1), 249.

Eagland, J. (1975). *In* "Water Relations of Food" (R. B. Duckworth, ed.), p. 73. Academic Press, New York.

El-Badawi, A. A., Anglemeier, A. F., and Cain, R. F. (1971). *Alexandria J. Agric. Res.* **19**, 89.

Empey, W. A., and Howard, A. (1954). *Food Preserv.* **14**(2), 33.

Ender, K., and Pfeiffer, H. (1975). *Fleisch* **29**, 136.

Engelke, F. (1961). Ph.D. Dissertation, University of Göttingen.

Ernst, E., and Hazlewood, C. F. (1978). *Inorg. Perspect. Biol. Med.* **2**, 27.

Evans, G. G. (1971). *Sci. Tech. Surv. — Br. Food Manuf. Ind. Res. Assoc.* **71**.

Evans, G. G., and Ranken, M. D. (1975). *J. Food Technol.* **10**, 63.

Fennema, O. R. (1966). *Cryobiology* **3**, 197.

Fennema, O. R. (1973). *In* "Low Temperature Preservation of Foods and Living Matters" (O. R. Fennema, W. D. Powrie, and E. H. Marth, eds.), pp. 101 and 504. Dekker, New York.

Fennema, O. R. (1977). *In* "Food Proteins" (J. R. Whitaker and S. R. Tannenbaum, eds.), p. 50. Avi Publ. Co., Westport, Connecticut.

Fewson, D., and Kirsammer, R. (1960). *Z. Tierphysiol., Tierernaehr. Futtermittelkd.* **15**, 46.

Fewson, D., Kirsammer, R., Seyfried, E., and Werhahn, E. (1964). *Z. Tierphysiol., Tierernaehr. Futtermittelkd.* **19**, 220.

Fischer, C., and Hamm, R. (1980). *Meat Sci.* **4**, 41.

Fischer, C., and Hofmann, K. (1978). *Fleischwirtschaft* **58**, 1763.

Fischer, C., and Honikel, K. O. (1980). *Fleischwirtschaft* **60**, 1703.

Fischer, C., Hofmann, K., and Hamm, R. (1976). *Fleischwirtschaft* **56**, 91.

Fischer, C., Scheper, J., and Hamm, R. (1977). *Fleischwirtschaft* **57**, 1826.

Fischer, C., Hamm, R., and Honikel, K. O. (1979). *Meat Sci.* **3**, 11.

Fischer, C., Honikel, K. O., and Hamm, R. (1980a). *Z. Lebensm.-Unters. -Forsch.* **171**, 105.

Fischer, C., Honikel, K. O., and Hamm, R. (1980b). *Z. Lebensm.-Unters. -Forsch.* **171**, 200.

Fischer, C., Honikel, K. O., and Hamm, R. (1982). *Z. Lebensm.-Unters. -Forsch.* **174**, 447.

Follett, M. J., Norman, G. H., and Ratcliff, P. W. (1974). *J. Food Technol.* **9**, 509.

Friberg, S. (1976). *In* "Food Emulsions" (S. Friberg, ed.), p. 1. Dekker, New York.

Froning, G. W., and Neelakantan, S. (1971). *Poult. Sci.* **50**, 839.

Fukazawa, T., Hashimoto, Y., and Yasui, T. (1961a). *J. Food Sci.* **26**, 541.

Fukazawa, T., Hashimoto, Y., and Yasui, T. (1961b). *J. Food Sci.* **26**, 550.

Fung, B. M. (1977). *Biochim. Biophys. Acta* **497**, 317.

Fung, B. M., Durham, D. L., and Wassil, D. A. (1975). *Biochim. Biophys. Acta* **399**, 191.

Garlid, K. D. (1979). *In* "Cell-Associated Water" (W. Drost-Hansen and J. S. Clegg, eds.), p. 293. Academic Press, New York.

George, A. R., Bendall, J. R., and Jones, R. C. D. (1980). *Meat Sci.* **4**, 51.

Girard, J. P. (1981a). *Sci. Aliments* **1**, 315.

Girard, J. P. (1981b). *Sci. Aliments* **1**, 329.

Girard, J. P., Dantchev, S., and Calderon, F. (1983). *Fleischwirtschaft* **63**, 909.

Godsalve, E. W., Davis, E. A., Gordon, J., and Davis, H. T. (1977). *J. Food Sci.* **42**, 1038.

Goldman, Y. E., Matsubara, I., and Simmons, R. M. (1979). *J. Physiol. (London)* **295**, 80P.

Golovkin, N. R., and Melusova, L. A. (1969). Bull. Int. Froid, Ann. **6**, 303.

Goma, M., Gyönös, K., and Biro G. (1970). *Fleischwirtschaft* **50**, 1681.

Gonzales, R. R. (1975). *Pilipp. J. Vet. Anim. Sci.* **1**, 197; *Food Sci. Technol. Abstr.* **11**, 5 S 909 (1979).

Gorska, I., and Duda, Z. (1981). *In* "Advances in Hot Meat Processing," Lect., IUFoSt Symp. Rydzyna, Poland.

Goussault, B. (1978). *Ind. Aliment. Agric.* **95**, 297.

Goutefongea, R. (1963). *Ann. Zootech.* **12**, 125.

Goutefongea, R. (1966). *Ann. Zootech.* **15**, 291.

Goutefongea, R. (1971). *Ann. Biol. Anim., Biochim., Biophys.* **11**, 233.

Grabowska, J., and Hamm, R. (1979). *Fleischwirtschaft* **59**, 1166.

Grau, R., and Hamm, R. (1953). *Naturwissenschaften* **40**, 29.

Grau, R., and Hamm, R. (1957). *Z. Lebensm.-Unters. -Forsch.* **105**, 446.

Grau, R., Hamm, R., and Baumann, A. (1953). *Biochem. Z.* **325**, 1.

Gravert, H. O. (1962). *Z. Tierz. Zuechtungsbiol.* **78**, 43.

Grosse, F. (1979). *Arch. Tierz.* **22**, 411.

Grosse, F., Otto, E., Guiard, V., Busch, K., and Herrendörfer, G. (1979). *Arch. Tierz.* **22**, 177.

Gumpen, S., and Martens, H. (1977). *Proc. Eur. Meet. Meat Res. Workers* **23**, N 6.

Gunar, E. V., Yakubow, G. Z., Kaminskaya, A. K., and Deberdeneva, Z. A. (1973). *Kholod. Tekh.* No. 11, p. 28; *Chem. Abstr.* **80**, 58529x (1974).

Halliday, D. A. (1978). *Process. Biochem.* July/Aug. 6.

Hamm, R. (1956). *Biochem. Z.* **328**, 309.

Hamm, R. (1957a). *Z. Lebensm.-Unters. -Forsch.* **106**, 281.

Hamm, R. (1957b). *Fleischwirtschaft* **9**, 477.

Hamm, R. (1958a). *Z. Lebensm.-Unters. -Forsch.* **107**, 1.

Hamm, R. (1958b). *Z. Lebensm.-Unters. -Forsch.* **107**, 423.

Hamm, R. (1958c). *Z. Lebensm.-Unters. -Forsch.* **108**, 280.

Hamm, R. (1958d). *Fleischwirtschaft* **10**, 700, 773.

Hamm, R. (1959a). *Z. Lebensm.-Unters. -Forsch.* **109**, 113.

Hamm, R. (1959b). *Z. Lebensm.-Unters. -Forsch.* **110**, 95.

Hamm, R. (1960). *Adv. Food Res.* **10**, 355.

Hamm, R. (1962a). *Z. Lebensm.-Unters. -Forsch.* **116**, 120.

Hamm, R. (1962b). *Z. Lebensm.-Unters. -Forsch.* **116**, 335.

Hamm, R. (1962c). *Z. Lebensm.-Unters. -Forsch.* **116**, 404.

Hamm, R. (1962d). *Z. Lebensm.-Unters. -Forsch.* **116**, 511.

Hamm, R. (1962e). *Z. Lebensm.-Unters. -Forsch.* **117**, 8.

Hamm, R. (1962f). *Z. Lebensm.-Unters. -Forsch.* **117**, 132.

Hamm, R. (1962g). *Z. Lebensm.-Unters. -Forsch.* **117**, 415.

Hamm, R. (1966). *In* "The Physiology and Biochemistry of Muscle as a Food, 1" (E. J. Briskey, R. G. Cassens, and J. C. Trautman, eds.), p. 363. Univ. of Wisconsin Press, Madison.

Hamm, R. (1971). *In* "Phosphates in Food Processing" (J. M. DeMan and P. Melnychin, eds.), p. 65. Avi Publ. Co., Westport, Connecticut.

Hamm, R. (1972). "Koloidchemie des Fleisches." Parey, Berlin.

Hamm, R. (1973). *Fleischwirtschaft* **53**, 73.

Hamm, R. (1975a). *In* "Meat" (D. J. A. Cole and R. A. Lawrie, eds.), p. 321. Butterworth, London.

Hamm, R. (1975b). *J. Text. Stud.* **6**, 281.

Hamm, R. (1977a). *Meat Sci.* **1**, 15.

Hamm, R. (1977b). *In* "Physical, Chemical and Biological Changes in Food Caused by Thermal Processing" (T. Høyem and O. Kvale, eds.), p. 101. Applied Science Publ., London.

Hamm, R. (1977c). *Fleischwirtschaft* **57**, 1502.

Hamm, R. (1978). *Proc. Meat Ind. Res. Conf.* p. 31.

Hamm, R. (1979a). *Fleischwirtschaft* **59**, 393, 561.

Hamm, R. (1979b). *Lebensmitteltechnol.* **12**(4), 19.

Hamm, R. (1980). *Dtsch. Lebensm.-Rundsch.* **76**, 263.

Hamm, R. (1981). *Dev. Meat Sci.* **2**, 93.

Hamm, R. (1982). *Food Technol.* **36**(11), 105.

Hamm, R. (1985). *In* "Proceedings of the Third Symposium on Properties of Water in Relation to Food Quality and Stability."

Hamm, R., and Deatherage, F. E. (1960). *Food Res.* **25**, 587.

Hamm, R., and Egginger, R. (1979). *Fleischwirtschaft* **59**, 1727.

Hamm, R., and Grabowska, J. (1978). *Fleischwirtschaft* **58**, 1345.

Hamm, R., and Grabowska, J. (1979). *Fleischwirtschaft* **55**, 1338.

Hamm, R., and Grabowska, J. (1980a). *Fleischerei* **31**, 930.

Hamm, R., and Grabowska, J. (1980b). *Fleischwirtschaft* **60**, 114.

Hamm, R., and Iwata, H. (1962). *Z. Lebensm.-Unters. -Forsch.* **117**, 20.

Hamm, R., and Neraal, R. (1977). *Z. Lebensm.-Unters. -Forsch.* **164**, 243.

Hamm, R., and Potthast, K. (1972). *Fleischwirtschaft* **52**, 206.

Hamm, R., and Rede, R. (1972). *Fleischwirtschaft* **52**, 331.

Hamm, R., and van Hoof, J. (1974). *Z. Lebensm.-Unters. -Forsch.* **156**, 87.

Hamm, R., Honikel, K. O., Fischer, C., and Hamid, A. (1980). *Fleischwirtschaft* **60**, 1567.

Hamm, R., Gottesmann, P., and Kijowski, J. (1982). *Fleischwirtschaft* **62**, 983.

Hamm, R., Honikel, K. O., and Kim, C. J. (1984). *Fleischwirtschaft* **64**, 1387.

Hansen, L. J. (1960). *Food Technol. (Chicago)* **14**, 565.

Hansen, J. R. (1971). *Biochim. Biophys. Acta* **230**, 482.

Haq, A., Webb, N. B., Whitfield, J. K., and Morrison, O. S. (1972). *J. Food Sci.* **37**, 480.

Hazlewood, C. F., Chang, D. C., Nichols, B. L., and Woessner, D. E. (1974). *Biophys. J.* **14**, 583.

Hearne, L. E., Penfield, M. P., and Goertz, G. E. (1978). *J. Food Sci.* **43**, 10.

Hedrick, H. B., Boillot, J. B., Brady, D. E., and Naumann, H. D. (1959). *Res. Bull. — M., Agr. Exp. Stn.* 717.

Heffron, J. J. A., and Hegarty, P. V. (1974). *Comp. Biochem. Physiol. A* **49A**, 43.

Helmer, R. L., and Saffle, R. L. (1963). *Food Technol.* **17**, 115.

Hermansson, A.-M. (1980). *Proc. Eur. Meet. Meat Res. Workers* **26** (Vol. 2), 134.

Hofmann, K. (1975). *Fleischwirtschaft* **55**, 25.

Hofmann, K. (1977). *Fleischwirtschaft* **57**, 727.

Hofmann, K., and Hamm, R. (1978). *Adv. Food Res.* **24**, 1.

Hofmann, K., Blüchel, E., Baudisch, K., and Klötzer, E. (1971). *Jahresber. Bundesanst. Fleischforsch.* **1**, 94.

Hofmann, K., Jolley, P., and Blüchel, E. (1980). *Fleischwirtschaft* **60**, 1717.

Hofmann, K., Hamm, R., and Blüchel, E. (1982). *Fleischwirtschaft* **62**, 87.

Holtz, A. (1966). Ph.D. Dissertation, University of Göttingen.

Honikel, K. O., and Fischer, C. (1980). *Fleischwirtschaft* **60**, 1709.

Honikel, K. O., and Hamm, R. (1978). *Meat Sci.* **2**, 181.

Honikel, K. O., and Hamm, R. (1983). *Fleischwirtschaft* **63**, 1320.

Honikel, K. O., and Kim, C. J. (1985). *Fleischwirtschaft* **65**, 1125.

Honikel, K. O., Fischer, C., and Hamm, R. (1980). *Fleischwirtschaft* **60**, 1577.

Honikel, K. O., Hamid, A., Fischer, C., and Hamm, R. (1981a). *J. Food Sci.* **46**, 1.

Honikel, K. O., Hamid, A., Fischer, C., and Hamm, R. (1981b). *J. Food Sci.* **46**, 23.

Honikel, K. O., Pyrcz, J., and Hamm, R. (1982). *Fleischwirtschaft* **62**, 507.

Honikel, K. O., Kim, C. J., Roncalés, P., and Hamm, R. (1985). *Meat Sci.* (in press).

Howard, A. (1960). *Proc. Eur. Meet. Meat Res. Workers* **6**.

Hunt, M. L., and Hedrick, H. B. (1977). *J. Food Sci.* **42**, 716.

Iles, N. A. (1973). *Sci. Tech. Surv. — Br. Food Manuf. Ind. Res. Assoc.* **81**.

Janicki, M. A., and Walczak, Z. (1954). *Przem. Rolny i Spozyw.* **8**, 197.

Jauregui, C. A., Regenstein, J. M., and Baker, R. C. (1981). *J. Food Sci.* **46**, 1271.

Johnson, H. R. (1976). *Proc. Meat Ind. Res. Conf.* p. 59.

Johnson, H. R., Aberle, E. D., Forrest, J. C., Haugh, C. C., and Judge, M. D. (1977). *J. Food Sci.* **42**, 522.

Jolley, P. D., Honikel, K. O., and Hamm, R. (1980–1981). *Meat Sci.* **5**, 99.

Jolley, P. D., Honikel, K. O., and Hamm, R. (1981). *Proc. Eur. Meet. Meat Res. Workers* **27** (Vol. 1), 101.

Jones, K. W., and Mandigo, R. W. (1982). *J. Food Sci.* **47**, 1930.

Joseph, R. L., and Connolly, J. (1977). *J. Food Technol.* **12**, 231.

Kastner, C. L., Henrickson, R. L., and Morrison, R. D. (1973). *J. Anim. Sci.* **36**, 484.

Khan, A. W., and Lentz, P. (1977). *Meat Sci.* **1**, 263.

Kim, C. J., Honikel, K. O., and Hamm, R. (1985a). *Fleischwirtschaft* **65**, 489.

Kim, C. J., Honikel, K. O., and Hamm, R. (1985b). *Fleischwirtschaft* **65**, 645.

Kijowski, J., and Niewiarowicz, A. (1978). *J. Food Technol.* **13**, 451.

Kijowski, J., and Niewiarowicz, A. (1980). *Fleischwirtschaft* **60**, 1236.

Kijowski, J., Niewiarowicz, A., and Kujawska-Biernat, B. (1981a). Sympos. *In* "Advances in Hot Meat Processing," Lect., IUFoST Symp. Rydzyna, Poland.

Kijowski, J., Pikul, J., and Niewiarowicz, A. (1981b). *In* "Advances in Hot Meat Processing," Lect., IUFoST Symp. Rydzyna, Poland.

Kotter, L. (1960). "Die Wirkung kondensierter Phosphate und anderer Salze auf tierisches Eiweiss." Verlag Schaper, Hannover.

Kotter, L., and Fischer, A. (1975). *Fleischwirtschaft* **55**, 360.

Kotter, L., and Palitzsch, A. (1970). *Fleischwirtschaft* **50**, 640.

Kotter, L., and Prändl, A. (1956). *Arch. Lebensmittelhyg.* **7**, 219.

Kramlich, W. (1971). *In* "The Science of Meat and Meat Products" (J. P. Price and B. S. Schweigert, eds.), p. 484. Freeman, San Francisco.

Labuza, T. P., and Lewicki, P. O. (1978). *J. Food Sci.* **43**, 1264.

Laroche, M. (1978). *Ann. Technol. Agric.* **27**, 849.

Law, H. M., Yang, S. P., Mullins, A. M., and Fiedler, M. M. (1967). *J. Food Sci.* **32**, 637.

Lawrie, R. A. (1958). *J. Sci. Food Agric.* **9**, 721.

Lawrie, R. A. (1968). *J. Sci. Food Agric.* **19**, 233.

Lawrie, R. A (1981). *In* "Advances in Hot Meat Processing," Lect., IUFoST Symp. Rydzyna, Poland.

Lawrie, R. A. (1983). *Int. J. Biochem.* **15**, 233.

Ledward, D. A. (1983). *Meat Sci.* **9**, 315.

Lee, C. M., Carroll, R. J., and Abdollahi, A (1981). *J. Food Sci.* **46**, 1789.

Lee, Y. B., Rickansrud, D. A., Hagberg, E. C., and Forsyther, R. H. (1978). *J. Food Sci.* **43**, 35.

Ling, G. N., and Walton, C. L. (1976). *Science* **191**, 293.

Locker, R. H., and Daines, G. J. (1973). *J. Sci. Food Agric.* **24**, 127.

Locker, R. H., and Daines, G. J. (1975). *J. Sci. Food Agric.* **26**, 1721.

Locker, R. H., and Daines, G. J. (1976). *J. Sci. Food Agric.* **27**, 193.

Love, R. M. (1966). *In* "Cryobiology" (H. T. Meryman, ed.), p. 313. Academic Press, New York.

Ludvigsen, J. (1954). *Beret. Forsoegslab. (Copenhagen)* p. 272.

Mahon, J. H., Schlamb, K., and Brotsky, E. (1971). *In* "Phosphates in Food Processing" (J. M. DeMan and P. Melnychin, eds.), p. 158. Avi Publ. Co., Westport, Connecticut.

Marsh, B. B., and Thompson, J. F. (1958). *J. Sci. Food Agric.* **9**, 417.

Martin, A. H., and Fredeen, H. T. (1974). *Can. J. Anim. Sci.* **54**, 127.

Meryman, H. (1971). *Cryobiology* **8**, 489.

Meyer, J. A., Brown, W. L., Giltner, M. J., and Guinn, J. R. (1964). *Food Technol. (Chicago)* **23**, 103.

Michels, P. (1968). "Verbindungen der Phosphorsäure in Lebensmitteln." Selbstverlag, Braunschweig.

Miller, A. J., Ackerman, S. A., and Palumbo, S. A. (1980). *J. Food Sci.* **45**, 1466.

Miller, W. O., Saffle, R. L., and Zirkle, S. B. (1968). *Food Technol. (Chicago)* **22**, 1139.

Millman, B. M. (1981). *J. Physiol. (London)* **320**, 118P.

Morrison, G. S., Webb, N. B., Blumer, T. N., Ivey, F. J., and Haq, A. (1971). *J. Food Sci.* **36**, 426.

Nakamura, R. (1973). *J. Food Sci.* **38**, 1113.

Nakayama, T., and Sato, Y. (1971a). *J. Text. Stud.* **2**, 75.

Nakayama, T., and Sato, Y. (1971b). *J. Text. Stud.* **2**, 475.

Neraal, R., and Hamm, R. (1977a). *Z. Lebensm.-Unters. -Forsch.* **163**, 18.

Neraal, R., and Hamm, R. (1977b). *Z. Lebensm.-Unters. -Forsch.* **163**, 123.

Nesterov, N., Grozdanov, A., Dilova, N., Petrova, P., and Tanzikov, M. (1977). *Proc. Eur. Meet. Meat Res. Workers* **23**, D4.

Niinivaara, F. P., and Rynnänen, T. (1953). *Fleischwirtschaft* **5**, 261.

Nilsson, R. (1969). *Proc. Eur. Meet. Meat Res. Workers* **15**.

Nusbaum, R. P. (1979). *Proc. Annu. Reciprocal Meat Conf.* **32**, 23.

Nyfeler, B., and Prabucki, A. L. (1976). *Schweiz. Landwirtsch. Monatsh.* **54**, 197.

Offer, G. (1984). *Meat Sci.* **10**, 155.

Offer, G., and Trinick, J. (1983). *Meat Sci.* **8**, 245.

Ohashi, R., Sera, H., and Ando, N. (1972). *Bull. Fac. Agric. Miyazaki Univ.* **19**, 261; *Food Sci. Technol. Abstr.* **5**, 9 S 1061 (1973).

Orr, H. L., and Wogar, W. G. (1979). *Poult. Sci.* **58**, 577.

Paardekooper, J. C., and Tinbergen, B. J. (1978). *Proc. Eur. Meet. Meat Res. Workers* **24** (Vol. 2), F1.

Pap, L. (1972). *Acta Aliment.* **1**, 371.

Park, S. W., Kang, T. S., Min, B. Y., Sun, K., and Yang, R. (1980). *Korean J. Food Sci. Technol.* **12** (1), 34; *Food Sci. Technol. Abstr.* **13**, 3 S 134 (1981).

Parrish, F. C., Rust, R. E., Popenhagen, G. R., and Miner, B. M. (1969). *J. Anim. Sci.* **29**, 398.

Partmann, W. (1968). *Fleischwirtschaft* **48**, 1317.

Partmann, W. (1971). *J. Text. Stud.* **2**, 328.

Partmann, W. (1973). *Fleischwirtschaft* **53**, 65.

Paul, P., Bratzler, L. J., Farwell, E. D., and Knight, K. (1952). *Food Res.* **17**, 504.

Penny, I. F. (1975). *J. Sci. Food Agric.* **26**, 1593.

Penny, I. F. (1977). *J. Sci. Food Agric.* **28**, 329.

Peschel, G., and Belouschek, F. (1979). *In* "Cell-Associated Water" (W. Drost-Hansen and J. S. Clegg, eds.), p. 3. Academic Press, New York.

Petrovic, L., Svrzic, G., Keleman-Masic, D., and Ibrocic, D. (1980). *Tehnol. Mesa* **21**, 102.

Pizza, A., Pedrielli, R., and Tagliavini, A. (1978). *Ind. Conserve* **53**, 17; *Food Sci. Technol. Abstr.* **10**, 11 S 1616 (1978).

Pohja, M. S. (1974). *Fleischwirtschaft* **54**, 1984.

Porteous, J. D. (1979). *Can. Inst. Food Sci. Technol. J.* **12**, 145.

Potthast, K., and Hamm, R. (1976). *Fleischwirtschaft* **56**, 978.

Powell, V. H. (1978). *Proc. Eur. Meet. Meat Res. Workers* **24** (Vol. 1) D 1.

Powrie, W. D. (1973). *In* "Low Temperature Preservation of Foods and Living Matters" (O. R. Fennema, W. D. Powrie, and E. H. Marth, eds.), p. 282. Dekker, New York.

Puolanne, E., and Matikkala, M. (1980). *Fleischwirtschaft* **60**, 1233.

Puolanne, E., and Ruusunen, M. (1978). *Fleischwirtschaft* **58**, 1543.

Puolanne, E., and Ruusunen, M. (1980). *Fleischwirtschaft* **60**, 1359.

Pyrcz, J., Honikel, K. O., and Hamm, R. (1981). *Fleischwirtschaft* **61**, 1875.

Radetic, P., Ristic, D., and Vuckovic, D. (1973). *Tehnol. Mesa* **14**, 291.

Rahelic, S., and Pribis, V. (1974). *Fleischwirtschaft* **54**, 491.

Randall, J. C. (1977). *Can Inst. Food Sci. Technol. J.* **10**, 147.

Randall, J. C., and Voisey, P. E. (1977). *Can. Inst. Food Sci. Technol. J.* **10**, 88.

Ranken, M. D. (1973). *Proc. Eur. Meet. Meat Res. Workers* **19** (Vol. 3), 1151.

Ranken, M. D. (1976). *Chem. Ind. (London)* **18**, 1052.

Rankow, M., and Skenderovic, B. (1976). *Pro. Eur. Meet. Meat Res. Workers* **22** (Vol. 1), D 7:1.

Ray, E. E., Stiffler, D. M., and Berry, B. W. (1980a). *Proc. Eur. Meet. Meat Res. Workers* **26** (Vol. 2), 26.

Ray, E. E., Stiffler, D. M., and Berry, B. W. (1980b). *J. Food Sci.* **45**, 769.

Reagan, J. O., Pirkle, S. C., Campion, D. R., and Carpenter, J. A. (1981). *J. Food Sci.* **46**, 838.

Regenstein, J. M., and Rank Stamm, J. (1979). *J. Food Biochem.* **3**, 213.

Riedel, L. (1961). *Kaeltetechnik* **13**, 122.

Ristic, M., Kijowski, J., and Schön, L. (1980). *Fleischwirtschaft* **60**, 105.

Rome, E. (1967). *J. Mol. Biol.* **27**, 591.

Rome, E. (1968). *J. Mol. Biol.* **37**, 331.

Rongey, E. H. (1965). *Proc. Meat Ind. Res. Conf.* p. 99.

Rydhag, L. (1977). *In* "Physical, Chemical and Biological Changes in Food Caused by Thermal Processing" (T. Høyem and O. Kvale, eds.), p. 224. Applied Science Publ., London.

Saffle, R. (1968). *Adv. Food Res.* **16**, 105.

Saffle, R. L., Christian, J. A., Carpenter, J. A., and Zirkle, S. B. (1967). *Food Technol. (Chicago)* **21**, 784.

Samejima, K., Ishiorishi, M., and Yasui, Y. (1981a). *J. Food Sci.* **46**, 1412.

Samejima, K., Tahara, S., and Yamamoto, K. (1981b). *J. Coll. Dairy, Nat. Sci. (Ebetsu, Japan)* **8**, 265; *Chem. Abstr.* **94**, 119624m (1981).

Scheper, J. (1972). *Fleischwirtschaft* **52**, 203.

Scheper, J. (1974). *Fleischwirtschaft* **54**, 1934.

Scheper, J. (1975). *Fleischwirtschaft* **55**, 1176.

Scheper, J. (1976). *Fleischwirtschaft* **56**, 970.

Scheper, J., Potthast, K., and Pfleiderer, U. E. (1979a.) *Fleischwirtschaft* **59**, 998.

Scheper, J., Pfleiderer, U. E., and Ehrhardt, S. (1979b). *Fleischwirtschaft* **59**, 1332.

Schmidt, G. R., Mawson, R. F., and Siegel, D. G. (1981). *Food Technol. (Chicago)* **35** (5), 235.

Schut, J. (1969). *Fleischwirtschaft* **49**, 67.

Schut, J. (1976). *In* "Food Emulsions" (S. Friberg, ed.), p. 385. Dekker, New York.

Schut, J. (1978). *Proc. Meat Ind. Res. Conf.* p. 1.

Schut, J., and Brouwer, F. (1971). *Proc. Eur. Meet. Meat Res. Workers* **17**, 775.

Schut, J., Visser, F. M. W., and Brouwer, F. (1978). *Proc. Eur. Meet. Meat Res. Workers* **24** (Vol. 3), W12.

Sec, M., and Modravy, V. (1970). *Cesk. Fysiol.* **19**, 106; *Chem. Abstr.* **74**, 38352g (1971).

Shults, G. W., and Wierbicki, E. (1973). *J. Food Sci.* **38**, 991.

Shults, G. W., Russelt, D. R., and Wierbicki, E. (1972). *J. Food Sci.* **37**, 860.

Singh, S. P., and Essary, E. J. (1971). *Poult. Sci.* **50**, 364.

Skenderovic, B., Rankow, M., and Nevescanin, S. (1975). *Proc. Eur. Meet. Meat Res. Workers* **21**, 227.

Smulders, F. J. M., Eikelenboom, G., and van Logtestijn, J. G. (1981). *Proc. Eur. Meet. Meat Res. Workers* **27** (Vol. 1), 151.

Sonn, H. (1964). Ph.D. Dissertation, University of Göttingen.

Steinhauf, D., Weniger, J. H., and Pahl, G. H. (1965). *Fleischwirtschaft* **45**, 29.

Sulzbacher, W. L. (1973). *J. Sci. Food Agric.* **24**, 589.

Swift, C. E. (1965). *Proc. Meat Ind. Res. Conf.* p. 78.

Swift, C. E., and Sulzbacher, W. L. (1963). *Food Technol. (Chicago)* **17**, 106.

Swift, C. E., Lokett, C., and Fryar, A. J. (1961). *Food Technol. (Chicago)* **15**, 468.

Szent-Györgyi, A. G. (1960). *In* "The Structure and Function of Muscle" (G. H. Bourne, ed.), Vol. 2, p. 1. Academic Press, New York.

Szent-Györgyi, A. (1973). *In* "From Theoretical Physics to Biology" (M. Marois, ed.), p. 113. Karger, Basel.

Tasker, P., Simon, S. E., Johnstone, B. M., Shankly, K. H., and Shwa, F. H. (1959). *J. Gen. Physiol.* **43**, 39.

Taylor, A. A., Shaw, B. D., and MacDougall, D. O. (1980–1981). *Meat Sci.* **5**, 109.

Theno, D. M., and Schmidt, G. R. (1978). *J. Food Sci.* **43**, 845.

Tinbergen, B. J., and Olsman, W. J. (1979). *J. Food Sci.* **44**, 693.

Toscano, P., and Autino, C. (1969). *Atti Soc. Ital. Sci. Vet.* **23**, 514.

Townsend, W. E., Witnauer, L. P., Riloff, J. A., and Swift, C. E. (1968). *Food Technol. (Chicago)* **22**, 319.

Tsai, T. C., and Ockerman, H. W. (1981). *J. Food Sci.* **46**, 697.

Ubertalle, A., and Mazzocco, P. (1979). *Ind. Aliment.* **18**, 533; *Food Sci. Technol. Abstr.* **12**, 5 S 570 (1980).

Valin, C., Goutefongea, R., and Kopp, J. (1971). *Rev. Gen. Froid* **62**, 923.

Valin, C., Palanska, O., and Goutefongea, R. (1975). *Ann. Technol. Agric.* **24**, 47.

Van den Oord, A. H. A., and Visser, I. P. R. (1973). *Fleischwirtschaft* **53**, 1427.

Van den Oord, A. H. A., and Wesdrop, J. (1978). *Proc. 24th Eur. Meet. Meat Res. Workers* **24** (Vol. 1), D 12.

Von Hippel, P. H. (1976). *In* "L'eau et les systèmes biologiques" (H. Alfsen and A. J. Bertraud, eds.), p. 19. CNRS, Paris.

Voyle, C. A. (1974). *Proc. Meat. Res. Inst. Symp.* **5**, 6.1.

Wardlaw, E. B., McKaskill, L. H., and Acton, J. C. (1973). *J. Food Sci.* **38**, 421.

Webb, N. B. (1974). *Proc. Meat Ind. Res. Conf.* p. 1.

Webb, N. B., Ivey, F. Y., Graig, H. B., Jones, V. H., and Monroe, R. J. (1970). *J. Food Sci.* **35**, 601.

Weber, H. H. (1925). *Biochem. Z.* **158**, 443, 473.

Wierbicki, E., and Deatherage, F. E. (1959). *J. Agric. Food Chem.* **6**, 387.

Wierbicki, E., Kunkle, L. E., and Deatherage, F. E. (1957). *Food Technol. (Chicago)* **11**, 69.

Wierbicki, E., Howker, J. J., and Shults, G. W. (1976). *J. Food Sci.* **41**, 1116.

Winger, R. J., and Pope, C. G. (1981a). *Meat Sci.* **5**, 355.

Winger, R. J., and Pope, C. G. (1981b). *Fleischwirtschaft* **61**, 1574.

Wirth, F. (1972). *Fleischwirtschaft* **52**, 212.

Wirth, F. (1976). *Fleischwirtschaft* **56**, 989.

Wismer-Pedersen, J., and Briskey, E. J. (1961). *Food Technol. (Chicago)* **15**, 232.

Woltersdorf, W., and Schraml, G. (1974). *Fleischwirtschaft* **54**, 942.

Yasui, T., Fukazawa, T., Takahashi, K., Sakanishi, M., and Hashimoto, Y. (1964). *J. Agric. Food Chem.* **12**, 399.

Yasui, T., Ishioroshi, M., Nakano, H., and Samejima, K. (1979). *J. Food Sci.* **44**, 1201.

Ziegler, G. R., and Acton, J. L. (1984). *Food Technol.* **38** (5), 77.

Znamiecki, P. M., Mellevitz, P., Sobina, L., and Mielnik, J. (1977). *Zesz. Nauk. Akad. Roln.-Tech. Olsztynie No. 178, p. 5; Food Sci. Technol. Abstr.* **11**, 4 S 675 (1979).

5

Processing and Fabrication

G. R. SCHMIDT

Department of Animal Sciences
Colorado State University
Fort Collins, Colorado

I. CURING

A. Ingredients

Meat curing historically has been defined as the addition of salt (sodium chloride) to meat for the purpose of preservation. The origin of curing meat by salting is lost in history, but it is known that the ancient Greeks and Romans had instructions for preservation of meat by packing it in salt (Binkerd and Kolari, 1975). As the art progressed, additional substances were added to meat for curing purposes. Eventually, meat curing came to be understood as the addition of salt, sodium or potassium nitrate, and, in some instances, other ingredients for the purpose of preserving and flavoring meat. The flavor-protecting properties and color-enhancing properties of the addition of sodium nitrate to meat were probably discovered by accident when this compound was an impurity in the crude rock salt employed in curing. Over the years there have been many changes in the production and marketing of cured meat (Cerveny, 1980).

Studies of the basic chemical reactions involved in producing the color of cured meat and of the role of microorganisms in reducing nitrate to nitrite eventually led to the use of a straight nitrite cure. Bacterial reduction of nitrate was impossible to control. The final nitrite level was quite variable, color

Copyright © 1986 by Academic Press, Inc.
All rights of reproduction in any form reserved.

development was irregular, and meat often spoiled during the curing process. Kerr *et al.* (1926) in the 1920s found that nitrite content of 54 cured meat samples ranged from 2 to 960 ppm (parts per million). A solution to this problem of variability in level of nitrite, with its resultant product variability, was direct addition of a controlled amount of nitrite either to the pickle or directly to the meat to be used for further processing. Permission for direct use of nitrite by packers under federal inspection was granted in January, 1923, by the Bureau of Animal Industry of the United States Department of Agriculture (USDA). In the next 2 years and 8 months experiments were conducted in 17 establishments with a variety of cured meat products. On the basis of these experiments, use of sodium nitrite in meat curing in federally inspected plants was formally authorized by the USDA in 1925 (USDA, 1925).

Ingoing levels of nitrite permitted for various cured meat products are shown in Table I. Finished products could not contain sodium nitrite in excess of 200 ppm. In a mixed cure, nitrate and nitrite were permitted but without a limit as to the amount of nitrate. In 1926, the Bureau of Animal Industry found that meat packers were using more nitrate than was needed for the curing process. A regulation was issued to limit the level of nitrate in pumping pickle to 1% (USDA, 1926).

Early curing practice relied on salt and nitrite penetration from the outside of the piece of meat to the center. Sixty to 70 days were allowed for diffusion to take place. New curing methods were developed and artery pumping of hams reduced the curing time to 1 to 4 days. Mild-cured hams were developed that could either be eaten cold or heated if baked ham was desired. Many sausage items that had been highly spiced and heavily cured for preservation were refined in flavor to meet consumer demands.

Although many meatpackers favored the straight nitrite cure, others believed that a mixed cure, containing both nitrate and nitrite, were more satisfactory. The arguments for a mixed cure were that the residual nitrate served as a constant source for nitrite and that it would be unnecessary to rely on microorganisms to reduce nitrate to nitrite. Accordingly in 1931, the Bureau of Animal Industry ruled that mixed curing solutions containing both nitrate and nitrite should contain not more than 0.25 oz (156 ppm) of nitrite and 2.75 oz (1716 ppm) of nitrate per 100 lb of meat (Tanner, 1944).

The USDA issued a regulation in 1978 that bacon be made with 120 ppm sodium nitrite and 550 ppm ascorbate or isoascorbate and that use of nitrate in pumped bacon be discontinued (USDA, 1978). The purpose of this regulation was to reduce the residual nitrite in processed product and thereby prevent the presence of nitrosamines in pumped bacon.

In its simplest form, the curing reaction that results in the cured meat pigment may be expressed as follows:

$$\text{Myoglobin} + \text{nitrite} \rightarrow \text{nitric oxide myoglobin}$$

Table I
Partial[a] List of Substances Acceptable for Use in the Preparation of Meat Products

Class of substance	Substance	Purpose	Products	Amount
Anticoagulants	Citric acid, sodium citrate	To prevent clotting	Fresh blood of livestock	0.2%—with or without water. When water is used to make a solution of citric acid or sodium citrate added to blood of livestock, not more than 2 parts of water to 1 part of citric acid or sodium citrate shall be used.
Antioxidants and oxygen interceptors	BHA (butylated hydroxyanisole)	To retard rancidity	Dry sausage	0.003% based on total weight; 0.006 percent in combination
	BHT (butylated hydroxytoluene)	To retard rancidity	Dry sausage	0.003% based on total weight
	Propyl gallate	To retard rancidity	Dry sausage	0.003% based on total weight
	TBHQ (tertiary butylhydroquinone)	To retard rancidity	Dry sausage	0.003% based on total weight; 0.006% in combination only with BHA and/or BHT
	BHA (butylated hydroxyanisole)	To retard rancidity	Fresh pork sausage, brown and serve sausage, Italian sausage products, pregrilled beef patties, and fresh sausage made from beef or beef and pork	0.01% based on fat content; 0.02% in combination based on fat content
	BHT (butylated hydroxytoluene)	To retard rancidity		0.01% based on fat content
	Propyl gallate	To retard rancidity		0.01% based on fat content
	TBHQ (tertiary butylhydroquinone)	To retard rancidity		0.01% based on fat content; 0.02% in combination only with BHA and/or BHT based on fat content
	BHA (butylated hydroxyanisole)	To retard rancidity	Dried meats	0.01% based on total weight; 0.01% in combination

(*continued*)

Table I (*Continued*)

Class of substance	Substance	Purpose	Products	Amount
	BHT (butylated hydroxytoluene)	To retard rancidity	Dried meats	0.01% based on total weight
	Propyl gallate	To retard rancidity	Dried meats	0.01% based on total weight
	TBHQ (tertiary butylhydroquinone)	To retard rancidity	Dried meats	0.01% based on total weight; 0.01% in combination only with BHA and/or BHT
Binders	Algin	To extend and stabilize product	Breading mix; sauces	Sufficient for purpose
	Carrageenan	To extend and stabilize product	Breading mix; sauces	Sufficient for purpose
	Carboxymethyl cellulose (cellulose gum)	To extend and stabilize product	Baked pies	Sufficient for purpose
	Enzyme (rennet)-treated calcium-reduced dried skim milk and calcium lactate	To bind and extend product	Sausages	3.5% total finished product (Calcium lactate required at rate of 10% of binder.)
	Enzyme (rennet)-treated sodium casseinate and calcium lactate	To bind and extend product	Imitation sausages; nonspecific loaves; soups, stews	Sufficient for purpose (Calcium lactate required at rate of 10% of binder.)
			Imitation sausages, nonspecific loaves, soups, stews	Sufficient for purpose (Calcium lactate required at rate of 25% of binder.)
	Methyl cellulose	To extend and to stabilize product (also carrier)	Meat and vegetable patties	0.15%
	Isolated soy protein	To bind and extend product	Sausage	2%
			Imitation sausage; nonspecific loaves; soups; stews	Sufficient for purpose

Sodium caseinate	To bind and extend product	Imitation sausage; nonspecific loaves; soups; stews	Sufficient for purpose
Dry or dried whey	To bind or thicken	Sausage, bockwurst	3.5% individually or collectively
Reduced lactose whey	To bind or thicken	Sausage, bockwurst	3.5% individually or collectively
Reduced minerals whey	To bind or thicken	Sausage, bockwurst	3.5% individually or collectively
Whey protein concentrate	To bind or thicken	Sausage, bockwurst	3.5% individually or collectively
Dry or dried whey	To bind or thicken	Imitation sausage; soups; stews; nonspecific loaves	Sufficient for purpose
Reduced lactose whey	To bind or thicken	Imitation sausage; soups; stews; nonspecific loaves	Sufficient for purpose
Reduced minerals whey	To bind or thicken	Imitation sausage; soups; stews; nonspecific loaves	Sufficient for purpose
Whey protein concentrate	To bind or thicken	Imitation sausage; soups; stews; nonspecific loaves	Sufficient for purpose
Dry or dried whey	To bind or thicken	Chili con carne; pork or beef with barbecue sauce	8% individually or collectively with other binders
Reduced lactose whey	To bind or thicken	Chili con carne; pork or beef with barbecue sauce	8% individually or collectively with other binders
Reduced minerals whey	To bind or thicken	Chili con carne; pork or beef with barbecue sauce	8% individually or collectively with other binders

(continued)

Table I (*Continued*)

Class of substance	Substance	Purpose	Products	Amount
	Whey protein concentrate	To bind or thicken	Chili con carne; pork or beef with barbecue sauce	8% individually or collectively with other binders
	Xanthan gum	To maintain uniform viscosity; suspension of particulate matter, emulsion stability, and freeze–thaw stability	Meat sauces, gravies or sauces and meats, canned or frozen and/or refrigerated meat salads, canned or frozen meat stews, canned chili or chili with beans, pizza topping mixes, and batter or breading mixes	Sufficient for purpose
Bleaching agent	Hydrogen peroxide	To remove color	Tripe (Substance must be removed from product by rinsing with clear water.)	Sufficient for purpose
Coloring agents (natural)	Alkanet, annatto, carotene, cochineal, green chlorophyl, saffron, and tumeric	To color casings or rendered fats; marking and branding product	Sausage casings; oleomargarine; shortening; marking or branding ink on product	Sufficient for purpose (may be mixed with approved artificial dyes or harmless inert material such as common salt and sugar)

Additive	Purpose	Products	Amount
Coloring agents (artificial)			
Coal tar dyes approved under the Federal Food, Drug, and Cosmetic Act (Operator must furnish evidence to inspector in charge that dye has been certified for use in connection with foods by the Food and Drug Administration.)	To color casings or rendered fats; marking and branding product	Sausage casings; oleomargarine; shortening; marking or branding ink on product	Sufficient for purpose (may be mixed with approved natural coloring matters or harmless inert material such as common salt or sugar)
Titanium dioxide	To color casings or rendered fats; marking and branding product	Canned ham salad spread and creamed type canned products	0.5%
Curing accelerators; must be used only in combination with curing agents			
Ascorbic acid / Erythorbic acid	To accelerate color fixing or preserve color during storage	Cured pork and beef cuts; cured comminuted meat food product	75 oz to 100 gal pickle at 10% pump level; 0.75 oz to 100 lb meat or meat byproduct; 10% solution to surfaces of cured cuts prior to packaging (The use of such solution shall not result in the addition of a significant amount of moisture to the product.)
Fumaric acid	To accelerate color fixing	Cured comminuted meat or meat food products	0.065% (or 1 oz to 100 lb) of the weight of the meat or meat byproducts, before processing
Glucono-δ-lactone	To accelerate color fixing	Cured, comminuted meat or meat food product / Genoa salami	8 oz to each 100 lb of meat or meat byproduct / 16 oz to 100 lb of meat (1.0%)

(continued)

Table I (*Continued*)

Class of substance	Substance	Purpose	Products	Amount
	Sodium acid pyrophosphate	To accelerate color fixing	Frankfurters, weiners, vienna, bologna, garlic bologna, knockwurst, and similar products	Not to exceed, alone or in combination with other curing accelerators the following: 8 oz in 100 lb of the meat, or meat and meat byproducts, content of the formula, nor 0.5% in the finished product
	Sodium ascorbate	To accelerate color fixing, or preserve color during storage	Cured pork and beef cuts, cured comminuted meat food product	87.5 oz to 100 gal pickle at 10% pump level; $\frac{7}{8}$ oz to 100 lb meat or meat byproduct; 10% solution to surfaces of cured cuts prior to packaging (The use of such solution shall not result in the addition of a significant amount of moisture to the product.)
	Sodium erythorbate Citric acid or sodium citrate	To accelerate color fixing, or preserve color during storage	Cured pork and beef cuts; cured comminuted meat food product	May be used in cured products or in 10% solution used to spray surfaces of cured cuts prior to packaging to replace up to 50% of the ascorbic acid, erythorbic acid, sodium ascorbate, or sodium erythorbate that is used
Curing agents	Sodium or potassium nitrate	Source of nitrite	Cured products other than bacon; nitrates may not be used in baby, junior, and toddler foods	7 lb to 100 gal pickle; 3.5 oz to 100 lb meat (dry cure); 2.75 oz to 100 lbs. chopped meat

Sodium or potassium nitrite (Supplies of sodium nitrite and potassium nitrite and mixtures containing them must be kept securely under the care of a responsible employee of the establishment. The specific nitrite content of such supplies must be known and clearly marked accordingly.)	To fix color	Cured products; nitrites may not be used in baby, junior, or toddler foods	2 lb to 100 gal pickle at 10% pump level; 1 oz to 100 lbs meat (dry cure); $\frac{1}{4}$ oz to 100 lb chopped meat and/or meat byproduct. The use of nitrites, nitrates, or combination shall not result in more than 200 ppm of nitrite, calculated as sodium nitrite, in finished product
Denuding agents; may be used in combination. Must be removed from tripe by rinsing with potable water. Lime (calcium oxide, calcium hydroxide) Sodium carbonate Sodium gluconate Sodium hydroxide Sodium persulfate Sodium silicates. (ortho, meta, and sesqui) Trisodium phosphate	To denude mucous membrane	Tripe	Sufficient for purpose
Flavoring agents; protectors and developers Program-approved artificial smoke flavoring	To flavor product	Various	Sufficient for purpose
Program-approved smoke flavoring	To flavor product	Various	Sufficient for purpose
Autolyzed yeast extract	To flavor product	Various	Sufficient for purpose
Harmless bacteria starters of the acidophilus type, lactic acid starter, or culture of	To develop flavor	Dry sausage, pork roll, thuringer, lebanon bologna, cervelat, and salami	0.5%
Pediococcus cerevisiae	To dissipate nitrite	Bacon	Sufficient for purpose

209

(continued)

Table I (*Continued*)

Class of substance	Substance	Purpose	Products	Amount
	Benzoic acid, sodium benzoate	To retard flavor reversion	Oleomargarine	0.1%
	Citric acid	To protect flavor Flavoring	Oleomargarine Chili con carne	Sufficient for purpose Sufficient for purpose
	Corn syrup solids, corn syrup, glucose syrup	To flavor	Chili con carne, sausage, hamburger meat loaf, luncheon meat, chopped or pressed ham	2.0% individually or collectively, calculated on a dry basis
	Dextrose	To flavor product	Sausage, ham, and cured products	Sufficient for purpose
	Malt syrup	To flavor product	Cured products	2.5%
	Milk protein hydrolysate	To flavor product	Various	Sufficient for purpose
	Monosodium glutamate	To flavor product	Various	Sufficient for purpose
	Sodium sulfoacetate derivative of mono- and diglycerides	To flavor product	Various	0.5%
	Sodium tripolyphosphate Mixtures of sodium tripolyphosphate and sodium metaphosphate, insoluble; and sodium polyphosphates, glassy	To help protect flavor	"Fresh Beef," "Beef for Further Cooking," "Cooked Beef," beef patties, meat loaves, meat toppings, and similar products derived from pork, lamb, veal, mutton, and goat meat that are cooked or frozen after processing	0.5% of total product

	Substance	Purpose	Product	Amount
	Sorbitol	To flavor, to facilitate the removal of casings from product and to reduce caramelization and charring	Cooked sausage labeled frankfurter, frank, furter, wiener, knockwurst	Nor more than 2% of the weight of the formula, excluding the formula weight of water or ice; not permitted in combination with corn syrup and/or corn syrup solids
Gases	Carbon dioxide solid (dry ice)	To cool product	Chopping of meat, packaging of product	Sufficient for purpose
	Nitrogen	To exclude oxygen	Sealed container	Sufficient for purpose
Miscellaneous	Potassium sorbate	To retard mold growth	Dry sausage	Two and one-half percent in water solution may be applied to casings after stuffing or casings may be dipped in solution prior to stuffing
	Propylparaben (propyl p-hydroxybenzoate)	To retard mold growth	Dry sausage	Three and one-half percent in water solution may be applied to casings after stuffing, or casings may be dipped in solution prior to stuffing
	Sodium bicarbonate	To neutralize excess acidity, cleaning vegetables	Rendered fats, soups, curing pickles	Sufficient for purpose
	Calcium propionate / Sodium propionate	To retard mold growth	Pizza crust	0.32% alone or in combination based on weight of the flour used
	Sodium hydroxide	To decrease the amount of cooked-out juices	Meat food products containing phosphates	May be used only in combination with phosphates in a ratio not to exceed one part sodium hydroxide to four parts phosphate; the combination shall not exceed 5% in pickle at 10%

(continued)

Table I (*Continued*)

Class of substance	Substance	Purpose	Products	Amount
Phosphates	Disodium phosphate Monosodium phosphate Sodium metaphosphate, insoluble Sodium polyphosphate, glassy Sodium tripolyphosphate Sodium pyrophosphate Sodium acid pyrophosphate Dipotassium phosphate Monopotassium phosphate Potassium tripolyphosphate Potassium pyrophosphate	To decrease the amount of cooked-out juices	Meat food products except where otherwise prohibited by the Federal meat inspection regulations	5% of phosphate in pickle at 10% pump level; 0.5% of phosphate in product (Only clear solution may be injected into product.)
Proteolytic enzymes	*Aspergillus oryzae* *Aspergillus flavus oryzae* group Bromelin Ficin Papain	To soften tissues	Raw meat cuts	Solutions consisting of water and approved proteolytic enzymes applied or injected into raw meat cuts shall not result in a gain of more than 3% above the weight of the untreated product

a A complete list is found in the Code of Federal Regulations, Animals and Animal Products.

The salts of nitrite are soluble in water and are considered to be active. In meat, the usually mild acid conditions lead to formation of only a small quantity of nitrous acid (Bard and Townsend, 1971). Nitrite is a weak reducing agent and is oxidized to nitrate only by strong chemical oxidants or by nitrifying bacteria. Nitrite oxidizes many reduced substances.

When nitrite is added to meat for curing, $<50\%$ of that added can be analyzed chemically after the completion of processing. Many cured meat items contain only a few to 50 ppm nitrite even though up to 156 ppm may have been added. It must be assumed that nitrite has been lost from the meat to the atmosphere or remains in the meat as a reaction product undetectable by current analytical methods for nitrite. Factors that affect the level of detectable nitrite are time and temperature employed during processing; amount of protein, fat, and carbohydrate; concentration of salt; concentration of nitrate; number and kind of microorganisms; and acidity (Cassens *et al.,* 1979).

Nitrite added to meat products reacts with myoglobin and hemoglobin of trapped red blood cells to form the cured meat color (Dryden and Birdsall, 1980). The general scheme of events is that when nitrite is first added to meat, the color is changed from the purple – red color of myoglobin to the brown of metmyoglobin. With time and reducing conditions, the color is converted to the rather dark red of nitric oxide myoglobin. Heat denaturation converts the pigment to a stable nitrosylhemachrome, which is pink. The heme group that gives myoglobin and hemoglobin its color has an organic part and an iron atom (Stryer, 1975). The organic part or protoporphyrin is made up of four pyrrole groups linked by methane bridges to form a tetrapyrrole ring. The iron atom binds to the four nitrogens in the center of the protoporphyrin ring and can form two additional bonds, one on either side of the heme plane. The iron atom can be in the ferrous (2+) or ferric (3+) state. In addition to the primary reaction with myoglobin, nitrite may also react with nonheme protein, lipids, and carbohydrates. Nitrite can also form nitrate, gases of nitrogen, and other reactions (Cassens *et al.,* 1979).

Nitrite is the critical agent in meat curing because it stabilizes color, produces flavor, and imparts a preservative effect. Nitrite is a reactive chemical known to participate in numerous reactions, with nitrosation being of greatest interest to meat curing. Concern has arisen about cured meat because of the possibility of *n*-nitrosamines being formed in it. Also, the residual or leftover nitrite ingested when the meat is consumed by individuals adds to the total body burden of nitrite (Cassens *et al.,* 1978).

ALKALINE PHOSPHATES

In addition to nitrite and/or nitrate, other compounds commonly included in a curing pickle are salt, ascorbates, alkaline phosphates, and sweeteners. The function of salt, ascorbates, and sweeteners in cured intact and sectioned and formed meats is similar to their function in sausage and are discussed in that section. Alkaline phosphates are often used in intact and sectioned and formed meat products. Therefore, alkaline phosphates are discussed in this section on curing.

The use of alkaline phosphates in the curing of primal cuts has been rather widely adopted by the meat industry in the United States. Various forms of phosphate have been used primarily to decrease the amount of shrinkage in smoked products and to reduce the degree of cookout in canned products (Hamm, 1971; Brotsky and Everson, 1973).

In the United States sodium tripolyphosphate; sodium metaphosphate, insoluble; sodium acid pyrophosphate; sodium polyphosphate, glassy; and disodium phosphate alone or in combination have been approved for use in certain products (Table I). Their use is restricted to an amount that will result in not more than 0.5% in the finished product.

Increased yields of primal cuts prepared with phosphate under commercial conditions have been reported to vary from 0 to 10%. The available data indicate that the approved phosphates are more effective in increasing yields as the final processing temperature is increased. Thus, smoked hams processed to 60°C (142°F) internally have shown smaller increases in yield than canned hams or fully cooked hams processed to 68°C (154°F) or higher. However, in smoked hams and canned hams alike, the effect of phosphates also may be noted in the decreased amount of readily expressible fluid in those hams containing added phosphate. In the late 1950s the USDA permitted use of phosphates in bacon. The legal limit was set at 0.5% in the finished product.

The use of some of the approved phosphates has raised problems not previously encountered in the processing of cured meats. The phosphate should be dissolved first in clear water before adding other ingredients so as to avoid precipitation of the phosphate in the more saturated pickles. Because of the corrosive action of alkaline phosphates, it has been found advisable to use plastic or stainless steel in handling the curing pickles and pumped primal cuts. Anodized cans, which provide for selective corrosion of an aluminum insert, have been helpful in reducing container corrosion in canned hams. Recrystallization of phosphate has been observed in and on the surface of cured meats prepared with polyphosphates. The crystals have been identified as disodium phosphate and are thought to result from the hydrolysis of the polyphosphate. The control of this condition is achieved by the use of reduced levels of phosphate in the cure.

B. Procedures

1. CURE APPLICATION

There are a number of methods of applying curing ingredients to intact cuts as well as to large chunks of meat. The rate of curing of these cuts depends on the rate of diffusion of the curing ingredient into the tissues. The lowest rate of curing results when the curing ingredients are applied externally in the form of a dry rub or in liquid form as a cover pickle. Such methods of curing are still practiced for such items as dry cured bacon and ham. These are generally specialty products processed for a regional market.

The internal injection of curing agents results in a much more rapid and uniform distribution of cure throughout the tissues. The most popular method for the application of curing materials to bacon and hams is by multi-needle injection. By the use of commercially available machines, the bellies or hams are injected simultaneously and automatically in a uniform pattern over the entire piece. A recent development is the simultaneous injection and tumbling of hams (Schmidt, 1981). The multiple injection of beef and pork cuts with a curing brine lends itself to rapid, mechanized handling of the product on the pumping lines.

The amount of pickle injected into a meat cut must be adjusted for a number of factors. The amount of the added ingredients are controlled for each type of product by USDA standards. "Ham" must have at least 20.5% meat protein on a fat-free basis, "ham with natural juices," 18.5%, and "ham water added," 17%. In "ham and water product," x% of weight is added ingredients for any product with <17% meat protein, where x represents the actual percent for that product. Cured and smoked bacon may not weigh more than the uncooked – uncured product (de Holl, 1978). The amount of pickle retained by the product through the entire processing procedure will depend on the method of injection, the dimensions of the product, and the weight loss before and during heat processing and during the chilling of the product. Intact meat chunks can be injected with pickle and processed directly after the injection. Some processors may want to immerse the intact cut in additional pickle, called a cover pickle, in order to allow time for equilibration of the curing ingredients.

2. MECHANICAL ACTION

A number of procedures have been introduced for the enhancement of water-binding capacities of cured – intact and sectioned and formed meat products. These procedures involve the use of machines that manipulate either bone-in or boneless meat products in such a way as to enhance the ability of the proteins within the product to bind the pickle during heat processing. These procedures are referred to as mixing, tumbling, or massaging. The tumbling and massaging of hams has been adequately reviewed by Theno *et al.* (1977), Schmidt and Siegel (1978), and Schmidt (1979). Depending on the program used, the intact or sectioned and formed meat cuts may be subjected to 10 min of fairly active mixing or up to 18 hr of intermittent massaging. All of these treatments act to disrupt the internal structure of the hams so that the salt and phosphates in the pickle act to extract the protein to enhance the binding of water within the muscle. In addition mechanical manipulation causes the protein on the surface of the muscle chunk to form an exudate. This exudate acts as a heat-set protein matrix to bind the chunks together once the product has been processed. The use of a vacuum mixer, massager, or tumbler prevents the incorporation of air into the exudate (Solomon *et al.*, 1980). This will enhance the strength of the binding material and prevent disruption of the internal structure of the product during heat processing in a closed container.

Rahelic *et al.* (1974) systematically studied the effect of tumbling and massaging on pasteurized canned hams. Massaging for up to 320 min decreased released juice in the cans while prolongation beyond 460 min increased it. Tenderness was similarly affected. Tumbling had similar effects. Histologically, increased duration of tumbling caused a swelling and loosening of the structure of the sarcomeres. After extensive tumbling, the sarcomere structure degraded completely. Actin filaments and Z disks were most rapidly broken down.

Theno *et al.* (1978b) collected samples of rectus femoris from cured ham muscles that had been massaged for 0, 1, 2, 4, 8, and 24 hr with 0, 1, 2, or 3% of added salt in the presence or absence of 0.5% phosphate for examination with a scanning electron microscope (SEM). After several hours of massaging, fiber disruption became evident. Further massaging resulted in longitudinal disruption of the fibers. Myofibrils were observed to separate and shred from the surface of the fibers. After 24 hr of massage, all treatments caused massive fiber disruption and loss of normal structural integrity. The effects of massaging were more pronounced in the presence of salts and phosphates at all time intervals.

Samples of the tacky exudate formed on meat surfaces as a result of massaging muscle in the presence of salt and/or phosphate were removed at intervals during 24 hr of massaging and observed using a light microscope. Samples without added salt or phosphate showed broken fibers and fragments from fiber disruption. Samples with salt or phosphate showed both solubilized protein and fragments from fiber disruption. Samples with salt and phosphate showed primarily clouds of solubilized protein. Length of massaging enhanced the effects in all samples (Theno *et al.*, 1978a).

Samples of the binding junction were removed from cooked ham rolls that had been stuffed at intervals during 24 hr of massaging in the presence of varying levels of salt (0, 1, 2, and 3%) and phosphate (0 and 0.5%) and examined using a light microscope to determine the microstructural characteristics of junctions exhibiting good and poor binding characteristics. Junctions in low-salt rolls (< 2%) were filled with fat and cellular fragments. Junctions from rolls with adequate salt (> 2%) and phosphate (0.5%) exhibited good binding characteristics (Theno *et al.*, 1978c).

Samples of the tacky exudate formed on the meat surfaces were also removed at time intervals and analyzed for fat, moisture, and protein. Results showed that as the massaging time increased, the percentage of fat and protein in the exudate increased in all treatments. The binding quality and cooking loss of the prepared ham rolls were also improved by the massaging process. A salt treatment of 2% appeared to be optimal for the development of adequate binding with decreased cooking loss. It was found that the presence of phosphate and absence of massaging resulted in the production of a product exhibiting cooking loss and binding properties superior to those of a product prepared in the absence of phosphate and presence of massaging, although the presence of both massaging and phosphate was beneficial for the production of an overall superior product (Siegel et al., 1978a).

The samples of the tacky exudate were also analyzed by sodium dodecyl sulfate (SDS)–polyacrylamide gel electrophoresis to determine the relative percentage of various myofibrillar proteins present. The results showed that phosphate exerts the greatest effect on the relative percentages of actin, myosin, and tropomyosin, and its action occurs primarily on meat surfaces before the massaging process is initiated. The massaging process involves a great degree of tissue destruction at the cellular level that aids in the extraction, solubilization, concentration, and distribution of the major myofibrillar proteins on surfaces and in interiors of muscle chunks, which is beneficial in the improvement of binding (Siegel et al., 1978b).

Krause et al. (1978) determined that tumbling significantely improved canned ham external appearance, color, sliceability, taste, aroma, and yield. Three-hr continuous tumbling resulted in less improvement in product quality and yield than did 18-hr intermittent tumbling. The addition of sodium tripolyphosphate and removal of most fat cover resulted in better sliceability and yield.

There are several points to remember regarding mechanical manipulation of pickle-injected meats. The appropriate amount of pickle must be injected. Mechanical manipulation should be completed as soon as possible after pickle injection. All of the injected pickle should be readily absorbed into the ham or other meat chunks shortly after mechanical manipulation begins regardless of the length of the mechanical action. After the last mechanical action, the product should be loaded into casings or cans as soon as possible. At this point, the muscles are very flexible and will conform to the shape of the container much more readily than if permitted to sit for a considerable length of time in a refrigerated room.

3. FORMING

There are many different types of equipment for filling either casings or cans with intact boneless cuts or with sectioned and formed meat chunks. Cuts may be placed by hand into a casing and the casing pulled tightly and clipped with a metal clip. More elaborate machines are available that draw a vacuum on the material prior to inserting it into casings, molds, or cans.

II. RESTRUCTURED MEATS

The class of products referred to as restructured meats includes not only sectioned and formed ham, but a growing list of cured and noncured products. The particles to be utilized in the finished product may be entire muscles, muscle chunks, finely comminuted batter, or coarse-ground, thinly sliced, flaked, or finely ground meat.

A considerable amount of work has been done on sectioned and formed meat products produced by flaking in a comitrol (Mandigo, 1974, 1975). For various products, grinders, choppers, flakers, and slicers are used to reduce the meat to the particle size preferred.

In some products, rather than using salt and phosphate to extract proteins to form an exudate, an emulsion of finely chopped material is added. This emulsion can be prepared by chopping meat that may contain reasonable levels of collagen and fat into a very fine batter. Some processors utilize various nonmeat extenders in this portion of the material (Vandenover and Yaiko, 1977). The addition of material into the chunks of the meat can be done by injecting isolated soy protein or other exogenous proteins along with a salt brine. It is important when injecting material, or when adding it as a separate emulsion to manipulate the product extensively mechanically after the addition of the nonmeat ingredient so as to incorporate the material into the muscle tissue evenly (Siegel et al., 1979).

The manipulation that is applied to cause protein extraction both to bind the moisture within the particles and to bind the particles together can be supplied by mixers, blenders, choppers, massagers, or tumblers.

Many of the products prepared for institutional use may be roasts or steaks that are either frozen, pasturized, or retorted to commercial sterility. These products are often served as sandwich items, breaded chopped steaks, or with a gravy sauce. Many of these items are served with considerable amounts of condiments along with a bread item to make a sandwich. Consumers generally prefer the product served in this manner (Ockerman and Crespo, 1977). Other ingredients that are commonly included in various sectioned and formed noncured meat items are salt, water, phosphate, and mechanically deboned tissue. Cross and Stanfield (1976) reported that consumers preferred restructured beef steaks with 30% fat and some added salt. Field et al. (1977) determined that the addition of 5% mechanically deboned meat improved the quality of restructured steaks. A possibility for the future is the inclusion of protein extracts from exogenous meat sources to be added as binders to aggregate meat particles into sectioned and formed products (Ford *et al.* 1978).

Considerable research is to be done in the area of increasing the yield, bind, and especially texture of various sectioned and formed products. The maintenance of meat texture, flavor, and color during various levels of heat processing and storage can be greatly improved. Work on tenderization and comminution equipment as well as various additives to the meat may enhance the ability to produce quality sectioned and formed uncured meat products, such as roasts, steaks, and various sliced items.

III. SAUSAGE

A. History

Sausage is a food that is prepared from comminuted and seasoned meat and is usually formed into a symmetrical shape. The word sausage is derived from the Latin *salsus,* which means salted or literally meat preserved by salting. Sausage making developed over a number of centuries, beginning with the simple process of salting and drying meat (Tauber, 1976). This was done to preserve fresh meat that could not be consumed immediately. The typical flavors, textures, and shapes of many sausages known today such as frankfurters, braunschweiger, and salami were named due to geographical location of their origin. Over 5 billion pounds of uncanned sausages were produced in the United States in 1979. The most popular sausages are frankfurters, fresh sausages, bologna, dry or semi-dry sausage, loaves, and liver sausages (Table II).

Table II
Several Meat Products Processed under Federal Inspection in 1979[a]

Meat product	Weight (10^6 lb)
Hams	1834
Bacon	1643
Beef, cooked and/or dried	303
Sausage	
Franks and wieners	1396
Fresh	1140
Bologna	679
Dried and semidried	333
Cured meat loaves and other cooked items	1078
Liver sausage	100
Other formulated products	254
Total Sausage	4980

[a] Source: USDA and American Meat Institute.

B. Protein Matrix Functionality

The manufacture of comminuted processed meat products is dependent on the formation of the functional protein matrix within the product. The properties of this matrix, which are different for each class of processed meats, give the product its characteristic texture and bite. These properties can be described in terms of emulsification or binding and gelation performance.

Saffle (1968) and Hansen (1960) described the basic structure of a meat emulsion as a mixture in which finely divided meat constituents are dispersed as a fat-in-water emulsion, with the discontinuous phase as fat and the continuous phase as water containing solubilized protein components. The water- and salt-soluble proteins of the meat emulsify fat globules by forming a protein matrix on their surfaces (Hansen, 1960; Swift *et al.*, 1961). These proteins accomplish this function of emulsion stabilization because of the presence of reactive groups that are oriented across the fat–water interface (Becher, 1965). Sulzbacher (1973) concluded that the binding quality of emulsion-type sausages is an outcome of the heat denaturation, and accompanying changes in solubility, of these emulsifying proteins. Once the fat is coated, the emulsion is stable for only a period of hours. However, heating the emulsion coagulates the protein and stabilizes the emulsion, so that the protein holds the fat in suspension for an unlimited period of time.

Emulsion stability is influenced by many factors (Schut, 1976) that include water-holding capacity of the meat selected; the level of meat, water, fat, salt, and nonmeat additives in the formulation; pH; mechanical treatment; and heat treatment. The means of formation of the protein matrix is one of the main factors contributing to the formation of a successful emulsion (Swift *et al.*, 1961). Photomicrographs of products show stronger cohesiveness of the protein matrix when it is thicker and the fat globules are more uniform (Froning and Neelakantan, 1971; Theno and Schmidt, 1978).

The mechanism of binding between meat chunks could be similar to the mechanism behind the heat-initiated emulsion stabilization in finely chopped sausages. The major difference would be that in emulsion sausages no large chunks of meat are present. The binding between chunks of meat is a phenomenon involving structural rearrangement of solubilized meat proteins. A loose protein structure is built up from previously dissolved protein. Siegel and Schmidt (1979) showed that the mechanism of binding between meat pieces involves the interaction of superthick synthetic filaments formed from intact myosin heavy chains in the extracted protein with heavy myofilaments located within muscle cells on or near sufaces of pieces of meat. An ionic strength of 0.6 μ and a pH of 6.0 are required to solubilize the myosin during mixing and comminution prior to heating to enable a strong protein matrix to form during cooking.

Protein gels are formed by intermolecular interactions resulting in a three-dimensional network of protein fibers that promotes structural rigidity (Fennema, 1976). The continuous intermeshing system of protein molecules holds or traps water. Critical parameters important to the type of gel formed include temperature, pH, salt, and protein concentration. These parameters alter the degree of cross-binding by changing the quarternary structure of the protein or the charged distribution of the polymer molecules. It is the flexibility of the polymer molecules and the number of connections between them that determine the elastic plastic nature of the gel structure (Paul and Palmer, 1972).

Myosin forms an irreversible gel that is initiated by heat. The gel exhibits a high water-binding capacity and very strong elastic qualities. The heat-induced gelatin of myosin is optimally developed at temperatures between 60 and 70°C at pH 6.0 and an ionic strength of 0.6 (Ishioroshi *et al.*, 1979; Yasui *et al.*, 1979). Samejima *et al.* 1979 studied some of the properties of gels formed by light meromyosin and heavy meromyosin and concluded that an intact myosin molecule offers the best functional properties. The properties of heat-induced myosin gels appear to be dependent upon the nature of the intact heavy chains of myosin (Siegel and Schmidt, 1979). Theno and Schmidt (1978) showed that the protein matrix and the fat droplet size in commercially manufactured frankfurters varied widely. The difference in appearance observed in comminuted meat products is highly affected by the mechanical treatments applied during comminution (Schut, 1976; Theno and Schmidt, 1978).

C. Ingredients

1. MEAT INGREDIENT FUNCTIONALITY

The abilities of various meat ingredients to provide extractable protein for matrix construction in the finished product are distinct measures of their value. Some meats have a very high myosin content and through appropriate technology this myosin can be extracted and utlized to enhance the quality of the finished product (Webb, 1974). Meats that are high in stromal protein or collagen have the lowest functionality. Meats that are high in fat also have a very low functionality. The amount of these lowly functional materials must be limited in commercial practice. When high levels of nonfunctional meat ingredients are improperly used in sausage manufacture, free fat shows up as fat caps on the ends of sausages. Another problem is the accruence of a thin film of grease on the surface of the sausages or the presence of pockets of jelly within the sausage.

Fat is an important constituent of meat products. Fat content affects the tenderness and juiciness of sausage. The degree of fat dispersion within the product affects the products' appearance and texture. Fat is limited to a maximum of 30% in cooked sausages and 50% in fresh sausages. The degree of saturation of the fat in the sausage will affect the palatability of the finished product.

Moisture accounts for 45 to 60% of the weight of processed meat products. Moisture content of cooked sausage products must not exceed four times the protein content plus 10%. In fresh sausages that are not heat-processed a maximum of 3% moisture may be added to facilitate processing. In order to obtain a level of moisture that is in compliance with the regulation and provide a product that is tender and juicy, the processor adds additional water to many products as part of the formulation. The amount of water added will depend on the amount of moisture that is lost during heat processing and chilling of the product prior to packaging.

Federal meat inspection regulations classify animal tissues used for preparation of comminuted meat products as either meat or meat by-products. To be classified as meat, tissues must be of skeletal origin and for purposes of labeling need be referred to only as beef, pork, veal, or mutton. Nonskeletal or smooth muscle tissues, such as lips, tripe, pork stomachs, and cardiac muscle are referred to as meat by-products and must be listed individually in the ingredient statement printed on packages.

Tissues vary in moisture–protein ratio, fat–lean ratio, and amount of pigment, as well as in their ability to bind moisture and fat. Sausage ingredients are classified as either binder or filler meats. Binder meats are further subdivided into high, medium, and low categories depending on their ability to bind water and fat. Meats with high binding properties are lean skeletal tissues, such as whole-carcass bull and cow meat. Beef shanks and veal are of medium value as binders. Low-binding meats contain a large proportion of fat, smooth muscle, or cardiac muscle tissue. These meats include regular pork trimmings, jowls, tongues, and hearts.

Fat content of meat used for comminuted meat products is influenced primarily by carcass grade and particular cut or type of trim. Variations in fat content greatly exceed those of moisture and protein. If moisture and protein are known, fat content may be approximated by difference, allowing ~ 0.8% for ash. When selecting meats for comminuted meat products, consideration should also be given to the percentage of myoglobin present in the raw material, principally because of the effect this pigment has on the color of the finished product. Heart and cheek meats are good sources of myoglobin and may be used to advantage in products that tend to be pale in color.

2. BEEF

The following cuts of boneless processing beef are usually quoted in market price sheets that serve the meat industry: whole-carcass bull meat (90% lean), whole-carcass cow meat (85% lean), lean trimmings (50, 75, or 85% lean), boneless chucks, and shoulder clods. In addition, the following meat by-products are sometimes used for the manufacture of comminuted meat products: head meat, hearts, lips, and weasand. When cheeks are trimmed of overlying glandular and connective tissue, the resultant product is called cheek meat, to differentiate it from the untrimmed product referred to only as cheeks. When cheek meat is used in a comminuted meat formulation, Federal inspection regulations require it to be labeled as either beef or pork cheeks. If, however, the cheeks are not trimmed, they must be listed separately as beef cheeks or pork cheeks.

3. PORK

Pork used in comminuted processed products comes from two sources: (1) boned primal cuts, usually from heavy logs, and (2) trimmings obtained during preparation of primal cuts for curing or merchandising as fresh pork. Pork trimmings may be either fresh or cured. When primal pork cuts that have already been injected with curing pickle are trimmed, the resultant meat is referred to as sweet pickled trim. Pork for sausage consists of 50 and 80% lean trimmings, boneless picnics, blade meat, skinned jowls, lips, snouts, cheek meat, and hearts.

4. VEAL AND MUTTON

Veal used in comminuted processed meats is either whole-carcass or veal trimmings. Both whole-carcass veal and veal trimmings are quite lean. On occasion, mutton is used in processed meat products. Mutton is usually quite dark in color and contributes desirable pigment to comminuted sausage or canned meat formulations. Mutton has good binding properties, but because of pronounced flavor, its usage is usually restricted to 20% or less of the total meat block.

5. VARIETY MEATS

Variety meats are used in many comminuted processed meat products. Government regulations specify that variety meats must be labeled specifically as to origin. Those finding greatest use in processed meat products are tongues, livers, hearts, tripe, and pork stomachs.

Federal inspection regulations require that all glandular and connective tissue obtained when long-cut tongues are converted to short-cut tongues be identified as tongue trimmings and further identified according to species. Trimmings from the tongue itself must be referred to as tongue meat and identified according to species. Tongue meat, however, may not include tongue trimmings. Tongues are scalded to remove the mucous membrane.

Most beef, calf, and lamb livers are sold fresh, but hog livers are used primarily for the manufacture of liver sausage and braunschweiger. Regardless of ultimate use, the gallbladder is always removed from the liver imediately after the liver is cut from the carcass. Livers are not generally scalded, but some processors of braunschweiger do use scalded livers.

When heart or heart meat is used for the manufacture of processed meats, the label must indicate species. Heart meat refers to the trimmed hearts with the cap removed.

Beef tripe is obtained from the paunch by trimming it free of adhering fat, opening it, removing the contents, washing, and scrubbing it. The mucous membrane covering the inner surface of the tripe is removed during the scrubbing operation. For use in processed meats, tripe is frequently scalded and cooked.

Pork stomachs are prepared similar to beef tripe. The stomachs are opened, washed, and may be scalded before being used.

6. PARTIALLY DEFATTED TISSUE

Partially defatted beef and pork tissues are subjected to low-temperature (49°C, 120°F) rendering to remove fat without denaturing the protein. Two types of partially defatted tissues are available: chopped and fatty. Chopped tissues can be used in some meat products up to 25% of the meat ingredient, but fatty tissues are limited to a level of 15% of the meat and meat by-product ingredient in sausage of the frankfurter class by Federal inspection regulations (de Holl, 1978).

7. MECHANICALLY DEBONED RED MEAT

Mechanically deboned red meat has been utilized in a number of sausage products (Field, 1974). The properties of mechanically deboned meat vary with percentage of bone in the original product, equipment design, and a wide variety of other factors. Yield affects the chemical and physical properties of mechanically deboned meat and probably affects flavor and color (Meiberg *et al.*, 1976).

8. SALT

Salt is the most common nonmeat ingredient added to sausages. Each batch of sausage produced contains from 1 to 5% salt, which serves the functions of giving flavor, preserving, and solubilizing proteins. The amount of salt used in sausage products varies depending on geographical location and the preference of individual sausage processors. Fermented sausages usually contain 3–5% salt, while fresh sausages have 1.5–2.0%. The vast majority of cooked sausage products contain 2–3% salt (Kramlich *et al.*, 1973).

Salt serves as a preservative by retarding bacterial growth, thereby functioning as a bacteriostatic rather than bactericidal agent. Bacteriostatic effectiveness is dependent on brine concentration in the sausage and is not a function alone of the total salt present. A 4–5% brine is generally sufficient to preserve properly handled sausages. Of vital importance to the successful manufacture of sausage is the ability of salt to solubilize muscle proteins.

Salt promotes the development of rancidity in fats, thus decreasing the useful storage life of both fresh and frozen, cured and uncured sausage products. Federal meat inspection regulations permit addition of certain antioxidants to dry and uncured frozen pork sausages to offset this. However, they require that packages be marked to indicate both their presence and purpose, such as "oxygen interceptor added to improve stability."

9. COLOR STABILIZERS

In 1955, sodium ascorbate or isoascorbate and their respective acids were legally permitted in cured meat. Regulations for the sodium salts permit addition of 0.875 oz (546 ppm) to 100 lb of meat or 87.5 oz (6570 ppm) to 100 gallons of pickle at 10% pump level (Table I). Ascorbate or isoascorbate accelerate formation of and improve the color stability of cured meat pigments, especially if the products are vacuum-packaged (Fox, 1974).

Phosphates may be used in the formulation of meat food products at a level not to exceed 8 oz in 100 lb of meat or meat and meat by-products or 0.5% in the finished product (Table I).

10. SWEETENING AGENTS

Only four sweeteners, sucrose, dextrose, lactose, and corn syrup or corn syrup solids, are widely used in the sausage industry. Sucrose and dextrose are used primarily for their sweetening ability.

The use of sucrose and dextrose is not restricted by federal regulations because their sweetness itself is self-limiting. Dextrose has been rated one-half to two-thirds as sweet as sucrose and is used at about the 1.0% level in sausages. Dextrose, a reducing sugar, aids in the manufacture of semi-dry sausages, particularly those prepared with a starter culture, by promoting the fermentation reaction responsible for the tangy flavor.

Nonfat dry milk is composed of 51% lactose. Nonfat dry milk is used widely as a nonmeat extender in the meat processing industry and, therefore, this sugar does find its way into sausage products (Rust, 1976).

Corn syrup and corn syrup solids are used extensively by sausage processors. Corn syrup consists of a mixture of dextrose, maltose, and higher saccharides. It possesses a bland flavor, being ~40% as sweet as sucrose. It is used chiefly as filler material but also has been shown to aid peelability of cellulosic casings from frankfurters. Corn syrups and solids may be classified according to their dextrose equivalent, DE. DE is a measure of the reducing sugar content calculated as dextrose and expressed as a percentage of the total dry substance. Federal regulations limit the amount of corn syrup permitted in sausage to 2.2% and corn syrup solids to 2.0%, but there are no limitations on their use in nonspecific loaves.

11. NONMEAT EXTENDERS

Meat processors have a wide variety of nonmeat products available to incorporate in sausages. These are often referred to as binder or extenders. They are added to meat formulations for one or more of the following reasons: to improve water and fat binding, to improve cooking yields, to improve slicing characteristics, to improve flavor, and to reduce formulation costs. The content of these materials permitted in sausage products is controlled by Federal meat inspection regulations. Individually or collectively, up to 3.5% of cereal, starch, vegetable flour, soy flour, soy protein concentrate, nonfat dry milk, and calcium-reduced nonfat dry milk are permitted in finished sausage products. Isolated soy protein, however, is restricted to 2%. Sausages containing > 3.5% of these nonmeat ingredients or more than 2% isolated soy protein are referred to as imitation, and must be labeled as such. For meat products known as loaves, Federal inspection regulations recognize two different types. Regulations governing the amount of extender materials allowed in finished products differ depending upon the type of loaf. Products identified as meat loaves are restricted to the percentage of extender materials allowed in other sausage products. Those referred to as nonspecific (the word meat does not appear in the name), such as pickle and pimento, macaroni and cheese, and luxury loaves are not restricted with respect to their content of extender materials (Roberts, 1974).

12. SEASONINGS

Seasoning is an inclusive term applied to any ingredient that, by itself or in combination, adds flavor to a food product. Sausage seasonings consist of mixtures of various spices. Other substances such as monosodium glutamate, hydrolyzed plant proteins, and flavor nucleotides are also used. Monosodium glutamate and flavor nucleotides enhance flavor while hydrolyzed plant proteins contribute a characteristic meaty flavor. Additionally, flavor nucleotides affect taste sensation centers on the tongue. In addition to imparting distinctive flavors and aromas to sausage, certain spices such as black pepper, cloves, ginger, mace, rosemary, sage, and thyme possess antioxidant properties.

Spices are dried aromatic vegetable substances. The term may be applied to all dried plant products, which include true spices, herbs, aromatic seeds, and dehydrated vegetables. Spices, being natural products, are liable to certain variations in flavor, strength, and quality because of changes in climatic conditions. Changing soil conditions, husbandry, and storage conditions all influence the spices' ultimate value. True spices such as allspice, ginger, nutmeg, and pepper are products of tropical plants. They may represent the root, bark, bud, flower, or fruit. Herbs are leaves of plants grown in both temperate and tropical zones and are relatively low in total oil content while true spices are relatively high. Marjoram, mint, sage, and thyme are examples of herbs. Aromatic seeds such as anise, coriander, dill, and mustard are derived from plants cultivated in both temperate and tropical areas. Vegetables such as garlic and onion are usually used in the dehydrated form.

Spices are used either whole or in one of the following processed forms: ground, essential oils, or oleoresins. Most spices are used in processed form. Whole peppercorns used in certain dry sausages are an example of a whole spice.

Spices are ground to various degrees of fineness. Ground spices are more uniformly dispersed in sausages but lose flavor more readily than whole spices. Standard ground spices pass through United States standard sieves from No. 20 to No. 60 mesh. Some sausage processors prefer spices that are ground exceedingly fine. These microground spice particles are twenty times smaller than No. 60 mesh ground particles. Extreme fineness of grind aids in rapid and complete dispersion of spices in sausage emulsions.

Flavor is largely the product of spice extractives: essential oils or oleoresins. Spice processors blend extractives to achieve a desired flavor. Several advantages such as the elimination of color specks, freedom from bacteria, reduced shipping costs, and less storage area are claimed for extractives. Essential oils are volatile oils removed from plants by one of several methods: steam distillation, absorption on neutral fat, or enzymatic action followed by steam distillation. Oleoresins are viscous, resinous materials obtained by extraction of ground spices with volatile solvents. The solvent is removed to arrive at the flavor and aromatic components.

Extractives are extended in compatible media because of their limited solubilities and the minute quantities needed to produce flavors and aromas equivalent to those of natural spices. Essential oils as well as oleoresins are available in either liquid or dry form. They are made water dispersible for use in sausage emulsions or liquids by blending with a solubilizing agent. For use as dry soluble spices they are blended into a dry salt or dextrose base. Individual salt or dextrose crystals are coated with the extractives, resulting in free-flowing spice products that blend easily with other dry ingredients. When added to sausages, salt or dextrose carriers dissolve, leaving spice extractives uniformly dispersed throughout the sausage.

13. ACIDULATION

Dry and semi-dry sausages have been produced for centuries. They represent a means of preservation by creation of low to intermediate moisture content, high salt concentration, and high acid content. The basic process is one of bacterial fermentation to produce lactic acid. Originally, wild bacteria were used to ferment the sausages. The technique of "back slopping" has been replaced to a large extent by the use of starter cultures. The starter cultures are usually a mixture of *Lactobacillus* and *Micrococcus* spp. Dextrose and glucose are usually used as the fermentable carbohydrate (Acton, 1978).

Chemical acidulants such as glucono-δ-lactone have been used alone or in combination with starter cultures to decrease pH and create the acid tangy taste (Terrell, 1978).

D. Casings

Casings determine sausage sizes and shapes. Casings may serve as processing molds, as containers during handling and shipping, and as merchandizing units for display. Casings must be sufficiently strong to contain the meat mass but have shrink and stretch characteristics that allow contraction and expansion of the meat mass during processing and storage. Casings must be able to withstand the forces produced during stuffing, linking, and closure.

Casings for the sausage industry are obtained from three basic materials: cellulose, collagen, and plastic. Five specific types: animal, regenerated collagen, cloth, cellulosic, and plastic casings are produced from these basic materials. The gastrointestinal tract from cattle, hogs, and sheep is used for casings. The structures are washed, scraped, treated with chemicals to remove soluble components, and salted. The various anatomical structures, such as the esophagus, stomach, small and large intestines, and rectum, are all separated, cleansed, salted, and graded as to size and condition.

Collagen casings are prepared by acquiring a suitable collagen source such as the corium layer of beef hides. The corium is extracted with an alkaline solution to remove soluble components and washed with potable water. The collagen is then swollen with acid to give a viscous mass of acid collagen that is pushed through an annular die to form a tube. The tube is fixed by moving through an alkaline bath and the neutralized collagen returns to a reasonable approximation of its original state. The tube is dried and cut to size. The small casings are edible and are used for fresh pork sausage links. Large collagen casings for other types of sausage are often treated with aldehydes to cross-link the collagen and increase the strength of the casing. This type of casing is removed from the sausage rather than eaten by the consumer.

Cellulosic casings include those made from cotton bags and those derived from processed cotton linters. These casings are composed of pure cellulose, food-grade glycerine, and water. These sausage casings are uniform, clean and easy to handle. They can be printed or pigmented to give an attractive appearance for retail display. Cellulosic casings are available in many sizes and types.

Sausages that are not smoked may be stuffed into impermeable plastic casings. These are formed from a copolymer of polyvinylidene and polyvinylchloride or polyethylene films. When further processing is required, as in the manufacture of liver sausage and various spreads, these products are cooked in water. Other meat products, such as fresh pork sausage or ground beef, require no cooking and are sold fresh or frozen in the plastic tubes (Kramlich *et al.*, 1973).

E. Processing Equipment

The type and extent of machinery and equipment required for meat product manufacturing depend upon the variety and volume of the operations. A sausage kitchen should include a grinder, mixer, cutter or mill, stuffer, and linker, with the necessary tables, cooking and smoking equipment, trucks, cages, and scales.

Points to consider in the selection of any equipment are appropriate materials to eliminate corrosion; rugged construction to minimize maintenance; satisfactory design to facilitate through cleaning; and appropriate capacity to meet the production requirements. All equipment also should be provided with necessary safeguards so that its operation will not be hazardous. Any equipment that is added to an operation must meet both Occupational Safety and Health Administration and USDA Food Safety and Inspection Service requirements (Rust, 1976).

The numerous procedures for the manufacture of the many different types of sausages are not discussed in this work. The reason for this decision is that many processors are manufacturing high-quality sausages with very different procedures. However, during the discussion of the equipment in this section, an attempt is made to highlight the most important constraints and processing variables associated with each type of equipment for various products.

1. GRINDERS

Grinders are used to cut the meats into small pieces so that the various meats may be throughly mixed with each other and with the curing materials and spices. Grinding is accomplished by worming the meats along a cylinder with sharp-edged ribs and then through perforated plates. As the meat is extruded through the holes in the plates, it is sliced away by revolving knives. The holes in the plates vary in both size and shape, depending upon whether the meat is to be cut into large pieces for canning or small pieces for sausage, whether the meat is tough and sinewy, or whether fat is to be cut for use in sausage or in a canned product. In general, plates with round holes ranging in diameter from $\frac{1}{8}$ to $1\frac{1}{4}$ in are used for sausage. The size of the pieces in the ground material varies with the diameter of the holes in the plate. In some types of plates the holes are tapered, being larger on the outside than on the cutting side. There are also a number of grinder – mixer combinations on the market, as well as tandem grinders with a transfer system between the two to permit a double grind in one operation. Some grinders are fitted with special devices to remove bone chips and sinew.

In order to maintain good lean and fat particle definition for the manufacture of fresh, dry, or semi-dry sausages, it is important to use a grinder with appropriately fitted blade and plate and to keep the meat very cold during grinding.

2. MIXERS AND BLENDERS

Mixers and blenders perform the needed mechanical agitation required to extract the salt-soluble contractile proteins needed for binding. Mixers are round-bottomed tanks equipped with two sets of parallel, wing-shaped paddles revolving in opposite directions and designed to work the meat back and forth and thoroughly mix it. End paddles scrape the meat off the ends of the hopper and force it to the center. The mixer should discharge the contents completely either through a bottom opening or by tilting the mixer on its side.

Most blenders, however, use open or closed, single or twin screws. They usually discharge through the end. Both mixers and blenders can be purchased of sufficient strength and fitted with tight covers to permit vacuumizing of the meat during mixing. Mixing units can be mounted on load cells coupled to weighing devices for accurate loading of various ingredients.

The duration, temperature, and sequence of adding ingredients to a mixer or blender has a profound effect on the amount of myofibrillar protein extracted and, therefore, on the properties of the sausage produced. If fat and lean particle definition is to be maintained, a short-time, low-temperature mix with the addition of salt near the end of the mixing time is utilized. If maximum protein extraction is the goal of mixing, then a high 16°C (60°F) or low −2°C (29°F) temperature, a longer mixing time, and addition of the salt at the start is used. In some operations mixing is utilized to form fat and lean cured pre-blends that can be analyzed for proximate composition, kept for several days, and utilized to provide materials for a least-cost formulation (Terrell, 1974).

3. CUTTERS

In the cutter comminution and mixing are accomplished by revolving the meat in a bowl past a series of knives mounted on a high-speed rotating arbor that is in a fixed position so that the knives pass through the meat as the bowl turns. The meat is guided to the knives by a plow arrangement that is in a fixed position inside the bowl.

In some cutters, bowl rotation speed and knife speed can both be controlled. Some machines are constructed to permit chopping under vacuum, which appears to be very desirable for removing air from a sausage batter. These units may be equipped with jackets for steam heating or brine chilling of the bowls. It is possible to chop and cook products such as braunschweiger or high-collagen mixtures in a single operation, stuff them hot, and subsequently chill them.

The final properties of sausages that are prepared in a chopper are affected by duration of chopping, final temperature of the batter after chopping, the composition of the meat, the presence of salt, the application of vacuum, the sharpness and number of knives, bowl speed, knife speed, and other factors. High-quality sausages are produced with various methods of bowl chopping. A given procedure is adapated for a product based on experimentation and processor preference. The meat that goes into the chopper may be intact muscles or finely ground. The final product may be coarsely chopped or a very finely chopped batter.

4. EMULSION MILLS

The operating principle of emulsion mills involves one or more rotating knives traveling at an extremely high rate of revolutions per minute so that the meat mixture is pulled from a hopper and forced through one or more perforated stationary plates. There are modifications of the mill principle for continuous production of fine batters. Meat is finely ground and preblended or prechopped in a cutter before it is passed through a mill. The mill has the function of reducing the meat and fat particles to very small size so as to produce a smooth batter of pastelike consistency. This type of consistency is often desired for frankfurters, bologna, liver sausage, and some portions of loaves.

5. STUFFERS

After the emulsion has been prepared, it is conveyed to an extruding device that places it in a casing, pan, or mold. Extruders are of the pump or piston type. Piston stuffers are used to force the meat mixture into casings or other containers. They are vertical cylinders with a capacity ranging from 50 to 500 lb, equipped with a cover that can be removed or tightened quickly and easily and contains a piston operated by means of air or hydraulic pressure. The piston moving upward in the cylinder forces meat out of the cylinder through openings in the side just below the cover and through quick-opening control cocks into stuffing tubes or horns and then into the containers. The cylinder and piston should be made of noncorrosive materials and the inside of the cylinder should be polished smooth. The piston should drop rapidly when the motivating pressure is released. Tubes or stuffing horns should be made of stainless metal.

Stuffers should be equipped with safety devices to shut off the pressure below the piston when the meat has been discharged from the cylinder. The cylinders should also have clean-up openings flush with the bottom of the stuffer. It is most important that the chamber under the piston be thoroughly sanitized daily. Failure to do so may result in serious bacterial contamination of the meat (Rust, 1976).

Continuous stuffers consist of a hopper with a continuous pump. In addition they may be designed for vacuumizing the chamber to remove air from the product. They work with fine, coarse, or sectioned and formed products.

6. PORTIONERS

Stuffers can be equipped with portioners, link twisters, or casing clipping units. They can be set for exact portion weights. There are a variety of linkers available ranging from link twisters with or without cutoffs for separating links to string tie type linkers to high-speed automatic linkers designed for production of finely cut sausages such as frankfurters.

7. PEELERS

When peelable cellulose casings are used they must be removed. Peelers are available that will do this at rates of 5000 lb or more per hour. In these units, the product is moistened and conditioned for peeling with steam. The casing is then slit and a jet of compressed air blows the casing open. It is subsequently sucked into a chamber much like a large vacuum cleaner. The peeled sausage is discharged and moves on to the packaging equipment.

8. HEATING

Air-conditioned smokehouses circulate air over heated coils and distribute smoke as well as humidity and temperature-controlled air by fans through ducts to the product. The smoke is generated in a separate compartment and drawn into the smokehouse. These installations may provide for regulation of temperature, humidity, and volume of circulated air and of density of smoke. The humidity is controlled by varying the amount of outside air introduced into the smokehouse or by injecting steam into the house.

Smokehouses also can perform either all or part of the cooking function, including cold or hot showering. They are sometimes called processing ovens. In all cases, smokehouses should be equipped with appropriate recording temperature monitors to record both product and house temperature.

Smokehouses are designed for manual or completely automated operation with various computer programming devices. Smokehouses can also be equipped with in-place cleaning systems.

Smoke can be generated from sawdust using some type of closed combustion chamber or applied as liquid or regenerated liquid smoke. Smokehouses using natural smoke generation require appropriate air pollution control equipment.

There are several continuous sausage or bacon production systems. In these systems the encased frankfurter emulsion is fed onto a conveyor that transports the product through a liquid smoke spray or a controlled smoke section followed by a high-velocity conditioned air stream for rapid cooking. Immediately afterward the frankfurters are chilled by air or brine, peeled, and packaged.

There are several other types of cooking equipment. In some plants boneless meat and large sausages in plastic or metal containers are cooked by submergence in water in steel tanks. Where a large volume is being processed, the cages on which the sausage is held are lowered into the cooking tanks. This eliminates the labor of placing the sausage in the tanks one stick at a time.

Tongues, cheek meat, livers, and similar products are cooked in steam-jacketed, round-bottomed kettles made of stainless steel. Steam-jacketed continuous agitator kettles are used for cooking products like pizza topping, taco mix, chili con carne, etc. Continuous broilers or fryers are used to cook some patties and sausage products.

F. Thermal Processing

1. SMOKE

The preservative effect of smoking is perhaps of less importance in modern processing operation than it once was. It remains important from the standpoint of reducing oxidative rancidity as well as reducing microbial populations. The ~ 20 different phenols contained in smoke are the principal preservative ingredients. Not only are they rather potent bacteriostats but they also serve as powerful antioxidants, which is extremely important in preserving meats. Since smoke is concentrated on the surface, the activity of these compounds is largely limited to that part of the product.

From the standpoint of color, smoking has several effects (Gilbert and Knowles, 1975). First, the heating stabilizes the cured meat color which, when coupled with surface drying, becomes a deep red. The brownish color contributed by smoke is due to the carbonyl compounds in smoke. These combine with free amino groups from the meat protein to form furfural compounds, which are brown in color. The extent of formation of these furfural compounds may explain why some smoked meat takes on a more brown than a red color.

The smoked flavor is largely due to the phenols, which give a characteristic smoky flavor, as do some of the carbonyl compounds (Sink, 1979). Most of the flavor-producing compounds in smoke appear to be steam-distillable and thus can easily be incorporated in liquid smoke flavorings.

Smoke also contains organic acids that are very important in developing surface coagulation of the meat protein. This is critical in the formation of the "skin" on skinless products, which is necessary for good peelability during processing.

Smoke contains some polycyclic hydrocarbons, which appear to be undesirable. Benz[a]pyrene and dibenz[a,b]anthracenes are potential carcinogens (Daun, 1979). These are present in extremely low concentrations in most smoked meats and can be eliminated in smoke flavorings.

Smoke is usually generated by burning wood or sawdust. In a situation where oxygen is restricted, the smoke is dark and high in carboxylic acids. This type of smoke is usually considered undesirable for meat. In most smoke generators, oxygen supplies are carefully controlled. During smoke generation, the greater the degree of oxidation of the smoke, the greater the quantity of organic acids and phenols produced and hence the better the quality of the smoke.

The temperature of smoke generation also has an effect. As temperatures increase over 302°C (575°F), the ratio of acids to phenols is reduced since acids are produced at lower temperatures and phenols at higher temperatures. From the information available, the best quality smoke is produced at a temperature of 340 to 400°C (650–750°F) with a subsequent oxidation temperature of 250°C (480°F). Temperatures around 400°C (750°F) encourage the production of the polycyclic hydrocarbons. Thus, a more practical temperature for smoke production might be closer to the 340°C (650°F) temperature. Liquid smoke may be added directly to products. For relatively low flavor level products such as franks or bologna, 2 to 4 oz/100 lb of 6% acid liquid smoke or 1 to 3 oz/100 lb of 10% acid liquid smoke are used. For various other products, such as dry and semi-dry sausage and smoke links, levels might go as high as 6 oz/100 lb for 6% acid material and 4 oz/100 lb for 10% acid material (Kramlich et al., 1973).

The liquid smoke flavoring should be diluted with some of the added water and added at the end of the chopping or blending cycle, leaving just sufficient time to distribute it uniformly throughout the mixture. The low pH of the liquid smoke solution could cause some protein denaturation and interfere with the gelling properties of the batter. When added in this manner, "smoke flavoring" must appear in the appropriate place in the ingredient statement plus the statement "Smoke Flavoring Added" in prominent letters contiguous to the product name.

Surface applications consist of spraying or heat regeneration of liquid smoke solution. Spray applications are preferred because of the ability of acids in the smoke solution to aid in surface coagulation of the protein, thus promoting skin formation on a product. Spray application is usually done at a temperature of $\sim 38°C$ (100°F) as the penetration of the liquid smoke is greatly increased at 38°C (100°F) as compared to room temperature. Higher temperatures (up to 60°C, 140°F) can be used where adequate exhaust systems are provided. Higher temperatures may cause casing breakage with animal casings.

The concentration used depends on the contact time available. The greatest part of the penetration or deposition occurs within the first few seconds of contact. Intermittent application is a practice being followed by many processors. "Smoke Flavoring Added" must appear on the label but need not appear in the list of ingredients.

There are heating systems available for developing smoke by heat regeneration of the liquid smoke. Usually these devices are used in place of the regular smoke generator.

Liquid solutions of smoke in oil are also available. These are used principally in such products as smoky links and braunschweiger. They are added to the product at rates of 2 to 4 oz per 100 lb of product. A more recent innovation is liquid smoke on a dextrin base, which is added in dry form at the rate of 0.25 to 0.5 lbs per 100 lb of meat (Wistreich, 1977).

2. HUMIDITY

Humidity in a smokehouse has a number of important effects. First, it affects smoke deposition. High humidities favor smoke deposition but they also tend to limit color development. With high humidities there is generally a greater degree of smoke penetration through the casing. When the surface is dry, the smoke is deposited on the casing surface, particularly if animal or collagen casings are used. The desirable red–brown color may not be developed with high humidities and the product surface will tend to be a dull brown or tan.

High humidities do not necessarily reduce shrinkage of the product. High humidity encourages fat rendering in the product. Humidities of $\sim 38\%$ for regular and 24% for the rapid peel type cellulose casings are preferred.

In the case of animal or collagen casings, somewhat higher humidities appear appropriate. It is advisable to start out the cycle with a dry-off period of low humidity to harden the casing surface and remove surface moisture, which can cause streaking.

With collagen casings particularly, the combination of smoke and low humidity hardens the casing. On the other hand, high humidities promote the softening of the collagen. This can be carried to the extent that the casings will disintegrate. Thus, developing the correct tenderness in a collagen or for that matter in an animal casing is a result of the critical use of a combination of smoke and drying at the beginning of the cycle with higher humidities at the end to promote the softening of the casing.

In a gravity house, the humidity is controlled by opening or closing the dampers. In an air-conditioned house the humidity is controlled by adding steam or water vapor and by the use of dampers. A wet bulb thermometer is a thermometer with a wick over the sensing bulb that measures the temperature as water evaporates from the wick. The lower the wet bulb temperature in comparison to the dry bulb, the greater the evaporation and hence the lower the humidity.

3. AIR CIRCULATION

Air circulation is critical as it promotes uniform heating of the product. Proper loading of the house is important because improper loading will restrict air circulation. In an air-conditioned house, the air circulation depends on circulating fans. Temperatures in the house should be checked at various points in the product load and variations over 6°C (10°F) should no be tolerated.

Air circulation has a substantial effect on heat transfer. In a static air situation, temperature zones will develop around the product and hence the temperature adjacent to the product surface may be considerably less than the house temperature. As a result the heating rate is reduced. If rapid heating is desired, air circulation must be increased. Many air-conditioned houses can give 10–12 air changes/min.

High air velocities have a tendency to produce more rapid drying along with more rapid heating. Thus, the air velocities must be controlled to a point where both the heating and the drying are in a desirable balance. It should also be recognized that it is more difficult to maintain high smoke densities with high air velocities unless the dampers are kept tightly closed to prevent loss of smoke.

4. COOKING

With controllable smokehouses, the entire smoking and cooking process can take place simultaneously. In addition, steam cooking is sometimes applied at the end of the cycle. Steam cooking is normally done only to smaller diameter products at temperatures of 77 to 82°C (170–180°F) for periods of 5 to 15 min.

Water cooking is used for many products in metal or plastic containers. When water is used, the temperature is usually held to ~71 to 77°C (160–170°F) or in the case of beef cuts 1–3°C (3–5°F) above the prescribed final product temperature. Higher temperatures promote the development of gelatin pockets, lower yields, and bursting of the plastic containers. Just as with air, maintaining uniform temperatures in the cook tank by means of water circulation is important.

REFERENCES

Acton, J. C. (1978). *Proc. Annu. Reciprocal Meat Conf.* 30, 49.

Bard, J., and Townsend, W. E. (1971). *In* "The Science of Meat and Meat Products" (J. F. Price and B. S. Schweigert, eds.), p. 452. Freeman, San Francisco, California.

Becher, P. (1965). *In* "Emulsions: Theory and Practice," 2nd ed., p. 2. Van Nostrand-Reinhold, Princeton, New Jersey.

Binkerd, E. F., and Kolari, O. E. (1975). *Food Cosmet. Toxicol.* 13, 655.

Brotsky, E., and Everson, C. W. (1973). *Proc. Meat Ind. Res Conf.* p. 107.

Cassens, R. G., Ito, T., Lee, M., and Buege, D. (1978). *BioScience* 28, 633.

Cassens, R. G., Greaser, M. L., Ito, T., and Lee, M. (1979). *Food Technol. (Chicago)* 33(7), 46.

Cerveny, J. G. (1980). *Food Technol. (Chicago)* 34(5), 240–243.

Cross, H. R., and Stanfield, M. L. (1976). *J. Food Sci.* 41, 1257.

Daun, H. (1979). *Food Technol. (Chicago)* 33(5), 66.

de Holl, J. C. (1978). "Encyclopedia of Labeling Meat and Poultry Products." Meat Plant Mag., St. Louis. p. 45.

Dryden F. D., and Birdsall, J. J. (1980). *Food Technol. (Chicago)* 34(7), 29.

Fenemma, O. R. (1976). "Principles of Food Science. Part I. Food Chemistry." Dekker, New York.

Field, R. A. (1974). *Proc. Meat Ind. Res. Conf.* p. 35.

Field, R. A., Booren, A., Larsen, S. A., and Kinnison, J. L. (1977). *J. Anim. Sci.* 45, 1289.

Ford, A. L., Jones, P. N., Macfarlane, J. J., Schmidt, G. R., and Turner, R. H. (1978). *J. Food Sci.* 43, 815.

Fox, J. B. (1974). *Proc. Meat Ind. Res. Conf.* p. 17.

Froning, G. W., and Neelakantan, S. (1971). *Poult. Sci.* 50, 839.

Gilbert, J., and Knowles, M. E. (1975). *J. Food Technol.* 10, 245.

Hamm, R. (1971). *In* "Phosphates in Food Processing" (J. M. DeMan and P. Melnychyn, eds.), p. 65. Avi Publ. Co., Westport, Connecticut.

Hansen, L. J. (1960). *Food. Technol.* 14, 565.

Ishioroshi, M., Samejima, K., and Yasui, T. (1979). *J. Food Sci.* 44, 1280.

Kerr, R. H., Marsh, C. T., Schroeder, W. F., and Boyer, E. A. (1926). *J. Agric. Res. (Washington, D.C.)* 6, 541.

Kramlich, W. E., Pearson, A. M., and Tauber, F. W. (1973). "Processed Meats." Avi Publ. Co., Westport, Connecticut.

Krause, R. J., Ockerman, H. W., Krol, B., Moerman, P. C., and Plimpton, R. F., Jr. (1978). *J. Food Sci.* 43, 853.

Mandigo, R. W. (1974). *Proc. Annu. Reciprocal Meat Conf.* 27, 403.

Mandigo, R. W. (1975). *Proc. Meat Ind. Res. Conf.* p. 43.

Meiburg, D. E., Beery, K. E., Brown, C. L., and Simon, S. (1976). *Proc. Meat Ind. Res. Conf.* p. 79.

Ockerman, H. W., and Crspo, F. L. (1977). *J. Food Sci.* 42, 1410.

Paul, P. C., and Palmer, H. H. (1972). *In* "Food Theory and Applications" (P. C. Paul and H. H. Palmer, eds.), p. 77. Wiley, New York.

Rahelic, S., Pribis, V., and Vicevic, Z. (1974). *Proc. Eur. Meeting Meat Res. Workers* 20, 133.

Roberts, L. H. (1974). *Proc. Meat Ind. Res. Conf.* p. 43.

Rust, R. E. (1976). "Sausage and Processed Meats Manufacturing." Am. Meat Inst., Arlington, Virginia.

Saffle, R. L. (1968). *Adv. Food Res.* 16, 105.

Samejima, K. Hasimoto, T., Yasui, T., and Fukazawa, T. (1969). *J. Food Sci.* 34, 242.

Schmidt, G. R. (1979). *Proc. Meat Ind. Res. Conf.* pp. 31–41.

Schmidt, G. R. (1981). *Proc. Meat Ind. Res. Conf.* p. 111.

Schmidt, G. R., and Siegel, D. G. (1978). *Proc. Annu. Reciprocal Meats Conf.* pp. 20–24.

Schut, J. (1976). *In* "Food Emulsions" (S. Friberg, ed.), p. 447. Dekker, New York.

Siegel, D. G., and Schmidt, G. R. (1979). *J. Food Sci.* **44,** 1129.

Siegel, D. G., Theno, D. M., and Schmidt, G. R. (1978a). *J. Food Sci.* **43,** 327.

Siegel, D. G., Theno, D. M., Schmidt, G. R., and Norton, H. W. (1978b). *J. Food Sci.* **43,** 431.

Siegel, D. G., Tuley, W. B., and Schmidt, G. R. (1979). *J. Food Sci.* **44,** 1272.

Sink, J. D. (1979). *Food Technol. (Chicago)* **33**(5), 72.

Solomon, L. W., Norton, H. W., and Schmidt, G. R. (1980). *J. Food Sci.* **45,** 438.

Stryer, L. (1975). "Biochemistry." Freeman, San Francisco, California.

Sulzbacher, W. L. (1973). *J. Sci. Food Agric.* **24,** 589.

Swift, C. E., Lockett, C., and Fryer, A. J. (1961). *Food Technol. (Chicago)* **15,** 468.

Tanner, F. W. (1944). "The Microbiology of Foods." Garrard, Champaign, Illinois.

Tauber, F. W. (1976). *Proc. Annu. Reciprocal Meat Conf.* **29,** p. 55.

Terrell, R. N. (1974). *Proc. Meat Ind. Res. Conf.* p. 23.

Terrell, R. N. (1978). *Proc. Annu. Reciprocal Meat Conf.* **30,** 39.

Theno, D. M., and Schmidt, G. R. (1978). *J. Food Sci.* **43,** 845.

Theno, D. M., Siegel, D. G., and Schmidt, G. R. (1977). *Proc. Meat Ind. Res. Conf.* pp. 53–69.

Theno, D. M., Siegel, D. G., and Schmidt, G. R. (1978a). *J. Food Sci.* **43,** 483.

Theno, D. M., Siegel, D. G., and Schmidt, G. R. (1978b). *J. Food Sci.* **43,** 488.

Theno, D. M., Siegel, D. G., and Schmidt, G. R. (1978c). *J. Food Sci.* **43,** U.S. Department of Agriculture (USDA) (1925). "Service and Regulatory Announcements," p. 102. Bur. Anim. Ind., USDA, Washington, D.C.

U.S. Department of Agriculture (USDA) (1926). "Service and Regulatory Announcements," p. 2. Bur. Anim. Ind., USDA, Washington, D.C.

U.S. Department of Agriculture (USDA) (1978). *Fed. Regist.* **43,** 20992.

Vandenover, R., and Yaiko, L. (1977). *Food Eng. Int.* **2,** 36.

Webb, N. B. (1974). *Proc. Meat Ind. Res. Conf.* p. 1.

Wistreich, H. E. (1977). *Proc. Meat Ind. Res. Conf.* p. 37.

Yasui, T., Ishioroshi, M., Nakano, H., and Samejima, K. (1979). *J. Food Sci.* **44,** 1201.

6

Meat Microbiology

A. A. KRAFT

Department of Food Technology
Iowa State University
Ames, Iowa

I. MICROBIAL SPOILAGE OF FRESH AND CURED MEATS

Changes in production, processing, marketing, and demands for meat in recent years have been described by Daly (1971), Ernst (1979), and Schmidt (1979). Also, Locker *et al.* (1975) reviewed preservation measures such as reduced water activity, low redox potentials, and carbon dioxide in conjunction with refrigeration temperatures for controlling microbial growth on meats. Other chapters in this book provide details on modern meat processing practices, all of which indicate that the meat industry is far from being a static enterprise. The technological developments in the meat industry present a continuing challenge in the area of meat microbiology to insure that such developments also result in a high level of wholesome products with improved shelf life. Consumer awareness has placed increased emphasis on quality and public health considerations.

Characteristics of meat that explain its great desirability for microorganisms and humans have been described by Ingram and Simonsen (1980) and are not detailed here. However, it must be emphasized that fresh meat is an ideal medium for bacterial growth and it is subject to rapid spoilage unless precau-

Copyright © 1986 by Academic Press, Inc.
All rights of reproduction in any form reserved.

tions are taken to retard microbial growth and activity. Organisms found on or in fresh meat come from many sources (Ayres, 1955). They can be transferred to the carcasses of meat animals from the feet and hides of the animals, from air, water, and soil, and from workers and processing equipment (Ayres, 1963). However, carcasses generally have few microorganisms on the surface immediately after slaughter; contamination results soon thereafter from the sources specified above. The interior of the animal is virtually free of organisms (in healthy animals) except for the lymph nodes and excluding the gastrointestinal and respiratory tracts (Ingram and Simonsen, 1980). Also, environmental conditions prior to slaughter affect the degree of contamination of the surfaces of the live animal because such conditions determine the cleanliness of the animal brought to slaughter. It is believed that contamination from the intestinal tract spreads throughout the meat if the animal is not eviscerated soon after slaughter. The effects of slaughtering, dressing, chilling, cutting, and boning, including hot boning, have been reviewed by Ingram and Simonsen (1980) and some of the operations are discussed in more detail in Section VII. In any event, a general result of handling during processing is to introduce more microorganisms onto the meat. When fresh meats are cut or ground, the microbial population can increase rapidly.

Before proceeding further into microbial spoilage of meats, it may be advisable to review basic concepts on growth of microorganisms in a medium such as meat.

Bacterial growth curves relate to increase in numbers of cells. When bacterial cells become introduced into meat, a period of adjustment usually occurs; this is known as the lag phase of bacterial growth. Growth, or multiplication, is offset by declining numbers of viable cells. The length of the lag phase depends on factors such as initial numbers, age of cells, and suitability of the environment, among other considerations. With large initial numbers, actively growing young cells, and a favorable environment, the lag phase is short, or possibly not even discernible, in comparison with the opposite conditions. Since initial populations are important in influencing the lag phase, sanitation and temperature control as means of restricting such populations also become very important.

Bacteria reproduce by fission: one cell becomes two, two become four, etc. The length of time needed for cell division to occur is the generation time, or time to double the population. Many of the bacteria that proliferate on meats can double their numbers in 20 to 30 min. With ideal conditions, such as fresh meat may provide, simple calculations may be made to determine how much time would be needed for bacterial numbers to reach spoilage populations. If, under ideal conditions, an organism divides every 20 min, populations may reach the millions with a matter of hours, depending on the initial contamination. Table I illustrates this point for ideal conditions. At levels of 10^7 to 10^8

Table I
Relation of Initial Loads of Bacteria to Time to Reach 10^7 Organisms with a 20-min Generation Time

Initial count	Approximate time to reach 10^7 bacteria per gram of meat (hr)
10,000	3.5
100,000	2.3
1,000,000	1.3

bacteria per gram of meat, off odors and possibly slime may be observed. Slime formation results from the bacterial colonies increasing in size and number to the point where they coalesce to form a sticky coating.

The relation between the initial bacterial load and time for slime formation on beef stored at different temperatures is shown in Fig. 1. The slopes of the curves indicate that the lower the temperature of storage, the longer the storage life for a given initial bacterial load. Another way to interpret this relationship is that for a given temperature, the effect on storage life is greater when initial numbers of bacteria are low than when they are high. With continual storage at higher temperature, such as 20°C, the effect of initial load becomes almost negligible. Only a few days difference for development of slime was observed when initial counts were about 10^4 compared with $10^2/cm^2$ at 20°C.

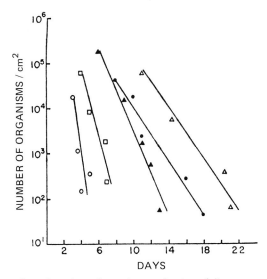

Fig. 1　Relation of initial numbers of organisms to the time of slime appearance on sliced beef held at 0 [△ and ● (Haines and Smith, 1933)], 5 (▲), 10 (□), and 20°C (○). From Ayres (1960, p. 10). Copyright by Institute of Food Technologists.

Several environmental conditions affect generation time of organisms on meats. Undoubtedly, the most important is temperature. In general, the higher the temperature, up to a point, the faster the growth rate. this relationship is illustrated in Fig. 2 for meat stored at temperatures from 0 to 20°C. Slopes of the growth curves decrease with decreasing temperature, as storage life increases, so that beef held at 0°C did not exhibit off odors until ~17 days later than similar beef kept at 20°C (Ayres, 1960).

When fresh meat is held at refrigeration temperatures (~0-10°C) under aerobic conditions, a type of microbial flora known as psychrotrophs will be favored over other organisms. Psychrotrophic bacteria can grow at such temperatures within a week or two, although their optimum temperatures are closer to room temperature (20-30°C). The typical bacterial flora that develops includes primarily *Pseudomonas, Acinetobacter, Moraxella, Alcaligenes,* and to a lesser degree, possibly Gram-positive spore formers, *Corynebacterium, Microbacterium, Arthrobacter,* Enterobacteriaceae, and other organisms. The predominant spoilage organisms on fresh refrigerated meat are *Pseudomonas* spp. (Ayres, 1960). These bacteria are aerobes and usually sensitive to low salt

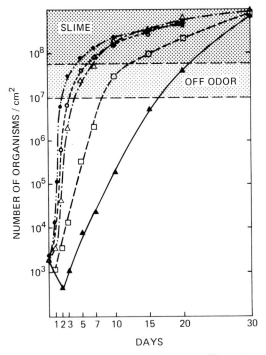

Fig. 2 Rates of growth of organisms on beef stored at 0 (▲), 5 (□), 10 (△), 15 (O), and 20°C (●). From Ayres (1960, p. 7). Copyright by Institute of Food Technologists.

concentrations, hence they are not typical spoilage organisms for cured meats. A relatively new genus, *Alteromonas,* also has been mentioned as a spoilage type, and includes the species formerly named *Pseudomonas putrefaciens.* Yeasts and molds also occur. *Pseudomonas,* however, dominates aerobic spoilage organisms by virtue of its rapid growth compared with competitive bacteria over a wide range of temperature and pH (Gill and Newton, 1977).

At temperatures $> 10°C$, problems can result from growth of pathogenic bacteria, or anaerobic spoilage organisms of the genus *Clostridium,* including *C. perfringens,* and fecal streptococci and Enterobacteriaceae (Ingram and Simonsen, 1980). These are mesophilic bacteria, with optimum growth temperatures of ~ 20 to 40°C. Although staphylococci are poor competitors, they may also be found on meats held at these temperatures. The importance of maintaining low temperatures for holding fresh meats cannot be sufficiently stressed.

Other environmental factors that influence generation time of microorganisms and keeping quality of meats are pH, salt concentration (water activity and osmotic pressure), oxidation–reduction potential, and production of metabolic wastes by the organisms themselves.

With regard to pH, meat provides favorable conditions for microbial growth when the pH is higher than ~ 5.5. Formation of lactic acid in muscle has been cited many times as a means for retarding microbial growth, with handling of animals prior to slaughter directed toward development of a low ultimate pH. Differences of pH in dark, normal, and pale pork influenced proportions of microorganisms recovered from the meat by Rey *et al.* (1976). Abnormal dark pork may be more susceptible to bacterial spoilage than normal meat, and normal pork more susceptible than pale pork because of the pH differences. As a consequence, while pork with a low pH has low water-holding capacity, soft texture, and poor salt retention, it may also harbor fewer bacteria than normal pork.

The oxidation–reduction properties of meat influence microbial survival or growth because of consumption of oxygen as tissue respires after slaughter. Since oxygen is no longer available from blood circulation after death of the animal, only the meat surfaces exposed to air may continue to provide aerobic conditions. In the deeper tissues of the meat, the flora that can grow is anaerobic or facultatively anaerobic. Most anaerobic bacteria are not psychrotrophic, so they do not readily develop inside refrigerated meat (Ingram and Simonsen, 1980). Cooling the deep tissues to inhibit anaerobic bacteria before the redox potential decreases sufficiently to favor these organisms is a necessary control measure. Bone taint, from growth of *Clostridium* spp., may be prevented by chilling the deep bone to ~ 15°C within 24 hr after slaughter (Locker *et al.,* 1975). Laboratory studies have been substantiated by actual experience with beef shipments in which temperature data and acceptability of the meat were recorded.

Gill (1980) reviewed invasion of deep tissue by bacteria after death of the animal, and from his work, concluded that it is not likely that postmortem invasion does occur. Other suggestions regarding agonal invasion (uptake of bacteria by the intestine after death) were also discounted. Gill also rejected the concept that contamination during slaughtering operations may be important in deeper tissue invasion, based on the bactericidal activity of the blood and lymph. He indicated that bacteria will be eliminated from the blood for 20 to 30 min after death when injected just before slaughter of the animal. However, survivors may grow within a few hours when carcasses are held at 30°C. Species differences existed in numbers eliminated, with *Clostridium perfringens* and salmonellae more resistant. Spores of *C. perfringens* were killed over the same time period as vegetative cells; bactericidal activity in carcass tissues was believed to be important in restricting growth of organisms in deep tissues. An initial inoculum in the order of 10^7 *C. perfringens* cells would be needed to cause contamination of beef at a level of 20 organisms/g. Earlier, Gill *et al.* (1976) showed that animals would not necessarily harbor live bacteria in the tissues after death even if the bacteria passed into the lymphatic system. Evisceration could be delayed up to 24 hr before deep tissue contamination might result from intestinal organisms.

The penetration of bacteria into meat was not considered by Maxcy (1981) to result from collagenolytic activity because most bacterial proteases have little effect on collagen and specific activity may therefore be lacking in bacterial proteases.

From these studies and conclusions, it is apparent that microbial spoilage of meats generally still may be attributed to the activity of the surface flora. Deep muscle penetration and growth are not as significant, although they may occur, particularly when chilling is inadequate.

A different type of flora is found on cured, processed meats than on fresh meats. Bacteria such as *Staphylococcus, Micrococcus, Lactobacillus, Microbacterium, Pedicoccus, Streptococcus, Clostridium,* and *Bacillus* are included in genera that have been isolated from cured meats. Salt-tolerant organisms may also include other types, such as *Vibrio* and *Sarcina.* Yeast and molds are not usually associated with freshly made cured meats, but after aging, sausages and country-cured hams may show growth of these fungi. Obviously, environmental factors such as temperature of processing, curing salts and other ingredients, smoking, and storage conditions all affect the flora that develops on cured meats.

While sodium nitrite at 200 ppm or less combined with salt and pH in the range 5.7–6.0 will inhibit many fresh meat spoilage organisms, the Gram-positive types mentioned above will survive or even proliferate under favorable conditions on cured meats. The effects of lower levels of nitrite such as 156 ppm and 40 ppm together with sorbate and ascorbate have received much attention since nitrite levels are now reduced compared with amounts used in earlier years.

According to Ingram and Simonsen (1980), an important distinguishing characteristic of cured meat spoilage is that putrefaction does not take place, differing from the spoilage associated with activity of psychrotrophs on fresh meats. With a water activity (a_w) of less than 0.96, psychrotrophic Gram-negative spoilage bacteria are inhibited. Enterobacteriaceae generally need an $a_w > 0.93$ for growth. At lower a_w levels, the cured meat organisms listed above become important. Spoilage evidenced by flavor change is not as objectionable as the proteolysis associated with fresh meats and often is characterized as sour or acid.

In general, the more acid the environment, the less salt needed for preservation. Salt is not as lethal at low temperatures as it is at higher temperatures. This has been shown by survival of salmonellae in curing brines held at ~5°C, but destruction of the bacteria at higher temperature brines. In most sausage products, 2.5 – 3% salt is used in curing; the concentration of salt in the aqueous phase of meat would be somewhat higher. Salt at this level is inhibitory to most fresh meat spoilage bacteria such as pseudomonads, so these are replaced by bacteria such as micrococci, streptococci, lactobacilli, and others previously mentioned.

Walker (1978) reviewed factors affecting growth of microorganisms in cured meats and some of the consequences of their growth.

Most bacteria have an optimum growth pH near 7.0, and as the pH becomes higher or lower the types of bacteria that can grow become more limited. The pH range of growth for the majority of bacteria found in fresh meats falls within the pH range 5.0 – 8.0. The pH of fresh meat will vary depending upon the amount of glycogen present at slaughter and subsequent changes in the meat. The pH will frequently fall in the range 5.7 – 6.0 but may on some occasions fall in a range as broad as 5.3 to 7.2. It has been shown that meat with a pH of 6.0 will spoil more rapidly than meat with a pH of 5.3. A number of bacteria exhibit a longer lag and a longer generation time at the lower pH. For example, the aerobic growth of food poisoning staphylococci is limited at pH 5.3 and anerobic growth is almost completely prevented at this pH. Thus the lower pH in meat products is desirable for delaying microbial spoilage. In some products the pH is lowered by addition of acids such as acetic, lactic, and citric, and in certain sausages the pH is lowered by acids produced by microbial fermentation. The growth of microorganisms that predominate on fresh meat is readily inhibited by pH 5.5 and by salt concentrations in the range of 5 to 8% and the combined effects that occur in processed meats are even greater than the two alone.

The role of nitrate in stabilizing of pasteurized cured meats has generally been accepted as being insignificant; any effect that it has is related to reducing the water activity.

Nitrite, of course, is an effective inhibitor of bacteria. The mechanism by which nitrite inhibits bacteria is not known in detail but it reacts via the undissociated nitrous acid form. Thus, its activity depends directly on pH. An optimum pH for the antimicrobial activity of nitrite is in the range 5.0 – 5.5. Nitrite blocks metabolic steps or reactions needed for outgrowth of spores, and it may act as a chelating agent (Tompkin et al., 1978a,b).

The extent to which nitrite inhibits growth of clostridia depends on such factors as pH, salt concentration, temperature, and numbers of cells. At low storage temperatures, less nitrite and salt are required to inhibit growth. Also, the smaller the number of susceptible bacteria, the smaller the amount of nitrite needed to prevent growth.

From the foregoing, it has been shown that a psychrotrophic association of microorganisms, predominantly pseudomonads and closely related organisms, originating from dirt on the animals, develops on fresh meat when it is cooled. These bacteria are adversely affected by conditions attained in cured meats (namely, pH, salt concentration, supply of oxygen, and presence of curing agents). They are also easily killed by heat; hence, they cause few problems in cured and processed meats.

Micrococci and streptococci, and microbacteria, which predominate on cured meats, resist salt and nitrite but are affected by pH in the practical range. Their heat resistance in the presence of fat may be relatively high. They may also be introduced into the product after heat processing.

Microbacteria and coryneforms are found in meats, both raw and cured, and various species possess individual properties (for example, resistance to pH and salt, ability to grow at low temperatures, and high heat resistance) that could cause problems if these characteristics are shared by a specific contaminant.

Lactobacilli are common in sausage, in which they sometimes cause greening. Some strains resist salt, acidity, and heating. Lactobacilli are common constituents in the gut of the pig and it is conceivable that the sausage strains originate there, but the most probable immediate source is unsanitary conditions in the processing plant.

Several defects may be caused by microorganisms on sausages. If substantial microbial growth occurs on the surface of a sausage or processed meat, a white or cream-colored or tannish slime will develop. This slime results from the accumulation of the cells of bacteria and yeasts. Slime is infrequently noted on the surfaces of vacuum-packaged products because there are fewer anaerobic bacteria likely to be contaminants on meat products. Also, organisms that grow under anaerobic conditions usually produce enough acid to inhibit growth of the aerobes that might form slime. Vacuum packaging prevents or inhibits the growth of molds, yeasts, micrococci, and others. Such facultative organisms as the streptococci, microbacteria, and lactobacilli can grow in vacuum-packaged meats. Even in vacuum packages, after a period of time, contaminants may multiply enough to produce visible growth or slime. A more common occurrence is growth of organisms in the liquid accumulated in the packages to make it appear milky. On nonvacuum-packaged products, before slime fully develops discrete colonies of growth can be observed; on frankfurters, for example, white, tan, or yellowish colonies can be observed, and as growth continues beads of growth develop and coalesce to produce a film that is sticky and slimy to the touch. As time goes on, yeasty and other off odors may develop. A mixture of micrococci, yeasts, Gram-positive rods, and streptococci may make up the flora of the surface film. The yeasts and micrococci frequently predominate under these circumstances. A vacuum-packaged luncheon meat that is properly refrigerated and has a low initial bacteria count can have a shelf life of several months.

Anaerobic metabolism of carbohydrates in meat products results in various fermentation products, primarily organic acids. Organisms such as the lactobacilli, streptococci, and microbacteria produce mainly lactic acid; its presence brings about a lowering of the pH of the meat and as a result the development of a sour flavor as described previously. The acid is not harmful but introduces an undesirable flavor into the product. Occasionally, certain bacteria are present that can produce carbon dioxide as well as acid; this condition may eventually produce gassy vacuum packages.

Additional deterioration in flavor of processed meat can occur as the result of enzymatic activity of microorganisms. These are lipolysis of fat and proteolytic metabolism of meat proteins. Ordinarily, these types of changes are not serious problems in vacuum-packaged processed meats. Microbial discoloration can occur, however, as green cores, green rings, and surface greening (Niven, 1951). Greening is well controlled by most sausage makers but it still occurs. Green cores are usualy associated with large sausages such as bologna. It occurs when greening bacteria are introduced into sausage emulsions and are not destroyed during cooking. The bacteria survive in the interior of the sausages but the color change does not occur until the product is cut and the surface is exposed to the air. These greening bacteria, principally *Lactobacillus viridescens*, produce hydrogen perioxide, which oxidizes the meat pigment to produce a grayish to greenish color. The change usually begins in the center of the meats and spreads to the edge. Spreading of the green area helps to distinguish microbiological greening from that caused by chemical or metallic sources. Cooking of processed meat products to temperatures of 67 to 68°C (152 to 155°F) destroys greening organisms; but occasionally if contamination by greening organisms is excessive, it may be necessary to cook to an internal temperature of 71°C (160°F). However, if good sanitation and proper refrigeration of the raw product are practiced to keep the numbers of organisms low, these problems should not develop.

The appearance of green rings in sausage rarely occurs. A combination of factors must be present for this problem to develop. Green rings appear at varying depths beneath the sausage, usually within 1 to 2 days after the sausage has been processed. This particular color change is noticeable as soon as the sausage is cut open, even though it has been properly refrigerated. Although greening bacteria are present throughout the sausages, discoloration develops in the form of rings, probably because the oxygen tension in this zone is conducive to oxidation of the pigment. This is the most frequently accepted theory for the occurrence of green rings. As with green cores, this condition is associated with the presence of large numbers of greening bacteria in the emulsion and subsequent undercooking of the product.

Surface greening can also be caused by similar bacteria. Surface greening caused by bacteria can be distinguished from that induced by chemicals or metals because of the length of time it takes for the defect to appear. Metallic discoloration may be discernible within a matter of hours, whereas the onset of microbial discoloration will not be noticeable much earlier than 5 days after processing. The time of onset varies with the numbers of organisms present. Surface greening does not occur on products that have been vacuum packaged.

Bacterial spoilage of braunschweiger and organisms involved was reviewed by Chyr *et al.* (1981). Gram-positive, catalase-negative cocci identified as *Streptococcus faecalis* and *Pedicoccus pentosaceus* were responsible. *Pedicoccus* produced souring with low pH and the *Streptococcus*' activity was characterized by a "perfumy or scented odor" under anaerobic conditions. Isolations were made from refrigerated ($5-7\,^{\circ}C$) products as well as braunschweiger held at $22\,^{\circ}C$. Previously, Steinke and Foster (1951) found differences in growth and types of microorganisms to be dependent on the kind of casing used for liver sausage. More permeable materials allowed for slime production consisting mainly of acid-forming micrococci, while less permeable materials did not permit slime formation. Chyr *et al.* (1980) had also determined that amount of contamination of pork liver was influential in determining numbers of organisms in the braunschweiger emulsion. With good bacterial quality of the raw emulsion, the product was capable of being held for more than 16 weeks at $5\,^{\circ}C$. The sausage was previously cooked to an internal temperature of $68\,^{\circ}C$, which caused a 97% reduction in bacterial numbers.

Effects of time and temperature of smoking of frankfurters was studied by Heiszler *et al.* (1972). They found that the major reduction in bacterial numbers occurred during the time the frankfurters were being heated to $60\,^{\circ}C$ ($140\,^{\circ}F$), although further reductions occurred in direct relationship to higher smokehouse temperatures up to $76.8\,^{\circ}C$ ($170\,^{\circ}F$). Also, with higher smoking temperatures, lower bacterial counts resulted during subsequent storage of the frankfurters at 5°C. Apparently, greater deposition of antimicrobial smoke constituents occurred with higher smokehouse temperatures, thus exerting a residual effect greater than that observed with lower temperatures of smoking. Heiszler et al. (1972) noted that incidence of coagulase-positive staphylococci on frankfurters after smoking was only ~1.5%, which is in agreement with the generally low incidence of Staphylococcus aureus recoveries from commercial frankfurters by Surkiewiez et al. (1976).

A survey of vacuum-packaged luncheon meats in Ontario by Duitschaever (1977) indicated that these cured meats did not present a public health problem although enterococci occurred in high incidence (all 159 samples were positive) and counts of Staphylococcus aureus were more than 1000/g in 20% of 30 positive samples. Luncheon meats examined were bologna, ham, chicken loaf, and macaroni cheese. Bacteriological quality varied considerably among different manufacturers' products. A noteworthy observation was that temperature abuse at the retail store contributed to high bacterial counts, with internal meat temperatures varying between 5 and 14°C at the time of purchase. Better sanitation practices in processing plants were considered to be necessary, as evidenced by the high incidence of enterococci.

For an excellent tabulation of cured meat defects and possible remedies, the reader is referred to Meat Industry, November, 1980.

Evidence has been obtained that modification of the curing salt mixture for frankfurters with 0.26% potassium sorbate, 40 or 140 ppm of sodium nitrite, and 500 ppm sodium isoascorbate did not greatly affect the natural microflora of the products or the pathogens Staphylococcus aureus and Clostridium perfringens (Hallerbach and Potter, 1981). With thuringer cervelat containing 78 ppm nitrite and other ingredients specified, similar results were observed. Wagner et al. (1982) found that vacuum-packaged bacon was protected from bacterial growth to a greater extent with 0.26% sorbate and 40 ppm nitrite than with 120 ppm nitrite and no sorbate in the cure. Reduction of nitrite from the traditional 200 ppm level, with other inhibitors such as sorbate, may be a feasible practice with regard to bacterial inhibition and related hazards from pathogenic organisms.

II. BIOCHEMICAL ACTIVITY OF SPOILAGE ORGANISMS IN MEATS

Activity of spoilage organisms on or in meats will be discussed primarily for psychrotrophs, since low-temperature spoilage is the main concern in keeping quality of most meat products (other than canned or dried meats).

Low temperatures are conducive to growth of psychrotrophs, which are characterized by their ability to produce off odors and slime on refrigerated meats (Ayres, 1960, 1963). Defects produced by low-temperature organisms run the gamut of odors and colors. Biochemical activity of psychrotrophs producing spoilage of meats has been the subject of much study. Lipolytic and proteolytic activity may not be as great at 30°C or higher temperatures as at refrigeration temperatures and room temperature < 30°C (Alford, 1960). Extracellular protease produced by a *Pseudomonas* from a frozen chicken pie was greater as test temperature was decreased from 30 to 0°C (Peterson and Gunderson, 1960). Biochemical activity of psychrotrophs was greater after chicken was frozen and thawed and then held in refrigerated storage than it was prior to freezing (Rey and Kraft, 1971). A similar situation may occur with other meats and may help explain why frozen – defrosted meats are still highly susceptible to spoilage after being held under refrigeration.

As Gill (1980) indicated, pseudomonads have an advantage over other meat microorganisms at low temperatures although initial number of pseudomonads may be low. Thus, *Pseudomonas* species readily attack proteins and eventually form breakdown products with characteristic off odors. This phenomenon has already been discussed. Lipolytic activity also contributes to off odors from the growth of pseudomonads on meats.

Again, it should be emphasized that environmental conditions and treatment of meats (packaging materials and methods, fresh versus cured, frozen, or dried meats) all affect microbial growth and subsequent types of spoilage that may occur. The influences of processing and packaging treatments were reviewed by Ingram and Simonsen (1980). The meat microbiologist must be familiar with these effects in dealing with particular products and their spoilage characteristics.

III. VIRUSES, PARASITES, FUNGI, AND OTHER ORGANISMS IN MEAT

Meats have been a subject of concern regarding transmission of viruses from foods to humans, but little evidence has been found to support this possibility (Cliver, 1980). Viruses that are infectious for animals do not usually infect man, but most outbreaks have been traced to human contamination through mishandling of food products derived from the animals. Viruses that contaminate meats are not likely to cause a human health hazard.

Viruses, nevertheless, may be transmitted through foods, whatever the original source, although they cannot multiply in meats. Transmission of viruses through foods was reviewed by Cliver (1971), who stated that intestinal viruses are sufficiently stable that they may be infectious in foods at the time the food is consumed. Heat processing inactivates viruses, but the time and temperature requirements vary considerably with the particular food item and different meats are no exception. Most viruses are inactivated at temperatures approximating those used for pasteurization (65°C) within 5 min (Sullivan *et al.*, 1970).

As indicated by Cliver (1980), viruses found in rare steak are probably of animal origin and not a health hazard, but viruses found in ground beef may be of human origin. A method developed by the U.S. Food and Drug Administration (Sullivan *et al.*, 1970) was used to examine market ground beef for viruses. No viruses were isolated from nine meat loaves in nine different markets, but poliovirus and echovirus were found in three packages of ground beef from three different markets.

Persistence of virus in inoculated sausage or ground beef was not significantly influenced by bacteria, although proteolytic bacteria were present in the meat tested (Hermann and Cliver, 1973). In the manufacture of Thuringer sausage, coxsackie virus type A9 (CA9) was able to survive bacterial fermentation by *Lactobacillus* as well as heat treatment at 49°C for as long as 6 hr. Indications were that enteroviruses are stable in ground meat products, which are liable to human contamination by excessive handling. Cliver's comments on meat as a source of recorded viral infections of humans relate to human origins; three outbreaks traced to meats were from human contamination, either by direct handling or from sewage (Cliver, 1980). Viral contamination of meats is still a potential hazard and requires the same precautions in sanitation as any other foodborne source. Larkin (1981) specified that foods should be heated to 70°C before consumption in areas where human intestinal content contamination is a possibility. Color change of red meats to brown in the center of the meat is an indication of sufficient heat treatment to inactivate most viruses.

The effects of food processing methods, institutional and home cooking procedures, and changes in technology are deserving of continuing study on transmission or survival of viruses in foods (Potter, 1973). For example, it may well be of significant public health interest to know the relationship of microwave cookery to virus survival, as well as changes in curing salt composition and addition or deletion of other food additives in present-day meats as a result of new regulations.

Several parasites may be transmitted to humans from meats, but current production and processing methods have decreased incidence of parasite infections from meats. Some parasites, their food source, human disease symptoms, and prevention of disease are listed in Table II.

Pork has been a common source of infection for trichinosis, which is probably the best-known food-borne parasitic disease. However, the causative agent, the nematode *Trichinella spiralis,* is distributed throughout the world and has been found in many types of wild animals in addition to the hog and the rat. Generally, trichinosis in humans is traced to consumption of undercooked pork, such as may be found in inadequately cured and undercooked sausage, or other pork products not certified as trichina-free. Obviously, such incidents of

Table II
Parsites Transmitted by Meats

Type of parasite	Food source	Symptoms	Prevention

Nematode
  Trichinella spiralis
	Pork
	Bear meat
	Walrus meat
		Occur in 3 stages:
		Gastroenteritis
		Muscle pain, fever,
		  other malaise
		Convalescence
			Heating (137°F,
			  58.3°C)
			Freezing — varies with
			  thickness of cut
			Curing (USDA, 1960)

Cestode *Taenia saginata*	Beef	Stomach pains, hunger	Thorough cooking
Taenia solium	Pork	Digestive disorder	Thorough cooking

trichinosis may be based on home processing operations. Decrease in infection in pigs because of greater care and a ban on feeding uncooked garbage to swine, as well as increased freezing of pork for home consumption, has contributed to decreased incidence of trichinosis in humans.

Federal inspection procedures may not detect trichinosis in swine carcasses, and uncooked pork products are regulated by the U.S. Department of Agriculture (1960) so that trichina are destroyed. Methods include heating, freezing, and curing. In heating, the entire meat cut must be heated to an internal temperature of 58.3°C (137°F) although some attempt has been made to change this minimum temperature. For freezing, times and temperatures depend on the thickness of the pork cut, and these conditions are as follows: for pieces not greater than 6 in., −15°C for 20 days, or −24°C for 10 days, or −29°C for 6 days. With larger thicknesses up to 27 inches, equivalent times at the temperatures specified are: 30 days, 20 days, and 12 days, respectively. Curing methods for destruction of viability of trichina are given in the federal regulations. Early symptoms of trichinosis resemble those of bacterial food poisoning, but later, when the larvae invade the muscles, muscle soreness, swelling, and fever may result. The majority of cases do not show clinical symptoms.

In some countries, inspection does attempt to detect cysts in the hog diaphragm by microscopic examination. At Iowa State University, Zimmerman *et al.* (1961) have devised a pepsin digestion – sedimentation procedure for counting larvae from diaphragm samples. Other methods for detection have been based on fluorescent antibody and agglutination procedures.

Fungi are important in meat microbiology not only because they are causes of spoilage, but because they may be potentially hazardous if they produce mycotoxins in meats.

As Jay (1978) indicated, bacteria usually outgrow molds and other fungi on meats and consume oxygen necessary for mold growth. Unless conditions become favorable for growth of molds as they become unfavorable for bacterial (e.g., lowering of water activity) growth, molds do not proliferate on meats.

On fresh meats, molds have been observed on beef quarters during transport on board ship in earlier times (Brooks and Hansford, 1923). "Black spot" and "whiskers" are characteristic signs of growth of molds such as *Thamnidium*, *Cladosporium*, and *Sporotrichum* spp. on the meat surfaces (Jay, 1978). Conditions conducive to mold growth on meats include reduction in water activity (a_w) to limit bacterial growth while still favoring molds; this may happen during long term storage or "ripening" (aging) of meats.

Jay (1978) listed 20 genera of molds isolated from different kinds of meats; *Aspergillus* and *Penicillium* species predominate. Among the meats given as sources were refrigerated beef, bacon, hams (particularly country-cured hams), salami, dry sausage, "cured meats," salted, minced pork, dried beef, and frankfurters.

In a study by Ayres *et al.* (1967), samples of cured and aged meats (fermented sausages, country-cured hams, and country-cured bacon) were analyzed for types of fungi present. A total of 670 strains were isolated from the meat products, which were collected from the United States and Europe. These strains included 456 molds and 214 yeasts, which were considered to be typical of the flora for the products. Fungi recovered from the same type of meats but in different locations were similar, indicating that a rather select flora exists regardless of where the meat is produced or marketed. Yeasts were more commonly associated with fermented sausages than with country-cured hams, mainly on the surface of the products, and contributed to spoilage when present in high numbers. Yeasts such as *Debaryomyces* and *Candida* spp. predominated. Ayres *et al.* (1967) concluded that fungi are important in retention of moisture and color of aged hams and sausages. Penicillia and aspergilli grow on hams after the meat loses ~ 25–30% of its original weight as moisture, and the a_w becomes reduced.

Yeasts have been recovered from bacon and sausage products, generally after some storage period (Cavett, 1962; Tonge *et al.*, 1964; Dowdell and Board, 1968).

Mycotoxins are compounds produced by molds and are toxic or have a biological effect on animals, including man (Bullerman, 1979). Many mycotoxins were first considered to be antibiotics, since they may have an antimicrobial action as well. However, their toxicity negates any potential as antibiotic substances. Mycotoxins of particular interest in foods are the aflatoxins, a group of chemically related substances which are metabolites of *Aspergillus flavus* and *A. parasiticus*. Because aflatoxins may present a significant public health hazard, the U.S. Food and Drug Administration, as well as regulatory agencies in other countries, has established limits to levels of aflatoxins permitted in foods or feeds. One possible route of infection with aflatoxins is by consumption of meat from animals that have been fed grains, particularly corn, contaminated with the toxins. Molds that may have a potential for producing mycotoxins in foods include not only *Aspergillus*, but also *Penicillium, Fusarium, Alternaria, Trichothecium, Cladosporium, Byssochlamys,* and *Sclerotinia* (Bullerman, 1979). When pigs were fed aflatoxin in their diets, the toxin was recovered from various tissues, with highest levels in the livers and kidneys (Jarvis, 1976). Ochratoxin, produced by species of *Aspergillus* and *Penicillium,* is a mycotoxin that has been found in tissues of pigs for bacon in Denmark (Krogh, 1976) and in pork meat in Yugoslavia (Krogh *et al.,* 1977).

Some molds, such as species of *Penicillium,* can grow and produce myocotoxins at refrigeration temperatures common to household refrigerators or supermarket display cases. Several types of mycotoxins were detected from culture extracts of molds collected from household refrigerators in a study by Torrey and Marth (1977). Some of the foods included meats or meat products.

Toxic molds have been recovered from cured smoked meats by several workers. Bullerman and Ayres (1968) studied 66 strains of molds from country-cured hams and sausages for aflatoxin-producing molds, and only *Aspergillus flavus* isolates were positive. Escher *et al.* (1973) were able to show production of ochratoxins from *A. ochraceus* on country-cured ham. Aspergilli and pencillia have been isolated from country-cured hams and sausages, *A. flavus* being responsible for production of aflatoxins and *Pencillium* spp. for producing penicillic acid. A temperature as low as 15°C was not inhibitory to aflatoxin production on cured meats in experiments designed to study temperature effects on toxin production. Aflatoxin production was observed on bacon, ham, and aged dry salami at 15°C (Bullerman *et al.,* 1969a,b). When bacterial spoilage was inhibited by antibiotics, aflatoxins were produced on inoculated fresh beef stored at 15, 20, and 30°C. However, at 10°C, bacterial and yeast spoilage resulted before aflatoxin could be detected. Other workers have indicated that aflatoxin was not produced on cured meats at temperatures below 15°C. Penicillic acid production is not favored by foods high in protein, but low in carbohydrates, such as meats (Bullerman, 1979).

Studies on the effects of curing salts on aflatoxin production by *Aspergillus parasiticus* have indicated that salt (NaCl) at levels of 2 to 3% or more was inhibitory, and that the inhibition depended on salt level, with or without nitrite in the curing salt mixture (Meier and Marth, 1977). It was believed that curing salt mixtures did not protect meat products from hazard of aflatoxin production if toxigenic molds were allowed to grow.

Aschenbrenner (1981) found that various levels of nitrite from 0 to 200 ppm had little effect on growth of *Aspergillus flavus* or *A. parasiticus* in sausage or media. Addition of 0.26% sorbate or 0.26% sorbate with 40 ppm nitrite was inhibitory to growth of *Aspergillus* and production of aflatoxin. Sorbate with no nitrite added had a greater inhibitory effect than did sorbate with nitrite. When NaCl at levels of 2.5 or 5.0% was included in media, inhibition of mold growth was more pronounced than when no salt was added. The possibility exists that sausages can support formation of aflatoxins when conditions are favorable for growth of toxigenic molds.

Persistence of mycotoxins in meats is important with regard to processing and storage treatments. Aflatoxins are considered to be stable in most food products (Bullerman, 1979). As Bullerman stated in his review, heat treatment does not generally reduce aflatoxin activity. Aflatoxin recovery decreases with storage time, with substantial reductions in recovery in frozen storage at −18°C. Koburger (1981) showed that yeast and mold populations decreased during storage of food samples, including meat, at freezer temperatures of −18, −26, and −34°C. Decreases were greatest at the highest storage temperature and not as pronounced as storage temperature was reduced. Plating medium influenced recoveries, with Antibiotic Plate Count agar (SPCA) believed to be more accurate than Potato Dextrose Agar (PDA) for enumeration.

Aflatoxins B_1 and G_1 were stable for 1 week in bologna, but patulin (a mycotoxin found in apples and apple products) and penicillic acid were not recovered (Lieu and Bullerman, 1977).

Information presented by Bullerman indicates that aflatoxins are the most stable mycotoxins in foods, but the public health significance of these findings is still not known.

IV. FOODBORNE DISEASES ASSOCIATED WITH MICROORGANISMS IN MEATS

A comprehensive work that provides valuable information on the subject of foodborne diseases caused by microorganisms is "Food-borne Infections and Intoxications," edited by Rieman (1969). The reader is referred to this book for details of the various organisms and diseases.

Bacterial foodborne disease is probably one of the most common forms of nonfatal illness occurring in the United States. Many of the bacteria responsible are so common that it is extremely difficult to protect foods completely from some of them. For example, staphylococci are common in the throat, on the skin, and in the air and can readily contaminate food. If these bacteria happen to be of the types that cause food poisoning, and conditions are right for their growth, such a food poisoning may result. There are probably a great many cases, particularly mild ones, that do not call for treatment by a physician. Probably most people have experienced at least a mild form of gastrointestinal upset at one time or another that is casually passed off by saying, "it must have been something I ate." This is quite true.

For many years, salmonellae, *Clostridium botulinum, C. perfringens,* and *Staphylococcus aureus* have been recognized as major types of pathogens involved in foodborne disease, with streptococci, *Bacillus cereus, Vibrio* spp., enteropathogenic *Escherichia coli,* and other organisms also of concern. Attention has also been given to the newer potential pathogens, *Yersinia enterocolitica* and *Campylobacter* species. Meats, since they are animal products, continue to be a focus of this attention.

Meats and poultry, and products containing these foods as ingredients, have been identified as the food vehicles responsible for more than half (~ 60%) of the reported foodborne disease outbreaks in the period 1968–1977 (Bryan, 1980). Food vehicles most commonly identified were ham, turkey, and roast beef. Cooked ground beef, pork, sausage, and chicken were also frequent food sources of outbreaks. Bryan cited data from the Communicable Disease Center (Center for Disease Control) of the U.S. Public Health Service. Of a total of about 1400 outbreaks, beef accounted for 35%, pork ~ 29%, and other meats, not including poultry, ~ 7%. Bryan reviewed the etiology of the foodborne diseases in relation to meats and poultry. As has been the case throughout many years, food service establishments were responsible for most of the mishandling resulting in the outbreaks (65%), followed by homes (31%), while processing plants accounted for the relatively small proportion of 4% of all outbreaks.

Probably the most characteristic feature of an outbreak of food poisoning is the explosive nature of the illness. Many people have had what is commonly known as the "GIs" at one time or another, in other words, a gastrointestinal upset. This is characteristic of food poisoning, perhaps only a minor set of symptoms in botulism, but generally typifies other foodborne disease. The term "ptomaine poisoning" is still sometimes used in error to describe food poisoning. This term has been in common use since it was first introduced about 1870 by an Italian toxicologist. It was taken from the word, "ptoma," meaning corpse. The symptoms were believed to be caused by food that had undergone putrefaction. Obviously, if an individual does come in contact with a food that is putrefied, changes are excellent that the individual will not consume such a food, provided, of course, that the olfactory sense is functioning. The term was a misnomer, and the more appropriate term is food poisoning. However, a distinction is made between types of food poisoning as food infections or food intoxications. In a food infection the illness is caused by eating a food containing disease-producing bacteria, such as *Salmonella* or possibly streptococci. True food poisoning refers to food intoxication, wherein the bacteria have already grown to produce a toxin in the food when it is eaten. This type of foodborne illness is caused by *Clostridium botulinum* or by *Staphylococcus aureus.*

Clostridium perfringens produces a mild intestinal disorder of short duration. Mishandling of foods during quantity cookery may result in perfringens poisoning. Meat cuts, stews, meat pies, meat sauces, and gravies prepared in large amounts are most frequently associated with this type of poisoning. If the food is kept hot enough or refrigerated properly, this organism can be held in check. This organism is mainly a problem in fresh meat dishes and not in sausages and processed meats. However, Bauer *et al.* (1981) reported that commercial pork sausage had an incidence of C. *perfringens* of ~39%, so it cannot be ignored as a potential pathogen in fresh pork sausage. Spores may survive when meats are cooked, which results in removing competitors for C. *perfringens.* Cooking produces a heat shock, allowing the spores to germinate to form vegetative cells in sufficient numbers to cause illness. As Bryan indicated (1980), outbreaks traced to meat may occur from the following causes: improper cooling, holding at warm temperatures, food preparation too far in advance of consumption, and inadequate reheating.

Salmonellae have been associated with a large variety of foods. Prevention of salmonellosis in meat products depends primarily upon good sanitation. The organisms are usually destroyed by heat (60°C, 140°F for 2.6 min) and are inhibited by moderate salt concentrations. If handling and sources of recontamination are avoided after meats have been heated, salmonellae should present no problem in processed meats. The generic name *Salmonella* is applied to a group of bacteria that once were called the paratyphoid bacteria. The genus also includes the organism that causes typhoid fever.

Salmonella infection resulting in food poisoning is caused by eating a large number of these bacteria in the food. This is not a true toxin type of poisoning; the bacteria have to grow, or multiply, in the food. Besides causing abdominal pain, diarrhea, chills, fever, vomiting, and possibly prostration, they may also produce bronchitis, pneumonia, endocarditis, meningitis, osteomyelitis, sinusitis, polynephritis, and abcesses. Food poisoning symptoms may come on within ~8 to 72 hr after swallowing the living organisms, which, although not characterized as producing true toxins, do produce an endotoxin in the body.

Salmonellae are widely distributed in animal species. Their usual habitat is the intestinal tract of animals and man. Animals have frequently been incriminated as a source of salmonellosis in man. Cattle have been known to carry the organisms for several years. There has been an increasing proportion of hogs that are infected as they move from farm to market to slaughter. Animals may become contaminated with salmonellae from feed, feed ingredients, particularly animal by-products, water, and other environmental sources. The bacteria can spread over meat surfaces during processing, and may cross-contaminate cooked meats by transfer from raw products when sanitation measures are inadequate. In a survey of retail stores, Swaminathan *et al.* (1978) found an overall incidence of salmonellae of ~15% in samples of pork, beef, and poultry. The highest contamination (21.5%) was from pork; ~11% of beef samples were contaminated.

With regard to the *Salmonella* problem, as it has been called, Silliker (1980) pointed out an interesting observation relating to the well-established fact that most foodborne diseases arise from food service establishments rather than processing plants. He advocated greater control over hotels, restaurants, food markets, institutions, and other preparation locations, and possibly less on on-line inspection of processing plants in order to make most advantageous use of inspectors.

Staphylococcus aureus, a Gram-positive coccus, is a cause of "staph" food poisoning. This organism can thrive in foods that are high in protein, sugar, and salt and that are held at a warm temperature for several hours. Generally, these organisms are associated with animal products that have been recontaminated after they have been heated. Meat products may contain staphylococcus enterotoxins after cooking, which destroys competing bacteria as well as staphylococci themselves (Bryan, 1980). Human contamination may occur, and then the infecting staphylococci may grow to produce enterotoxin if temperature conditions are favorable. While the bacteria are readily destroyed by heat, the toxin or enterotoxins they produce are not. The generation time for staphylococci is very short at a temperature of $10°C (50°F)$ or higher, and when growth conditions are ideal, a few organisms can change to millions within a matter of hours.

Staphylococcus food poisoning is not an infection, but is an intoxication that is characterized by having an abrupt, and often violent, onset of nausea, cramps, vomiting, diarrhea, and occasionally prostration.

While the toxin produced by a few cells is insignificant, the accumulated enterotoxin elaborated by millions of staphylococci is sufficient to induce symptoms within 2 to 4 hr. The illness lasts only a day or two and rarely results in death.

Staphylococcus aureus is abundant in the environment and it has been estimated that one-third to one-half of the human population has some of these organisms in nasal passages, mouth and throat, skin, and hair. When enterotoxin is detectable, it is likely that one or more of several things occurred: the ingredients were heavily contaminated, raw materials, including meat, were mishandled, and/or the processed product was cross-contaminated after processing. Raw and processed meats should be maintained at low temperatures, and any leftover materials should be promptly refrigerated or disposed of immediately. In past years, several meat processors experienced considerable difficulty from the occurrence of *S. aureus* and enterotoxin in Genoa salami. Factors affecting growth of staphylococci and production of enterotoxins were investigated by Lee *et al.* (1977). Bryan (1980) reviewed the possible causes of outbreaks of staphylococcal intoxication following consumption of any fermented sausage such as Genoa salami. Microorganisms, particularly staphylococci, may grow in products that are not smoked or cooked and are held at temperatures $> 18°C$ before drying. Staphylococci may be favored, as previously indicated, by the salt, low a_w and oxidation – reduction state, in comparison with competing organisms. Enterotoxin may be formed during early fermentation before inhibition results from processing treatments.

Botulism is a rare food toxemia caused by the anaerobic, Gram-positive, spore-forming bacillus, *Clostridium botulinum*. Even though it is rare, this type of food poisoning is of great concern because of its high mortality rate. The organism grows in the food and produces a neurotoxin. Symptoms develop normally within 8 to 72 hr, usually 12–36 hr, after the contaminated food is eaten. Early signs of botulism are weakness, fatigue, dizziness, blurred vision, and difficulty in speaking and swallowing. When sufficient toxin has been ingested, the diaphragm and chest muscles become paralyzed and the patient dies of asphyxiation.

Most botulism outbreaks are caused by home-preserved foods. While the danger posed by *Clostridium botulinum* in processed meat products is minimal, it still remains a hazard in meats. Nitrite and salt (as well as refrigeration) play important roles in controlling botulism in cured meats. Commercial meat products implicated in botulism outbreaks during the period 1968–1977 were spaghetti with meatballs, beef stew, and beef pot pie; inadequate thermal processing was a contributory factor in some instances (Bryan, 1980).

Yersinia enterocolitica has been isolated from meat foods and has become of increasing concern as a cause of gastroenteritis and other syndromes in humans. *Y. enterocolitica* has been recovered from vacuum-packaged beef and lamb after long storage (21–35 days) at 1 to 3°C. Freezing caused large decreases in numbers of bacteria on inoculated beef held at -23°C (Hanna *et al.*, 1976, 1977). The epidemiology of the organism is not distinct but study has been intensified on its isolation, identification, and pathogenicity.

V. DETECTION AND EVALUATION OF MICROORGANISMS IN MEATS

A concise but informative description of microorganisms and methods for their analysis is given by Johnston and Elliott (1976). The reader is referred to this work for details on sources of meat microorganisms and methods for their determination on ready-to-eat red meat (cooked cured and uncured meats, canned cured meats, fermented sausage, dried meats, and vinegar-pickled meats) and raw meat.

Guidelines for the microbiological evaluation of meat were presented by Kotula *et al.* (1980). The importance of meat microbiology was stressed with regard to monitoring microbial contamination of meat and processing equipment. The guidelines were formulated to serve as a reference for research and extension workers as well as teachers. Sampling, enumeration, and identification procedures were the main areas of consideration. These guidelines therefore are recommended reading for those individuals fitting the professional categories named, but students also should be in the reader audiences. Sampling methods for airborne organisms in processing plants include the use of exposure plates, millipore filters, and various types of aerosol sampling devices. For sampling meat surfaces, the contact plate, filter, swab, sponge, rinse, core, and free juice (drip) methods are suggested. All these methods are given in some detail in the article as well as in various source books and will not be detailed here. Sampling for internal portions involves cutting and excising followed by subsequent dilution, plating, incubation, and enumeration. Similarly, comminuted meat and ingredients are sampled on a gravimetric basis for making appropriate dilutions with standard bacteriological techniques for obtaining an estimate of microbial populations.

Transport of samples has the objectives of providing the samples to the analyst with a minimum of change in the population between the time of sampling and the time of receipt in the laboratory. Low temperature is used with insulated containers to protect from external temperature influences. On-site evaluation eliminates the need for transport, but this is not possible without a well-equipped mobile laboratory.

Tompkin (1976) reviewed sample transport, preparation, media, and incubation procedures for meats and emphasized that the choices made should be specified because of their influence on results obtained. If samples are frozen, they must be defrosted to permit analysis. Various methods of preparing initial suspensions include the use of a blender with different size containers such as canning jars for the Osterizer, the Stomacher, or simply a jar or bottle with diluent and sample mixed by shaking. The last method, as Tompkin indicated, is best suited for liquid samples such as drip from a package of meat or brine.

The Stomacher is a relatively recently accepted device in the United States for blending meat samples, although it has been more popular in Europe for some years. Several workers have compared the Stomacher to the blender for making a meat suspension for microbial analysis; both methods have advantages and disadvantages. The Stomacher essentially consists of a machine that has two paddles that beat a plastic bag containing the samples against a metal door. The beating action causes a shearing of the sample, intended to release bacteria while forming the meat suspension. One advantage of the Stomacher is that the bag and the residual sample may be easily discarded after analysis, in comparison with the cleanup work involved with reusable glassware. Emswiler et al. (1977) concluded that recoveries of bacteria from meat with either the Stomacher or blender were not significantly different. They therefore favored the more labor-saving Stomacher system for microbiological examination of meat products. A word of caution should be given here. This author's experience with the Stomacher has occasionally been a harrowing one because of broken plastic bags in the machine with subsequent salmonellae or other hazard not easily controlled. A recommendation is to use two bags, polyethylene or Cryovac, instead of one for holding the sample.

Several methods available for sampling meat surfaces were listed by Vanderzant et al. (1976). They stated that it should be understood that much variation in bacterial members and types may occur at different sites on a carcass or cut of meat. The ability of the meat surface to retain microorganisms affects the usefulness of the method for sampling or removal of those organisms, and all of these mechanisms also depend on the processing treatment given to the meat. These variables were reviewed by Vanderzant et al. (1976).

Diluents for bacteriological analysis have been the subject of much research, since they may affect bacterial populations. Once the meat sample is blended, however, the meat itself assumes a major role in influencing numbers of organisms recovered. Most methods texts provide adequate information on types of diluents that should be used, such as 0.1% peptone water, phosphate-buffered water, or saline solutions. Media that are employed for meat microorganisms also are listed in the appropriate texts and by the authors cited above.

Tompkin (1976) listed media for aerobic plate counts as plate count agar and APT agar, the latter for lactic acid producers. In our laboratory, trypticase soy agar (TSA) and LBS agar are used for the same purposes. For staphylococci, Baird–Parker agar is the medium of choice. Coliforms may be enumerated routinely with violet red bile (VRB) agar. Liquid (broth) media such as lauryl tryptose broth followed by brilliant green lactose bile broth for coliforms or EC broth for *Escherichia coli* are also used for enumeration. A good medium for analysis of *Clostridium perfringens* is sulfite–polymyxin–sulfadiazene (SPS) agar with D-cycloserine (Walker and Rey, 1972). Several media have been advocated for detecting salmonellae in meat products. Procedures recommended by the Association of Official Analytical Chemists (1976) are considered as "official" and are recognized by the Food and Drug Administration. Also, a good working procedure for anlaysis of salmonellae in meats is given by Galton et al. (1968). Again, all these methods are discussed or described in the "Compendium of Methods for the Microbiological Examination of Foods," edited by M. L. Speck (1976).

Many new methods have been applied to microbiological analysis of foods, including meats. These have been reviewed by Goldschmidt and Fung (1978). Such methods include the use of diagnostic kits, such as the Minitek system, the R/B system, the Pathotec system, the API system, Enterotube, and Auxotab among others. These commercial kits have been studied quite thoroughly; in general, they argree with conventional methods with an accuracy of 90% or more (Fung, 1976). Methods have been devised for automating media preparation, plating, and staining. For further information, the reader is referred to the paper by Goldschmidt and Fung (1978).

Several rapid analytical tests were evaluated by Strange et al. (1977) for their relationship to bacterial contamination of intact meat. Tests included color value, thiobarbituric acid number (TBA value) extract release volume [ERV, a predictor of microbial quality of beef described by Jay (1964)], pH, tryrosine value, and redox potential. Of these, the color measured by reflectance and the tryrosine value were the most closely related to bacterial populations.

Undoubtedly, many other methods could be described relating to detection and evaluation of microorganisms in meats. The subject has been widely researched, and the reader should recognize that there is a wide variety of procedures available. However, the references cited will serve well to present commonly accepted methods.

VI. CONTROL OF MICROORGANISMS IN MEATS AND QUALITY ASSURANCE

Control of microorganisms in meats has already been discussed to some extent with relation to methods of preservation such as high temperature (cooking and canning), low temperature (refrigeration and freezing), curing and smoking, dehydration, and fermentation. Packaging methods will be described in a later section, as will fermentation (with regard to use of microorganisms in processing). All of these operations serve to control microorganisms in or on meats. They involve many complex reactions relating to microorganisms and their meat substrates, as described in the section on spoilage.

Quality assurance programs for meat inspection and processing operations have been described by Angelotti (1978). The Meat and Poultry Inspection Acts require that the U.S. Department of Agriculture provide inspection of meat and poultry for sale in commerce after slaughter and processing. The amendments of 1967 to the Meat Inspection Act required that states provide inspection equal to federal inspection for those products remaining within the state. Many states have requested federal inspection even for such products because of difficulties encountered in conducting state inspection programs.

Because of the costs involved in implementing such large-scale inspection programs, the federal government has been seeking less drastic means than continuous inspection at slaughter. However, as Angelotti indicated, the demands of the consumer and responsibility of the federal and state governments are such that continuous inspection at slaughter of animals is necessary. For further processing operations, which have increased considerably in the last few decades, the additional inspection required is very difficult because of the large number of products involved.

Quality assurance programs have provided longer shelf life and consistency of quality and have benefited consumers as well as processors. Angelotti (1978) proposed that all processing be subject to voluntary quality assurance programs and that intensity of inspection be geared toward degree of health hazard as well as economic loss for different product–process combinations.

Regulations concerning cooking of beef have been issued in the Federal Register (Angelotti, 1977) because of outbreaks of salmonellosis from consumption of commercially prepared roast beef. The regulations stated that all such beef be processed to an internal temperature of 63°C (145°F). Goodfellow and Brown (1978) conducted research to find safe processing schedules that would still allow for production of rare roast beef while destroying salmonellae. They developed time–temperature processes that utilized sufficiently low internal temperatures to maintain the characteristic red color of rare beef, but with destruction of salmonellae. Processes that were based on a temperature below 57.6°C (136°F) provided desirable color. The regulations amending requirements to cook roast beef to a minimum internal temperature of 63°C (145°F) were published in 1978 (Federal Register, 1978).

Survival of several *Salmonella* serotype species and *Staphylococcus aureus* was studied by Smith *et al.* (1977), who inoculated beef snack sausages with the organisms and heated the products to temperatures of 51.1 to 52.2 or 57.8 to 58.9°C. The sausages were dried, and no visible salmonellae or staphylococci were detected with the higher heat treatment. The low internal temperature treatment was insufficient to eliminate food poisoning bacteria.

Research on thermal destruction of microorganisms in meat cooked by microwave or conventional means was reported by Crespo and Ockerman (1977). In general, as noted by earlier investigators, these workers believed that low-temperature (149 ± 6°C) oven cooking was more effective than microwave cooking in destroying meat microorganisms. High-temperature (232 ± 6°C) oven cooking was still more effective than the low-temperature oven cooking procedure for meat cooked to internal temperatures of 34, 61, and 75°C.

Many other reports have been concerned with control of meat microorganisms by thermal processing as well as other preservation methods. It is not the purpose of the author to review all these processes in this section, since much of this information is related to other sections in this chapter. Furthermore, the subject is such an extensive one that it would require a volume of its own. Much research in control of microbial growth on meats has dealt with use of hypochlorite solutions for sanitizing beef. Generally, bacterial contamination has been reduced by hypochlorite treatment (Titus *et al.*, 1978). However, results are variable and depend on many factors that influence the effectiveness of the hypochlorite.

VII. EFFECTS OF PROCESSING AND PACKAGING
METHODS ON MICROBIOLOGY OF MEATS

Several new processes for handling meats have been adapted or are being considered for use by the meat industry. Each innovation requires that attention be given to the microbiology of the product from the two main aspects of development of pathogens and growth of spoilage organisms.

New methods in meat processing described by Schmidt (1979) included vacuum chopping, the production of sectioned and formed meats with massaging and tumbling, making restructured meat products, controlled atmosphere smokehouses, and modified atmosphere packaging, which for this purpose might broadly include vacuum packaging and packaging in an atmosphere of carbon dioxide and other gases. Additional methods include high-temperature aging of intact or excised muscle for increased tenderness, hot processing, electrical stimulation for tenderness, and use of cryogenic gases for rapid chilling. Some of these operations may be used in combination. Since all of these techniques change the nature of the meat substrate from its original condition immediately after slaughter and initial processing, it is obvious that they may also affect the microbiology of the products.

Quality of restructured beef steaks after refrigerated and frozen storage was evaluated by Ockerman and Organisciak (1979). Tumbled beef tissue containing 2% salt was made into patties, vacuum packaged, and stored at refrigerator $(4° \pm 2°C)$ and freezer temperatures $(-23 \pm 3°C)$. Aerobic and anaerobic bacteria counts increased during refrigerated storage but remained relatively constant during frozen storage for 90 days, although thiobarbituric acid (TBA) values increased during both holding conditions. These investigators believed that antioxidants or phosphates as such would be needed for storage stability. No direct relation was indicated by the authors between bacterial counts and quality deterioration.

Vacuum packaging has been discussed to some extent in an earlier section. Vacuum packaging has been shown to enhance storage life of fresh pork (Lulves, 1981). The flora that normally develops as aerobic spoilage bacteria with conventional packaging is suppressed, and facultative anaerobes are favored (Holland, 1980). Lactic acid producers, Enterobacteriaceae, and *Microbacterium* may develop on vacuum packaged meats. Baran *et al.* (1970) observed that, in general, growth of aerobes was more rapid on fresh meat packaged in air than in vacuum, but the converse was true for anaerobic bacteria. It is believed that potential pathogens such as salmonellae, *Clostridium perfringens,* and *Staphylococcus aureus* do not present a hazard in vacuum-packaged meats since refrigeration will inhibit their growth (Holland, 1980). Vacuum-packaged luncheon meats, including sliced ham, did not harbor potential pathogens in a recent study of retail store products by Steele and Stiles (1981).

Modified atmospheres, in which a predetermined gaseous atmosphere is flushed into a gas-impermeable package, have been used to extend keeping time by inhibiting microbial growth on meats. Gases include carbon dioxide (CO_2), oxygen (O_2), and nitrogen (N_2). Modified atmosphere packaging provides a complex environment in the package. Respiration by the meat and consumption of oxygen by microorganisms depletes oxygen in the package, while carbon dioxide levels increase (one reason for the inhibitory effect of vacuum packaging is believed to be accumulation of carbon dioxide in the barrier package). Carbon dioxide may also diffuse into the meat (Ogilvy and Ayres, 1951a; Seideman et al., 1979). The gas extends the lag phase and generation time of bacteria, thus inhibiting growth. Much research has been done on the use of carbon dioxide for preserving meats and effects on the microbial flora. Ogilvy and Ayres (1951b) found that frankfurters had improved storage life with CO_2 concentrations up to 50%; greater concentrations did not improve keeping time any further. Use of carbon dioxide on fresh meat is limited to concentrations no higher than ~20 to 25% in the atmosphere because of discoloration at higher levels (Brooks, 1933). Materials having low gas permeability are best adapted for retarding spoilage of CO_2-packaged fresh meat (Kraft and Ayres, 1952).

Christopher et al. (1979) tested several gas mixtures of O_2, CO_2, and N_2 in comparison with vacuum packaging of beef roasts. The modified atmospheres generally allowed greater growth of psychrotrophic bacteria than did vacuum packaging after storage. Steaks prepared from the roasts showed similar trends, but differences were not significant.

An excellent review of modified atmospheres for fresh meats was presented by Holland (1980).

A system related to modified atmosphere packaging is hypobaric storage, which is storage in a controlled environment of low pressure, low temperature, and high humidity (Restaino and Hill, 1981). In studies in which a hypobaric trailer (a cargo container) was used at −2.2 to 0°C (28–31°F) 85–95% relative humidity, and pressure of 5 to 20 mm Hg, meats had extended shelf life of ~7 days for pork, and 2 weeks for bone-in processed hams compared with conventional refrigerated storage. Lamb, beef, and veal were also transported over a 42-day journey in excellent condition in the hypobaric container. Pseudomonas predominated at the time of spoilage of the meats.

High-temperature–short-time aging of beef has not only a beneficial effect on tenderness, but also on delay of spoilage compared with beef held at low temperature for sufficient time to produce equivalent tenderization (Rey et al., 1970). When beef muscle is excised after low-temperature holding of sides and then holding the excised muscle in the cooler, retail cuts may spoil more rapidly than if the muscle remains intact (Rey et al., 1978).

The use of cryogenic gases, particularly liquid nitrogen (LN_2) and liquid carbon dioxide (LCO_2) for refrigerants for meats is a fairly recent development. The rapid freezing rate provided by cryogens has advantages of reducing tissue damage and associated quality losses. Studies on the bacteriological quality of retail cuts from beef loins shell-frozen with LN_2 indicated the loins could be held at room temperature in insulated styrofoam boxes for 2 or 3 days without undue health hazard (Rey et al., 1971). Even with fluctuations in external temperature representing environmental conditions during holding and air transport, the chances of increasing occurrence of foodborne disease organisms on subsequently prepared retail cuts did not exist (Rey et al., 1972). In other work, beef patties frozen by LN_2, LCO_2, or by mechanical freezing had significantly greater reductions in bacterial counts with cryogenic freezing than with conventional freezing (Kraft et al., 1979). Survival in frozen storage for 5 months at $-18°C$ was greatest for Moraxella–Acinetobacter (42%), followed by Pseudomonas (32%), Staphylococcus (10%), Micrococcus (10%), and Streptococcus (5%). Staphylococci were protected by high fat content in beef patties, but populations remained relatively stable with all freezing methods. Cryogenic freezing is considered to be a feasible method for reducing microbial populations on ground meat and may be of practical significance for the fast-food industry.

Hot-processed, cryogenically frozen ground beef has greater sensory quality characteristics than the conventionally processed product (Jacobs and Sebranek, 1980). Hot-boned ground beef did not differ in microbiological quality from conventionally processed beef when the meat was examined 1, 2, 4, or 8 hr after slaughter and compared with control beef chilled for 24 to 48 hr (McMillin et al., 1981). Similar results were reported earlier by Emswiler and Kotula (1979) for ground beef from hot-boned or cold-boned carcasses.

Hot-boning combined with electrical stimulation for beef carcasses, primal cuts, ground beef, pork, and lamb cuts did not result in unusual increases in bacterial numbers on the meat (Kotula, 1981; Stern, 1980). Kotula indicated that electrical stimulation does not produce important effects on microbial populations on meat and can be used to advantage with hot-boning. Fung et al. (1981) recommended chilling hot-boned beef to 21°C within 3 to 9 hr after fabrication.

Because ground beef has been extended with textured soy protein (TSP) in several applications, the microbiology of ground beef–soy products has received considerable attention. Combinations of meat and soy protein have generally been considered to have shorter keeping times and higher bacterial loads than ground beef with no soy. Harrison et al. (1981) found that bacterial counts in a 20% TSP–ground beef blend were higher than in ground beef alone and spoilage occurred faster in the beef–soy product. Differences in bacterial flora were not believed to cause differences in shelf life, since both types of products had *Pseudomonas* as the predominant spoilage organism. In frozen beef–soy products, survival rates of aerobic and anaerobic organisms were higher than in frozen beef with no soy added after storage at $-29°C$ for as long as 56 days (Obioha and Kraft, 1976).

Another innovation in meat processing is the production of mechanically deboned meat (MDM) as a means of utilizing more meat protein to alleviate the shortage of protein for nutritional purposes and to make use of meat combinations that might otherwise be lost. The terminology for MDM has undergone some changes, but in this chapter, the simple MDM is used. Palatability is not decreased when MDM is added at levels up to 20% to ground beef (Cross *et al.*, 1977). Ground beef prepared with MDM at levels from 0 to 30% was comparable in microbiological safety in studies done by Emswiler *et al.* (1978). Chub packs of the ground beef were stored at $-1.1°C$ for 24 days and $-23.3°C$ for 6 months and analyzed at intervals for aerobic organisms, coliforms, *Escherichia coli, Staphylococcus aureus, Clostridium perfringens,* and salmonellae. These investigators indicated that with sound manufcturing practices, MDM can be produced as a wholesome product in ground beef.

VIII. MICROBIOLOGICAL CRITERIA FOR MEATS

In considering microbiological criteria for meat, it should be understood that several definitions may apply to the term *microbiological criteria*. The Food Protection Committee of the National Academy of Sciences (1964) defined a microbiological criterion as "any microbiological specification, recommended limit, or standard," and all three of those terms also have specific definitions. A specification usually applies to a value for buying and selling, or for in-house quality control, a limit is a suggested maximum number of microorganisms, and a standard is a legal maximum number of microorganisms or specific types of organisms. Criteria also may be considered as advisory (specifications) or mandatory (legally binding).

There has been much activity among meat industry groups, trade associations, and regulatory agencies concerning such criteria for meats in recent years. However, this type of activity is of long standing but continues to receive attention because of the many interests involved. Quality assurance programs have been based on microbiological criteria; this was reviewed briefly in Section VI. The status of microbiological standards for meat up to 1977 was reviewed by Tompkin (1977). Earlier, Elliott (1970, 1973) described the U.S. Department of Agriculture's regulatory programs for meat and poultry and for inspection under the programs, as applied to processing and equipment sanitation. As long ago as 1968, the USDA announced that it would establish microbiological criteria for meat and poultry products (Federal Register, 1968). Since then, several advisory committees and trade association committees, as well as the National Academy of Sciences and National Research Council, have reviewed microbiological criteria for meats.

Sampling plans have also been included in microbiological criteria, and the term *microbiological guideline* has come into the picture in connection with sampling. A microbiological guideline is, to quote Elliott (1970), "that level of bacteria in a final product, or in a shipped product, that requires identity and correction of causative factors in current and future production, or in handling after production." The microbiological guideline is based on a sampling plan that accounts for the commodity, the kind of test performed, and the relative hazard involved. Elliott stated that the program was correctional, citing the outbreaks of *Staphylococcus* food poisoning from Genoa salami as an example of remedial help by the USDA. This monitoring operation involved much work by the USDA and industry. Good manufacturing practice for salami products required that numbers of staphylococci should be low in the meat ingredients and that an active lactic acid starter culture be used or acidification be employed to inhibit growth of *Staphylococcus*.

Ground beef has received considerable attention with regard to standards, and since Oregon established standards for meat products, additional review has been done. According to Carl (1975), Oregon's establishment of microbiological standards for meat products at the retail level resulted in improvement in sanitary conditions in retail meat markets. Winslow (1975, 1979) questioned the effectiveness of the Oregon standards, which were repealed in 1977 as a result of another review by a committee appointed by the Oregon State Director of Agriculture.

The Center for Disease Control issued a report in 1975 that described an analysis of foodborne disease outbreaks from meat products during the period 1966–1973. The report indicated that ground beef, cold cuts, and frankfurters were relatively infrequently associated with food poisoning outbreaks, and that they were not "high-risk foods." Surveys for salmonellae substantiated this conclusion. McClory (1976) also stated that use of microbiological standards for ground beef was not an effective means of protection of the public health.

In 1970, the American Meat Science Association held a conference in which microbiological standards for fresh, canned, and cured meats were discussed. As Sulzbacher (1970) stated in his summary remarks, microbiological guidelines may serve to improve plant sanitation and quality control, but official standards cannot guarantee absolute safety of meat products. The purposes of microbiological criteria, as stated by Kotula (1970), will serve to close this section: to increase wholesomeness in meats, and to provide increased confidence and satisfaction in meat as a food.

The situation regarding microbiological criteria for meats is still under consideration, and may well occupy the attention of industry, government, and consumer groups for some time to come.

IX. USE OF MICROORGANISMS IN MEAT PROCESSING

The use of microbial cultures in meat processing was reviewed by Bacus and Brown (1981). They described the role of starter cultures in meat fermentations and their relation to inhibition of pathogenic bacteria. *Pedicoccus cerevisiae* was the first commercial starter culture for meats, partly because of its ability to survive lyophilization and obviously because of its ability to produce acid. *Pediococcus, Micrococcus,* and *Lactobacillus* are used as starter cultures for fermented meat products today. Some organisms considered to have been micrococci may actually be coagulase-negative staphylococci, as stated by Bacus and Brown. These organisms may also reduce nitrate and have catalase activity, which are desirable traits. Cultures may be marketed for fermentations as freeze-dried or frozen concentrates. Starter cultures for meat fermentations were developed after 1940 and have been improved considerably since then.

The major objectives of the use of starter cultures for meat products are to provide sufficient numbers of cells to compete successfully against pathogens and then produce sufficient acidity to inhibit any potentially hazardous organisms, while at the same time to promote quality and prolong shelf life. The problem of staphylococcus food poisoning in dry or semi-dry fermented sausage has been described earlier. Current recommendations specify controlled rapid acid formation in fermented sausage, either by starter cultures or addition of acidulants to control staphylococci (National Academy of Sciences, 1975).

Lactobacillus plantarum and *Pedicoccus cerevisiae* have been shown to restrict growth of staphylococci in country-style ham (Bartholomew and Blumer, 1980).

Lactic acid bacteria are also inhibitory to salmonellae; this has been demonstrated for skim milk and sausage. Gilliland (1980) demonstrated inhibition of *Clostridium perfringens* by *Lactobacillus* and *Pedicoccus*. Antagonisms against spoilage bacteria in ground beef have also been observed.

Lactic acid starter cultures have the ability to lower residual nitrite levels in cured meats, thus decreasing the potential for formation of nitrosamines (Bacus and Brown, 1981). Lactic cultures, by virture of their acid production, will inhibit growth of *Clostridium botulinum*. With rapid acid production in the presence of carbohydrates, lactic cultures may prevent toxin formation by *C. botulinum*.

Stimulation of acid production by starter cultures in Lebanon bologna as a result of action of the spice mix of black pepper, allspice, and nutmeg has been observed by Kissinger and Zaika (1978). Stimulation is not due to increased numbers of organisms, but to actual production of acid.

The future of use of microbial cultures for meat processing is a promising one, according to Bacus and Brown (1981). Meat cultures may become more specialized for particular products and processes for enhanced consistency and insurance of safety. Such application has the appeal of natural control and is a biological means for product development and maintenance of wholesomeness of meat products.

X. SUMMARY AND RESEARCH NEEDS

The meat industry has made technological advances in recent years that require that research be done on microbiology in relation to new processing methods and newly developed products or products in stages of commercial development. Improvement in shelf life and maintenance of a wholesome meat supply for the consuming public have been primary goals in the quality assurance programs of the industry. Developments in meat handling and processing have not only presented a continuing challenge to meat microbiologists, but have also stimulated greater interest in microbiological standards for meats.

Since meat originates from a live animal, much of the microbiology of meat products is associated with organisms found on or in the animal. However, processing operations, handling by workers, and plant sanitation have a great influence on the microorganisms found on the products. Research in prevention of contamination from live animal to finished meat product has been a major activity for many years, and undoubtedly, it will continue to be so for many years to come. Awareness of consumers and pressure of consumer groups has spurred regulatory agencies and industry to take increased action in providing wholesome meat products at an affordable price. The industry also needs to be profitable and continues to seek more economical means of production, utilization, and distribution to provide high-quality products. All of these factors make interesting research possibilities for meat microbiologists. Relationships between microbiology of meats and types of products as well as innovations in processing have been discussed in this chapter.

Increased emphasis is needed on research directed toward greater utilization of preservation measures that are relatively low in energy and water consumption. Effective sanitation measures are a continuing need in processing plants. Low-temperature treatment of meats, such as by use of cryogenic gases to provide inhibition of surface microbial growth, may warrant further investigation. A definite area for continued future research is that of modified atmosphere packaging, from the aspects of gaseous atmospheres and the most applicable packaging films. Vacuum packaging or packaging in different gaseous environments modifies the spoilage flora from aerobic types producing rather offensive off odors to more facultative souring types of bacteria. These findings should be studied in more detail so that greater advantage may be taken of controlled packaging systems to modify, inhibit, or perhaps eliminate spoilage types and potential pathogens.

The effectiveness of substitutes for nitrites, such as sorbates, in cured meats as inhibitors of microbial growth has been shown. However, this type of work is another area that demands continuous attention because of possible undesirable side effects on humans, and possibilities of development of more resistant strains of microorganisms over a time period.

Differences in types of microbial flora proliferating on fresh versus cured or smoked meats have been described and the reasons for these differences have been explained on the basis of salts and other ingredients added, pH, water activity, oxidation – reduction potential, and smoking, heat treatment, or refrigerated holding conditions. Additional research is needed on cured meat microbiology from aspects of prevention of growth of staphylococci, particularly in dry or semi-dry fermented sausage, and rapid means for detecting staphylococcal enterotoxins. New kinds of bacterial pathogens and viruses and parasites have also required special attention in meats, particularly regarding new processing and cooking methods, including institutional and home microwave cookery. Research may prove that this is a potential problem, or that no undue hazard exists.

Since mycotoxins have been shown to be carcinogenic as well as to produce other harmful effects in animals, continued mycotoxin research is needed on meat products that are subject to mold contamination. Salt at levels > 2% have been shown to inhibit aspergilli. Reduction of salt levels in cured meats as a possible means of reducing hypertension needs investigation with regard to mold growth and aflatoxin production. Sorbate is inhibitory to *Aspergillus flavus* and *A. parasiticus* and should be considered in future research if levels of NaCl are decreased in cured meats.

In addition to pathogens that have been found in different meat products over the years, attention is being given to more recently recognized organisms that potentially could cause disease by infecting meat products. *Yersinia enterocolitica* and *Campylobacter* species are included in this category. However, salmonellae, staphylococci, *Clostridium perfringens,* and C. *botulinum* are the major types of pathogens involved in food borne disease from meats. Food service establishments continue to be the main sources of illness from all foods, while processing plants account for only a small proportion of disease outbreaks. *Staphylococcus* food poisoning in dry or semi-dry sausages has been a problem in the past but can be controlled by controlled and rapid acidulation. Lactic acid starter cultures may be used to advantage for this purpose. Future research may well be directed toward development of a greater variety of fermented products and to antagonisms among microorganisms that may be used to advantage in decreasing risk from pathogens in meats.

New methods for microbiological analysis should be investigated further for application to meats or designed specifically for more direct application to meats. Time-saving, reproducible, accurate methods of analysis still have high priority in quality control operations. Guidelines have been published for microbiological evaluation of meat, and several references are available for procedures to be followed.

New methods of processing meats have been described in other chapters. Recent practices include aging meat at elevated temperatures, vacuum chopping, tumbling and massaging meat products, restructuring meat products, modified atmosphere packaging, hot processing and deboning, mechanically deboning meat, use of cryogens for freezing and refrigeration, electrical stimulation, and combinations of some of these practices. The microbiology of these operations has been studied to some extent, but by no means completely for all these methods. For example, hot processing can be varied considerably in conjunction with electrical stimulation. Research still should be done on the microbiology of the different procedures and should accompany any new developments in processing meats. As has been described earlier, additional work needs to be performed on effects of new formulations of curing salts and other ingredients in cured meats on potential pathogens, including toxigenic molds and spoilage organisms. Survival of microorganisms in meats cooked by microwaves, either for institution or home consumption, is a subject for additional research.

Soy protein–beef combinations generally promote greater growth of bacteria than beef with no soy extender, hence keeping quality may be decreased in the extended product. Research is needed to elucidate reasons for differences and to make practical use of the findings. With good manufacturing practices, ground beef containing mechanically deboned meat can be comparable in wholesomeness to meat products prepared with no mechanically deboned meat.

Because of interest by industry, regulatory agencies, and consumers in wholesomeness of meat products, increased attention has been given to microbiological criteria for meats. The meat industry has effectively used microbiological criteria for in-house quality control and for meeting other specifications. These programs may need to be expanded to include more products with one objective being to decrease the demand for federal and state inspection. Voluntary quality assurance programs as well as the regulatory and advisory criteria now in existence are systems for microbiological control of meat products at the processing plant level. Research should be continued on the relation between numbers and types of various organisms, their significance, and problems of interpretation with regard to microbiological criteria. Certainly, additional work on the relationship of criteria to product safety is needed and undoubtedly this problem will remain a subject of research, debate, and controversy in the future.

REFERENCES

Alford, J. A. (1960). Effect of incubation temperature on biochemical tests in the general *Pseudomonas* and *Achromobacter*. *J. Bacteriol.* **79**, 591–593.

Angelotti, R. (1977). Cooking requirements for cooked roast beef. *Fed Regist.* **42** (171), 44217.

Angelotti, R. (1978). Quality assurance programs for meat and poultry inspection and processing. *Food Technol. (Chicago)* **32**, 48–50.

Aschenbrenner, C. R. (1981). The effect of various levels of nitrite, sorbate, and erythorbate on *Aspergillus flavus* and *Aspergillus parasiticus*. M.S. Thesis, Iowa State University, Ames.

Association of Official Analytical Chemists (1976). "Bacteriological Analytical Manual for Foods," 4th ed. AOAC, Washington, D.C.

Ayres, J. C. (1955). Microbiological implications in handling, slaughtering, and dressing meat animals. *Adv. Food. Res.* **6**, 109–161.

Ayres, J. C. (1960). Temperature relationships and some other characteristics of the microbial flora developing on refrigerated beef. *Food Res.* **25**, 1–18.

Ayres, J. C., Lillard, D. A., and Leistner, L. (1967). Mold ripened meat products. *Proc. Annu. Reciprocal Meat Conf.* **20**, 156–168.

Bacus, J. N., and Brown, W. L. (1981). Use of microbial cultures: Meat products. *Food Technol. (Chicago)* **35**, 75–83.

Baran, W. L., Kraft, A. A., and Walker, H. W. (1970). Effects of carbon dioxide and vacuum packaging of color and bacterial count of meat. *J. Milk Food Technol.* **33**, 77–82.

Bartholomew, D. T., and Blumer, T. N. (1980). Inhibition of *Staphylococcus* by lactic acid bacteria in country-style hams. *J. Food Sci.* **45**, 420–425, 430.

Bauer, F. T., Carpenter, J. A., and Reagan, J. O. (1981). Prevalence of *Clostridium perfringens* in pork during processing. *J. Food Prot.* **44**, 279–283.

Brooks, F. T., and Hansford, C. G. (1923). Mould growth upon cold-store meat. *Trans. Br. Mycol. Soc.* **8**, 113–141.

Brooks, J. (1933). The effect of carbon dioxide on the colour changes or bloom of lean meat. *J. Soc. Chem. Ind., London* **52**, 17T–19T.

Bryan, F. L. (1980). Foodborne diseases in the United States associated with meat and poultry. *J. Food Prot.* **43**, 140–150.

Bullerman, L. (1979). Significance of mycotoxins to food safety and human health. *J. Food Prot.* **42**, 65–86.

Bullerman, L. B., and Ayres, J. C. (1968). Aflatoxin-producing potential of fungi isolated from cured and aged meats. *Appl. Microbiol.* **16**, 1945–1946.

Bullerman, L. B., Hartman, P. A., and Ayres, J. C. (1969a). Aflatoxin production in meats. I. Stored meats. *Appl. Microbiol.* **18**, 714–717.

Bullerman, L. B., Hartman, P. A., and Ayres, J. C. (1969b). Aflatoxin production in meats. II. Aged dry salamis and aged country cured hams. *Appl. Microbiol.* **18**, 718–722.

Carl, K. E. (1975). Oregon's experience with microbiological standards for meat. *J. Milk Food Technol.* **38**, 483–486.

Cavett, J. J. (1962). The microbiology of vacuum packed sliced bacon. *J. Appl. Bacteriol.* **25**, 282–289.

Center for Disease Control (1975). "Morbidity and Mortality Weekly Report," Vol. 24 (27). U.S. Dept. of Health, Education and Welfare, Public Health Serv., Atlanta, Georgia.

Christopher, F. M., Seideman, S. C., Carpenter, Z. L., Smith, G. C., and Vanderzant, C. (1979). Microbiology of beef packaged in various gas atmospheres. *J. Food Prot.* **42**, 240–244.

Chyr, C. Y., Walker, H. W., and Sebranek, J. C. (1980). Influence of raw ingredients, nitrite levels, and cooking temperatures on the microbiological quality of braunschweiger. *J. Food Sci.* **45**, 1732–1735.

Chyr, C. Y., Walker, H. W., and Sebranek, J. G. (1981). Bacteria associated with spoilage of braunschweiger. *J. Food Sci.* **46**, 468–470, 474.

Cliver, D. O. (1971). Transmission of viruses through foods. *CRC Crit. Rev. Environ. Control* **1**, 551–573.

Cliver, D. O. (1980). Viral hazards in meat. *Proc. Annu. Reciprocal Meat Conf.* **33**, 63–64.

Crespo, F. L., and Ockerman, H. W. (1977). Thermal destruction of microorganisms in meat by microwave and conventional cooking. *J. Food Prot.* **40**, 442–444.

Cross, H. R., Stroud, J., Carpenter, Z. L., Kotula, A. W., Nolan, T. W., and Smith, G. C. (1977). Use of mechanically deboned meat in ground beef patties. *J. Food Sci.* **42**, 1496–1499.

Daly, M. J. (1971). Trends in the U.S. meat economy. *Food Technol. (Chicago)* **25**, 826–829.

Dowdell, M. J., and Board, R. G. (1968). A microbiological survey of British fresh sausage. *J. Appl. Bacteriol.* **31**, 378–396.

Duitschaever, C. L. (1977). Bacteriological evaluation of some luncheon meats in the Canadian retail market. *J. Food Prot.* **40**, 382–384.

Elliott, R. P. (1970). Microbiological criteria in USDA regulatory programs for meat and poultry. *J. Milk Food Technol.* **33**, 173–177.

Elliott, R. P. (1973). Red meat and poultry inspection: microbiology of equipment and processing. *J. Milk Food Technol.* **36**, 337–339.

Emswiler, B. S., and Kotula, A. W. (1979). Bacteriological quality of ground beef prepared from hot and chilled beef carcasses. *J. Food Prot.* **42**, 561–562, 576.

Emswiler, B. S., Pierson, C. J., and Kotula, A. W. (1977). Stomaching vs. blending. *Food Technol. (Chicago)* **31**, 40–42.

Emswiler, B. S., Pierson, C. J., Kotula, A. W., and Cross, H. R. (1978). Microbiological evaluation of ground beef containing mechanically deboned beef. *J. Food Sci.* **43**, 158–161.

Ernst, L. J. (1979). Changing demands for meat. *Proc. Meat Ind. Res. Conf.* pp. 1–10.

Escher, F. E., Koehler, P. E., and Ayres, J. C. (1973). Production of ochratoxins A and B on country cured ham. *Appl. Microbiol.* **26**, 27–30.

Federal Register (1968). U.S. Department of Agriculture, Consumer and Marketing Service. Establishment of microbiological criteria to supplement regular sanitary inspections. *Fed. Regist.* **33** (247), 19042, 19043.

Federal Register (1978). Cooking requirements for cooked beef and roast beef. Title 9. Ch. III, Parts 310 and 320. *Fed. Regist.* **43** (138), 30791-30793.

Food Protection Committee (1964). An evaluation of public health hazards from microbiological contamination of foods. *N.A.S.-N.R.C., Publ.* **1195**, 1–64.

Fung, D. Y. C. (1976). New methods for pathogens: A review. *Proc. Annu. Reciprocal Meats Conf.* **29**, 284–300.

Fung, D. Y. C., Kastner, C. L., Lee, C., Hunt, M. C., Dikeman, M. E., and Kropf, D. H. (1981). Initial chilling rate effects on bacterial growth on hot-boned beef. *J. Food Prot.* **44**, 539–544.

Galton, M. M., Morris, G. K., and Martin, W. T. (1968). "Salmonellae in Foods and Feeds. Review of Isolation Methods and Recommended Procedures." Communicable Disease Center, Atlanta, Georgia.

Gill, C. O. (1980). Total and intramuscular bacterial populations of carcasses and cuts. *Proc. Annu. Reciprocal Meat Conf.* **33**, 47–53.

Gill, C. O., and Newton, K. G. (1977). The development of aerobic spoilage flora on meat stored at chill temperatures. *J. Appl. Bacteriol.* **73**, 189–195.

Gill, C. O., Penney, N., and Nottingham, P. M. (1976). Effects of delayed evisceration on the microbial quality of meat. *Appl. Environ. Microbiol.* **31**, 465–468.

Gilliland, S. E. (1980). Use of lactobacilli to preserve fresh meat. *Proc. Annu. Reciprocal Meat Conf.* **33**, 54–62.

Goldschmidt, M. C., and Fung, D. Y. C. (1978). New methods for microbiological analysis of food. *J. Food Prot.* **41**, 201–219.

Goodfellow, S. J., and Brown, W. L. (1978). Fate of *Salmonella* inoculated into beef for cooking. *J. Food Prot.* **31**, 598–605.

Hallerbach, C. M., and Potter, N. N. (1981). Effects of nitrite and sorbate on bacterial populations in frankfurters and thuringer cervelat. *J. Food Prot.* **44**, 341–346.

Hanna, M. O., Zink, D. L., Carpenter, Z. L., and Vanderzant, C. (1976). *Yersinia enterocolitica*-like organisms from vacuum packaged beef and lamb. *J. Food Sci.* **41**, 1254–1256.

Hanna, M. O., Stewart, J. C., Carpenter, Z. L., and Vanderzant, C. (1977). Effect of heating, freezing, and pH on *Yersinia enterocolitica*-like organisms from meat. *J. Food Prot.* **40**, 689–692.

Harrison, M. A., Melton, C. C., and Draughon, F. A. (1981). Bacterial flora of ground beef and soy extended ground beef during storage. *J. Food Sci.* **46**, 1088–1090.

Heiszler, M. G., Kraft, A. A., Rey, C. R., and Rust, R. E. (1972)., Effect of time and temperature of smoking on microorganisms on frankfurters. *J. Food Sci.* **37**, 845–849.

Hermann, J. E., and Cliver, D. O. (1973). Enterovirus persistence in sausage and ground beef. *J. Milk Food Technol.* **36**, 426–428.

Holland, G. C. (1980). Modified atmospheres for fresh meat distribution. *Proc. Meat. Ind. Res. Conf.* pp. 21–39.

Ingram, M., and Simonsen, B. (1980). Meat and meat products. *In* "Microbial Ecology of Foods" (J. H. Silliker, Chairman), Vol. 2, pp. 333–409. Academic Press, New York.

Jacobs, D. K., and Sebranek, J. G. (1980). Use of prerigor beef for frozen ground beef patties. *J. Food Sci.* **45**, 648–651.

Jarvis, B. (1976). Mycotoxins in food. *In* "Microbiology in Agriculture, Fisheries and Food" (F. A. Skinner and J. G. Carr, eds.), pp. 251–267. Academic Press, New York.

Jay, J. M. (1964). Beef microbial quality determined by extract-release volume (ERV). *Food Technol. (Chicago)* **18**, 1637–1641.

Jay, J. M. (1978). Meats, poultry and seafoods. *In* "Food and Beverage Mycology" (L. Beuchat, ed.), pp. 129–144. Avi Publ. Co., Westport, Connecticut.

Johnston, R. W., and Elliott, R. P. (1976). *In* "Compendium of Methods for the Microbiological Examination of Foods" (M. L. Speck, ed.), pp. 540–548. Am. Public Health Assoc., Washington, D.C.

Kissinger, J. C., and Zaika, L. L. (1978). Effect of major spices in Lebanon bologna on acid production by starter culture organisms. *J. Food Prot.* **41**, 429–431.

Koburger, J. A. (1981). Effect of frozen storage on fungi in foods. *J. Food Prot.* **44**, 300–301.

Kotula, A. W. (1970). Microbiological criteria for fresh meat. *Proc. Annu. Reciprocal Meats Conf.* **23**, 121–138.

Kotula, A. W. (1981). Microbiology of hot-boned and electrostimulated meat. *J. Food Prot.* **44**, 545–549.

Kotula, A. W., Ayres, J. C., Huhtanen, C. N., Stern, N. J., Stringer, W. C., and Tompkin, R. B. (1980). Guidelines for microbiological evaluation of meat. *Proc. Annu. Reciprocal Meats Conf.* **33**, 65–70.

Kraft, A. A., and Ayres, J. C. (1952). Post-mortem changes in stored meats. IV. Effect of packaging materials on keeping quality of self-service meats. *Food Technol. (Chicago)* **6**, 8–12.

Kraft, A. A., Reddy, K. V., Sebranek, J. G., Rust, R. E., and Hotchkiss, D. K. (1979). Effect of composition and method of freezing on microbial flora of ground beef patties. *J. Food Sci.* **44,** 350–354.

Krogh, P. (1976). Mycotoxic nephropathy. *Adv. Vet. Sci. Comp. Med.* **20,** 147–170.

Krogh, P., Hald, B., Plestina, R., and Ceovic, S. (1977). Balkan (endemic) nephropathy and foodborne Ochratoxin A: Preliminary results of a survey of foodstuffs. *Acta Pathol. Microbiol. Scand., Sect. B:* **85B,** 238–240.

Larkin, E. P. (1981). Food contaminants—viruses. *J. Food Prot.* **44,** 320–325.

Lee, I. C., Harmon, L. G., and Price. J. F. (1977). Growth and enterotoxin production by staphylococci in Genoa salami. *J. Food Prot.* **40,** 325–329.

Lieu, F. Y., and Bullerman, L. B. (1977). Production and stability of aflatoxins, penicillic acid and patulin in several substrates. *J. Food Sci.* **42,** 1222–1224, 1228.

Locker, R. H., Davey, C. L., Nottingham, P. M., Haughey, D. P., and Law, N. H. (1975). New concepts in meat processing. *Adv. Food Res.* **21,** 158–221.

Lulves, W. J. (1981). Quality changes in vacuum packaged fresh pork. Ph.D. dissertation, Iowa State University, Ames.

McClory, M. J. (1976). Update: Microbiology of fresh meat. A retail perspective. *Proc. Annu. Reciprocal Meats Conf.* **29,** 317–335.

McMillin, D. J., Sebranek, J. G., and Kraft, A. A. (1981). Microbial quality of hot-processed ground beef patties processed after various holding times. *J. Food Sci.* **46,** 488–490.

Maxcy, R. B. (1981). Surface microenvironment and penetration of bacteria into meat. *J. Food Prot.* **44,** 550–552.

Meier, K. R., and Marth, E. H. (1977). Production of aflatoxin by *Aspergillus parasiticus* NRRL-2999 during growth in the presence of curing salts. *Mycopathologia* **61,** 77–84.

National Academy of Sciences (1975). Prevention of microbial and parasitic hazards associated with processed foods. *In* "Fermented Foods, Fermented Sausages," Chapter 7. Natl. Acad. Sci., Washington, D.C.

Niven, C. F. (1951). Sausage discolorations of bacterial origin. *Am. Meat Inst. Found., Bull.* **13.**

Obioha, I. W., and Kraft, A. A. (1976). Bacteriology of ground beef and soy-extended ground beef. *J. Milk Food Technol.* **39,** 706 (abstr.).

Ockerman, H. W., and Organisciak, C. S. (1979). Quality of restructured beef steaks after refrigerated and frozen storage. *J. Food Prot.* **42,** 126–130.

Ogilvy, W. S., and Ayres, J. C. (1951a). Postmortem changes in stored mets. II. The effect of atmospheres containing carbon dioxide in prolonging the storage life of cut-up chicken. *Food Technol. (Chicago)* **5,** 97–102.

Ogilvy, W. S., and Ayres, J. C. (1951b). Postmortem changes in stored meats. III. The effect of atmospheres containing carbon dioxide in prolonging the storage life of frankfurters. *Food Technol. (Chicago)* **5,** 300–303.

Peterson, A. C., and Gunderson, M. F. (1960). Some characteristics of proteolytic enzymes from *Pseudomonas fluorescens. Appl. Microbiol.* **8,** 98–104.

Potter, N. (1973). Viruses in foods. *J. Milk Food Technol.* **36,** 307–310.

Restaino, L., and Hill, W. M. (1981). Microbiology of meats in a hypobaric environment. *J. Food Prot.* **44,** 535–538.

Rey, C. R., and Kraft, A. A. (1971). Effect of freezing and packaging methods on survival and biochemical activity of spoilage organisms on chicken. *J. Food Sci.* **36,** 454–458.

Rey, C. R., Kraft, A. A., Walker, H. W., and Parrish, F. C., Jr. (1970). Microbial changes in meat during aging at elevated temperature and later refrigerated storage. *Food Technol. (Chicago)* **24,** 67–71.

Rey, C. R., Kraft, A. A., and Rust, R. E. (1971). Microbiology of beef shell frozen with liquid nitrogen. *J. Food Sci.* **36,** 955–958.

Rey, C. R., Kraft, A. A., and Rust, R. E. (1972). Effect of fluctuating storage temperatures on microorganisms on beef shell frozen with liquid nitrogen. *J. Food Sci.* **37**, 865–868.

Rey, C. R., Kraft, A. A., and Parrish, F. C., Jr. (1978). Microbiological studies on aging of intact and excised beef muscle. *J. Food Prot.* **41**, 259–262.

Rieman, H., ed. (1969). "Food-Borne Infections and Intoxications." Academic Press, New York.

Schmidt, G. R. (1979). New methods in meat processing. *Proc. Meat Ind. Res. Conf.* pp. 31–40.

Seideman, S. C., Carpenter, Z. L., Smith, G. C., Dill, C. W., and Vanderzant, C. (1979). Physical and sensory characteristics of beef packaged in modified gas atmospheres. *J. Food Prot.* **42**, 233–239.

Silliker, J. M. (1980. Status of *Salmonella*—ten years later. *J. Food Prot.* **43**, 307–313.

Smith, J. L., Huhtanen, C. N., Kissinger, J. C., and Palumbo, S. A. (1977). Destruction of *Salmonella* and *Staphylococcus* during processing of a nonfermented snack sausage. *J. Food Prot.* **40**, 465–467.

Speck, M. L., ed. (1976). "Compendium of Methods for the Microbiological Examination of Foods." Am. Public Health Assoc., Washington, D.C.

Steele, J. E., and Stiles, M. E. (1981). Microbial quality of vacuum packaged sliced ham. *J. Food Prot.* **44**, 435–439.

Steinke, P. K. W., and Foster, E. M. (1951). Effect of different artificial casings on the microbial changes in refrigerated liver sausage. *Food Res.* **16**, 289–293.

Stern, N. J. (1980). Effect of boning, electrical stimulation and medicated diet on the microbiological quality of lamb cuts. *J. Food Sci.* **45**, 1749–1752.

Strange, E.D., Benedict, R. C., Smith, J. L., and Swift, C. E. (1977). Evaluation of rapid tests for monitoring alterations in meat quality during storage. *J. Food Prot.* **40**, 843–847.

Sullivan, R., Fassolitis, A. C., and Read, R. B., Jr. (1970). Method for isolating viruses from ground beef. *J. Food Sci.* **35**, 624–626.

Sulzbacher, W. L. (1970). Microbiological standards, summary and projections. *Proc. Annu. Reciprocal Meat Conf.* **23**, 146–153.

Surkiewicz, B. F., Johnston, R. W., and Carosella, J. M. (1976). Bacteriological survey of frankfurters produced at establishments under federal inspection. *J. Milk Food Technol.* **39**, 7–9.

Swaminathan, B., Link, M. A. B., and Ayres, J. C. (1978). Incidence of salmonellae in raw meat and poultry samples in retail stores. *J. Food Prot.* **41**, 518–520.

Titus, T. C., Acton, J. C., McCaskill, L., and Johnson, M. G. (1978). Microbial persistence on inoculated beef plates sprayed with hypochlorite solutions. *J. Food Prot.* **41**, 606–612.

Tompkin, R. B. (1976). Microbiological techniques: Sample transport, sample preparation, media, and incubation. *Proc. Annu. Reciprocal Meat Conf.* **29**, 270–277.

Tompkin, R. B. (1977). Meat microbiology update. *Proc. Meat Ind. Res. Conf.* pp. 129–132.

Tompkin, R. B., Christiansen, L. N., and Shaparis, A. B. (1978a). Enhancing nitrite inhibition of *Clostridium botulinum* with isoascorbate in perishable canned cured meat. *Appl. Environ. Microbiol.* **35**, 59–61.

Tompkin, R. B., Christiansen, L. N., and Shaparis, A. B. (1978b). Effect of prior refrigeration on botulinal outgrowth in perishable canned cured meat when temperature abused. *Appl. Environ. Microbiol.* **35**, 863–866.

Tonge, R. J., Baird-Parker, A. C., and Cavett, J. J. (1964). Chemical and microbiological changes during storage of vacuum packaged sliced bacon. *J. Appl. Bacteriol.* **27**, 252–264.

Torrey, G. S., and Marth, E. H. (1977). Isolation and toxicity of molds from foods stored in homes. *J. Food Prot.* **40**, 187–190.

U.S. Department of Agriculture (1960). "Regulations Governing the Meat Inspection of the United States Department of Agriculture". U.S. Govt. Printing Office, Washington, D.C.

Vanderzant, C., Carpenter, Z. L., and Smith, G. C. (1976). Sampling carcasses and meat products. *Proc. Annu. Reciprocal Meat Conf.* **29**, 258–269.

Wagner, M. K., Kraft, A. A., Sebranek, J. G., Rust, R. E., and Amundson, C. M. (1982). Effect of pork belly-type on the microbiology of bacon cured with or without potassium sorbate. *J. Food Prot.* **45,** 29–32.

Walker, H. W. (1978). "Meat Microbiology" (presented at the Sausage and Processed Meats Short Course). Iowa State University, Ames.

Walker, H. W., and Rey, C. R. (1972). Methodology for *Staphylococcus aureus* and *Clostridium perfringens. Proc. Annu. Reciprocal Meat Conf.* **25,** 342–359.

Winslow, R. L. (1975). A retailer's experience with the Oregon bacterial standards for meat. *J. Milk Food Technol.* **38,** 487–489.

Winslow, R. L. (1979). Bacterial standards for retail meats. *J. Food Prot.* **42,** 438–442.

Zimmerman, W. J., Schwarte, L. H., and Biester, H. E. (1961). On occurrence of *Trichinella spiralis* in pork sausage available in Iowa (1953–60). *J. Parasitol.* **47,** 429–432.

7

Sensory Qualities of Meat

H. R. CROSS, P. R. DURLAND, and S. C. SEIDEMAN
USDA-ARS Roman L. Hruska
U.S. Meat Animal Research Center
Clay Center, Nebraska

I. INTRODUCTION

In recent years, economic pressures have challenged the livestock and meat industry to seek ways of producing meat products that will enable the consumer to receive maximum palatability benefits at the lowest cost. Such appeal depends on factors such as meat color, flavor, aroma, tenderness, and method of cookery. Variation in these factors influences consumer acceptance of the meat product. Thus, this chapter emphasizes the growing importance and scope of these sensory properties of meat as they are related to consumer preference.

II. MEAT COLOR

It is important to the meat industry that the physical appearance of meat offered to the consumer at the retail level be of a high degree of acceptability. The only quality characteristic that consumers have to base their initial purchase decision on is physical appearance. Additionally, a consumer expects a cut of meat to look as appealing after it is cooked as it appears in the retail meat case. In considering the specific features contributing to physical appearance, researchers agree that meat color is one of the most influential quality factors in consumer selection. Adams and Huffman (1972) claimed that consumers relate the color of lean to freshness. Urbain (1952) reported that consumers have learned through experience that the color of fresh meat is bright red and any deviation from this bright red color is unacceptable. For example, discoloration of packaged fresh meat in the retail case is known as "loss of bloom" to the meat industry, but to the consumer it may mean spoilage. There are many factors that could be responsible for "loss of bloom." When discussing meat color, factors affecting discoloration should be emphasized as well as meat color intensity.

The principal pigments in muscle tissue associated with color are the blood pigment hemoglobin, which is primarily in the blood stream, and a smaller molecule, myoglobin, which is located within the muscle cell. Other pigments are also found in muscle and meat but are present in such small amounts that their contribution to meat color is minor. Between 20 and 30% of the total pigment present in the live animal is hemoglobin (Fox, 1966). The biological function of hemoglobin is to transport oxygen from the lungs to the muscle cells via the circulatory system. At the cell wall, hemoglobin surrenders the oxygen molecule to myoglobin. Myoglobin then functions to bind oxygen in muscle cells for use in the subsequent metabolic breakdown of various metabolites, such as in the tricarboxylic acid cycle.

Myoglobin (MW 16,000) is a conjugated protein consisting of the iron–porphyrin compound, heme, combined with a simple globin protein. The heme functions to bind oxygen while the globin protein serves to surround and protect the heme. Hemoglobin (MW 67,000) is a conjugated protein containing four heme molecules and four globin proteins. The heme portion of these two molecules is identical. The slight differences lie in the globin protein that surrounds the hemes. Essentially, hemoglobin and myoglobin are nearly identical and undergo the same color reactions. The hemoglobin molecules can, therefore, be thought of as four myoglobin molecules bound together. The abbreviated formula for the heme molecule is shown in Fig. 1. The iron atom of heme is bound covalently to four nitrogens, each constituting part of the pyrole ring of the porphyrin structure. In myoglobin, six bond orbitals are present. The iron atom is bound to four nitrogen atoms in the porphyrin ring

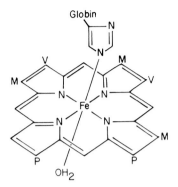

Fig. 1 Schematic representation of the heme complex of myoglobin. Adapted from Bodwell and McClain (1971).

and one nitrogen atom of a histidine molecule of the globin protein. The sixth bond orbital is open for the formation of complexes with several compounds (Bodwell and McClain, 1971). Variations at this sixth position are, in part, responsible for differences in the color of meat.

When an animal is stunned and bled, the blood and associated hemoglobin are removed from the animal. This leaves myoglobin as the primary pigment responsible for meat color. In living tissue, the reduced form of myoglobin, which is purple in color, exists in equilibrium with its oxygenated form, oxymyoglobin, which is bright red. The iron atom in both of these pigment forms is in the reduced state (Fe^{2+} or ferrous). When the interior of the meat is exposed to oxygen from the air, the oxygen combines with the heme of myoglobin to produce oxymyoglobin. Thus, the meat color changes from purplish-red to bright red. If oxygen is then excluded from the cut meat, the color will eventually turn purplish-red again because the pigment is deoxygenated back to myoglobin (Fig. 2).

Changes in meat color are due to two chemical reactions within the heme portion of myoglobin. One reaction is dependent upon the molecule bound at

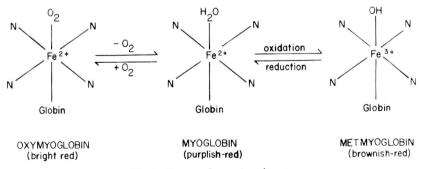

Fig. 2   Pigment changes in red meat.

the free binding site of the heme molecule. The other reaction involves the oxidation – reduction state of the iron molecule in the heme. The conversion of purple reduced myoglobin to red oxymyoglobin is caused by oxygenation or merely the covalent binding of molecular oxygen from the atmosphere to the free binding site of the heme of myoglobin (Clydesdale and Francis, 1971). This oxygenation reaction is usually noticeable on fresh meat in <0.5 hr (Johnson, 1974) and is called *blooming* in the meat industry. The resulting red oxymyoglobin is stable as long as the heme remains oxygenated (Clydesdale and Francis, 1971) and the iron in the heme remains in the reduced state.

Another form of myoglobin characterized by the oxidation of the iron of the heme in myoglobin from the ferrous (Fe^{2+}) to the ferric (Fe^{3+}) state is called metmyoglobin and is brown in color. Metmyoglobin is the major pigment noticed on discolored meat and results from the oxidation of the iron atom. It appears as an undesirable brownish-red pigment (Fig. 2). This reaction can be reversible so long as there are reducing substances such as NADH (nicotinamide adenine dinucleotide) present in the meat. When the reducing ability of the muscle is lost, however, the color of the meat remains brown because the oxidized iron atom in the heme cannot be reduced. The palatability of such meat, however, will usually still be satisfactory after cooking (Mangel, 1951). After the formation of metmyoglobin, further oxidative changes in myoglobin caused by enzymes and bacteria produce a series of brown, green, and faded-appearing compounds (Watts, 1954).

Some muscles within a carcass discolor at a faster rate than others. This is caused by differences in the metmyoglobin reducing ability (MRA) of the various muscles. Some muscles have more reducing equivalents present, in which the iron in the heme of the myoglobin molecule is in the reduced state for a longer period of time, resulting in either reduced myoglobin or oxymyoglobin. This explains why some cuts will last 4 – 5 days in a retail case while others will discolor after only 1 – 2 days. The MRA of a muscle is based upon the amount of glucose and reducing enzymes within that muscle. MRA can be affected by such antemortem factors as state of nutrition and amount of exercise an animal receives prior to slaughter.

Factors Affecting the Discoloration of Meat Pigments

Discoloration of fresh meat results from those factors that cause a change in the state of the iron atom in the heme of myoglobin (Fe^{2+} or Fe^{3+}) and/or a change in the molecule at the free binding site of the heme. The primary color states of myoglobin are shown in Fig. 3. Reduced myoglobin, oxymyoglobin, and metmyoglobin are the most important to meat color with respect to their proportions and distribution. Variation in the color of meat depends on the

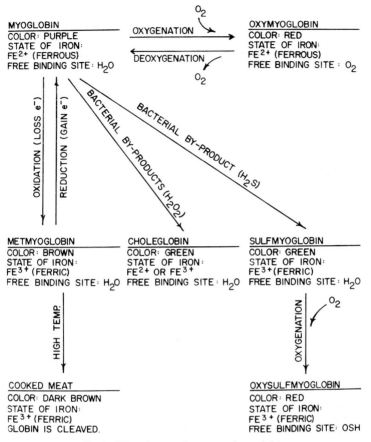

Fig. 3   The primary color states of myoglobin.

reversibility of the chemical states of myoglobin related to oxygen availability and binding.

The typical color of the interior of a piece of fresh meat that is devoid of oxygen is dark purple. This purple color is characteristic of myoglobin in its reduced state and is the predominant form present in meat. When exposed to oxygen, reduced myoglobin is oxygenated to oxymyoglobin, which is red in color. This red color of oxymyoglobin is the typical red color of meat that the consumer is most likely to prefer. The pigment associated with the undesirable brownish color of meat is termed metmyoglobin. Its presence results in a discoloration problem and occurs when the iron molecule in the heme of myoglobin is oxidized from the ferrous (Fe^{2+}) to the ferric (Fe^{3+}) state.

The color of meat seen in the retail case depends on the proportion of these pigments in relation to one another. Although two or more pigments may be present, the color of meat depends on the pigment that is present in the largest quantity. For example, the typical bright red color observed on the exterior surface of a meat cut is characteristic of oxymyoglobin because it is the pigment that predominates quantitatively. Signs of brownish discoloration occur when at least 60% of the unstable reduced myoglobin pigments in a particular area become oxidized to metmyoglobin (Lawrie, 1966).

Associated with meat discoloration is oxygen tension. High oxygen tensions are characteristic of oxymyoglobin and are found on the external reddish surface of meat. Low oxygen tension favors metmyoglobin formation and thus results in a brown discolored appearance (Taylor, 1972).

The oxidation reaction of purple myoglobin to brown metmyoglobin is caused by factors such as high temperature, low pH, salt, low oxygen atmospheres, aerobic bacteria, and low oxygen permeability of wrapping film. These factors are important in that they cause deoxygenation of oxymyoglobin to unstable reduced myoglobin and a decrease in available oxygen, which causes low oxygen tension.

High temperatures ($> 3°C$) cause a denaturation effect on the globin moiety. Its function of protecting the heme is reduced, thus deoxygenating oxymyoglobin to unstable reduced myoglobin. Unstable reduced myoglobin is then oxidized to metmyoglobin (Walters, 1975).

At pH values < 5.4, oxidation of myoglobin will occur. Low pH will cause denaturation of the globin protein moiety that protects the heme and causes the subsequent disassociation of oxygen from the heme as well as the oxidation of the iron molecule. Acids are well-known oxidizing agents and therefore oxidize reduced myoglobin to metmyoglobin. As the pH continues to decrease, the rate at which oxidation occurs will increase (Brooks, 1937).

Salt, an oxidizing agent of myoglobin, has two mechanisms of promoting oxidation. First, salt lowers the pH of any buffers in meat, thus oxidizing reduced myoglobin to metmyoglobin. Second, salt decreases the uptake of oxygen, thus promoting low oxygen tension resulting in oxidation (Brooks, 1937).

Meat discoloration may also be attributed to bacterial contamination, which may be either aerobic or anaerobic depending upon the conditions under which it occurs. The high oxygen demand of aerobic bacteria in their logarithmic phase of growth leads to the formation of metmyoglobin, resulting in the discoloration effect. Aerobic bacteria such as *Pseudomonas geniculata, P. aeruginosa, P. fluorescens,* and *Achromobacter faciens* have been found to cause the formation of brown metmyoglobin by reducing the oxygen tension on the meat surface (Robach and Costilow, 1962). Facultative anaerobic bacteria such as *Lactobacillus planterum* do not cause this discoloration problem, because these bacteria consume low amounts of oxygen (Robach and Costilow, 1962). Also evident is the fact that as aerobic bacterial counts increase, the surface of the meat changes from red oxymyoglobin to brown metmyoglobin and then to purple reduced myoglobin. The high counts of bacteria actually prevent oxygen from reaching the meat surface, thus permitting rapid enzymatic reduction of brown metmyoglobin to purple reduced myoglobin by either meat reducing enzymes or bacterial reductants. Some bacteria are also capable of producing by-products that oxidize the iron molecule of the heme portion of myoglobin and become attached to the free binding site of heme. The two most common bacterial oxidizing by-products are hydrogen sulfide (H_2S) and hydrogen peroxide (H_2O_2). When they become attached to the free binding site of the heme of unstable myoglobin, they produce sulfmyoglobin and choleglobin, respectively (Fig. 3). Hydrogen sulfide production causes green discoloration on vacuum-packaged meat (anaerobic atmospheres). Meat that has a high pH (> 6.0) and low oxygen tension is susceptible to bacterial hydrogen sulfide production. An example of a bacterium that causes this green discoloration of meat is *Pseudomonas mephitica*. When a vacuum-packaged cut that has green discolorations on it is removed from the vacuum package, the green sulfmyoglobin will become oxygenated to oxysulfmyoglobin, which is red in appearance but has the odor of rotten eggs. The bacterial by-product hydrogen peroxide can produce green areas of discoloration on aerobically packaged meat. It causes the greening through the oxidation of heme pigments to form choleglobin (Figure 3) or degradation beyond porphyrins to bile pigments (Walters, 1975).

The type of packaging film used also plays an important role in oxidation and bacterial growth. Relatively oxygen-impermeable films such as those used in vacuum packaging will usually cause metmyoglobin formation initially due to low oxygen tensions, but this will revert to purple reduced myoglobin when a substantial amount of the oxygen has been utilized and converted to carbon dioxide (Pirko and Ayres, 1957).

The color of meat as seen in the retail case is a very influential factor affecting consumer acceptance. Although it appears stable to the naked eye, the color of meat is continuously undergoing reactions related to oxygen (Watts *et al.*, 1966). For example, a cut of meat that is purple in color due to the presence of reduced myoglobin will undergo a color change if exposed to oxygen. This color change is due to one of two opposing reactions related to oxygen. These reactions are autoxidation and enzymatic reduction (MRA) and will remain active for different periods of time depending on factors such as species and type of muscle. If the reducing enzymatic activity decreases, for example, brown metmyoglobin will result.

The retail case life of meat is also influenced by temperature conditions.

Low temperatures are desirable because they slow the deoxygenation process and meat will stay in the desirable form of red oxymyoglobin. It also tends to inhibit bacterial growth and subsequent discoloration activity.

The lighting in a retail case can also influence consumer acceptance of meat color. Meat exposed to a reddish illuminated atmosphere will appear to be redder (Urbain, 1952), thus improving its consumer appeal. Soft white fluorescent lighting will cause discoloration but at a rate relatively less than other forms of lighting. The photochemical effect of lighting also causes meat color deterioration.

In essence, low light intensities that do not increase the temperature of the meat surface will not cause discoloration. High-intensity lights, however, increase the surface temperature of meat and cause the globin protein to denature. Subsequent oxidation of the iron in the heme occurs. Additionally, the irradiation from these lights may promote lipid oxidization because the iron oxidation within the myoglobin catalyzes such changes in lipids.

The rainbowed, multi-hued discoloration occasionally seen on meat surfaces is termed iridescence. It is seen when white light strikes the fibrous character of a cut surface of meat and projects a prism effect (Ockerman, 1973). Although this effect is very common on cooked meat products, it is only occasionally found on fresh meat.

Differences in the color intensity of meat are influenced by factors such as animal species, sex and chronological age of the animal, muscle differences, activity of the animal, rate of postmortem pH decline, and the ultimte pH of the meat.

The variation in color intensity of meat between animals of different species is caused by differing concentrations of myoglobin present in their muscle tissues (Table I). As the concentration of myoglobin in muscle tissues increases, the meat becomes darker. Because of the high concentration of myoglobin in beef muscle tissue, it is the darkest and produces cuts that are bright cherry red. The color of lamb meat is a light red to brick red because of its intermediate myoglobin content. Pork, which contains the lowest concentrations of myoglobin, is a grayish pink color.

Chronological age of animals within a species also affects color intensity. As

Table I
Effect of Species on Myoglobin Concentration

Species	Color	Myoglobin concentration (mg/g) wet tissue
Pork	Grayish-pink	1–3
Lamb	Bright red	4–8
Beef	Bright cherry red	4–10

Table II
Effect of Chronological Age on Myoglobin
Concentration[a]

Chronological age	Myoglobin concentration (mg/g) wet tissue
Veal	1–3
Young beef	4–10
Mature beef	16–20

[a] Adapted from Clydesdale and Francis (1971).

chronological age increases, the concentration of myoglobin in muscle tissues also increases (Table II) and makes the resulting meat darker in color. A youthful animal possesses high quality meat that appears light, bright red in color (Urbain, 1952).

The color intensity of meat is also believed to be related to the activity of the animal. Animal muscles used for locomotion appear darker in color because they usually contain a higher concentration of myoglobin than support muscles. This difference in muscle type is due to the increased amount of oxygen required for the production of energy in locomotor muscles (Table III). Animals that are less active thus produce meat containing lower concentrations of myoglobin and lighter in color than meat from more active animals. Red muscle fibers contain higher concentrations of myoglobin than white muscle fibers. Red muscle fibers have predominately aerobic metabolism and thus utilize more oxygen than white muscle fibers, which have predominately anaerobic metabolism.

Another factor influencing the color intensity of meat is pH. Ultimately, the pH of muscle is affected by immediate antemortem and postmortem conditions of the animal and carcass, respectively. If an animal is subjected to short-term, violent excitement immediately before slaughter, its meat will have a low ultimate pH. The result is a condition known as *PSE* (pale, soft, and exudative) in swine. PSE occurs when there is a rapid antemortem or postmortem pH decline while the carcass temperature is still high. A buildup of lactic acid

Table III
Effect of Muscle Type on Myoglobin Concentration

Type of muscle	Color	Myoglobin concentration (mg/g) wet tissue
Extensor carpi radialis (locomotor)	Dark	12
Longissimus dorsi (support)	Light	6

Table IV
Effect of pH on Muscle Color Intensity[a]

	Muscle glycogen concentration	Muscle color	Muscle lactic acid concentration	Muscle pH
At death				
Normal	1%	Purple	Very low	7.0
Dark-cutting beef	0.5–0.2%	Purple	Very low	7.0
PSE pork	0.6%	Purple	Medium	6.4
24 hrs postmortem				
Normal	0.1%	Bright red (beef) Pink (pork)	High	5.6
Dark-cutting beef	0.1%	Shady dark or black	Low	6.3–6.7
PSE pork	0.1%	Pale, soft, and exudative	Very high	5.1

[a] Adapted from G. C. Smith (personal communication, 1977).

causes the low pH and is the metabolic end product of the anaerobic break-down of glucose and glycogen. The color of meat is paler because the low pH causes the muscle structure (fibrils) to "open" and scatter light (Walters, 1975). The myoglobin fraction is also more readily oxidized to metmyoglobin, which has a low color intensity, as a result of the low pH (Walters, 1975).

DFD (dark, firm, and dry) is a condition opposite to PSE in that it occurs as a result of a high ultimate pH. It is found in beef and pork when the animals are exposed to long-term depletion of nutrients, such as during an extended transit haul. The meat appears dark and has a texture that is very firm and dry. Because the meat is dark, carcasses with this condition are known as dark-cutters. Their muscle fibers are swollen and tightly packed together, forming a barrier to the diffusion of oxygen and the absorption of light (Walters, 1975). Table IV summarizes the effect of pH on muscle color intensity.

III. FLAVOR AND AROMA

Flavor and aroma are terms used to describe the senses of taste and smell and thus need defining for the purposes of this chapter. Taste is defined as the sensory attribute of certain soluble substances perceptible by the taste buds, usually the four sensations (sweet, sour, salty, or bitter). Odor represents the sensory attribute of certain volatile substances perceptible by the olfactory organ. Flavor is the complex combination of the olfactory and gustatory attributes perceived during tasting, which may be influenced by tactile, thermal, pain, and even kinaesthetic effects. Flavor can be affected by such factors as texture, temperature, and pH (Lawrie, 1966). The mechanism by which odor and taste affect our perception of flavor is thought to be that odor and taste substances act by interfering with enzyme-catalyzed reactions in the olfactory receptors or taste buds, respectively (Haagen-Smit, 1952). The active part of a taste or odorous molecule fits some part of a protein structure in the olfactory receptors or taste buds, thereby altering reactions and giving rise to characteristic odors and flavors (Haagen-Smith, 1952). Under ideal conditions, the sense of smell is said to be 10,000 times more sensitive than that of taste (Lawrie, 1966). This acute sensitivity of olfactory receptors in the nasal cavity accounts for the innumerable odor sensations a human is able to perceive as opposed to taste sensations.

Various taste and odor sensations are developed in meat exposed to heat and subsequent mastication. The breakdown of the fibrous cooked meat texture results in the release of flavorous juices and volatile aroma compounds responsible for the characteristic flavor of meat (Patterson, 1974). Fresh raw meat flavor has been characterized as having either a mild serumlike taste or a commercial lactic acid taste. The serumlike taste is thought to be due to blood salts in the meat and products of pyrolysis and saliva. The odor of fresh raw meat tends to intensify during maturity and exposure to heat. Cooked meat odor has been described with terms such as "animal," "brothy," "metallic," "sour," "sweet," "nosefilling," and "fatty." The ultimate odor of meat is influenced by the cooking method, the kind of meat, and the treatment of the meat prior to cooking (Price and Schweigert, 1971).

Chemical compounds thought to contribute to the distinctive flavor and intensity of flavor in meat include nitrogenous extractives such as creatine, creatinine, purines, and pyrimidines. Aroma condensates from cooked meat have been shown to contain chemical compounds such as ammonia, amines, indoles, hydrogen sulfide, and short-chain aliphatic acids (Weir, 1960). The relationship of these compounds to specific cooked meat aroma descriptions, however, ("animal," "brothy," "metallic," "sour," "sweet," "nosefilling," and "fatty") has not been established (Weir, 1960).

Studies of volatile compounds from cooked meat have suggested that ring compounds (e.g., oxazoline and trithiolan), hydrocarbons, aldehydes, ketones, alcohols, acids, esters, ethers, lactones, aromatics, sulfur-containing compounds (e.g., mercaptans and sulfides), and nitrogen-containing compounds (e.g., amines and ammonia) are of importance in creating the characteristic flavor of meat (Hornstein, 1971).

In 1912 a French chemist, Maillard, observed that when he heated glucose and glycine, a brown pigment, or melanordin, was produced. This reaction resulted in a characteristic meaty flavor and subsequently was referred to as the Maillard reaction. Flavor production via the Maillard reaction proceeds from reactions of amines, peptides, amino acids, and proteins with reducing sugars. The primary step involved is a condensation reaction between the α-amino groups of the amino acids or protein and the carbonyl groups of reducing sugars and gives rise to the carbonyl – amino reaction product, which can be converted into a glycosylamine.

The N-glycosylamine undergoes an Amadori rearrangement to form a 1-amino-1-deoxy-2-ketose. At that time, two pathways are possible. In the first, 2-furaldehydes are formed and in the second pathway, C-methyl aldehydes, ketoaldehydes, ketols, and reductones are formed as end products (Hodge, 1967). These end products provide reactive reagents that will undergo a third pathway or Strecker degradation. This pathway is not concerned primarily with producing the resulting pigments or melanordins, but it does act to provide reducing compounds essential for their production. In addition, it is responsible for the production of a characteristic flavor and aroma. The flavor compounds formed as a consequence of this reaction are found to be aldehyde derivatives formed through the oxidative degradation of amino acids.

IV. TENDERNESS

Muscles within the same animal differ greatly in tenderness. The characteristic differences in tenderness occur as a result of the collective effects of numerous traits (Smith *et al.*, 1978). Smith *et al.* (1978) suggested that these traits can be broadly classified as actomyosin effects, background effects such as the amount of connective tissue and the chemical state of collagen, and bulk density and lubrication effects. These effects are discussed in greater detail in other chapters so they are not discussed here.

V. MEAT COOKERY

Meat cookery and the effect of heat on specific muscle components is one of the most important yet least understood aspects of meat research. Through proper meat cookery practices, beneficial treatments or management styles imposed on cattle and/or carcasses to improve palatability can be emphasized. Examples of such treatments or management styles include preslaughter management and nutrition, sex condition, physiological maturity, fat content, connective tissue state, postmortem aging and contraction, and storage.

Historically, meat was cooked for its pasteurization effect; however, in more recent centuries, the application of heat to meat has been conducted for the development of a characteristic color, odor, flavor and texture. Heldman (1975) suggested that the understanding of meat cookery may be slow due to complexities introduced by size, shape, composition of meat, and difficulties encountered with measuring thermal properties.

Table V shows the progressive changes that occur in sarcoplasmic proteins, myofibrillar proteins, collagen, tenderness, and flavor as affected by cooking temperature and can be used as a reference throughout this section.

Numerous studies have been conducted in the area of meat cookery based on the structural components of meat that are primarily responsible for its tenderness. Of meat's structural components, the connective tissue proteins and the myofibrillar protein fractions are of prime importance because they perform opposing actions that ultimately affect meat texture and tenderness. Lawrie

Table V
Changes in Proteins During Cooking

Change	Temperature (°C)
Sarcoplasmic proteins	
Sarcoplasmic proteins lose solubility and denature	40–60
Maillard reaction	40–50
Myofibrillar protein	
Denaturation of myofibrillar proteins	40–50
Coagulation of myofibrillar proteins	57–75
Unfolding of actomyosin molecules	<70
Myosin loses solubility	45–50
Myosin denaturation	53–65
Actin loses solubility	<80
Tropomyosin and troponin denaturation	30–70
Collagen	
Collagen solubilization	60–70
Collagen shrinkage	60–75
Collagen conversion to gelatin	65
Collagen fiber disintegration	60–80
Structural changes	
Sarcomere shortening and decrease in fiber diameter	40–50
Birefringence begins to diminish	54–56
Loss of M-line structure, distintegration of thin and thick filaments	60–80
Tenderness changes	
Denaturation of contractile proteins (first toughening step)	40–50
Collagen shrinkage (second toughening step)	65–70
Tenderization begins	54
Flavor changes	
Meat flavor development	>70

(1966) and Visser *et al.* (1960) summarized this effect of heat on meat structure as producing a softening of the connective tissue by conversion of the collagen to gelatin, accompanied by a toughening of the muscle fibers due to heat coagulation of the myofibrillar proteins.

A. Effect of Heat on Meat Color

When meat is heated, dramatic changes occur within its chemical structure that affect its color. Specifically, when muscle is heated, the globin moiety of myoglobin becomes denatured; thus, its function of protecting the heme diminishes. This results in a change in the valence state of iron in the heme and a change of molecules at the free binding site. In essence, what occurs is that all reduced myoglobin and oxymyoglobin molecules are converted to the met-myoglobin form, which is brown in color. Once this reaction is completed, muscle pigments, as well as other proteins within the muscle, continue to denature. Myoglobin and other sarcoplasmic proteins denature in the range 40–50°C. In addition, the amine groups of amino acids that make up muscle proteins will react with available reducing sugars, such as free glucose, and undergo a Maillard browning reaction. It may be important that the consumer be aware of the changes that occur in meat upon exposure to heat so that he or she may prepare a desirable product in the home.

B. Effect of Heat on Water-Holding Capacity

The water-holding capacity (WHC) of meat can be defined as the ability of meat to hold fast to its own or added water during the application of forces (pressing, heating, or centrifugation) (Hamm, 1969). Lean red meat contains ~75% water (Hamm, 1969).

Hydration water represents 4–5% of this total water content and is water that is tightly bound to either myosin or actin, depending on the shape and charge of the protein (Hamm, 1969). Free water is water that represents the major percentage of the total water content and is immobilized within the muscle tissues. Its state is influenced by the molecular arrangement of the myofibrillar proteins. Myofibrils are well suited to retain water because the three-dimensional network of their filaments provides an open space for water to be immobilized (Price and Schweigert, 1971). Tightening this space between the myofibrillar protein network by contraction or protein denaturation by heating results in a decrease of immobilized water and an increase in expressible water (Hamm, 1969). The amount of water muscle tissue retains can determine how various heat treatments will effect the protein samples. It has been observed that heating protein samples for a long period of time can have the same effect on water retention as does heating at high temperatures for short periods of time (Mann and Briggs, 1950; Rence *et al.,* 1953). Ritchey and Hostetler (1965) cooked beef to various internal temperatures of 61, 68, 74, and 80°C. Their results indicated that as the internal temperature of the meat increased, the meat became less juicy and cooking loss percentages increased. Bengtsson *et al.* (1976) found that up to the temperature of 65 to 70°C any apparent water loss occurred almost entirely by evaporation from a wet surface and at temperature > 65°C weight loss by liquid drip was significant. Marshall *et al.* (1960) found that during low-temperature roasting, as the internal temperature increased, evaporative losses also increased. He also noted that as the oven temperature was increased and internal temperature increased, drip losses increased, although total losses were the greatest at the lowest oven temperature.

Heating muscle tissue causes denaturation of myofibrillar proteins followed by their coagulation, shrinkage of the myofilaments, and a tightening of the microstructure of myofibrils (Cheng, 1976). All of these factors can increase the amount of free water in muscle tissue when temperatures and cooking times are increased. This increase in the amount of free water is evidenced by increased cooking loss percentages.

Hostetler and Landmann (1968) found that as the myofibrillar proteins coagulate after they are denatured by heat, they lose their water-holding ability. Upon denaturation, the immobilized water in the proteins of the myofilaments is freed and when it escapes from the intermyofibrillar spaces carries with it some soluble sarcoplasmic proteins. This combination of water and sarcoplasmic proteins appears as an amorphous substance and is often seen when meat is cooked to 53°C or higher.

Davey and Gilbert (1974) reported that when water in its hydrated form is released and makes its transition to water in its free form, there is a decrease in the rate of temperature rise between 65 and 75°C and attributed this decrease to the massive movement of water from a hydrated to a free form. The process of collagen shrinkage is presumed to be responsible for squeezing hydrated water from muscle tissue to its free form.

Since the A-band region is more thermally stable than the I-band region of the sarcomere, the loss of water-holding capacity may be more severe in the I-band region than in the A-band region (Cheng and Parrish, 1976).

The sarcolemma isolates the myofibril from surrounding substances and thus acts as a barrier to the sideways movement of water perpendicular to the meat fibers (Locker and Daines, 1974). When meat is cooked, the sarcolemma loses its envelope effect and exhibits perforations in its structure.

C. Effect of Heat on Muscle Proteins

Many studies have been conducted to investigate the mechanism of heat denaturation of proteins. The initial changes that take place in muscle tissue when exposed to heat are the unfolding of peptide chains and coagulation of the proteins. The unfolding step exposes side chains that attract one another and form aggregates of increasing size (Price and Schweigert, 1971). Ultimately, these aggregates reach a size such that they can no longer remain in solution and thus precipitate (Price and Schweigert, 1971). The overall result is coagulation of the protein. Protein coagulation is generally assumed to take place between 57 and 75°C. A specific coagulation temperature, however, has not been established since muscle fibers consist of several different proteins, each with a different coagulation temperature.

1. THE MYOFIBRILLAR PROTEINS

The myofibrillar proteins that undergo significant structural changes that affect the ultimate tenderness of meat when exposed to heat include myosin, actin, tropomyosin, and troponin.

The myosin protein fraction undergoes denaturation upon heating to 53°C and has been found to form a leathery gel within the muscle system (Locker, 1956). This leathery gel is quite tough and contains mostly coagulated myosin. It gradually loses solubility between 45 and 50°C and is almost completely insoluble at 55°C (Cheng, 1976). The myosin molecule remains intact until temperatures of 70 to 80°C are reached.

Actin alone has been found to have no effect on tenderness but does exert a significant effect on heat gelation when combined with myosin by establishing rigor type linkages (Samejima *et al.*, 1969). When temperatures reach 80°C, actin begins to lose its solubility (Cheng, 1976).

The regulatory proteins tropomyosin and troponin appear to denature in the temperature range of 30 to 50°C and cause a rapid loss of calcium sensitivity of the myofibrillar system (Hartshorne *et al.*, 1972). This calcium sensitivity loss does not affect the total process of myosin or actomyosin denaturation but it does affect the aggregation of the myofibrillar system. Even though these regulatory proteins lose their calcium binding properties and regulatory action on actomyosin ATPase at temperatures of 40 to 50°C (Hartshorne *et al.*, 1972), they are not entirely denatured until heated to temperatures in excess of 70°C.

An unfolding of the actomyosin molecules occurs at temperatures up to 70°C. Higher temperatures cause the formation of S—S bonds between peptide chains of the actomyosin and result in an increase in meat toughness (Hamm and Hofmann, 1965).

2. SARCOPLASMIC PROTEINS

The majority of the sarcoplasmic proteins gradually decrease in solubility between the temperatures of 40 and 60°C (Hamm, 1966) and are almost completely denatured at temperatures of 62°C (Bendall, 1964). Myoglobin has been found to be one of the most heat-stable sarcoplasmic proteins. Many researchers have found that prolonged heating or high-temperature heating of muscle was essential before a significant coagulation of myoglobin was evidenced.

3. CONNECTIVE TISSUE

At temperatures of approximately 60 to 75°C collagen fibers suddenly contract and undergo a toughening phase called collagen shrinkage (Davey and Gilbert, 1974). This collagen shrinkage is accompanied by an increase in the ultimate tenderness of the meat and is due to the denaturation of the myofibrillar proteins that occurs in the 58–62°C temperature range (Draudt, 1972). However, at temperatures > 60°C, collagen shrinkage causes a contracture of the collagen sheath and an exudation of moisture from the myofibrils, resulting in a toughening effect (Davey and Gilbert, 1974).

Collagen solubilization begins at ~ 60 to 70°C and continues to increase as temperature increases (McCrae and Paul, 1974). Penfield and Meyer (1975) indicated that accompanying an increase in collagen solubilization was a decrease in tenderness at different end-point temperatures. However, at very high temperatures (e.g., 98°C), even though collagen solubilization increases, the predominant factor causing increased toughness at this temperature is a hardening of the myofibrillar proteins (McCrae and Paul, 1974).

D. Effect of Heat on Muscle Structure

An important change caused by heat that occurs in the structure of muscle is a shortening effect of the muscle or a decrease in the sarcomere length. Sarcomere shortening occurs when myofibrillar proteins are denatured and when collagen shrinkage occurs. Sarcomere shortening in isolated fibers begins when the temperature reaches 40–50°C (Hegarty and Allen, 1972). Also occurring in the 40–50°C range is a decrease in fiber diameter thought to be attributed to the heat coagulation of myofibrillar proteins. Since it has been shown that shortened sarcomeres are associated with the actomyosin toughness of meat (Herring *et al.*, 1967), any changes in myofibril length or diameter, such as those caused by surface shrinkage, ultimately effects the tenderness of meat.

Myofibril shortening begins when the temperatures of individual heated fibers reach 50°C, which is slightly before birefringence is noticeably diminishing (54–56°C) (Hostetler and Landmann, 1968). The loss of birefringence coincides with the temperature at which myosin is coagulated and sarcoplasmic proteins are rapidly beginning to coagulate. Fiber shortening is thus associated with the actual coagulation processes of the various proteins in the muscle. When birefringence is diminished, the A-band structure is disturbed in that the coagulation of the myofibrillar proteins places them out of alignment (Hostetler and Landmann, 1968).

Schmidt and Parrish (1971) found from electron and phase contrast microscopy that the myofibrillar proteins compressed and sarcomeres shortened at 50°C. At 60°C, heat caused the loss of the M-line structure, disintegration of the thick and thin filaments, and coagulation and further myofibrillar shrinkage. Heating to 70 and 80°C caused further disintegration of the thin filaments and further coagulation of the thick filaments. In an ultrastructural study, Dahlin *et al.* (1976) found that heating turkey meat from red (semitendinosus) and white (pectoralis major) muscles to 82°C resulted in Z-disk thickening and a decrease in A-band length in both muscle types. I bands were found to be removed from white muscle but remained intact in red muscle. Heating to 82°C completely removed the thin filaments in the pectoralis major muscle of the turkey (Hegarty and Allen, 1975). To summarize, the following progressive changes occur in muscle ultrastructure at temperatures of 60 to 80°C: collagen fiber disintegration, myofibril fragmentation, endomysial sheath swelling, and coagulation and shrinkage of myofibrils. Also evident is a transition from a fibrous to a granular structure, which can be attributed to heat-induced changes in the sarcoplasmic proteins (Schaller and Powrie, 1972).

E. Effect of Heat on Palatability

Practically all meat is heated prior to consumption so as to obtain a characteristic meaty flavor, texture, and aroma. The flavor of raw meat is weak, salty, and bloodlike. The true meaty flavor develops when raw meat is exposed to temperatures exceeding 70°C.

Texture, or tenderness, of meat is probably the most important palatability factor affecting consumer acceptance of meat. Two structures in meat are responsible for its tenderness: muscle fibers and connective tissue. In review, the effects of cooking on meat structure can be summarized as producing a softening of the connective tissue by conversion of the collagen to gelatin, accompanied by a toughening of the muscle fibers due to heat coagulation of the myofibrillar proteins. Exposure to temperatures up to 50°C has little effect on meat tenderness (Machlik and Draudt, 1963). However, at temperatures of 54°C and above, the tenderness of meat increases and at 60 to 64°C tenderness reaches a maximum. These changes are thought to be caused by collagen shrinkage, which is observed at 60°C. Davey and Gilbert (1974) found that as meat is heated, two distinctly separate phases of toughening occur at different temperatures. The first phase occurs between 40 and 50°C and is believed to be due to the denaturation of the contractile system. The second phase occurs between 65 and 70°C and results in a doubling toughness effect believed to be associated with collagen shrinkage. In this 65–70°C temperature range the percentage of fiber shrinkage and weight loss increases.

Numerous studies have been conducted to relate the effect of internal temperature on the tenderness of meat. In fact, it has been found that the final internal temperature may have a greater effect on palatability than maturity or marbling. As the internal temperature of meat increases, the meat becomes less tender, as indicated by increased shear force values and sensory panel scores, and becomes drier and harder. If the internal temperature is below that of collagen shrinkage (60–65°C), then the major decrease in tenderness for cuts high in connective tissue does not occur. If the temperature is higher than that of collagen shrinkage, however, the coagulation effects will result in a greater weight loss percentage and a more tightly packed, less tender muscle tissue texture (Laakkonen et al., 1970).

Factors such as oven temperature, relative humidity, sample dimension, and initial sample temperature are also thought to play an important role in the final internal temperature and weight loss yield during oven cooking of beef (Bengtsson et al., 1976).

Method of cookery, as well as the rate of heat penetration, has a pronounced effect on the ultimate palatability of meat. A suitable cookery method for meat cuts is essential because myofibrils toughen by extensive heating, while the collagenous connective tissue requires a moist atmosphere and a long cooking period, such as that used in stewing, so as to soften the connective tissue through the conversion of collagen to gelatin (McCrae and Paul, 1974). In general, there is little or no softening of connective tissue at 60°C even after 64 min of heating; the critical temperature for softening is around 65°C, at which point collagen shrinkage is observed. If meat is held at this temperature for a lengthy period, the collagen begins to soften.

The rate of heat penetration may also affect various palatability properties of meat. The rate at which the interior of meat is heated is influenced by the rate at which energy is supplied, the rate at which energy is transmitted to the meat, the shape and size of the sample being heated, the composition of the sample, the spatial distribution of areas of lean, fat, connective tissue, and bone, the characteristics of the meat surface, and any changes induced in the meat by heat, including protein denaturation, loss of water, and melting of fat (Paul and Palmer, 1972).

McCrae and Paul (1974) found that the rate of heat penetration was fastest between 30 and 50°C and slowest between 60 and 70°C or 50 and 60°C. They concluded that these results were due to an increase in energy utilization for protein denaturation and water evaporation. Microwave cookery resulted in the fastest rate of heat penetration, followed by oven-broiling, braising, and roasting (in order of descending penetration rate).

Another variable that affects meat palatability is the time required to reach the final internal temperature. These time–temperature interactions cause both physical and chemical changes in the connective tissues and muscle fibers. Haughey (1968) reported that cooking time was dependent upon heat transfer to the surface of the meat, internal heat transfer, which, by raising the temperature, denatures the proteins and produces the cooked meat state, and mass transfer, which results from a procedural reaction beginning with a release of water originally immobilized by the proteins. This water diffuses through the meat because of concentration gradients and results in evaporation or drip loss from the meat surface.

Increased tenderization of meat resulting from slower rates of heating may be attributed to collagen becoming more soluble without extensive hardening of muscle fibers (Paul, 1963). Laakkonen *et al.* (1970) suggested that the retention of meat juices along with the shrinkage of collagen may explain the tenderization effect of cooking meat for a long period of time at low temperatures.

VI. METHODS FOR THE EVALUATION OF MEAT QUALITY

A. Sensory Panel Methods

The food industry has relied on sensory evaluation as a "guidance tool" in the areas of product development, quality control, and marketing. With the continuing expansion and diversity of the food industry, the sensory evaluation field has implemented techniques that produce valid and reliable data.

The field of sensory evaluation emphasizes the utilization of subjective procedures (measurements made by humans) rather than objective procedures (measurements made by instruments). The reason for this is that some instruments take minutes to respond yet measure only one attribute whereas the human senses can respond within seconds yet measure the composite "mouthfeel" (Amerine *et al.*, 1965). Instrumental devices, however, can supplement subjective procedures. For instance, correlations can be obtained from the measurements made between the trained taste panel and the instrument.

This field has classified subjective panel methods into two divisions: affective and analytical. The affective method measures consumer reactions of a particular product in terms of either acceptance or preference. Preference is defined as an expression of higher degree of liking, a choice of one object over others, and/or a psychological continuum of affectivity upon which choices are made [IFT], (Institute of Food Technologists, 1981). This method usually consists of a 50 to 100-member consumer panel. The panelists are chosen at random from the population to be sampled and do not receive any formal training. The objective of the affective method is to determine the future performance of a product in the marketing field.

There are three types of affectives: paired-preference, ranking and rating. They are presented in outline in Tables VI and VII.

Table VI
Classification of Sensory Evaluation Methods and Panels[a]

Classification of methods by function	Appropriate methods[b]	Type and number of panelists[c]
Analytical: Evaluates differences or similarity, quality, and/or quantity of sensory characteristics of a product		Screened for interest, ability to discriminate differences, and reproduce results
		Trained to function as a human analytical instrument
Discriminative:		Normal sensory acuity
Difference: Measures simply whether samples are different	Paired-comparison Duo–trio Triangle Ranking Rating difference/scaler difference from control	Periodic requalification Panel size depends on product variability and judgment reproducibility No recommended "magic number"—a number often used is 10; a recommended minimum number is generally 5, since any fewer could represent too much dependence upon one individual's responses
Sensitivity: Measures ability of individuals to detect sensory characteristic(s)	Threshold Dilution	

Descriptive: Measures qualitative and/or quantitative characteristic(s)	Attribute rating
	  Category scaling
	  Ratio scaling (magnitude estimation)
	Descriptive analysis
	  Flavor profile analysis
	  Texture profile analysis
	  Quantitative descriptive analysis	

Affective: Evaluates preference and/or acceptance and/or opinions of product	Paired-preference
	Ranking
	Rating
	  Hedonic (verbal or facial) scale
	  Food action scale
		Randomly selected
		Untrained
		Representative of target population
		Consumers of test product
		No recommended "magic number"—minimum is generally 24 panelists, which is sometimes considered rough product screening; 50–100 panelists usually considered adequate

[a] Adapted from the Institute of Food Technologists (1981).
[b] See Table VII.
[c] Ignoring these considerations increases the possibility of bias and error in the results.

Table VII
Information Guide for Sensory Evaluation Methods[a]

Method	Number of samples per test	Analysis of data	Selected references for analysis of data[a]
1. Paired-comparison (or paired-preference)	2	Binomial distribution	Gridgeman (1955, 1961); Peryam (1958); Roessler et al. (1978); Scheffe (1952)
2. Duo-trio	3 (2 identical, 1 different)	Binomial distribution	Gridgeman (1955); Lockhart (1951); Peryam (1958); Roessler et al. (1978)
3. Triangle	3 (2 identical, 1 different)	Binomial distribution	Byer and Abrams (1953); Gridgeman (1955); Kramer and Twigg (1962); Peryam (1958); Roessler et al. (1978)
4. Ranking	2–7	Rank analysis Analysis of variance	Kahan et al. (1973); Kramer (1960, 1963); Snedecor and Cochran (1967); Tompkins and Pratt (1959)
5. Rating difference/scaler difference from control	1–18 (the larger number only if mild-flavored or rated for texture only)	Analysis of variance Rank analysis	Mahoney et al. (1957); Peryam (1958); Snedecor and Cochran (1967); Tukey (1951); Wiley et al. (1957)
6. Threshold	5–15	Sequential analysis	American Society for Testing and Materials (ASTM) (1968); Green and Swets (1971); Gregson (1962); Guilford (1954); Pilgrim et al. (1955); Wald (1947)
7. Dilution	5–15	Sequential analysis	Bohren and Jordan (1953); Tilgner (1962a,b); Wald (1947)

Method	Scale	Statistical analysis	References
8. Attribute rating (category scaling and ratio scaling or magnitude estimation)	1–18 (the larger number only if mild-flavored or rated for texture only)	Analysis of variance Rank analysis Regression analysis Factor analysis Graphic presentation	ASTM (1972a); Carlin et al. (1956); Duncan (1955); Kramer and Twigg (1962); Mahoney et al. (1957); Snedecor and Cochran (1967); Wiley et al. (1957); Moskowitz (1974, 1975, 1978); Moskowitz and Barbe (1976); Moskowitz and Wehrly (1972); Moskowitz and Sidel (1971); Stevens (1962); Winer (1971)
9. Flavor profile analysis	1	Graphic presentation Principal components and multivariate analysis of variance	Caul (1957); Little (1958); Sjöström and Cairncross (1954); Sjöström et al. (1957); Kendall and Stuart (1968); Morrison (1976)
10. Texture profile analysis	1–5	Graphic presentation Principal components and multivariate analysis of variance	Brandt et al. (1963); Civille and Szczesniak (1973); Szczesniak et al. (1963); Kendall and Stuart (1968); Morrison (1976)
11. Quantitative descriptive analysis	1–5	Analysis of variance Factor analysis Regression analysis Graphic presentation	Stone et al. (1974)
12. Hedonic (verbal or facial) scale rating	1–18 (the larger number only if mild-flavored or rated for texture only)	Analysis of variance Rank analysis	ASTM (1972b); Ellis (1966); Gridgeman (1961); Hopkins (1950); Peryam and Pilgrim (1957)
13. Food action scale rating	1–18 (the larger number only if mild-flavored or rated for texture only)	Analysis of variance Rank analysis	Schutz (1965)

[a] Adapted from the Institute of Food Technologists (1981).

The paired-preference test is conducted in such a manner that the panelist is asked to express a preference based on a specific attribute of the sample. The method may also be used to make multiple paired comparisons within a sample series. Results are obtained in terms of the relative frequencies of choice of the two samples represented.

The ranking test is an extension of the paired preference test. Three or more samples are presented simultaneously. The total number of samples tested will depend on the product and the ability of the panelist. The panelist is asked to rank the samples according to his or her preference. The amount or degree of preference cannot be determined by this method (Tables VI and VII).

Rating tests permit the evaluation of intensity of a particular attribute. A number of rating scales are being used:

1. *Food Action Scale.* This scale measures the general level of acceptance of a meat product rather than a specific attribute. Nine rating categories ranging "I would eat this every opportunity I had" to "I would eat this only if I were forced to" are represented. One or more samples may be tested.

2. *Hedonic Scale.* This scale is used to measure the level of liking for meat products by a population. Several variations of the traditional nine-point word hedonic scale have been used. These include (1) a reduced number of rating categories (fewer than five is not recommended); (2) a larger number of "like" rating categories than "dislike;" (3) omission of the neutral rating category; (4) substitution for the verbal categories by caricatures representing degrees of pleasure and displeasure (Fig. 4); and (5) use of a nonstructured, nonnumerical line scale anchored with "like" and "dislike" on each end. A hedonic rating test can yield both absolute and relative information about the test samples. Absolute information is derived from the degree of liking (or disliking) indicated for each sample, and relative information is derived from the direction and degree of difference between the sample scores.

CONSUMER SCORE SHEET
SAMPLE QUESTIONNAIRE FACIAL HEDONIC METHOD
ACCEPTANCE QUESTIONNAIRE

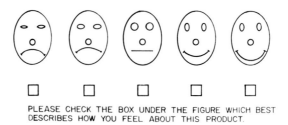

PLEASE CHECK THE BOX UNDER THE FIGURE WHICH BEST
DESCRIBES HOW YOU FEEL ABOUT THIS PRODUCT.

Fig. 4   A facial hedonic rating scale system. Adapted from Cross *et al.* (1978b).

"Analytical tests are used for laboratory evaluations of products in terms of differences or similarities and for identification and quantification of sensory characteristics (IFT, 1981)." There are two major types of analytical tests: discriminative and descriptive. Both tests use trained and experienced panelists.

There are two types of discriminative tests: difference and sensitivity. Difference tests measure whether samples are different at some predetermined α level, while sensitivity tests measure the ability of an individual to detect sensory characteristics (Table VI). There are several types of difference tests outlined in Table VI. These tests are described in greater detail in IFT's "Sensory Evaluation Guide for Testing Food and Beverage Products (IFT, 1981)." There are two types of sensitivity tests: threshold and dilution (Table VI). The recommended uses are outlined in Table VIII.

Analytical descriptive tests attempt to identify sensory characteristics and quantify them. Panelists are selected for their ability to perceive differences between test products and verbalize perceptions. Long and specialized training is required to perform profile and quantitative descriptive analyses. There are two types of descriptive tests: attribute rating and descriptive analysis. There are two types of attribute rating tests: category scaling and ratio scaling (Tables

Table VIII
Recommended Sensory Test Methods for Specific Types of Applications[a]

Type of application	Appropriate test methods listed in Table VII
New product development	1, 2, 3, 4, 5, 8, 9, 10, 11, 12, 13
Product matching	1, 2, 3, 5, 8, 9, 10, 11, 12, 13
Product improvement	1, 2, 3, 5, 8, 9, 10, 11, 12, 13
Process change	1, 2, 3, 5, 8, 9, 10, 11, 12, 13
Cost reduction and/or selection of a new source of supply	1, 2, 3, 5, 8, 9, 10, 11, 12, 13
Quality control	1, 2, 3, 5, 8, 9, 10, 11
Storage stability	1, 2, 3, 4, 5, 8, 9, 10, 11, 12, 13
Product grading or rating	8
Consumer acceptance and/or opinions	12, 13
Consumer preference	1, 4, 12
Panelist selection and training	1, 2, 3, 4, 5, 6, 7, 8
Correlation of sensory with chemical and physical measurements	5, 8, 9, 10, 11

[a] Adapted from the Institute of Food Technologists (1981).

VI and VII). There are several types of descriptive analysis tests: flavor profile analysis, quantitative descriptive analysis, and texture profile analysis (IFT, 1981).

It is very important to select the correct sensory method for specific types of applications. Recommendations are presented in Table VIII (IFT, 1981).

B. Selection and Training of Sensory Panelists

With proper experience and training, sensory panelists are expected to become familiar enough with certain aspects of sensory characteristics that when performing actual analysis they are able to refer to specific sensations and identify them. This type of trained testing requires that the human subject possess considerable sensitivity and precision. The type of individual who has this potential as a panelist should have the following characteristics: interest in the sensory program; motivation to perform a selected task; time available for panel participation; normal sensory acuity; good health, being free from allergies, frequent head colds, and sickness (Amerine *et al.*, 1965); and capability of discriminating differences and reproducing results. Factors such as age and sex are not as critical.

At present, trained sensory panels are being utilized to the greatest extent in industry, universities, and government agencies. The selection and training process is initiated when the organization establishes a perspective toward the program's purposes, role, and objectives. Following this, the program begins a recruitment process (Fig. 5). Potential taste panelists are recruited through newsletters, personal contact, and seminars. At this time, the candidates are told about the testing procedures, the testing objectives, the importance of the research, and the duration of the testing program. Also, the candidates are informed of the performance criteria that regulate the panel selection procedures. Prior to screening tests, personal interviews are conducted in an attempt to eliminate unqualified candidates. The interviewer determines if the candidate is interested in the sensory program and if he or she is available to participate in the program on a regular basis. In addition to those factors considered when selecting a sensory panel, there are minor ones that are related in a peripheral manner to the panelist's qualifications. These include educational level, work classification, technical background, work experience, sex, age, smoking status, and sensory experience. All the information gathered in the personal interview provides the basis for disqualifying candidates who are neither interested nor available, classifying the candidates as potential panelists for routine or for special tests, and selecting panelists who are to be screened and trained in descriptive analysis (Cross *et al.*, 1978a,b). Training for the inexperienced panelists stops here.

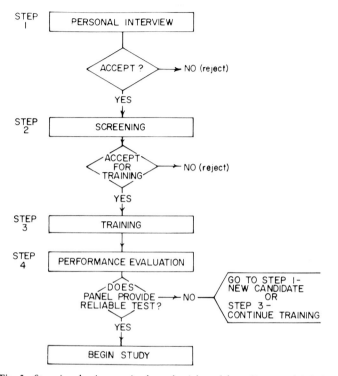

Fig. 5 Steps in selecting a trained panel. Adapted from Cross *et al.* (1978a).

For the experienced and trained panelists, a screening process is next. The process differs for the two types of panelists in that it is more complex for the trained panelists. Both categories of panelists are required to possess discriminatory skills; however, the trained panelists must also be capable of characterizing specific attributes of a food product.

Screening methodology for experienced panelists is accomplished in a selective manner. The two basic methods are screening with triangle tests as described by Cross *et al.* (1978a,b) (Fig. 6); and screening with a sequential analysis procedure as described by Bradley (1953). The Cross *et al.* (1978a,b) screening process is more comprehensive because the candidates are required to participate in a total of 10 sessions, whereas the second method allows unqualified panelists to withdraw at any time. These tests, however, have the same objective of screening panelists into experienced and inexperienced pools. When this process is completed, actual training sessions begin. Panelists are required to participate in comprehensive training programs (duration period of 2 to 6 months or longer) and must display stable personalities and adequate verbal

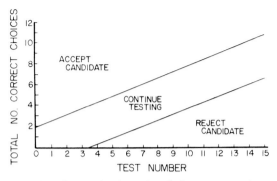

Fig. 6 Sequential analysis chart used to determine whether to accept, reject, or continue testing a panelist during the screening step. Adapted from Cross *et al.* (1978a).

skills. Other selective criteria can be applied depending on the test methodology.

Performance evaluation can begin soon after training is initiated to help the panel leader identify problems among individual panelists. In general, however, the panel can be evaluated after 10 or 12 training sessions. The performance evaluation can identify specific problem areas for individual panelists and often confirm the panel leader's observations. Additional training sessions should be held to concentrate on the problem areas identified. After 2 or 3 weeks of additional training, another evaluation can be conducted. These evaluations can help the panel leader to judge the results of training.

Cross *et al.* (1978a,b) reported on methods for testing the performance of the potential panel. Panel evaluation can be spread over 4 days, with three sessions per day and three samples per session. The 4-day test can be carried out with nine samples that are prepared to have wide ranges in tenderness, juiciness, and connective tissue. The data for each candidate are treated by a one-way analysis of variance (ANOVA) with nine treatments and four observations per cell (treatment). The design is treated as a balanced-lattice design (Cochran and Cox, 1957), so that effects of day and session can be studied. From the ANOVA table, the F ratio, mean square treatments/mean square error, is calculated. Used in this context, the F ratio is a measure of a panelist's ability to award different scores to different samples while being able to repeat himself on the same sample a day or more later. The degree to which a person discriminates among samples and is consistent in his replicate judgements is reflected in his F ratio (American Society for Testing and Materials [ASTM], 1968). The larger the F ratio, the better the panelist. Candidates can be ranked on the basis of their F ratios.

Results of a test for the 11 panelists and the panel leader as reported by Cross *et al.* (1978a) are presented in Table IX. The panel leader, or expert, partici-

Table IX
Panel Performance Evaluation: F Ratios by Palatability Attribute[a]

| | | F ratio | | | | | | | | | | |
| | | | | | | | Panelist | | | | | |
Attribute	Panel leader	1	2	3	4	5	6	7	8	9	10	11
Tenderness	14.40	2.76	7.12	7.54	8.29	8.87	8.67	6.12	5.37	6.77	8.78	3.64
Juiciness	5.33	2.37	3.59	8.03	3.58	8.78	6.94	6.14	4.59	1.89	7.50	3.18
Connective tissue	8.78	4.02	4.39	4.07	5.37	5.32	8.32	3.76	0.95	6.84	12.32	4.03

[a] Adapted from Cross et al. (1978a).

Table X
Ranking of Panelists Based on F Ratios in Table VII[a]

| | | Ranking Panelist | | | | | | | | | | |
Attribute	Panel leader	1	2	3	4	5	6	7	8	9	10	11
Tenderness	1	12	7	6	5	2	4	9	10	8	3	11
Juiciness	6	11	8	2	9	1	4	5	7	12	3	10
Connective tissue	2	10	7	8	5	6	3	11	12	4	1	9
Sum	9	33	22	16	19	9	11	25	29	24	7	30
Overall ranking	2	12	7	5	6	3	4	9	10	8	1	11

[a] Adapted from Cross et al. (1978a).

pated in the sample scoring but had no prior knowledge of the samples being evaluated. The *F* ratios should be compared between the panel leader and each individual panelist. It was not unusual for an individual panelist to have a higher *F* ratio than the panel leader.

Table X is a ranking of the *F* ratios (Cross *et al.*, 1978a). The *F* ratios in Table IX and the rankings in Table X can help the panel leader make training decisions. For example, Panelist 8 had a problem with connective tissue and Panelist 1 with tenderness, juiciness, and connective tissue.

A single evaluation, as summarized in Tables IX and X, is not a conclusive basis for the decision of who should or should not be on the panel. Subsequent tests are useful to evaluate the panelist's performance throughout the study. This panel (Cross *et al.*, 1978a) was trained for 4 months and was tested four times. Panelist 1 was consistently rated last and ultimately excused.

The number of panelists selected for the final panel should be based on the test results. Inclusion of a person with less than satisfactory results only to achieve a predetermined panel size is a mistake. ASTM (1968) requires a minimum of five panelists, since fewer would represent too much dependence upon any one individual's response.

C. Selection of Research Cooking Methods

The selection of a cookery method is an integral part of organizing and guiding a research project, because preparation has a significant influence on the sensory characteristics (e.g., tenderness, juiciness, and flavor) of meat as well as related mechanical measurements. The method of cookery chosen should be based on the objectives of the experiment.

Three methods of cookery — broiling, roasting, and braising — are discussed in this section because they are determined by the American Meat Science Association (Cross *et al.*, 1978b) as being the most common meat research cookery methods while also being quite common in the home. The methods of meat cookery are categorized as either dry heat cookery or moist heat cookery procedures. The former applies to broiling and roasting while the latter applies to braising.

Dry heat cookery is characterized by its relatively short heating period and high temperature. This procedure is inadequate in solubilizing connective tissue (collagen), but it does create a unique flavor due to a sugar–amine browning action. For the roasting method, the transmittance of heat to the meat is accomplished by air convection in a closed oven preheated to 165°C (Cross *et al.*, 1978b). This method is used for large tender meat cuts that are cooked fat side up on an oven rack in an open roasting pan to catch drippings and to permit self-basting. Recommended degrees of doneness for roasting meat include beef 70°C, lamb 75°C, and pork 75°C (Cross *et al.*, 1978b).

For the broiling method, the meat cut is cooked either above or below the source of radiant heat. The oven door is opened slightly and the cut is turned over halfway through the cooking period. The meat is removed when the desired internal temperature is reached. When comparing these two types of dry heat cookery, Cross et al. (1979) found that palatability parameters were similar in beef steaks that varied in marbling and maturity. Cooking loss values, however, were lower for steaks that were roasted than for those broiled.

Methods of moist heat cookery include braising, pressure cooking, and/or any type of cookery that utilizes water. Although practically all meat cuts can be satisfactorily subjected to braising, beef cuts high in connective tissue content are traditionally cooked in this manner. For this form of cookery, meat is browned in a small amount of fat, placed on a rack or in a pan, and cooked slowly in a covered utensil containing a small amount of liquid. The recommended final internal temperature of the meat is 85°C (beef) at an oven temperature of 165°C (Cross et al., 1978b).

In general, both cooking time and cooking temperature are important to consider relative to the tenderizing and toughening changes that occur in the structure of meat as caused by heat. Time is especially influential on the softening process and temperature on the toughening process (Weir, 1960). To ensure uniformity, all meat cooked for purposes of analyses should be cooked to the same internal endpoint temperature. A recording potentiometer serves to accurately measure and record the internal temperature of the meat. Although chemical thermometers (alcohol and mercury) may also be employed for measuring internal temperature, household thermometers should not be used because they lack the degree of accuracy required for research purposes (Carlin and Harrison, 1978).

To evaluate the consistency of temperature endpoints at a specific time after cooking, the meat sample is cut down the center to expose its interior. A degree of doneness score is obtained by comparing the internal color of the sample to a set of standardized photographs that show different degrees of doneness. The recommended internal endpoint temperatures for degrees of doneness in beef and lamb are rare (60°C), medium (70°C), and well done (80°C) (Cross et al., 1979).

D. Presentation of the Samples

There are many factors to consider when presenting evaluation samples to panel members. The testing area must be free from distracting pictures, colors, noises, and odors. It is essential to position panelists away from one another either in booths, rooms, or with partitions. Controlled temperature and humidity of the area is favorable for maximum performance. Lighting in the testing area must be uniform, adequate, and adjustable. Although special lighting (i.e., red fluorescent) is sometimes employed to cover obvious color distractions on a sample, natural fluorescent lights are suitable for most sample evaluations. The panelists are provided with this comfortable setting to encourage maximum concentration on the sample testings.

When the testing area is sufficiently prepared to permit an appropriate evaluation performance, the taste panelists are seated and the tasting procedure begins. A "warmup" sample is initially presented to the panelists to prevent any first-sample bias. The number of samples depend on the type of product, the sensory characteristic being evaluated, and the experience of the panelists. Enough samples (usually no fewer than two) should be available to the panelists to ensure evaluation scores that are representative of the entire product under study. Sample sizes should be uniform and large enough that the panelists are able to evaluate several different characteristics (Carlin and Harrison, 1978). In meat evaluation, some scientists recommend 2.5-cm or 1.3-cm diameter cores for evaluation but others recommend using cubes or slices (Cross *et al.*, 1978b).

Another factor influencing panel evaluation scores is the order in which the samples are presented to the judges. Eindhoven *et al.* (1964) concluded that when samples are judged simultaneously, as in visual comparisons, the later samples are rated lower. This phenomenon is termed *position effect* and is independent of two other effects termed *contrast* and *convergence*. These two principles refer to sample presentations when the quality of one sample influences the panelist's opinion of the following sample. Situations such as these are impossible to avoid but should be considered when conclusions are made.

The temperature of the samples should be constant and consistent throughout the evaluation performance. If the cooking intervals are appropriately spaced, samples will not reach their temperature endpoints at corresponding times (Cross *et al.*, 1979) and thus can be individually evaluated before they cool.

Samples are presented in sanitary, white, odorless, and tasteless containers. With the proper evaluation sheet, the panelist rates each sample individually by the method and order learned in the training session. Between tasting samples, agents such as room temperature water or crackers, are used by the panelists to remove aftertastes.

E. Color Evaluation

Meat color can be defined by characteristics such as hue, value, and chroma, with hue pertaining to the actual tint or color (i.e., red, green, etc.) of the meat, value being the degree of lightness or darkness of meat color, and chroma referring to the intensity of the meat color. These characteristics can be evaluated by either subjective or objective methods of analyses.

Subjectively, meat color evaluation utilizes a trained panel that establishes standards and criteria of analysis. Subjective methods of evaluating meat color involve the use of highly trained panelists. Scoring systems used to describe color intensity (i.e., 8 ≡ dark red, 1 ≡ pale) are referred to as visual color standards and are usually photographs. Visual color standards enable the panelists to analyze the whole meat surface visually. In addition, only humans can evaluate percentage surface discoloration, whereas no machine is capable of this.

An objective method of determining the percentages of myoglobin, oxymyoglobin, and metmyoglobin on the meat surface is spectrophotometry. To calculate these percentages, the total amount of light reflected from a sample at various wavelengths is read. From these data, a ratio of the amount of light absorbed to the amount of light transmitted is calculated.

Although objective color analysis helps to determine the amount of pigment in muscle and its relationship to muscle color, it is not capable of determining the density of muscle fibers or other factors viewed only by the human eye that affect a meat cut's appearance.

F. Odor Evaluation

Odor is the sensory attribute of certain volatile substances perceived by the olfactory organ. Presently, there is no instrument available that compares with the olfactory systems in the nasal cavity of humans to evaluate odorous substances. The most popular subjective method of evaluating odor is the sniff test. It involves the use of a numerical scale based on factors such as intensity (4 ≡ intense off odor, 1 ≡ no off odor) or verbal description (rancid, ammonia-like, sulfurous, etc.). The first scale described determines the extent of the odor (intensity) while the second scale gives a specific description of the odor. Another important factor to consider that affects the results of the sniff test is the temperature of the meat. Many odoriferous substances in meat are noticeable only at very high temperatures. For example, although boar odor cannot be detected in either the live animal or the carcass, it is apparent and can be evaluated by heating a piece of fat from the carcass. A number of other odoriferous compounds may result from the breakdown of long-chain fatty acids (which have no odor) to short-chain fatty acids, which are very volatile when exposed to high temperatures.

Objective methods of evaluating meat odors are rarely used because they do not correlate well with subjective human evaluation. However, an example of an objective method for evaluating meat odor would be to heat a meat sample in a boiling flask and capture and condense the volatiles in a distillation column. The odoriferous volatiles would then be run through a gas chromatograph. In order to determine the type of gas chromatograph column and procedure to use, a subjective descriptive panel must first be used to describe the odor in an effort to estimate logically what compound measurable in the gas chromagraph it is desirable to quantify (i.e., volatile fatty acids, esters, amines, etc.).

G. Flavor Evaluation

As in the case of odor, there is no known instrument available to measure adequately the components normally associated with the various attributes of meat flavor. The major problem scientists encounter with meat flavor evaluation is trying to correlate analytical results accurately with sensory descriptions.

In descriptive sensory methodology trained sensory panelists are used to describe the flavor characteristics of a given meat sample. For each specific flavor characteristic, a scale is used to indicate the intensity of the flavor characteristic.

Objective methods employed to evaluate meat flavor include gas chromatography, mass spectrometry, and liquid chromatography. Continuing research in this area has led to the discovery of hundreds of flavor compounds in meat, therefore making analysis difficult.

H. Juiciness Evaluation

Juiciness of meat can be directly related to the fat and moisture content of the meat. It is also influenced by such factors as the flavor and texture of the meat. For example, some flavor components of meat can cause the rapid release of saliva, thus giving an impression of juiciness. Fat present in meat may also act as a means of lubrication and sustained juiciness. Texture determines the ease with which moisture can be expressed from meat when it is masticated, also contributing an impression of juiciness.

A subjective method of evaluating juiciness is through the use of the trained sensory panel that evaluates meat on a numerical scale ($8 \equiv$ extremely juicy, $1 \equiv$ extremely dry). The evaluators are conscious of an initial and final release of juices in the meat samples as parameters for its evaluation.

Objective methods to measure meat juiciness are related to the water-holding capacity of meat. Bound and immobilized water in meat is calculated after a meat sample is exposed to high amounts of pressure, such as with the Carver Laboratory Press. These values are correlated to the juiciness characteristics of meat. Pressure applied to meat samples is accomplished by placing the meat sample on filter paper between two metal plates and applying pressure to the plates. The wetted areas on the filter paper are measured and are thus proportional to the amount of free water in the meat sample (see Chapter 4 for a more extensive discussion).

Another method of objectively evaluating juiciness is the water extraction of muscle protein procedure (Sybesma, 1965). This procedure involves powdering the meat in a Waring blender followed by centrifugation and filtration of the homogenate. The filtrate is then mixed with a buffer solution and measured in a spectrophotometer as percentage transmission at a standard wavelength. A control with a given percentage transmission value is used to separate acceptable and unacceptable results.

I. Tenderness Evaluation

Variation in tenderness between meat cuts can be caused by variations in fat content, amount, and chemical state of connective tissue and actomyosin effects. Subjective methods of evaluating tenderness include both the trained and consumer panels. Trained panelists evaluate texture on factors such as the amount of sensory detectable connective tissue, ease of fragmentation, overall tenderness, and texture classification.

Amount of sensory detectable connective tissue is based on the panelist's ability to detect rubberlike strands of connective tissue in the cooked meat towards the end of mastication. Ease of fragmentation is evaluated by the panelist, who works the meat between the tongue, cheek and teeth in order to determine how easily the meat fragments. Tenderness can initially be evaluated after three or five chews and then again after 10 to 15 chews, depending on the methodology needed to satisfy the objective of the experiment.

Some objective methods of evaluating meat tenderness include: the Warner–Bratzler Shear Device, Lee–Kramer Shear Press, Instron, Universal Testing Machine, and fragmentation index. The Warner–Bratzler Shear Device, Lee–Kramer Shear Press, and the Instron are used to evaluate the tenderness of cooked meat while the fragmentation index is used to predict tenderness based on raw meat measurements.

The most commonly used instrument for objectively measuring meat tenderness is the Warner–Bratzler Shear Device. It measures the amount of force necessary to shear a cylindrical sample of meat with a 1-mm-thick blade. A dynamometer is used to calibrate the force as read in pounds or kilograms. The greater the force value in pounds, the tougher the meat sample (Szczesniak and Targeson, 1965). Factors found to affect the accuracy of the Warner–Bratzler Shear Device that may need to be considered include doneness of the cooked meat, uniformity of cylindrical sample size, direction of the muscle fibers, amount of connective tissue and fat deposits present, temperature of the sample, and the speed at which the sample is sheared (Bratzler, 1949). This instrument usually correlates well with trained taste panel evaluation scores. The Instron texture testing machine is similar to the Warner–Bratzler Shear Device, except that it can also measure cohesion and elasticity.

The Lee – Kramer Shear Press is a mechanical instrument designed to measure the firmness or hardness of cooked and uncooked food. It contains cell teeth that act in the same basic manner that a human's teeth do in that food is first compressed then sheared. A hydraulic drive system moves the cell teeth at a constant rate to compress the food sample against a stationary test cell. This in turn results in fast return of the cell teeth and a time – force curve recording on a strip chart. An automatic digital readout records the area under the force curve, which is the work performed on the sample under test. Results measured with this instrument, like the Warner – Bratzler Press, correlate well with sensory panel tenderness values.

Fragmentation index is a procedure that measures the breaking strength of muscle tissues (Calkins and Davis, 1980; Davis *et al.*, 1980; Reagan *et al.*, 1975). Raw meat samples are homogenized, filtered, and centrifuged. The supernatant is removed and the excess fat is collected from the centrifuge tube. The tube plus the remaining pellet is then weighed at a constant time of 5 to 10 min. The fragmentation index equation equals average pellet weight multiplied by 100. It may vary from values of 50 (very tough) to 300 (very tender). This procedure is relatively new and is thus in the experimental stages of development.

VII. SUMMARY AND RESEARCH NEEDS

The various attributes of fresh and cooked meat that influence consumer acceptance have been summarized. Many antemortem and postmortem factors can influence the sensory qualities of meat. This chapter has briefly outlined some of the effects of those factors on sensory traits. Various methods of measuring or evaluating these traits have also been discussed. Since such great emphasis is placed on sensory panel evaluations, much care should be taken in selecting and training sensory panels. Data may be misinterpreted if poorly or improperly trained sensory panels are used. The livestock and meat industry in the United States has relied heavily on feeding high-energy rations for long periods of time to assure the production of a high-quality product. With ever-increasing energy and production costs, we can no longer afford to feed animals to assure a high level of palatability. We must continue to conduct research that will develop postmortem handling systems that will assure the consumer of a uniformly acceptable product. Research must continue in the areas of postmortem handling of meat to maintain or improve meat quality. Increased research is needed on more objective means of measuring meat quality traits. In addition, research is needed to develop more suitable methods for selecting, training, and evaluating sensory panels.

REFERENCES

Adams, J. R., and Huffman, D. L. (1972). Effect of controlled gas atmospheres and temperature on quality of packaged pork. *J. Food Sci.* **37**, 869–872.

American Society for Testing and Materials (1968). *In* "Manual on Sensory Testing Methods," STP 434. ASTM Committee E-18, Philadelphia, Pennsylvania.

American Society for Testing and Materials (1972a). Standard recommended practice for determining effect of packaging on food and beverage products during storage, E460. *In* "1975 Annual Book of ASTM Standards," Part 46, p. 257. ASTM Committee E-18, Philadelphia, Pennsylvania.

American Society for Testing and Materials (1972b). Standard recommended practice for sensory evaluation of industrial and institutional food purchases, E461. *In* "1975 Annual Book of ASTM Standards," Part 46, p. 260. ASTM Committee E-18, Philadelphia, Pennsylvania.

Amerine, M. A., Pangborn, R. M., and Roessler, E. B. (1965). "Principles of Sensory Evaluation of Food." Academic Press, New York.

Bendall, J. R. (1964). Meat proteins. *In* "Proteins and Their Reactions" (H. W. Schultz and A. F. Anglemier, eds.), p. 225. Avi Publ. Co., Westport, Connecticut.

Bengtsson, N. E., Jakobsson, B., and Dagerskog, M. (1976). Cooking of beef by oven roasting: A study of heat and mass transfer. *J. Food Sci.* **40**, 1047.

Bodwell, C. E., and McClain, P. E. (1971). Chemistry of animal tissues. Proteins. *In* "The Science of Meat and Meat Products" (J. F. Price and B. W. Schweigert, eds.), 2nd ed., p. 78. San Francisco.

Bohren, B. B., and Jordan, R. (1953). A technique for detecting flavor changes in stored dried eggs. *Food Res.* **18**, 583.

Bradley, R. A. (1953). Some statistical methods in taste testing and quality evaluation. *Biometrics* **9**, 22–38.

Brandt, M. A., Skinner, E. Z., and Coleman, J. A. (1963). Texture profile methods. *J. Food Sci.* **28**, 404.

Bratzler, L. J. (1949). Determining the tenderness of meat by the use of the Warner–Bratzler method. *Proc. Annu. Reciprocal Meat Conf.* **2**, 117–120.

Brooks, J. (1937). Color of meat. *Food Ind.* **9**, 707.

Byer, A. J., and Adrams, D. (1953). A comparison of the triangle and two-sample taste-test methods. *Food Technol.* **7**, 185.

Calkins, C. R., and Davis, G. W. (1980). Fragmentation index of raw muscle as a tenderness predictor of steaks from USDA Good and Standard steer and bullock carcasses. *J. Anim. Sci.* **50**, 1067–1072.

Carlin, A. F., and Harrison, D. L. (1978). "Cooking and Sensory Methods used in Experimental Studies on Meat." National Live Stock and Meat Board, Chicago.

Carlin, A. F., Kempthorne, O., and Gordon, J. (1956). Some aspects of numerical scoring in subjective evaluation of foods. *Food Res.* **21**, 273–281.

Caul, J. F. (1957). The profile method of flavor analysis. *Adv. Food Res.* **7**, 1–40.

Cheng, C. S. (1976). Biochemical studies of postmortem aging, calcium ion and heating of bovine skeletal muscle. Ph.D. Thesis, Iowa State University, Ames.

Cheng, C. S., and Parrish, F. C., Jr. (1976). Scanning electron microscopy of bovine muscle: Effect of heating on ultrastructure. *J. Food Sci.* **41**, 1449–1454.

Civille, G. V. (1979). Descriptive analysis — flavor profile, texture profile and quantitative descriptive analysis. *In* "Sensory Evaluation Methods for the Practicing Food Technologist," pp. 6-1–6-22. Inst. Food Technol., Chicago.

Civille, G. V., and Szczeniak, A. S. (1973). Guidelines to training a texture profile panel. *J. Text. Stud.* **4**, 204.

Clydesdale, F. M., and Francis, F. J. (1971). The chemistry of meat color. *Food Prod. Dev.* **5**, 81–84.

Cochran, W. G., and Cox, G. M. (1957). "Experimental Design," 2nd ed. Wiley, New York.

Cross, H. R., Moen, R., and Stanfield, M. S. (1978a). Training and testing of judges for sensory analysis of meat quality. *Food Technol. (Chicago)* **32**, 48–54.

Cross, H. R., Bernholdt, H. F., Dikeman, M. E., Greene, B. E., Moody, W. G., Staggs, R., and West, R. L. (1978b). "Guidelines for Cookery and Sensory Evaluation of Meat." *Am. Meat Sci. Assoc.* Chicago, Illinois.

Cross, H. R., Stanfield, M. S., and Elder, R. S. (1979). Comparison of roasting versus broiling on the sensory characteristics of beef *longissimus. J. Food Sci.* **44**, 310–311.

Dahlin, K. J., Allen, C. E., Benson, E. S., and Hegarty, P. V. J. (1976). Ultrastructural differences induced by heat in red and white avian muscles in rigor mortis. *J. Ultrastruct. Res.* **55**, 96.

Davey, C. L., and Gilbert, K. V. (1974). Temperature-dependent cooking toughness in beef. *J. Sci. Food Agric.* **25**, 931–938.

Davis, G. W., Dutson, T. R., Smith, G. C., and Carpenter, Z. L. (1980). Fragmentation procedure for bovine longissimus muscle as an index of cooked steak tenderness. *J. Food Sci.* **45**, 880–884.

Draudt, H. N. (1972). Changes in meat during cooking. *Proc. Annu. Reciprocal Meat Conf.* **25**, 243.

Duncan, D. B. (1955). Multiple range and multiple F tests. *Biometrics* **11**, 1.

Eindhoven, J., Peryan, D., Heiligman, F., and Baker, G. A. (1964). Effect of sample sequence on food preference. *J. Food Sci.* **29**, 4.

Ellis, B. H. (1966). "A Guide Book for Sensory Testing." Continental Can Co., Chicago, Illinois.

Fox, J. B., Jr. (1966). The chemistry of meat pigments. *J. Agric. Food Chem.* **14**, 207.

Green, D., and Swets, J. (1971). "Signal Detection Theory in Psychophysics," p. 30. Krieger, Huntington, New York.

Gregson, R. A. M. (1962). A rating-scale method for determining absolute taste thresholds. *J. Food Sci.* **27**, 376.

Gridgeman, N. T. (1955). Taste comparisons: Two samples or three? *Food Technol. (Chicago)* **9**, 148–150.

Gridgeman, N. T. (1961). A comparison of some taste-test methods. *J. Food Sci.* **26**, 171.

Gridgeman, N. T. (1964). Sensory comparisons: The 2-stage triangle test with sample variability. *J. Food Sci.* **29**, 112–117.

Guilford, J. P., (1954). "Psychometric Methods," 2nd ed. McGraw-Hill, New York.

Haagen-Smit, A. J. (1952). Smell and taste. *Sci. Am.* **186**, 28–32.

Hamm, R. (1966). Heating of muscle systems. *In* "The Physiology and Biochemistry of Muscle as a Food, 1" (E. J. Briskey, R. G. Cassens, and J. C. Trautman, eds.), p. 363. Univ. of Wisconsin Press, Madison.

Hamm, R. (1969). Properties of meat proteins. *In* "Proteins as Human Food" (R. A. Lawrie, ed.), pp. 167–185. Avi Publ. Co., Westport, Connecticut.

Hamm, R., and Hofmann, K. (1965). Changes in the sulphydryl and disulphide groups in beef muscle proteins during heating. *Nature (London)* **207**, 1269.

Hartshorne, D. J., Barns, E. M., Parker, L., and Fuchs, F. (1972). The effect of temperature on actomyosin. *Biochim. Biophys. Acta* **267**, 190.

Haughey, D. P. (1968). A physical approach to the cooking of meat. *Proc. Meat Ind. Res. Conf.* No. 10 (mimeo.). Meat Ind. Res. Inst., New Zealand, Hamilton.

Hegarty, P. V. J., and Allen, C. E. (1972). Rigor-stretched turkey muscles: Effect of heat on fiber dimensions and shear values. *J. Food Sci.* **37**, 652.

Hegarty, P. V. J., and Allen, C. E. (1975). Thermo effects on the length of sarcomeres in muscles held at different tensions. *J. Food Sci.* **40**, 24.

Heldman, D. R. (1975). Heat transfer in meat. *Proc. Annu. Reciprocal Meat Conf.* **28**, 314.

Herring, H. K., Cassens, R. G., and Briskey, E. J. (1967). Factors affecting collagen solubility in bovine muscles. *J. Food Sci.* **32**, 534–538.

Hodge, J. E. (1967). Nonenzymatic browning reactions. *In* "Chemistry and Physiology of Flavors," (H. W. Schultz, ed.), 465–491.

Hopkins, J. W. (1950). A procedure for quantifying subjective appraisals of odor, flavor and texture of foodstuffs. *Biometrics* **6**, 1–16.

Hopkins, J. W. (1954). Some observations on sensitivity and repeatability of trial taste difference tests. *Biometrics* **10**, 521–530.

Hornstein, I. (1971). Chemistry of meat flavor. *In* "The Science of Meat and Meat Products" (J. F. Price and B. S. Schweigert, eds.). Freeman, San Francisco.

Hostetler, R. L., and Landmann, W. A. (1968). Photomicrographic studies of dynamic changes in muscle fiber fragments. 1. Effect of various heat treatments on length, width and birefringence. *J. Food Sci.* **33**, 468.

Institute of Food Technologists, Sensory Evaluation Division (IFT) (1981). Sensory evaluation guide for testing food and beverage products. *Food Technol. (Chicago)* **35**, 11.

Johnson, B. Y. (1974). Chilled vacuum-packed beef. *CSIRO Food Res.* **34**, 14–21.

Kahan, G., Cooper, D., Papavasiliou, A., and Kramer, A. (1973). Expanded tables for determining significance of differences for ranked data. *Food Technol. (Chicago)* **27** (5), 61.

Kendall, M. G., and Stuart, A. (1968). "Advanced Theory of Statistics," 2nd ed., Vol. 3. Wiley, New York.

Kraft, A. A., and Ayres, J. C. (1954). Effect of display case lighting on color and bacterial growth on packaged fresh beef. *Food Technol. (Chicago)* **8**, 290–295.

Kramer, A. (1960). A rapid method for determining significance of differences from rank sums. *Food Technol. (Chicago)* **14**, 576–581.

Kramer, A. (1963). Revised tables for determining significance of differences. *Food Technol. (Chicago)* **17**, 124.

Kramer, A., and Twigg, B. A. (1962). "Fundamentals of Quality Control for the Food Industry." Avi Publ. Co., Westport, Connecticut.

Laakkonen, E., Wellington, G. H., and Sherbon, J. (1970). Low temperature, long-time heating of bovine muscle. 1. Changes in tenderness, water-binding capacity, pH and amount of water soluble components. *J. Food Sci.* **35**, 175.

Lawrie, R. A. (1966). The eating quality of meat. *In* "Meat Science" (R. A. Lawrie, ed.). 286–342. Pergamon, Oxford.

Little, A. D., Inc. (1958). "Flavor Research and Food Acceptance." Van Nostrand-Reinhold, Princeton, New Jersey.

Locker, R. H. (1956). The disassociation of myosin by heat coagulation. *Biochim. Biophys. Acta* **20**, 514.

Locker, R. H., and Daines, G. J. (1974). Effect of mode of cutting on cooking loss in beef. *J. Sci. Food Agric.* **25**, 939.

Lockhart, E. E. (1951). Binomial systems and organoleptic analysis. *Food Technol. (Chicago)* **5**, 428–431.

McCrae, S. E., and Paul, P. C. (1974). Rate of heating as it affects the solubilization of beef muscle collagen. *J. Food Sci.* **39**, 18.

Machlik, S. M., nd Draudt, H. N. (1963). The effect of heating time and temperature on the shear of beef semitendinosus muscle. *J. Food Sci.* **28**, 711.

Mahoney, C. H., Stier, H. L., and Crosby, E. A. (1957). Evaluating flavor differences in canned foods. I. Genesis of a simplified procedure for making flavor difference tests. II. Fundamentals of the simplified procedure. *Food Technol. (Chicago)* **11**, Insert 29, 37.

Mangel, M. (1951). The determination of methemoglobin in beef muscle extracts. *Res. Bull. — Mo., Agric. Exp. Stn.* **474.**

Mann, R. L., and Briggs, D. R. (1950). Effect of solvent and heat treatments on soybean proteins as evidenced by electrophoretic analysis. *Cereal Chem.* **27,** 258.

Marshall, N., Wood, L., and Patton, M. B. (1960). Cooking choice grade, top round beef roasts. Effect of internal temperature on yield and cooking time. *J. Am. Diet. Assoc.* **36,** 341.

Morrison, D. F. (1976). "Multivariate Statistical Methods." McGraw-Hill, New York.

Moskowitz, H. R. (1974). Sensory evaluation by magnitude estimation. *Food Technol. (Chicago)* **28,** 16.

Moskowitz, H. R. (1975). Applications of sensory measurement to food evaluations. II. Methods of ratio scaling. *Lebensm.-Wiss. Technol.* **8,** 249.

Moskowitz, H. R. (1978). Magnitude estimation: Notes on how, what, where and why to use it. *J. Food Qual.* **1,** 195.

Moskowitz, H. R., and Barbe, C. (1976). Psychometric analysis of food aromas by profiling and multidimensional scaling. *J. Food Sci.* **41,** 567.

Moskowitz, H. R., and Sidel, J. L. (1971). Magnitude and hedonic scales for food acceptability. *J. Food Sci.* **36,** 677.

Moskowitz, H. R., and Wehrly, T. (1972). Economic applications of sweetness scales. *J. Food Sci.* **37,** 411.

Ockerman, H. W. (1973). "Chemistry of Meat Tissue," pp. XI-3, XII-7. Ohio State Univ. Press, Columbus.

Patterson, R. L. S. (1975). The flavor of meat. *In* "Meat" (D. J. A. Cole and R. A. Lawrie, eds.), p. 359. Butterworth, London.

Paul, P. C. (1963). Influence of methods of cooking on meat tenderness. *In* "Proceedings of the Meat Tenderness Symposium," pp. 225–241. Campbell Soup Co., Camden, New Jersey.

Paul, P. C., and Palmer, H. H., eds. (1972). "Food Theory and Application." Wiley, New York.

Penfield, M. P., and Meyer, B. H. (1975). Changes in tenderness and collagen of beef semitendinosus muscle heated at two rates. *J. Food Sci.* **40,** 150.

Peryam, D. R. (1958). Sensory difference tests. *Food Technol. (Chicago)* **12,** 231–236.

Peryam, D. R., and Pilgrim, F. J. (1957). Hedonic scale method of measuring food preferences. *Food Technol. (Chicago)* **11,** Suppl., 9–14.

Pilgrim, F. J., Schultz, H. G., and Peryam, D. R. (1955). Influence of monosodium glutamate on taste perception. *Food Res.* **20,** 310–314.

Pirko, P. C., and Ayres, J. C. (1957). Pigment changes in packaged beef during storage. *Food Technol. (Chicago)* **11,** 461.

Prell, P. A. (1976). Preparation of reports and manuscripts which include sensory evaluation data. *Food Technol. (Chicago)* **30,** 11.

Price, J. F., and Schweigert, B. S., eds. (1971). "The Science of Meat and Meat Products," 2nd ed. Freeman, San Francisco.

Rainey, B. D. (1980). Sensory evaluation methods. IFT short course. *In* "Selection and Training of Panelists for Sensory Testing," Chapter 7. *Inst. Food Technol.,* Chicago, Illinois.

Reagan, J. O., Dutson, T. R., Carpenter, Z. L., and Smith, G. C. (1975). Muscle fragmentation indices for predicting cooked beef tenderness. *J. Food Sci.* **40,** 1093.

Rence, J. W., Mohammad, A., and Mecham, D. K. (1953). Heat denaturation of gluten. *Cereal Chem.* **30,** 115.

Ritchey, S. J., and Hostetler, R. L. (1965). The effect of small temperature changes on two beef muscles as determined by panel scores and shear-force values. *Food Technol. (Chicago)* **19,** 93.

Robach, D. L., and Costilow, R. N. (1962). Role of bacteria in the oxidation of myoglobin. *Appl. Microbiol.* **9,** 529–533.

Roessler, E., Pangborn, R., Sidel, J., and Stone, H. (1978). Expanded statistical tables for estimating significance in paired-preference, paired-difference, duo–trio, and triangle tests. *J. Food Sci.* 43, 940.

Samejima, K., Hashimoto, Y., Yasui, T., and Fukazawa, T. (1969). Heat gelling properties of myosin, actin, actomyosin and myosin-subunits in a saline model system. *J. Food Sci.* 34, 242.

Schaller, D. R., and Powrie, W. D. (1972). Scanning electron microscopy of heated beef, chicken and rainbow trout muscle. *J. Inst. Can. Sci. Technol. Aliment.* 5, 184.

Scheffe, H. (1952). An analysis of variance for paired comparisons. *J. Am. Stat. Assoc.* 47, 381.

Schmidt, J. G., and Parrish, F. C., Jr. (1971). Molecular properties of postmortem muscle. 10. Effect of internal temperature and carcass maturity on structure of bovine longissimus. *J. Food Sci.* 36, 110.

Schutz, H. G. (1965). A food action rating scale for measuring food acceptance. *J. Food Sci.* 30, 365.

Sjöström, L. B., and Cairncross, S. E. (1951). Flavor profiles, a method of judging dairy products. *Am. Milk Rev.* 12, 42–44.

Sjöström, L. B., and Cairncross, S. E. (1954). The descriptive analysis of flavor. *In* "Food Acceptance Testing Methodology," Symposium sponsored by Quartermaster Food and Container Institute for the Armed Forces, Chicago.

Sjöström, L. B., Cairncross, S. E., and Caul, J. F. (1957). Methodology of the flavor profile. *Food Technol. (Chicago)* 11, Suppl., 20–25.

Smith, G. C., King, G. T., and Carpenter, Z. L. (1978). "Laboratory Manual for Meat Science." Texas A&M University, College Station.

Snedecor, G. W., and Cochran, W. G. (1967). "Statistical Methods," 6th ed. Iowa State University, Ames.

Stevens, S. S. (1962). The surprising simplicity of sensory metrics. *Am. Psychol.* 17, 29.

Stone, H., Sidel, J., Oliver, S., Woolsey, A., and Singleton, R. C. (1974). Sensory evaluation by quantitative descriptive analysis. *Food Technol. (Chicago)* 28, 24.

Sybesma, W. (1965). "Exudative Meat 'Meat Degeneration' in the Netherlands," Data about genetic and environmental influence on the frequency of this quality defect. Research Institute for Animal Husbandry, "Schoonoord," Zeist, The Netherlands (mimeo.).

Szczesniak, A. S., and Farkas, E. (1962). Objective characterization of the mouthfeel of gum solutions. *J. Food Sci.* 27, 381–385.

Szczesniak, A. S., and Targeson, K. W. (1965). Methods of meat texture measurement viewed from the background of factors affecting tenderness. *Adv. Food Res.* 14, 134.

Szczesniak, A. S., Brandt, M. A., and Friedman, H. H. (1963). Development of standard rating scales for mechanical parameters of texture and correlation between the objective and the sensory methods of texture evaluation. *J. Food Sci.* 28, 397.

Taylor, A. A. (1972). Gases in fresh meat packaging. *Meat World* 5, 3.

Tilgner, D. J. (1962a). Dilution tests for odor and flavor analysis. *Food Technol. (Chicago)* 16, 26.

Tilgner, D. J. (1962b). Anchored sensory evaluation tests—A status report. *Food Technol. (Chicago)* 16, 47.

Tompkins, M. D., and Pratt, G. B. (1959). Comparison of flavor evaluation methods for frozen citrus concentrate. *Food Technol. (Chicago)* 13, 149–152.

Tukey, J. W. (1951). Quick and dirty methods in statistics. II. Simple analyses for standard designs. *Proc. Conf. Am. Soc. Qual. Control* 5, 189–197.

Urbain, W. M. (1952). Oxygen is key to the color of meat. *Natl. Provis.* 127, 140–141.

Visser, R. Y., Harrison, D. L., Goertz, G. E., Bunyan, M., Skelton, M. M., and Mackintosh, D. L. (1960). The effect of degree of doneness on the tenderness and juiciness of beef cooked in the oven and in deep fat. *Food Technol. (Chicago)* 14, 193.

Wald, A. (1947). "Sequential Analysis." Wiley, New York.

Walters, C. L. (1975). *In* "Meat" (D. J. A. Cole and R. A. Lawrie, eds.), pp. 385–401. Butterworth, London.

Watts, B. M. (1954). Oxidative rancidity and discoloration in meat. *Adv. Food Res.* **5**, 1.

Watts, B. M., Kendrick, J., Zipser, M. W., Hutchins, B., and Saleh, B. (1966). Enzymatic reducing pathways in meat. *J. Food Sci.* **31**, 855–862.

Weir, C. E. (1960). Palatability characteristics of meat. *In* "The Science of Meat and Meat Products" (J. F. Price and B. S. Schweigert, eds.). Freeman, San Francisco.

Wiley, R. C., Briant, A. M., Fagerson, I. S., Sabry, J. H., and Murphy, E. F. (1957). Evaluation of flavor changes due to pesticides—a regional approach. *Food Res.* **22**, 192–205.

Winer, B. J. (1971). "Statistical Principles in Experimental Design," 2nd ed. McGraw-Hill, New York.

8

Nutritional Composition and Value of Meat and Meat Products

C. E. BODWELL

Energy and Protein Nutrition Laboratory
Beltsville Human Nutrition Research Center
Agriculture Research Service, U.S. Department of Agriculture
Beltsville, Maryland

B. A. ANDERSON

Nutrition Monitoring Division
Human Nutrition Information Service, U.S. Department of Agriculture
Hyattsville, Maryland

I. INTRODUCTION

Meat and meat products made from the muscle tissues of beef, pork, lamb, poultry, and fish constitute a major staple food, on a nutritional basis, in the diets of the large majority of the United States population. In addition, the highly positive organoleptic qualities of meat and the sense of satiety that follows a meat meal serve to make meats, in a general sense, a unique dietary component.

In this chapter, data that indicate the dietary importance of meats are presented first. Then, representative data describing the nutritional composition of the edible portion of foods from beef, pork, lamb, veal, turkey, and fish are reported and discussed. Next is the consideration of nutritional quality or bioavailability, which is important in fully appreciating the role of meats in our daily diets. Other factors that may affect nutritional value are then discussed. The relation of nutrient content to energy (calorie) content is compared between meats or meat products and other principal protein sources. Lastly, the question of the possible deleterious effects of high-meat diets on health is briefly discussed.

II. IMPORTANCE

Household consumption surveys (Table I) indicate that weekly consumption of all types of meats in the United States has increased markedly during the past several decades. Between 1948 and 1977, consumption of red meats increased ~ 1½ times, of poultry by about 2 times, and of fish by ~ 1¼ times. When data

Table I

Quantity of Meats Consumed per Person[a] in the United States during a Week in Spring[b]

| | Food group[c] | | |
Year	Red meats	Poultry	Fish
1942	1.762[d]	0.311	0.273
1948[e]	2.418	.462	.327
1955	3.033	.709	.396
1965	3.359	.854	.368
1977	3.513	.941	.408

[a] 21 meals at home in a week equals one person.

[b] Quantities in pounds. Data for 1942 from U.S. Department of Agriculture (1944); for 1948, U.S. Department of Agriculture (1954); for 1955, U.S. Department of Agriculture (1956); for 1965, U.S. Department of Agriculture (1968); for 1977, U.S. Department of Agriculture (1982a).

[c] Food groups in 1942 and 1948 were more similar to the nutrition food groups rather than marketing food groups used in 1955, 1965, and 1977.

[d] Excludes bacon and salt pork in 1942.

[e] Urban households only with two or more persons in families — excludes one-person households.

Table II
Per Capita Consumption of Red Meats, Poultry, and Fish[a]

Year	Beef	Veal	Pork	Lamb and mutton	Edible offals	Total red meat[b]	Poultry	Fish
1910	55.6	6.6	57.9	5.8	10.3	136.2	18.0	—
1920	46.7	7.3	59.1	4.8	10.2	128.1	16.0	—
1930	48.9	5.8	62.4	6.0	8.9	121.7	17.9	12.2
1940	43.4	6.7	68.4	5.9	9.7	134.1	17.5	13.0
1950	50.1	7.3	64.4	3.6	10.1	135.5	25.1	13.8
1960	64.2	5.2	60.3	4.3	10.1	144.1	34.4	13.7
1970	84.1	2.4	62.0	2.9	10.9	162.3	48.9	15.8
1979	79.6	1.6	64.6	1.3	10.2	157.3	62.0	17.4

[a] In pounds retail weight equivalent. Includes processed meats on a fresh weight basis and game fish (based on disappearance data). From U.S. Department of Agriculture (1965, 1981).
[b] Excludes game.

based on disappearance of fresh retail cuts in the market place are used, the long-term trend was similar (Table II), but consumption of red meats decreased slightly and consumption of poultry increased markedly between 1970 and 1979. From those and similar data, the consumption of cooked, edible meats can be calculated (van Arsdall, 1979). For cooked, edible meat, per capita

Table III
Percentage Contribution of the Meat Group[a] to the Average Nutrient Intake Per Day Per Individual, Based on 3 Days of Dietary Reports, April 1977–March 1978[b]

Nutrient	(%)
Energy	28.4
Protein	48.7
Fat	42.4
Carbohydrate	5.1
Calcium	7.5
Iron	34.5
Magnesium	17.4
Phosphorus	29.0
Vitamin A	12.0
Thiamin	23.6
Riboflavin	24.2
Niacin	44.3
Vitamin B_6	39.9
Vitamin B_{12}	51.1
Ascorbic acid	5.2

[a] Meat group includes red meats, poultry, fish, edible offal, and some mixed meat dishes.
[b] From U.S. Department of Agriculture (1984).

consumption in 1977 was estimated (in pounds) at 44.9 for beef and veal, 21.8 for pork (excluding lard), 0.6 for lamb and mutton, and 27.6 for poultry. Comparable estimates for fish are not available.

The importance of meats in the United States diet is also indicated by the percentage contribution of meats to the average daily intakes of various nutrients (Table III). Based on 1977–1978 dietary survey data, meats provided 40–50% of the protein, fat, niacin, vitamin B_6, and vitamin B_{12}, more than 30% of the iron and $\sim$25% of the thiamin and riboflavin. Red meats provided a considerably higher proportion of those nutrients than did either poultry or fish (Table II).

III. NUTRIENT COMPOSITION AND ENERGY VALUE

In the following tables, representative data are given for cuts that were selected because their values are indicative of the usual range of values between cuts within the carcass for each species and because they are also those cuts that are consumed at the highest level. Because space is limited, data are given for a single USDA quality grade (e.g., beef) or class (i.e., poultry). Likewise, data for cooked products are limited to one or two methods of cooking. The proximate composition and energy value of various products are discussed first, mineral and vitamin content next, and lipid content last. For the most part, the values for cooked products are emphasized because they indicate consumed nutrients. Except as otherwise indicated, values are for *lean only*, without bone and with the intermuscular, seam, and external fat trimmed. Data for the composition of cuts not listed, cuts with fat included, and cuts cooked by alternative methods are available in various referenced handbooks.

As with any composition data, they constitute the best estimates of the expected nutritional composition of a particular cut cooked by the method specified. In practice, the values for any cut may differ due to variation among animals, production practices, and other factors, some of which are discussed in this chapter.

A. Proximate Composition and Energy Value

Values for water, food energy (kcal; 1 kcal = 4.186 joules), protein ($N \times 6.25$), total lipids, carbohydrate, and ash for raw and cooked red meats, poultry, and fish are given in Tables IV–X. In the meats listed, dietary fiber is absent, carbohydrate is usually absent, and the amount of total ash is low. (In fresh meats, some glycogen may be present but the level is variable and usually

Table IV

Proximate, Vitamin, and Mineral Composition of Raw Hamburger and Raw and Cooked Beef Round and Rib[a]

Nutrients (per 100 g edible portion)	Hamburger (raw) Nominal fat level (%)			Round (Choice grade)			Rib (Choice grade)		
					Cooked			Cooked	
	17	21	27	Raw	Braised	Roasted	Raw	Broiled	Roasted
Proximate									
Water (g)	63.19	60.18	56.06	71.25	56.89	61.28	69.41	58.43	57.90
Food energy (kcal)	234	264	310	146	225	222	163	225	243
Protein (N × 6.25) (g)	18.70	17.69	16.62	21.87	31.59	28.99	20.13	28.04	26.74
Total lipids (g)	17.06	20.67	26.55	6.28	9.96	10.84	8.52	11.64	14.27
Carbohydrate (g)	0	0	0	0	0	0	0	0	0
Ash (g)	0.93	0.94	1.00	1.03	1.14	1.15	1.01	1.23	1.09
Minerals									
Calcium (mg)	6	8	8	4	5	5	10	13	13
Copper (µg)	72	73	62	83	134	126	67	100	89
Iron (mg)	1.95	1.77	1.73	2.38	3.46	3.15	2.18	2.57	2.30
Magnesium (mg)	20	18	16	25	25	28	22	27	25
Manganese (µg)	17	15	17	14	18	16	14	16	16
Phosphorus (mg)	141	136	130	214	272	239	196	208	219
Potassium (mg)	284	261	228	371	308	397	373	394	404
Sodium (mg)	66	69	68	59	51	65	63	69	75
Zinc (mg)	4.14	3.86	3.55	3.51	5.48	5.06	4.66	7.00	6.19
Vitamins									
Ascorbic acid (mg)	0	0	0	0	0	0	0	0	0
Thiamin (mg)	0.06	0.05	0.04	0.12	0.08	0.09	0.09	0.11	0.08
Riboflavin (mg)	0.25	0.21	0.15	0.20	0.26	0.26	0.15	0.22	0.20
Niacin (mg)	4.53	4.51	4.48	4.13	4.08	4.10	3.74	4.80	3.63
Pantothenic acid (mg)	0.39	0.37	0.35	0.44	0.42	0.57	0.34	0.34	0.44
Vitamin B_6 (mg)	0.26	0.25	0.24	0.57	0.36	0.50	0.42	0.40	0.37
Folacin (µg)	8	7	7	10	11	12	6	8	8
Vitamin B_{12} (µg)	2.06	2.34	2.65	2.93	2.47	2.80	3.57	3.32	3.37
Vitamin A (IU)	0	0	0	0	0	0	0	0	0

[a] From U.S. Department of Agriculture (1986); roasted values calculated from raw values by use of retention levels.

Table V
Proximate, Vitamin, and Mineral Composition of Raw and Cooked Pork[a]

Nutrients (per 100 g edible portion)	Rib			Fresh Ham		Cured Ham		Sausage		
	Raw	Cooked		Raw	Roasted	Unheated	Roasted	Fresh	Cooked	Smoked Linked (Cooked)
		Broiled	Roasted							
Proximate										
Water (g)	69.56	54.47	56.96	72.90	59.70	68.26	65.02	44.52	44.57	39.27
Food energy (kcal)	161	258	245	136	220	147	159	417	369	389
Protein ($N \times 6.25$) (g)	21.80	28.82	28.21	20.48	28.32	22.32	25.70	11.69	19.65	22.17
Total lipids (g)	7.53	14.94	13.80	5.41	11.03	5.71	5.50	40.29	31.16	31.75
Carbohydrate (g)	0	0	0	0	0	0.05	0.04	1.02	1.03	2.10
Ash (g)	0.98	1.25	1.04	1.05	1.11	3.66	3.74	2.49	3.06	4.72
Minerals										
Calcium (mg)	9	15	11	6	7	7	9	18	32	30
Copper (μg)	60	80	78	75	108	79	102	70	140	70
Iron (mg)	0.77	0.81	1.0	1.01	1.12	0.81	0.84	0.91	1.25	1.16
Magnesium (mg)	22	30	21	25	24	18	17	11	17	19
Manganese (μg)	10	20	11	29	37	35	30	—	71	—
Phosphorus (mg)	241	266	256	229	281	232	227	118	184	162
Potassium (mg)	423	439	423	369	373	371	329	205	361	336
Sodium (mg)	45	67	46	55	55	1516	1259	804	1294	1500
Zinc (mg)	1.62	2.38	2.22	2.27	2.27	2.04	2.62	1.59	2.50	2.82
Vitamins										
Ascorbic acid (mg)	0.3	0.3	0.4	0.9	0.9	0	0	2	2	2
Thiamin (mg)	0.86	0.89	0.64	0.88	0.69	0.93	0.80	0.545	0.741	0.700
Riboflavin (mg)	0.27	0.32	0.31	0.23	0.35	0.23	0.21	0.164	0.254	0.257
Niacin (mg)	5.04	5.23	5.36	5.34	4.94	5.25	5.04	2.835	4.518	4.532
Pantothenic acid (mg)	0.78	0.75	0.58	0.81	0.67	0.54	0.50	0.40	0.72	0.78
Vitamin B_6 (mg)	0.30	0.31	0.31	0.30	0.32	0.53	0.47	0.25	0.33	0.35
Folacin (μg)	7	9.0	9.0	9	12	4	4.0	4	—	—
Vitamin B_{12} (μg)	0.53	0.70	0.55	0.71	0.72	0.87	0.67	1.13	1.73	1.63
Vitamin A (IU)	6	6	8.0	6	9.0	0	0	—	—	—

[a] From U.S. Department of Agriculture (1983).

Table VI

Proximate, Vitamin, and Mineral Composition of Raw and Cooked (Roasted) Lamb and Veal[a]

Nutrients (per 100 g edible portion)	Lamb (Choice Grade)				Veal			
	Leg		Rib–loin		Round		Rib	
	Raw	Cooked	Raw	Cooked	Raw	Cooked	Raw	Cooked
Proximate								
Water (g)	72.92	60.74	69.26	56.49	74.51	67.01	74.73	64.86
Food energy (kcal)	134	208	172	265	104	145	115	161
Protein ($N \times 6.25$) (g)	19.99	28.33	19.26	24.37	22.20	30.39	22.74	29.18
Total lipids (g)	5.35	9.62	9.94	17.82	1.12	1.88	1.54	4.03
Carbohydrate (g)	0	0	0	0	0	0	0	0
Ash (g)	1.02	1.00	0.97	.99	1.10	1.23	1.20	1.26
Minerals								
Calcium (mg)	5	7	11	13	12	15	13	15
Copper (μg)	143	151	108	137	103	140	69	88
Iron (mg)	1.70	2.16	1.50	1.71	0.64	0.88	0.60	0.77
Magnesium (mg)	22	24	21	20	23	23	24	22
Manganese (μg)	28	35	27	30	—	—	—	—
Phosphorus (mg)	167	187	165	167	150	176	140	154
Potassium (mg)	299	296	280	275	389	411	415	411
Sodium (mg)	62	59	79	75	49	54	38	40
Zinc (mg)	3.00	4.8	2.90	4.0	1.95	2.82	2.17	2.94
Vitamins								
Ascorbic acid (mg)	0	0	0	0	0	0	0	0
Thiamin (mg)	0.15	0.10	0.12	0.10	0.10	0.08	0.11	0.08
Riboflavin (mg)	0.28	0.36	0.26	0.27	0.21	0.26	0.20	0.23
Niacin (mg)	6.23	5.92	5.81	6.30	7.44	8.36	7.62	8.02
Pantothenic acid (mg)	0.55	0.65	0.53	0.55	1.06	1.15	1.09	1.11
Vitamin B_6 (mg)	0.28	0.34	0.26	0.28	0.40	0.44	0.41	0.43
Folacin (μg)	4	3	4	3	7	4	7	4
Vitamin B_{12} (μg)	2.15	2.59	2.07	2.18	1.75	1.92	1.80	1.85
Vitamin A (IU)	0	0	0	0	0	0	0	0

[a] USDA, Human Nutrition Information Service, Nutrient Data Research Branch (unpublished data); values for cooked lamb for calcium, zinc, manganese, pantothenic acid, vitamin B_6, and vitamin B_{12} calculated from values for raw lamb and values for cooked veal calculated from values for raw veal; dashes denote lack of reliable data for a constituent believed to be present in measurable amount.

Table VII

Proximate, Vitamin, and Mineral Composition of Raw, Cooked (Roasted), and Canned Chicken[a]

Nutrients (per 100 g edible portion)	Combined[b]				Light meat	
	Flesh and skin		Flesh only		Flesh and skin	
	Raw	Cooked	Raw	Cooked	Raw	Cooked
Proximate						
Water (g)	65.99	59.45	75.46	63.79	68.60	60.51
Food energy (kcal)	215	239	119	190	186	222
Protein ($N \times 6.25$) (g)	18.60	27.30	21.39	28.93	20.27	29.02
Total lipids (g)	15.06	13.603	3.08	7.41	11.07	10.85
Carbohydrate (g)	0.00	0.00	0.00	0.00	0.00	0.00
Ash (g)	0.79	0.92	0.96	1.02	0.86	0.93
Minerals						
Calcium (mg)	11	15	12	15	11	15
Iron (mg)	0.90	1.26	0.89	1.21	0.79	1.14
Magnesium (mg)	20	23	25	25	23	25
Phosphorus (mg)	147	182	173	195	163	200
Potassium (mg)	189	223	229	243	204	227
Sodium (mg)	70	82	77	86	65	75
Zinc (mg)	1.31	1.94	1.54	2.10	0.93	1.23
Copper (mg)	0.048	0.066	0.053	0.067	0.040	0.053
Manganese (mg)	0.019	0.020	0.019	0.019	0.018	0.018
Vitamins						
Ascorbic acid (mg)	1.6	0.0	2.3	0.0	0.9	0.0
Thiamin (mg)	0.060	0.063	0.073	0.069	0.059	0.060
Riboflavin (mg)	0.120	0.168	0.142	0.178	0.086	0.118
Niacin (mg)	6.801	8.487	8.239	9.173	8.908	11.134
Pantothenic acid (mg)	0.910	1.030	1.058	1.104	0.794	0.926
Vitamin B_6 (mg)	0.35	0.40	0.43	0.47	0.48	0.52
Folacin (μg)	6	5	7	6	4	3
Vitamin B_{12} (μg)	0.31	0.30	0.37	0.33	0.34	0.32
Vitamin A (IU)	140	161	52	53	99	110

[a] U.S. Department of Agriculture (1979); data for broilers or fryers except for canned chicken (all classes); dashes denote lack of reliable data for a constituent believed to be present in measurable amount.

[b] Includes both light and dark meat (whole carcass).

very low; thus, in composition tables, technically traces of carbohydrate should be listed, but since this is not quantifiable, zero values are listed.) Therefore, the sum of moisture, protein, and fat (lipid) is relatively constant, and the loss of moisture during cooking or processing increases the amount of protein or fat per 100 g (~ 3 oz) of edible portion (without bone). The amounts of moisture, protein, and fat, however, vary considerably among cuts and species. For the red meats (Tables IV–VI), cuts from the round, ham, or leg are much leaner than those from the rib, loin, or shoulder. Both in the raw and cooked states,

Light meat		Dark meat				Canned, boned, with broth
Flesh only		Flesh and skin		Flesh only		
Raw	Cooked	Raw	Cooked	Raw	Cooked	
74.86	64.76	65.42	58.63	75.99	63.06	68.65
114	173	237	253	125	205	165
23.20	30.91	16.69	25.97	20.08	27.37	21.77
1.65	4.51	18.34	15.78	4.31	9.73	7.95
0.00	0.00	0.00	0.00	0.00	0.00	0.00
0.98	1.02	0.76	0.92	0.94	1.02	1.81
12	15	11	15	12	15	14
0.73	1.06	0.98	1.36	1.03	1.33	1.58
27	27	19	22	23	23	12
187	216	136	168	162	179	—
239	247	178	220	222	240	138
68	77	73	87	85	93	503
0.97	1.23	1.58	2.49	2.00	2.80	—
0.040	0.050	0.054	0.077	0.063	0.080	—
0.018	0.017	0.019	0.021	0.021	0.021	—
1.2	0.0	2.1	0.0	3.1	0.0	2.0
0.068	0.065	0.061	0.066	0.077	0.073	0.015
0.092	0.116	0.146	0.207	0.184	0.227	0.129
10.604	12.421	5.211	6.359	6.246	6.548	6.329
0.822	0.972	0.994	1.111	1.249	1.210	0.850
0.54	0.60	0.25	0.31	0.33	0.36	0.35
4	4	7	7	10	8	—
0.38	0.34	0.29	0.29	0.36	0.32	0.29
28	29	170	201	72	72	—

lean cuts have lower levels of fat, generally higher levels of protein, and slightly higher levels of moisture than do fat cuts. In general, energy content parallels fat content. That is particularly true for ground meats (hamburger, pork sausage, ground lamb) in which energy content depends directly upon the relative amounts of fat and lean used in their manufacture.

For chicken and turkey, values are listed for light plus dark meat, light meat alone and dark meat alone (Tables VII – VIII), with or without skin. Light meat (e.g., breast, wings) contains less fat and more protein than dark meat (e.g., legs,

Table VIII

Proximate, Vitamin, and Mineral Composition of Raw and Cooked (Roasted) Turkey[a]

Nutrients (per 100 g) edible portion	Combined			
	Flesh and skin		Flesh only	
	Raw	Cooked	Raw	Cooked
Proximate				
Water (g)	70.40	61.70	74.16	64.88
Food energy (kcal)	160	208	119	170
Protein ($N \times 6.25$) (g)	20.42	28.10	21.77	29.32
Total lipids (g)	8.02	9.73	2.86	4.97
Carbohydrate (g)	0.00	0.00	0.00	0.00
Ash (g)	0.88	1.00	0.97	1.05
Minerals				
Calcium (mg)	15	26	14	25
Iron (mg)	1.43	1.79	1.45	1.78
Magnesium (mg)	22	25	25	26
Phosphorus (mg)	178	203	195	213
Potassium (mg)	266	280	296	298
Sodium (mg)	65	68	70	70
Zinc (mg)	2.20	2.96	2.37	3.10
Copper (mg)	0.103	0.093	0.109	0.094
Manganese (mg)	0.020	0.021	0.021	0.021
Vitamins				
Ascorbic acid (mg)	0.0	0.0	0.0	0.0
Thiamin (mg)	0.064	0.057	0.072	0.062
Riboflavin (mg)	0.155	0.177	0.168	0.182
Niacin (mg)	4.085	5.086	4.544	5.443
Pantothenic acid (mg)	0.807	0.858	0.907	0.943
Vitamin B_6 (mg)	0.41	0.41	0.47	0.46
Folacin (μg)	8	7	9	7
Vitamin B_{12} (μg)	0.40	0.35	0.43	0.37
Vitamin A (IU)	6	0	0	0

[a] From U.S. Department of Agriculture (1979); values are for all classes (toms, hens, etc.); combined values include both light and dark meat (whole carcass).

thighs). The most striking differences among poultry cuts, however, are related to the association of fat with the skin. Removal of the skin markedly lowers fat content and correspondingly increases protein content and decreases energy density.

For all species, heart and gizzard (organ meats) are largely muscle tissue and are low in fat and high in protein (Table IX).

Values representative of fish species with fat contents that are low (cod, flounder, halibut), intermediate (mackerel, tuna) or high (herring, salmon) are given in Table X. The general belief that fish are lower in fat than red meats is

Light meat				Dark meat			
Flesh and skin		Flesh only		Flesh and skin		Flesh only	
Raw	Cooked	Raw	Cooked	Raw	Cooked	Raw	Cooked
69.83	62.83	73.82	66.27	71.13	60.23	74.48	63.09
159	197	115	157	160	221	125	187
21.64	28.57	23.56	29.90	18.92	27.49	20.07	28.57
7.36	8.33	1.56	3.22	8.80	11.54	4.38	7.22
0.00	0.00	0.00	0.00	0.00	0.00	0.00	0.00
0.90	1.02	1.00	1.08	0.86	0.97	0.93	1.02
13	21	12	19	17	33	17	32
1.21	1.41	1.19	1.35	1.69	2.27	1.75	2.33
24	26	27	28	20	23	22	24
184	208	204	219	170	196	184	204
271	285	305	305	261	274	286	290
59	63	63	64	71	76	77	79
1.57	2.04	1.62	2.04	2.95	4.16	3.22	4.46
0.075	0.048	0.075	0.042	0.137	0.150	0.147	0.160
0.018	0.020	0.019	0.020	0.021	0.023	0.022	0.023
0.0	0.0	0.0	0.0	0.0	0.0	0.0	0.0
0.056	0.056	0.064	0.061	0.073	0.058	0.081	0.063
0.115	0.132	0.122	0.129	0.202	0.235	0.221	0.248
5.137	6.289	5.844	6.838	2.855	3.530	3.075	3.649
0.615	0.626	0.688	0.677	1.033	1.160	1.155	1.286
0.48	0.47	0.56	0.54	0.32	0.32	0.36	0.36
7	6	8	6	10	9	11	9
0.42	0.35	0.45	0.37	0.38	0.36	0.40	0.37
6	0	0	0	5	0	0	0

not necessarily true. In particular, fish canned in oil provide a high level of calories per 100 g of edible portion.

Representative values for a variety of sausage and luncheon meats manufactured from beef, pork, chicken, and/or turkey are given in Table XI. The compositions of these products vary widely among meats, amounts of fat, salts, spices, and other nonmeat components added, and the processing procedures used in their manufacture. Among the products listed, moisture content varies from ~27% (pepperoni) to ~72% (turkey breast meat), fat from ~2% (turkey breast meat) to ~44% (pepperoni), protein from ~10% (Vienna sausage) to

Table IX

Proximate, Vitamin, and Mineral Composition of Raw and Cooked Heart and Gizzard[a]

	Heart							
Nutrients (per 100 g edible portion)	Beef		Calf		Pork		Lamb	
	Raw	Braised	Raw	Braised	Raw	Braised	Raw	Braised
Proximate								
Water (g)	74.75	65.27	76.2	60.3	76.2	68.1	71.6	54.1
Food energy (kcal)	102	173	124	208	118	148	162	260
Protein (g)	15.96	28.79	15.0	27.8	17.3	23.6	16.8	29.5
Total lipid (g)	3.8	5.6	5.9	9.1	4.4	5.0	9.6	14.4
Carbohydrate (g)	0.5	0.0	1.8	1.8	1.3	0.4	1.0	1.0
Ash (g)	1.0	1.1	1.1	1.0	0.8	1.0	1.0	0.9
Minerals								
Calcium (mg)	2	6	3	4	5	7	11	14
Copper (mg)	0.253	0.740	—	—	0.408	0.508	0.48	—
Iron (mg)	4.6	7.5	3.0	4.4	4.68	5.8	1.09	—
Magnesium (mg)	23	25	—	—	19	24	38	—
Manganese (μg)	19	59	—	—	63	73	36	—
Phosphorus (mg)	173	250	160	148	169	178	249	231
Potassium (mg)	266	233	208	250	294	206	348	—
Sodium (mg)	63	63	94	113	56	35	103	—
Zinc (mg)	2.38	3.13	—	—	2.8	3.1	1.97	—
Vitamins								
Ascorbic acid (mg)	8.0	6.0	1	Trace	5	2.0	1.0	Trace
Thiamin (mg)	.19	.14	0.63	0.29	0.61	0.56	0.45	0.21
Riboflavin (mg)	1.02	1.54	1.05	1.44	1.19	1.70	0.74	1.03
Niacin (mg)	9.460	4.070	2.600	8.1	6.76	6.05	6.3	6.4
Pantothenic acid (mg)	2.320	.87	—	—	2.515	2.47	3.0	—
Vitamin B$_6$ (mg)	.43	.21	—	—	0.39	0.39	—	—
Folacin (μg)	2.1	1.94	—	—	4	4	—	—
Vitamin B$_{12}$ (μg)	13.66	14.300	—	—	3.80	3.79	5.2	—
Vitamin A (IU)	—	—	30	40	8	22	70	100

[a] Values for beef from U.S. Department of Agriculture (in press); values for pantothenic acid, vitamin B$_6$, and vitamin B$_{12}$ for calf and lamb heart from U.S Department of Agriculture (1969); other values for calf and values for lamb from U.S Department of Agriculture (1963); values for chicken and turkey from U.S. Department of Agriculture (1979); values for pork from U.S Department of Agriculture (1983).

~24% (corned beef loaf), and carbohydrate from none (corned beef loaf, turkey breast) to ~9% (olive loaf, pork).

Values for several baby foods are given in Table XII. All contain between 13 and 16% protein. With some exceptions (chicken and turkey sticks), most contain <10% fat and between 75 and 80% water.

B. Vitamins and Minerals

In general, meats and meat products are excellent sources of thiamin, riboflavin, and niacin, and good to excellent sources of vitamin B$_6$ and vitamin B$_{12}$

				Gizzard			
Chicken		Turkey		Chicken		Turkey	
Raw	Simmered	Raw	Simmered	Raw	Simmered	Raw	Simmered
73.56	64.85	73.27	64.23	76.19	67.32	75.58	65.40
153	185	143	177	118	153	117	163
15.55	26.41	18.05	26.76	18.19	27.15	19.10	29.43
9.33	7.92	6.99	6.10	4.19	3.66	3.70	3.88
0.71	0.10	0.65	2.05	0.58	1.14	0.62	0.60
0.85	0.72	1.05	0.87	0.85	0.73	1.01	0.70
12	19	9	13	8	10	9	15
0.346	0.502	0.317	0.627	0.096	0.110	0.141	0.173
5.96	9.03	4.80	6.89	3.51	4.15	3.80	5.44
15	20	22	22	16	20	17	19
89	107	52	92	65	62	64	98
177	199	190	205	135	155	143	128
176	132	277	183	236	179	343	211
74	48	87	55	76	67	80	54
6.59	7.30	3.52	5.27	3.00	4.38	2.01	4.16
3.2	1.8	3.2	1.7	3.2	1.6	3.2	1.6
0.152	0.070	0.199	0.071	0.034	0.027	0.057	0.033
0.728	0.741	1.088	0.882	0.188	0.244	0.231	0.327
4.883	2.803	4.093	3.253	4.725	3.975	3.435	3.072
2.559	2.654	2.674	2.723	0.752	0.714	0.909	0.848
0.36	0.32	0.36	0.32	0.14	0.12	0.14	0.12
72	80	72	79	52	53	52	52
7.29	7.29	7.29	7.15	2.12	1.94	2.12	1.90
30	28	30	28	217	188	217	185

(Tables IV – XII). They are also excellent sources of zinc, highly available (see below) iron, and copper and provide significant amounts of phosphorous, potassium, magnesium, selenium, and chromium. Especially high levels of thiamin are provided by pork, high levels of niacin and vitamin B_6 by chicken (light meat), and high levels of vitamins B_6 and B_{12} by beef. Beef also provides especially high levels of iron and zinc.

Some muscle organ meats (heart, gizzard) are also excellent sources of riboflavin, niacin, pantothenic acid, and vitamin B_6. Some fish provide high levels of niacin, pantothenic acid, and vitamin B_{12}.

Table X

Proximate, Vitamin, and Mineral Composition of Selected Raw, Cooked, or Canned Fish[a]

Nutrients (per 100 g edible portion)	Cod		Flounder		Halibut (Atlantic and Pacific)	
	Raw[b]	Broiled[c]	Raw	Baked	Raw	Broiled[c]
Proximate						
Water (g)	80.89	72.2	80.8	58.1	76.5	66.6
Food energy (kcal)	82	141	83	202	100	171
Protein (g)	17.61	20.2	17.3	30.0	20.9	25.2
Total lipid (g)	0.73	6.5	1.0	8.2	1.2	7.0
Carbohydrate (g)	0	0.4	0	0	0	0
Ash (g)	1.16	—	1.3	2.2	1.4	1.7
Minerals						
Calcium (mg)	18	14	27	23	13	16
Copper (μg)	87	—	170	—	—	—
Iron (mg)	0.43	0.5	0.79	1.4	0.7	0.8
Magnesium (mg)	30	32	24	—	—	—
Manganese (μg)	—	—	40	—	—	—
Phosphorus (mg)	195	223	182	344	211	248
Potassium (mg)	392	444	316	587	449	525
Sodium (mg)	58	147	61	237	54	134
Zinc (mg)	1.69	0.5	0.46	—	—	—
Vitamins						
Ascorbic acid (mg)	0	3	1.0	2	—	—
Thiamin (mg)	0.07	0.05	0.10	0.07	0.07	0.05
Riboflavin (mg)	0.09	0.07	0.08	0.08	0.07	0.07
Niacin (mg)	2.02	2.14	2.3	2.5	8.3	8.3
Pantothenic acid (mg)	0.17	—	0.86	—	0.275	—
Vitamin B$_6$ (mg)	0.21	0.21	0.19	—	0.43	—
Folacin (μg)	2	—	0.5	—	—	—
Vitamin B$_{12}$ (μg)	0.57	0.69	5.4	—	1.0	—
Vitamin A (IU)	0	294	48	—	440	680

[a] Values for cod from U.S. Department of Agriculture, Human Nutrition Information Service, Nutrient Data Research Branch (unpublished data); values for raw flounder, herring, and mackerel from Sidwell et al. (1974, 1977, 1978a,b); values for vitamins B$_6$ and B$_{12}$ and pantothenic acid for halibut, salmon, tuna, and herring (canned), from U.S. Department of Agriculture (1969); all other values from U.S. Department of Agriculture (1963); dashes denote lack of reliable data for a constituent believed to be present in measurable amount.

[b] Atlantic, fillet.

[c] With margarine.

Herring (Atlantic)		Mackerel (Pacific)		Salmon (Atlantic)		Tuna	
Raw	Canned (with liquid)	Raw	Canned (with liquid)	Raw	Canned (with liquid)	Raw (bluefin)	Canned (in oil, with liquid)
60.1	62.9	71.7	66.4	63.6	64.2	70.5	52.6
219	208	142	180	217	203	145	288
18.2	19.9	22.0	21.1	22.5	21.7	25.2	24.2
15.7	13.6	5.3	10.0	13.4	12.2	4.1	20.5
0	0	0	0	0	0	0	0
1.7	3.7	1.5	2.5	1.4	1.6	1.3	2.4
142	147	58	260	79	—	—	6
170	—	90	—	—	—	—	—
1.09	1.8	2.93	2.2	0.9	—	1.3	1.1
38	—	30	—	—	—	—	—
10	—	0	—	—	—	—	—
324	297	267	288	186	—	—	294
348	—	320	—	—	—	—	301
103	—	90	—	—	—	—	800
0.17	—	0.10	—	—	—	—	—
9.0	—	3	—	9	—	—	—
0.05	—	0.09	0.03	—	—	—	0.04
0.26	0.18	0.26	0.33	0.08	—	—	0.09
3.8	—	12.4	8.8	7.2	—	—	10.1
2.43	0.700	0.44	—	1.30	0.550	0.550	0.325
0.31	0.160	0.65	—	0.70	0.300	0.80	0.425
10	—	—	—	—	—	—	—
11.4	8.00	4.2	—	4.00	6.89	3.00	2.20
814	—	107	30	—	—	—	90

Table XI

Proximate, Vitamin, and Mineral Composition of Sausage Products and Luncheon Meats[a]

Nutrients (per 100 g edible portion)	Beerwurst (beer salami)		Bologna			
	Beef	Pork	Beef	Beef and pork	Pork	Turkey
Proximate						
Water (g)	53.70	61.46	54.84	54.30	60.60	65.08
Food energy (kcal)	324	238	313	316	247	199
Protein ($N \times 6.25$) (g)	12.33	14.24	11.69	11.69	15.30	13.73
Total lipid (g)	29.40	18.80	28.36	28.26	19.87	15.21
Carbohydrate (g)	1.73	2.06	1.95	2.79	0.73	0.97
Ash (g)	2.83	3.44	3.16	2.97	3.50	3.30
Minerals						
Calcium (mg)	9	8	12	12	11	84
Copper (mg)	0.05	0.05	0.03	0.08	0.08	0.03
Iron (mg)	1.36	0.76	1.40	1.51	0.77	1.53
Magnesium (mg)	12	13	10	11	14	14
Manganese (mg)	—	0.032	0.028	0.039	0.036	—
Phosphorus (mg)	105	103	82	91	139	131
Potassium (mg)	184	253	155	180	281	199
Sodium (mg)	932	1240	1001	1019	1184	878
Zinc (mg)	2.63	1.72	2.00	1.94	2.03	1.74
Vitamins						
Ascorbic acid (mg)	14[c]	29[c]	19[c]	21[c]	35[c]	—
Thiamin (mg)	0.113	0.554	0.056	0.172	0.523	0.055
Riboflavin (mg)	0.123	0.192	0.128	0.137	0.157	0.165
Niacin (mg)	2.870	3.254	2.631	2.580	3.900	3.527
Pantothenic acid (mg)	0.33	0.49	0.28	0.28	0.72	—
Vitamin B_6 (mg)	0.21	0.35	0.18	0.18	0.27	—
Folacin (μg)	3	3	5	5	5	—
Vitamin B_{12} (μg)	2.11	0.87	1.41	1.33	0.93	—
Vitamin A (IU)	—	—	—	—	—	—

[a] From USDA (1980); dashes denote lack of reliable data for a constitutent believed to be present in measurable amount.
[b] Values for all-pork smoked link sausage are given in Table V.
[c] Product contains added sodium ascorbate.

C. Amino Acids

Representative data for the amino acid compositions of selected cooked meats, sausage and luncheon meat products, and baby foods are given in Tables XIII–XV. In general, all meat products are excellent sources of essential amino acids that, nutritionally, are highly available. The role of meat, as the major source of high quality protein in the United States diet is discussed in a subsequent section.

Bratwurst, pork, cooked	Chicken roll	Corned beef loaf	Frankfurter			
			Beef	Beef and pork	Chicken	Turkey
56.13	68.60	67.32	54.00	53.87	57.53	62.99
301	159	163	322	320	257	226
14.08	19.53	23.70	11.29	11.28	12.93	14.28
25.87	7.38	6.80	29.42	29.15	19.48	17.70
2.07	2.45	0.00	2.39	2.55	6.79	1.49
1.85	2.05	2.80	2.90	3.15	3.28	3.53
44	43	11	12	11	95	106
0.09	0.04	0.06	0.06	0.08	—	—
1.29	0.97	2.03	1.33	1.15	2.00	1.84
15	19	10	10	10	—	—
0.046	—	0.031	0.033	0.032	—	—
149	157	64	82	86	—	134
212	228	89	159	167	—	1.79
557	584	1037	1024	1120	1370	1426
2.30	0.72	3.83	2.12	1.84	—	—
1	0	8[c]	25[c]	26[c]	—	—
0.505	0.065	0.010	0.051	0.199	0.066	0.041
0.183	0.130	0.117	0.102	0.120	0.115	0.179
3.200	5.291	1.637	2.527	2.634	3.089	4.132
0.32	—	0.19	0.29	0.35	—	—
0.21	—	0.14	0.11	0.13	—	—
—	—	—	4	4	—	—
0.95	—	1.8	1.64	1.30	—	—
—	—	—	—	—	—	—

(continued)

D. Lipids

The fats of meats contain several types of lipids, including triglycerides (glycerol esters of fatty acids), phospholipids (e.g., lecithin, cephalin, sphingomyelin), and cholesterols. The concentration of cholesterol and phospholipid is relatively constant in skeletal muscle. Their proportion of the total lipid varies as a function of the total lipid in the muscle. As total lipid content of the meat decreases from 12 to 1%, the percentage phospholipid of the fat can increase

Table XI *(Continued)*

Nutrients (per 100 g edible portion)	Salami (cooked)			Salami (dry)	
	Beef	Beef and pork	Turkey	Pork	Pork and beef
Proximate					
Water (g)	59.34	60.40	65.86	36.18	34.70
Food energy (kcal)	254	250	196	407	418
Protein ($N \times 6.25$) (g)	14.70	13.92	16.37	22.58	22.86
Total lipid (g)	20.10	20.11	13.80	33.72	34.39
Carbohydrate (g)	2.49	2.25	0.55	1.60	2.59
Ash (g)	3.38	3.32	3.42	5.92	5.47
Minerals					
Calcium (mg)	9	13	20	13	8
Copper (mg)	0.09	0.23	0.05	0.16	0.08
Iron (mg)	2.00	2.67	1.61	1.30	1.51
Magnesium (mg)	13	15	15	22	17
Manganese (mg)	0.047	0.057	—	0.070	0.038
Phosphorus (mg)	101	115	106	229	142
Potassium (mg)	225	198	244	—	378
Sodium (mg)	1158	1065	1004	2260	1860
Zinc (mg)	2.14	2.14	1.81	4.20	3.23
Vitamins					
Ascorbic acid (mg)	15[c]	12[c]	—	—	26[c]
Thiamin (mg)	0.127	0.239	0.065	0.930	0.600
Riboflavin (mg)	0.257	0.376	0.175	0.330	0.285
Niacin (mg)	3.412	3.553	3.527	5.600	4.867
Pantothenic acid (mg)	0.96	0.85	—	—	1.06
Vitamin B_6 (mg)	0.22	0.21	—	0.55	0.50
Folacin (μg)	2	2	—	—	—
Vitamin B_{12} (μg)	4.85	3.65	—	2.80	1.90
Vitamin A (IU)	—	—	—	—	—

from <10 to >80% (Weihrauch and Son, 1983), the cholesterol from <1% to >5%, and the triglycerides can decrease from >90% to <20%. The compositions of fats are usually reported as fatty acid compositions. For cooked meats, sausage and luncheon meat products, and baby foods, fatty acids and representative total cholesterol values appear in Tables XVI–XVIII. In the original analysis, the fatty acids of triglycerides and phospholipids were released by hydrolysis and then methylated and determined; data were expressed as percentages of the total free fatty acid methyl esters. The values listed in Tables XVI–XVIII were calculated by converting the data for fatty acid methyl esters to grams of free fatty acids in the total fat per 100 g of edible portion of meat by use of appropriate conversion factors (0.869–0.907 for beef, 0.910 for pork,

Smoked linked sausage, pork and beef[b]	Thuringer	Turkey breast meat	Turkey ham	Turkey Roll Light and dark	Light	Vienna Sausage (Canned)
52.16	48.00	71.85	71.38	70.15	71.55	59.93
336	347	110	128	149	147	279
13.40	16.04	22.50	18.93	18.14	18.70	10.29
30.33	29.93	1.58	5.08	6.99	7.22	25.20
1.43	2.29	0.00	0.37	2.13	0.53	2.04
2.69	3.74	4.18	4.23	2.60	2.00	2.53
10	7	7	10	32	40	10
0.06	0.09	0.05	—	0.07	0.04	0.03
1.45	2.04	0.40	2.76	1.35	1.28	0.88
12	12	20	—	18	16	7
0.037	0.031	—	—	—	—	0.029
107	100	229	191	168	183	49
189	231	278	325	270	251	101
945	1453	1431	1996	586	489	953
2.11	2.02	1.13	—	2.00	1.56	1.60
19[c]	23[c]	0	—	—	—	0
0.260	0.169	0.040	0.052	0.091	0.089	0.087
0.170	0.299	0.107	0.247	0.284	0.226	0.107
3.227	4.088	8.322	3.527	4.800	7.000	1.613
0.44	0.55	0.59	—	—	—	—
0.17	0.30	0.36	—	—	—	0.12
—	—	—	—	—	—	—
1.51	4.61	2.02	—	—	—	1.02
—	0	0	0	—	—	—

(continued)

0.878 for lamb, 0.726 for veal, 0.890 for chicken with skin, 0.862 for chicken without skin, and 0.686 for cod).

Reflecting their higher lipid content, the fattest cut of red meats (rib) and chicken with skin or dark chicken meat, contain much higher levels of fatty acids per gram of edible lean tissue than do the leaner cuts of red meats and chicken without skin or light chicken meat. The major saturated fatty acids are myristic (14:0), palmitic (16:0), and stearic (18:0). The major monounsaturated acids are palmitoleic (16:1) and oleic (18:1); the principal polyunsaturated acids are linoleic (18:2), linolenic (18:3), and arachidonic (20:4). Both linoleic and arachidonic acids are essential fatty acids of which meats in general, provide appreciable amounts.

Table XI *(Continued)*

Nutrients (per 100 g edible portion)	Ham, chopped (canned)	Ham, sliced Extra lean	Regular	Italian sausage (cooked)	Kielbasa (kolbassy)
Proximate					
Water (g)	60.77	70.52	64.64	49.95	53.95
Food energy (kcal)	239	131	182	323	310
Protein ($N \times 6.25$) (g)	16.06	19.35	17.56	20.02	13.26
Total lipid (g)	18.83	4.96	10.57	25.70	27.15
Carbohydrate (g)	0.28	0.96	3.11	1.50	2.14
Ash (g)	4.08	4.21	4.11	2.82	3.50
Minerals					
Calcium (mg)	7	7	7	24	44
Copper (mg)	0.05	0.07	0.99	0.08	0.10
Iron (mg)	0.95	0.76	0.10	1.50	1.45
Magnesium (mg)	13	17	19	18	16
Manganese (mg)	0.025	0.033	0.031	0.082	0.040
Phosphorus (mg)	139	218	247	170	148
Potassium (mg)	284	350	332	304	271
Sodium (mg)	1365	1429	1317	922	1076
Zinc (mg)	1.83	1.93	2.14	2.38	2.02
Vitamins					
Ascorbic acid (mg)	2	26[c]	28[c]	2	21[c]
Thiamin (mg)	0.535	0.932	0.863	0.623	0.228
Riboflavin (mg)	0.165	0.223	0.252	0.233	0.214
Niacin (mg)	3.200	4.838	5.251	4.165	2.879
Pantothenic acid (mg)	—	0.47	0.45	0.45	0.82
Vitamin B_6 (mg)	0.32	0.46	0.34	0.33	0.18
Folacin (μg)	—	4	3	—	—
Vitamin B_{12} (μg)	0.70	0.75	0.83	1.30	1.61
Vitamin A (IU)	—	—	—	—	—

Less than one-half of the fatty acids in meats are saturated. However, in the calculation of the P/S ratios (the ratio of polyunsaturated to saturated fatty acids), the levels of monosaturated fatty acids are not considered. Thus, the P/S ratios for the fats of beef and lamb are quite low (<0.15) and for fats of pork and veal are intermediate (0.34–0.39). The P/S ratios are high for chicken (0.77–0.85; all classes) and for cod (2.09) and other fish.

The fatty acid compositions of sausage and luncheon meat products (Table XVII) and of baby foods (Table XVIII) are highly variable and reflect the meats (or meats and fats) used in their manufacture. Among the data presented, P/S ratios vary from 0.08 to 0.87 for sausage and luncheon meats and from 0.06 (beef junior) to 0.95 (chicken, strained) for baby foods.

The cholesterol levels listed in Tables XVI and XVII vary from 41 to 103 mg

| Knockwurst (knackwurst) | Luncheon Meat | | Olive loaf, pork | Pepperoni | Polish sausage |
	Beef loaf	Pork beef			
55.50	52.53	49.28	58.18	27.06	53.15
308	308	353	235	497	326
11.88	14.37	12.59	11.84	20.97	14.10
27.76	26.20	32.16	16.50	43.97	28.72
1.76	2.90	2.33	9.16	2.84	1.63
3.10	4.00	3.64	4.33	5.17	2.40
11	11	9	109	10	12
0.06	0.12	0.04	0.05	0.07	0.09
0.91	2.32	0.86	0.54	1.40	1.44
11	14	14	19	16	14
—	0.045	0.029	0.035	—	0.049
98	119	86	127	119	136
199	208	202	297	347	237
1010	1329	1293	1484	2040	876
1.66	2.53	1.66	1.38	2.50	1.93
27[c]	13[c]	13[c]	9[c]	—	1
0.342	0.110	0.314	0.295	0.320	0.502
0.140	0.220	0.152	0.260	0.250	0.148
2.734	3.660	2.830	1.835	4.960	3.443
0.32	0.52	0.63	0.77	1.87	0.45
0.17	0.19	0.20	0.23	0.25	0.19
—	—	6	—	—	—
1.18	3.89	1.28	1.26	2.51	0.98
—	—	—	—	—	—

per 100 g of edible portion. Among the cooked meats (Table XVI), cured ham and cod both contain relatively low levels. Among the sausage and luncheon meat products, several have cholesterol levels of 50 mg or less per 100 g edible portion. Data for the cholesterol levels of baby foods are not available.

IV. PROTEIN NUTRITIONAL QUALITY

Proteins differ in their nutritional value because of differences in their content and distribution of indispensable (essential) amino acids and because of differences in the nutritional availability of their amino acids. The nine amino acids that are considered indispensable (National Research Council, 1980) are

Table XII
Proximate, Vitamins, and Mineral Composition of Baby Foods[a]

Nutrients (per 100 g edible portion)	Beef			Pork			Lamb	
	Strained	Junior	With beef heart, strained	Strained	Ham, strained	Ham, junior	Strained	Junior
Proximate								
Water (g)	80.6	79.9	82.5	78.4	79.4	78.5	80.3	79.6
Food energy (kcal)	107	106	94	124	111	125	103	112
Protein ($N \times 6.25$) (g)	13.6	14.5	12.7	14.0	13.9	15.1	14.1	15.2
Total lipid (g)	5.4	4.9	4.4	7.1	5.8	6.7	4.7	5.2
Carbohydrate (g)	0.0	0.0	0.0	0.0	0.0	0.0	0.1	0.0
Fiber (g)	—	—	0.1	—	—	—	—	—
Ash (g)	0.8	0.8	0.9	1.1	1.0	1.0	0.8	0.9
Minerals								
Calcium (mg)	7	8	4	5	6	5	7	7
Copper (mg)	0.043	0.092	0.129	0.072	0.065	—	0.055	0.057
Iron (mg)	1.47	1.65	1.99	1.00	1.03	1.01	1.49	1.66
Magnesium (mg)	17	9	12	10	13	11	13	10
Manganese (μg)	—	—	—	—	—	—	—	—
Phosphorus (mg)	84	72	94	94	81	89	97	91
Potassium (mg)	220	190	200	223	204	210	205	211
Sodium (mg)	81	66	63	42	41	64	62	73
Zinc (mg)	2.455	2.000	1.835	2.270	2.246	1.700	2.756	2.600
Vitamins								
Ascorbic acid (mg)	2.1	1.9	2.1	1.8	2.1	2.1	1.2	1.7
Thiamin (mg)	0.010	0.010	0.022	0.146	0.139	0.142	0.019	0.019
Riboflavin (mg)	0.142	0.159	0.357	0.203	0.154	0.194	0.200	0.191
Niacin (mg)	2.849	3.283	3.885	2.269	2.633	2.840	2.925	3.193
Pantothenic acid (mg)	0.340	0.353	—	—	0.510	0.531	0.410	0.424
Vitamin B_6 (mg)	0.140	0.120	0.118	0.205	0.252	0.200	0.152	0.183
Folacin (μg)	5.5	5.7	5.0	1.9	2.0	2.1	2.3	2.0
Vitamin B_{12} (μg)	1.420	1.474	—	0.990	—	—	2.190	2.269
Vitamin A (IU)	185	103	126	38	38	32	86	27

[a] From U.S. Department of Agriculture (1978); dashes denote lack of reliable data for a constituent believed to be present in measureable amount.

tryptophan, threonine, leucine, isoleucine, lysine, methionine, phenylalanine, valine, and histidine. In meeting the requirements for phenylalanine and methionine, tyrosine can spare or replace part of the phenylalanine and cystine can replace part of the methionine. Thus, those two amino acids are also included in any assessment of protein quality. In general, all meats and meat products contain high levels of the indispensable amino acids plus tyrosine and cystine (Tables XIII–XV). Furthermore, the amino acids of meats and meat products are highly available. That is, the protein is highly digestible and, except after severe heat treatment (experimental conditions), no appreciable amounts of the amino acids are unavailable following digestion. In general, meats and meat products are 95–100% digestible whereas the digestibility of protein from some plant sources is as low as 65 to 75% (Hopkins, 1981).

	Veal		Chicken			Turkey		
	Strained	Junior	Strained	Junior	Sticks, junior	Strained	Junior	Sticks, junior
	79.8	79.8	77.5	76.0	68.3	78.9	77.5	69.8
	101	110	130	149	188	114	129	182
	13.5	15.3	13.7	14.7	14.6	14.3	15.4	13.7
	4.8	5.0	7.9	9.6	14.4	5.8	7.1	14.2
	0.0	0.0	0.1	0.0	1.4	0.1	0.0	1.4
	0.1	0.2	—	—	0.2	—	—	0.5
	0.9	0.9	0.8	0.8	1.2	0.9	0.8	0.9
	7	6	64	55	73	23	28	72
	0.040	0.075	0.045	0.045	—	0.040	—	—
	1.27	1.25	1.40	0.99	1.56	1.20	1.35	1.24
	12	11	13	11	—	14	12	—
	—	—	—	—	—	—	—	—
	98	98	97	90	121	126	95	103
	216	236	141	122	106	231	180	91
	69	69	47	51	479	55	72	483
	2.000	2.518	1.210	1.008	—	1.830	1.800	—
	2.3	2.1	1.7	1.5	1.7	2.2	2.4	1.5
	0.016	0.018	0.014	0.014	0.017	0.018	0.016	0.010
	0.159	0.183	0.152	0.163	0.197	0.209	0.250	0.155
	3.552	3.808	3.255	3.420	2.005	3.667	3.483	1.752
	0.430	0.454	0.680	0.726	—	0.570	0.609	—
	0.149	0.118	0.200	0.188	0.103	0.180	0.166	0.076
	5.9	6.7	10.4	11.1	—	11.3	12.1	—
	—	—	—	—	—	1.000	1.069	—
	46	50	135	186	3.178	560	568	228

Whether evaluated by chemical methods, animal bioassays, or studies with human subjects, the proteins of meats and meat products, together with other foods from animals (eggs and milk), are generally of the highest quality (Bodwell, 1977; Bodwell et al., 1981). Even with proper processing, however, quality is not maintained in some manufactured products. As a test of protein quality for various meats or meat products, the amount of each indispensable amino acid is expressed as a ratio to the amount specified for an ideal reference protein or amino acid pattern (Table XIX). A ratio of 1.0 (a value of 100%) denotes equivalence, in amino acid composition alone, to the reference (whole egg). Among the examples shown, a few have marginal or deficient levels of tryptophan. In general, however, all meats or meat products, including those listed in Table XIX, have an excess of total essential amino acids.

Table XIII
Amino Acid Compositions of Cooked Meats[a]

Amino acid (g/100 g edible portion)	Beef round	Beef rib	Lamb leg	Lamb rib	Ham (fresh)	Ham (cured)	Veal	Chicken				Cod
								Dark (with skin)	Dark (flesh only)	Light (with skin)	Light (flesh only)	
Tryptophan	.325	.300	0.363	0.312	0.381	0.308	0.399	0.289	0.320	0.326	0.361	.224
Threonine	1.267	1.168	1.301	1.119	1.332	1.143	1.318	1.071	1.156	1.202	1.305	.884
Isoleucine	1.304	1.203	1.446	1.244	1.368	1.127	1.605	1.288	1.445	1.458	1.632	.930
Leucine	2.292	2.114	2.185	1.879	2.302	2.040	2.227	1.883	2.053	2.119	2.319	1.643
Lysine	2.413	2.226	2.384	2.051	2.791	2.179	2.538	2.105	2.325	2.374	2.626	1.863
Methionine	.742	.685	0.698	0.600	0.698	0.678	0.695	0.688	0.757	0.776	0.855	.599
Cystine	.325	.300	0.335	0.289	0.367	.386	0.360	0.348	0.350	0.385	0.396	.217
Phenylalanine	1.132	1.044	1.151	0.990	1.133	1.110	1.235	1.007	1.086	1.130	1.226	.787
Tyrosine	.974	.900	0.979	0.842	1.010	0.843	1.094	0.832	0.924	0.940	1.043	.680
Valine	1.411	1.301	1.419	1.220	1.518	1.114	1.570	1.258	1.357	1.412	1.533	1.040
Arginine	1.833	1.691	1.863	1.603	1.966	1.669	1.979	1.634	1.651	1.811	1.864	1.212
Histidine	.993	.916	0.793	0.682	1.436	0.921	0.977	0.759	0.849	0.858	0.959	.593
Alanine	1.749	1.614	1.781	1.532	1.676	1.517	1.804	1.523	1.493	1.679	1.686	1.225
Aspartic Acid	2.649	2.444	2.643	2.273	2.628	2.434	2.995	2.316	2.439	2.587	2.754	2.074
Glutamic Acid	4.357	4.019	4.383	3.771	4.399	4.190	4.741	3.787	4.098	4.254	4.629	3.016
Glycine	1.582	1.459	1.568	1.349	1.291	1.336	1.454	1.718	1.344	1.823	1.518	.972
Proline	1.281	1.181	1.292	1.111	1.074	1.098	1.225	1.277	1.125	1.381	1.271	.716
Serine	1.109	1.023	1.188	1.022	1.165	1.053	1.337	0.918	0.941	1.021	1.063	.820
% Protein (N × 6.25)	29.0	26.7	28.3	24.4	28.3	25.7	30.4	26.0	27.4	29.0	30.9	20.2

[a] Values for roasted cuts except for cod (broiled with margarine); values for beef, lamb, veal, and cod from U.S. Department of Agriculture, Human Nutrition Information Service, Nutrient Data Research Branch (unpublished data); for pork from U.S. Department of Agriculture (1983); for poultry from U.S. Department of Agriculture (1979); all values for lean only, except for chicken, with skin.

The Recommended Dietary Allowance (RDA) for protein from a mixed diet is based on the assumption that the protein would be provided by animal plus plant sources and would be utilized about 75% as well as the reference protein (whole egg). Because most meats and meat products are utilized as efficiently as egg protein, the amounts of protein from meats required to meet the RDAs would actually be less than the amounts required from a mixed diet (Table XX). An 85-g serving (~3 oz) of cooked meat provides between 20 and 25 g of high quality protein that meets 50–100% of the various RDAs specified for different age groups.

Complementation is another important aspect in assessing the protein nutritive value of meat and meat products (Bressani, 1977). Most legumes are deficient (low) in the sulfur amino acids (methionine and cystine), and most cereals are deficient (low) in lysine. When foods from either source are consumed together with meats, the excess of sulfur amino acids or of lysine in meats (Table XIX) compensates for the lack of those amino acids in the legume or cereal protein. The centuries-old Oriental practice of adding a small amount of beef, pork, chicken, fish, or egg to either cereal-based or legume-based foods markedly improves the nutritional value of the protein from the predominant plant sources.

V. BIOAVAILABILITY OF MINERALS AND VITAMINS

The minerals and vitamins of meats and meat products are readily available for meeting nutritional needs. In contrast to many plant protein sources, meats contain no substances (such as tannins, caffeine, or phytate) that could interfere with the digestion or absorption of specific minerals. In addition, inclusion of meats in the diet is of particular importance in relation to nutritional iron status.

In foods, iron is present as either heme iron or nonheme iron. Heme iron is readily absorbed and its absorption is not affected by other dietary components. The iron in meats (red meats, poultry, fish) is generally about two-thirds heme iron and one-third nonheme iron. However, dependent on the age of the animal, the proportion of iron present as heme iron may be lower (e.g., muscle from an aged cow may be only 40–45% heme iron). Nonheme iron is present in nonmeat foods, as well as in meat, but it is less readily absorbed than heme iron. The absorption (and hence, utilization) of nonheme iron is greatly enhanced by the consumption of meat or ascorbic acid in the same meal (Monsen et al., 1978). A meal with low iron availability contains <30 g of meat and <25 mg of ascorbic acid while a meal with high iron availability contains 90 g or more of meat or >75 mg of ascorbic acid or a combination of 30–90 g meat and 25–75 mg ascorbic acid (National Research Council, 1980).

Table XIV
Amino Acid Composition of Sausage and Luncheon Meat Products[a]

				Bologna			Corned beef loaf
	Beerwurst			Beef and			
Amino acid	Beef	Pork	Beef	pork	Pork	Bratwurst	
Tryptophan	0.112	0.112	0.107	0.105	0.149	0.113	0.169
Threonine	0.466	0.565	0.442	0.511	0.641	0.556	0.927
Isoleucine	0.533	0.485	0.505	0.507	0.663	0.514	0.890
Leucine	0.906	0.943	0.859	0.898	1.168	0.944	1.638
Lysine	0.945	1.032	0.896	0.883	1.204	1.070	1.820
Methionine	0.286	0.355	0.271	0.277	0.412	0.342	0.525
Cystine	0.158	0.107	0.149	0.136	0.171	0.142	0.248
Phenylalanine	0.444	0.451	0.421	0.462	0.585	0.471	0.838
Tyrosine	0.402	0.403	0.381	0.359	0.482	0.406	0.638
Valine	0.543	0.510	0.514	0.621	0.737	0.566	1.026
Arginine	0.762	0.809	0.722	0.699	1.004	0.831	1.642
Histidine	0.393	0.419	0.372	0.318	0.482	0.406	0.623
Alanine	0.888	0.770	0.841	0.729	0.979	0.789	1.631
Aspartic acid	1.207	1.198	1.144	1.028	1.403	1.172	2.016
Glutamic acid	2.009	2.141	1.904	1.874	2.296	1.947	3.278
Glycine	1.030	0.781	0.976	0.864	1.077	0.854	2.244
Proline	0.888	0.640	0.841	0.748	0.773	0.656	1.554
Serine	0.497	0.569	0.472	0.509	0.634	0.545	0.947
% Protein (N × 6.25)	12.33	14.24	11.69	11.69	15.30	14.08	23.70

[a] In g per 100 g meat product. From U.S. Department of Agriculture (1980).

VI. OTHER FACTORS AFFECTING NUTRITIONAL VALUE

A. Fatness Level

A comprehensive discussion of various other factors that may affect the nutritional value of meats and meat products is beyond the scope of this chapter. For the most part, only general comments or summaries are included. More detailed data and references to specific studies have been provided in several reviews or texts (Harris and von Loesecke, 1960; Price and Schweigert, 1971; Harris and Karmas, 1975; Lawrie, 1979; Sommers and Hagen, 1981).

USDA grades for red meats include both quality grades [primarily dependent upon the amount of marbling (visible intramuscular fat)] and yield grades (dependent upon percentage of trimmed, boneless major retail cuts that can be derived from the carcass). Yield grades apply only to whole carcasses or wholesale cuts and thus are not used at the retail level. Except for pork and most veal,

Beef	Frankfurter Beef and pork	Ham, chopped, canned	Ham, sliced Extra lean	Regular	Italian sausage	Kielbasa (kolbassy)	Knockwurst
0.103	0.082	0.182	0.236	0.214	0.161	0.138	0.108
0.426	0.406	0.717	0.861	0.781	0.792	0.432	0.479
0.487	0.485	0.691	0.873	0.792	0.731	0.639	0.466
0.829	0.819	1.249	1.554	1.411	1.343	0.874	0.821
0.865	0.904	1.380	1.730	1.571	1.522	1.010	0.933
0.262	0.228	0.419	0.510	0.464	0.486	0.276	0.287
0.145	0.130	0.190	0.291	0.264	0.201	0.225	0.147
0.407	0.359	0.622	0.780	0.708	0.670	0.501	0.407
0.368	0.314	0.527	0.601	0.545	0.577	0.490	0.361
0.496	0.471	0.719	0.904	0.821	0.804	0.639	0.515
0.697	0.849	0.997	1.359	1.233	1.182	0.942	0.709
0.359	0.350	0.637	0.786	0.714	0.577	0.314	0.361
0.812	0.769	0.917	1.126	1.022	1.121	0.848	0.675
1.105	1.115	1.485	1.875	1.703	1.666	1.218	1.011
1.838	1.852	2.371	2.931	2.661	2.768	1.618	1.619
0.942	0.824	0.786	0.935	0.849	1.214	1.039	0.703
0.812	0.542	0.683	0.786	0.714	0.932	0.687	0.539
0.455	0.462	0.627	0.761	0.692	0.775	0.530	0.477
11.29	11.28	16.06	19.35	17.56	20.02	13.26	11.88

(continued)

quality grades are used at the retail level and, in general, the higher quality grades have a higher fat content in the lean portion of the cuts. As noted in Section III,A, the sum of protein, fat, and moisture is relatively constant, and an increase in the content of fat results in a decrease in the content of protein and/or moisture. This is most applicable when entire retail cuts are considered (both lean and external fat). Rice (1971) has summarized the earlier data on the effects of fat level on the composition of entire retail cuts of beef, pork, and lamb.

A recent study (Cross *et al.*, 1980) was designed to evaluate the separable yields of four USDA quality grades (Prime, Choice, Good, Standard) of beef. When all retail cuts were combined, quality grade did not significantly affect the percentages of lean, intermuscular fat or subcutaneous fat in the raw or cooked meat. Individual cuts also were analyzed. Among cooked rib roasts from carcasses from the four grades, yields of lean varied from 53.6 (Choice) to 58.5% (Good), dissectable fat from 20.6 (Good) to 36.6% (Prime), and bone plus connective tissue from 14.4 (Prime) to 19.8% (Standard). Similar values for bottom round roasts (cooked) were less variable, and maximum differences between grades were only 2% or less for percentages of lean and dissectable

Table XIV (Continued)

Amino acid	Luncheon meat, pork and beef	Olive loaf	Pepper-oni	Polish Sausage	Pork Sausage Fresh	Pork Sausage Cooked	Salami, cooked Beef	Salami, cooked Beef and pork
Tryptophan	0.133	0.104	0.200	0.138	0.094	0.157	0.134	0.114
Threonine	0.544	0.472	0.848	0.591	0.462	0.777	0.555	0.521
Isoleucine	0.643	0.420	0.908	0.611	0.427	0.717	0.635	0.675
Leucine	1.048	0.869	1.578	1.076	0.784	1.317	1.080	0.929
Lysine	1.185	0.809	1.635	1.110	0.889	1.494	1.127	1.107
Methionine	0.288	0.301	0.536	0.379	0.284	0.478	0.341	0.301
Cystine	0.230	0.144	0.247	0.157	0.118	0.198	0.188	0.196
Phenylalanine	0.510	0.419	0.785	0.539	0.391	0.657	0.529	0.481
Tyrosine	0.502	0.383	0.660	0.444	0.337	0.566	0.480	0.552
Valine	0.713	0.508	0.978	0.679	0.469	0.789	0.647	0.668
Arginine	0.925	0.583	1.346	0.925	0.691	1.160	0.908	0.855
Histidine	0.405	0.292	0.664	0.444	0.337	0.566	0.468	0.359
Alanine	0.806	0.617	1.403	0.902	0.655	1.101	1.058	0.880
Aspartic acid	1.312	0.951	1.970	1.293	0.973	1.635	1.439	1.285
Glutamic acid	1.769	1.919	3.245	2.117	1.617	2.716	2.394	1.929
Glycine	0.985	0.680	1.577	0.992	0.709	1.192	1.228	1.189
Proline	0.729	0.830	1.225	0.712	0.544	0.914	1.058	0.831
Serine	0.528	0.517	0.860	0.584	0.452	0.761	0.593	0.537
% Protein ($N \times 6.25$)	12.59	11.84	28.97	14.10	11.69	19.65	14.70	13.92

fat.   In a related study, Fox and Ono (1980) analyzed samples from 168 beef carcasses with marbling scores equivalent to the Good, Choice, and Prime grades.   Total protein ($N \times 6.25$), niacin, and riboflavin tended to decrease only slightly as marbling score increased while thiamin and vitamin B_6 levels did not follow any pattern in relation to marbling score.   When Wolf and Ono (1980) analyzed samples of raw separable lean from cuts of the same 168 beef carcasses (50 Good grade, 70 Choice grade, 48 Prime grade), differences between quality grades in the levels of iron, magnesium, potassium, sodium, zinc, and copper were negligible.

Those data indicate that the nutritive value for a particular cut of beef is generally similar, regardless of quality grade, if external fat is removed.   Although few recent data are available, similar findings would be expected for lamb and pork.   For chicken and turkey, as noted earlier (Section III,A), the single most important factor in nutrient content is the presence or absence of the skin and its associated fat.

| | Salami, dry | | Smoked linked sausage | | | | | |
| | | Pork and beef | Pork | Pork and beef | Thuringer | Turkey breast meat | Turkey ham | Vienna sausage, canned |
Pork								
0.253	0.210	0.216	0.107	0.154	0.256	0.215	0.109
1.012	0.959	0.929	0.465	0.688	1.001	0.842	0.357
1.084	0.971	0.951	0.485	0.768	1.170	0.984	0.557
1.625	1.732	1.692	0.810	1.048	1.793	1.508	0.797
1.878	1.824	1.745	0.903	1.383	2.120	1.784	0.791
0.470	0.593	0.596	0.364	0.352	0.652	0.548	0.265
0.289	0.262	0.248	0.105	0.197	0.234	0.197	0.175
0.940	0.868	0.848	0.405	0.577	0.893	0.751	0.425
0.686	0.710	0.699	0.386	0.542	0.889	0.748	0.341
1.120	1.083	1.068	0.464	0.804	1.195	1.006	0.573
1.373	1.516	1.454	0.765	0.993	1.570	1.321	0.707
0.614	0.699	0.699	0.395	0.469	0.702	0.591	0.273
1.336	1.480	1.419	0.774	1.014	1.393	1.172	0.651
2.095	2.073	2.032	1.136	1.438	2.185	1.839	1.005
3.829	3.383	3.327	1.994	2.205	3.672	3.090	1.304
1.553	1.641	1.561	0.857	1.103	1.116	0.939	1.013
1.336	1.188	1.121	0.673	0.860	0.936	0.788	0.606
0.903	0.942	0.919	0.510	0.678	1.001	0.842	0.432
22.58	22.86	22.17	13.40	16.04	22.50	18.93	10.29

B. Storage (Fresh Meats and Luncheon Meats)

In general, fresh fish, poultry, and ground meats should be stored for no longer than 1–2 days in the refrigerator without freezing. Fresh cuts of red meats, cured pork, and luncheon meats usually should be stored for no longer than 3–5 days. Under those recommended conditions, nutritive value should remain essentially unchanged.

C. Cooking

Of the changes in meat caused by cooking (other than denaturation, etc. of protein), the losses of specific nutrients are nutritionally important. Minerals, vitamins, or fat can be lost into the cooking medium (oil or water), and vitamins can be lost by destruction or alteration by heat to a biologically nonuseful

Table XV
Amino Acid Composition of Baby Foods[a]

Amino Acid (g amino acid/100 g)	Beef			Pork		
	Strained	Junior	Beef heart, strained	Strained	Ham, strained	Ham, junior
Tryptophan	0.137	0.146	0.126	0.136	0.138	0.149
Threonine	0.597	0.635	0.511	0.617	0.603	0.656
Isoleucine	0.619	0.659	0.619	0.680	0.661	0.718
Leucine	1.091	1.162	1.010	1.127	1.111	1.208
Lysine	1.133	1.206	1.051	1.158	1.180	1.283
Methionine	0.418	0.445	0.329	0.398	0.355	0.386
Cystine	0.159	0.169	0.130	0.154	0.171	0.186
Phenylalanine	0.527	0.561	0.523	0.570	0.530	0.576
Tyrosine	0.453	0.482	0.365	0.510	0.466	0.506
Valine	0.688	0.733	0.690	0.702	0.716	0.779
Arginine	0.928	0.988	0.832	0.948	0.940	1.022
Histidine	0.462	0.492	0.339	0.447	0.472	0.514
Alanine	0.858	0.914	0.866	0.879	0.838	0.911
Aspartic acid	1.198	1.275	1.173	1.310	1.326	1.442
Glutamic acid	2.069	2.203	2.017	2.133	2.089	2.271
Glycine	0.869	0.925	0.822	0.760	0.796	0.866
Proline	0.682	0.726	0.657	0.747	0.645	0.702
Serine	0.499	0.531	0.442	0.541	0.503	0.547
% Protein ($N \times 6.25$)	13.6	14.5	12.7	14.0	13.9	15.1

[a] From U.S. Department of Agriculture (1978).

form. Data from earlier studies were summarized elsewhere (Sommers and Hagen, 1981; Harris and Karmas, 1975). Recent data that illustrate the effects of various cooking methods on nutrient retention are summarized in Table XXI. For beef and pork, retentions of the various nutrients were similar across all cuts within each species; only cooking method significantly affected nutrient retention. Retention approached or was greater than 100% for protein ($N \times 6.25$), total fat, iron, zinc, copper, riboflavin, and cholesterol. Thiamin and vitamin B_6 were the most labile nutrients.

For beef, braising resulted in the lowest retentions for most nutrients. For pork, braised cuts, in general, also had lower retentions of the B vitamins and the water-soluble minerals than did broiled or roasted cuts, but losses of iron, zinc, and protein were slightly greater in broiled samples.

For chicken, stewing caused greater losses of calcium, magnesium, potassium, thiamin, and niacin than did roasting (Table XXI). Retention values for roasted turkey and chicken were similar except for copper, which was markedly lower for turkey.

In compiling the vitamin compositions of foods in the United Kingdom, Paul and Southgate (1978) used a value of 20% for losses of thiamin, riboflavin,

	Lamb		Veal		Chicken			Turkey		
							Sticks, junior			Sticks, junior
	Strained	Junior	Strained	Junior	Strained	Junior	junior	Strained	Junior	junior
	0.140	0.151	0.156	0.176	0.156	0.167	0.117	0.149	0.160	0.101
	0.642	0.693	0.564	0.638	0.615	0.660	0.572	0.633	0.681	0.546
	0.662	0.714	0.610	0.689	0.645	0.693	0.729	0.717	0.772	0.629
	1.112	1.200	1.044	1.180	1.058	1.136	1.142	1.138	1.225	1.071
	1.249	1.349	1.085	1.227	1.144	1.228	1.160	1.184	1.274	1.176
	0.441	0.476	0.298	0.337	0.367	0.394	0.323	0.443	0.477	0.304
	0.194	0.209	0.175	0.198	0.180	0.193	0.127	0.174	0.187	0.122
	0.554	0.598	0.523	0.591	0.558	0.599	0.673	0.590	0.634	0.610
	0.493	0.532	0.432	0.489	0.439	0.471	0.497	0.503	0.541	0.483
	0.714	0.770	0.657	0.743	0.689	0.740	0.764	0.727	0.782	0.651
	0.923	0.996	0.897	1.014	0.957	1.028	1.003	0.909	0.979	0.883
	0.356	0.384	0.415	0.469	0.415	0.445	0.462	0.366	0.393	0.367
	0.882	0.953	0.845	0.956	0.871	0.936	0.792	0.905	0.974	0.704
	1.324	1.429	1.193	1.349	1.269	1.362	1.263	1.426	1.534	1.272
	2.156	2.328	1.939	2.192	1.982	2.128	2.201	2.267	2.439	2.196
	8.819	0.885	0.927	1.048	0.913	0.981	0.811	0.941	1.013	0.682
	0.788	0.851	0.746	0.843	0.707	0.759	0.902	0.704	0.757	0.651
	0.540	0.583	0.480	0.543	0.498	0.535	0.534	0.562	0.605	0.505
	14.1	15.2	13.5	15.3	13.7	14.7	14.6	14.3	15.4	13.7

nicotinic acid (niacin), and vitamin B_6 for red meats and poultry cooked by roasting, frying and grilling (all dry heat cooking methods). For losses after stewing or boiling, they used values of 60% (thiamin), 30% (riboflavin), 50% (nicotinic acid and vitamin B_6), and 40% (pantothenic acid). They used average losses of 20% and 0% for vitamin B_{12} and vitamin A, respectively, for all cooking methods.

Few data are available for the losses of nutrients from fish during cooking. Paul and Southgate (1978), however, assumed some tentative values for losses of vitamins in fish cooked by different methods. Respectively, according to cooking method (poaching, baking, frying/grilling), the values were thiamin: 10, 30, and 20%; riboflavin: 0, 20, and 20%; nicotinic acid: 10, 20, and 20%; vitamin B_6: 0, 10, and 20%; pantothenic acid: 20, 20, and 20%; free folic acid: 50, 30, and 0%; total folic acid: 0, 20, and 0%; and vitamin B_{12}: 0, 10, and 0%.

For most meats, retentions of protein are high regardless of the method of cooking (Table XXI). Considerable other data (not reported) also indicate that cooking does not significantly affect the protein nutritional value of meats and meat products. Limited data also suggest that infrared or microwave cooking has little effect on protein nutritive value (Bodwell and Womack, 1978).

Table XVI
Lipid Composition of Cooked Meats[a]

Fatty acids (g/100 g) edible portion	Beef		Ham		Lamb		Veal		Chicken				Cod
	Round	Rib	Fresh	Cured	Leg	Rib	Leg	Rib	Dark (with skin)	Dark (flesh only)	Light (with skin)	Light (flesh only)	
Saturated, total	3.93	6.05	3.80	1.84	4.118	7.627	.573	1.229	4.37	2.66	3.05	1.27	1.18
10:0	0.004	0.009	0.01	0.02	.019	.036	—	—	—	—	—	—	—
12:0	0.005	0.010	0.13	0.02	.019	.036	—	—	0.03	0.03	0.01	0.01	—
14:0	0.290	0.450	0.13	0.07	.212	.393	.043	.093	0.12	0.07	0.09	0.04	0.05
15:0	—	—	—	—	.039	.072	—	—	—	—	—	—	—
16:0	2.363	3.417	2.43	1.17	2.001	3.707	.291	.625	3.19	1.84	2.25	0.87	0.93
17:0	—	—	—	—	.115	.213	—	—	—	—	—	—	—
18:0	1.087	1.890	1.22	0.56	1.289	2.388	.190	.407	0.90	0.63	0.62	0.32	0.18
20:0	—	—	—	—	.077	.142	—	—	—	—	—	—	—
Monounsaturated, total	5.054	6.251	4.96	2.53	3.743	6.933	.565	1.213	6.19	3.56	4.26	1.54	0.80
14:1	—	—	—	—	.068	.125	.013	.028	—	—	—	—	—
16:1	0.491	0.484	0.34	0.21	.193	.357	.049	.105	0.86	0.49	0.57	0.18	0.12
18:1	4.346	5.536	4.51	2.32	3.406	6.308	.468	1.004	5.11	2.97	3.51	1.30	0.57
20:1	0.002	0.006	0.08	—	—	—	—	—	0.15	0.05	0.12	0.03	0.0
Polyunsaturated, total	0.453	0.433	1.34	0.63	.587	1.087	.226	.484	3.49	2.26	2.31	0.98	2.47
18:2	0.324	0.341	0.95	0.50	.356	.659	.105	.226	3.04	1.87	1.98	0.74	0.04
18:3	0.029	0.031	0.32	0.06	.096	.178	.023	.048	0.14	0.09	0.09	0.04	0.01
20:4	0.072	0.043	0.06	0.07	.125	.232	.062	.133	0.14	0.14	0.09	0.08	0.22
20:5	0.026	0.018	—	—	—	—	—	—	0.02	0.01	0.01	0.01	0.71
22:5	—	—	—	—	—	—	—	—	0.03	0.03	0.02	0.02	0.08
22:6	—	—	—	—	—	—	—	—	0.05	0.05	0.03	0.03	1.36
P/S Ratio	0.12	0.07	0.35	0.34	0.14	0.14	0.39	0.39	0.80	0.85	0.76	0.77	2.09
Total lipids (%)	10.8	14.3	11.0	5.5	9.6	17.8	1.9	4.0	15.8	9.7	10.9	4.5	6.5
Cholesterol (mg/100 g)	81	80	94	55	103	90	94	88	91	93	84	85	57

[a] Values for roasted cuts except for cod (broiled with margarine); values for beef round, lamb, veal, and cod from USDA, Human Nutrition Information Service, Nutrient Data Research Branch (unpublished data); for beef from U.S. Department of Agriculture (1986); for pork from U.S. Department of Agriculture (1983b); for poultry from U.S. Department of Agriculture (1979); all values for lean only, except for values for chicken, with skin; dashes denote lack of reliable data for a constituent believed to be present in measurable amount; P/S ratio of polyunsaturated to saturated fatty acids.

D. Thermal Processing

In general, the protein in canned or other thermally processed meats is not significantly altered by the common procedures (Sommers and Hagen, 1981). Exceptions may occur with prolonged heating ($>$ 1 hr) at elevated temperatures ($>$ 100°C). Loss of minerals is small, except for some leaching from foods processed in liquids.

Considerable data describe the losses of various vitamins during thermal processing. In general, levels are not significantly reduced of niacin and riboflavin but are reduced of vitamin B_6, vitamin B_{12}, pantothenic acid, and thiamin. Losses of thiamin and vitamin B_{12}, in particular, may range from 20 to 80%, depending upon the length and severity of the processing conditions.

Data are limited on the effects of long-term storage of canned meats. In a few studies, however, losses (up to 50%) of thiamin have been reported (Sommers and Hagen, 1981).

E. Freezing

Freezing per se of fresh cuts of meats probably causes little change in their nutritive value. During frozen storage and upon thawing, however, vitamins, in particular, may be lost. For any specific vitamin, estimates of the magnitude of these losses vary widely among studies. The data have recently been summarized (Harris and Karmas, 1975; Lawrie, 1979; Sommers and Hagen, 1981).

In most studies, only slight losses of niacin have been attributed to frozen storage. For thiamin, riboflavin, vitamin B_6, and vitamin B_{12}, estimates of losses attributed to storage and to thawing (drip) range from none to as high as 70 to 80%. In general, quick freezing, low-temperature frozen storage, and thawing at refrigerated temperatures minimize possible losses.

F. Curing and Smoking

Results from a few detailed studies on the effects of curing and smoking on protein nutritional value and on the retention of vitamins have been reported, but few data on mineral retention are available (Sommers and Hagen, 1981). No strong evidence indicate that either curing or smoking significantly affects the protein nutritive value of hams. Retentions of 70 to 100% have been reported for niacin, riboflavin, and thiamin in cured smoked hams. Corresponding values for bacon sides are similar. Losses during storage may further decrease thiamin by 15 to 30%, and prolonged storage could cause increases in free fatty acid content and some oxidation of fats.

Table XVII

Lipid Composition of Sausage and Luncheon Meat Products[a]

Fatty acids (g/100 g)	Beerwurst Beef	Beerwurst Pork	Bologna Beef	Bologna Beef and pork	Bologna Pork
Saturated, total	12.00	6.28	11.66	10.70	6.88
10:0	0.03	0.09	0.07	0.06	0.02
12:0	0.03	0.09	0.04	0.04	0.02
14:0	0.94	0.27	0.82	0.62	0.24
15:0	—	—	—	—	—
16:0	7.49	3.89	6.55	6.27	4.33
17:0	—	—	—	—	—
18:0	3.51	1.94	4.19	3.70	2.27
20:0	—	—	—	—	—
Monounsaturated, total	14.15	8.98	13.28	13.39	9.78
14:1	—	—	—	—	—
16:1	2.31	0.77	1.56	1.38	0.72
18:1	11.84	8.21	11.72	12.01	9.07
20:1	—	—	—	—	—
Polyunsaturated, total	0.96	2.36	1.05	2.40	2.12
18:2	0.71	1.94	0.81	1.99	1.84
18:3	0.25	0.42	0.24	0.41	0.28
20:4	—	—	—	—	—
20:5	—	—	—	—	—
22:5	—	—	—	—	—
22:6	—	—	—	—	—
P/S Ratio	0.08	0.38	0.09	0.22	0.31
Total lipids (%)	29.40	18.80	28.36	28.26	19.87
Cholesterol (mg/100 g)	56	59	56	55	59

[a] From U.S. Department of Agriculture (1980); dashes denote lack of reliable data for a constituent believed to be present in measurable amount; P/S ratio, ratio of polyunsaturated to saturated fatty acids.

G. Dehydration

Some nutrients are lost during freeze-drying and the subsequent storage of dehydrated meats, but protein nutritional value generally is not affected. Depending on time and temperature of storage, significant amounts ($>70\%$) of thiamin and from slight to moderate levels of riboflavin, pantothenic acid, and niacin may be lost (Labuza, 1972; Sommers and Hagen, 1981).

H. Irradiation Preservation

Under conditions appropriate for commercial application, the preservation of uncooked meats by exposure to ionizing radiation does not significantly

Bratwurst	Chicken roll, light meat	Corned beef loaf	Frankfurter		
			Beef	Beef and pork	Chicken
9.32	2.02	2.68	11.96	10.76	5.54
0.03	—	0.02	0.03	0.08	—
0.01	0.01	0.02	0.03	0.06	0.05
0.30	0.06	0.18	0.89	0.53	0.18
—	—	—	—	—	—
5.85	1.51	1.39	6.91	6.45	4.12
—	—	—	—	—	—
3.13	0.39	1.08	4.09	3.65	1.10
—	—	—	—	—	—
12.19	2.96	3.03	14.35	13.67	8.48
—	—	—	—	—	—
0.82	0.41	0.47	1.79	1.31	1.20
11.37	2.43	2.56	12.56	12.36	7.14
—	—	—	—	—	—
2.74	1.60	0.38	1.16	2.73	4.04
2.48	1.36	0.29	0.86	2.34	3.74
0.26	0.06	0.09	0.30	0.39	0.15
—	0.08	—	—	—	—
—	0.00	—	—	—	—
—	0.01	—	—	—	—
—	0.02	—	—	—	—
0.29	0.79	0.14	0.10	0.26	0.73
25.87	7.38	6.80	29.42	29.15	19.48
60	50	43	48	50	101

(continued)

decrease the nutritional quality of protein, lipid, or carbohydrate. Some destruction of vitamins, thiamin in particular, occurs but is no more severe than that from conventional methods of preservation. Various aspects of irradiation preservation and its effects on the nutritive value and wholesomeness of foods are discussed by Josephson *et al.* (1978), Thomas *et al.* (1981), the World Health Organization (1981), and GAMMASTER (1981).

I. Mechanical Deboning

Mechanically deboned red meats, officially termed "mechanically separated (species)" and mechanically deboned poultry contain higher levels of calcium, unsaturated fats, and fluoride than do conventionally hand-boned meats. The

Table XVII (Continued)

Fatty acids (g/100 g)	Ham, Chopped, (canned)	Ham, sliced Extra Lean	Regular	Italian sausage	Kielbasa (kolbassy)	Knockwurst (knackwurst)
Saturated, total	6.28	1.62	3.39	9.07	9.91	10.20
10:0	0.03	0.02	0.02	0.07	0.03	0.11
12:0	0.01	0.01	0.02	0.05	0.02	0.09
14:0	0.24	0.08	0.15	0.45	0.36	0.58
15:0	—	—	—	—	—	—
16:0	4.02	1.01	2.13	5.39	6.13	5.97
17:0	—	—	—	—	—	—
18:0	1.99	0.51	1.08	3.07	3.36	3.46
20:0	—	—	—	—	—	—
Monounsaturated, total	9.17	2.35	4.95	11.94	12.94	12.81
14:1	—	—	—	—	—	—
16:1	0.67	0.22	0.43	1.18	0.97	1.28
18:1	8.49	2.13	4.52	10.77	11.97	11.53
20:1	—	—	—	—	—	—
Polyunsaturated, total	2.05	0.48	1.21	3.29	3.08	2.92
18:2	1.86	0.43	1.04	2.84	2.68	2.48
18:3	0.20	0.05	0.17	0.44	0.40	0.44
20:4	—	—	—	—	—	—
20:5	—	—	—	—	—	—
22:5	—	—	—	—	—	—
22:6	—	—	—	—	—	—
P/S Ratio	0.33	0.30	0.36	0.36	0.31	0.29
Total lipid (%)	18.83	4.96	10.57	25.70	27.15	27.76
Cholesterol (mg/100 g)	49	47	57	78	67	58

products of the mechanical deboning process have been well characterized [Goldstrand, 1975; Chant et al., 1977; Murphy et al., 1979; Council for Agricultural Science and Technology (CAST), 1980].

VII. NUTRITIONAL VALUE VERSUS CALORIC DENSITY

Calories are of major dietary concern to large segments of the United States population. Given the low levels of exercise that are common, high, or even moderate, caloric intakes can cause overweight. For many individuals, a reduction in energy intake (calories) would be prudent if the reduced intake provides adequate amounts of all essential nutrients. Rather than a crash diet, a dietary regime should be established as the basis for a lifelong pattern of eating for both weight control and good health.

Beef loaf	Luncheon Meat Pork and beef	Olive loaf	Pepperoni	Polish sausage	Pork sausage Fresh	Cooked
11.18	11.59	5.85	16.13	10.33	14.47	10.81
0.03	0.08	0.01	—	0.07	0.11	0.12
0.01	0.05	0.01	—	0.07	0.08	0.09
0.75	0.48	0.27	0.86	0.40	0.57	0.45
—	—	—	—	—	—	—
6.32	7.09	3.54	10.29	6.40	8.87	6.53
—	—	—	—	—	—	—
4.08	3.89	2.02	4.69	3.38	4.85	3.62
—	—	—	—	—	—	—
12.25	15.09	7.87	21.11	13.52	18.53	13.90
—	—	—	—	—	—	—
1.54	1.35	0.66	1.77	0.95	1.45	1.09
10.71	13.74	7.21	18.92	12.58	17.09	12.81
—	—	—	—	—	—	—
0.87	3.74	1.94	4.37	3.08	5.24	3.81
0.65	3.16	1.82	3.74	2.79	4.39	3.28
0.21	0.58	0.11	0.41	0.29	0.85	0.54
—	—	—	0.14	—	—	—
—	—	—	—	—	—	—
—	—	—	—	—	—	—
—	—	—	—	—	—	—
0.08	0.32	0.33	0.27	0.30	0.36	0.35
26.20	32.16	16.50	43.97	28.72	40.29	31.16
64	55	38	—	70	68	83

(continued)

In such an approach, meats and meat products would still be major dietary staples for most individuals. Selected meats and meat products provide a high return in terms of nutrients provided per 1000 kcal or per usual serving (Tables XXII, XXIII). Many meats provide higher levels of protein per 1000 kcal than do some other sources of protein. Furthermore, the content of saturated fats, when expressed in terms of protein, in meats is either lower than or equal to that in many other common protein sources.

VIII. DIETARY MEATS AND HEALTH

During the past few years, various governmental and other organizations have published recommendations for prudent diets or for national dietary goals (American Heart Association, 1978; U.S. Senate, 1977a,b; Health, Education and Welfare, 1979; U.S. Department of Agriculture and Department of Health,

Table XVII *(Continued)*

Fatty acids (g/100 g)	Salami (cooked)		Salami (dry)		Smoked link sausage	
	Beef	Beef and pork	Pork	Pork and beef	Pork	Pork and beef
Saturated, total	8.43	8.09	11.89	12.20	11.32	10.62
10:0	0.06	0.10	—	0.06	0.04	10.05
12:0	0.05	0.07	—	0.04	0.04	0.04
14:0	0.58	0.47	0.52	0.51	0.40	0.41
15:0	—	—	—	—	—	—
16:0	4.71	4.53	7.64	7.60	7.07	6.57
17:0	—	—	—	—	—	—
18:0	3.04	2.93	3.56	4.00	3.76	3.55
20:0	—	—	—	—	—	—
Monounsaturated, total	9.30	9.19	16.00	17.10	14.64	14.19
14:1	—	—	—	—	—	—
16:1	1.20	0.89	1.22	1.69	1.01	1.13
18:1	8.10	8.30	14.67	15.40	13.63	13.06
20:1	—	—	—	—	—	—
22:1	—	—	—	—	—	—
Polyunsaturated, total	0.88	2.02	3.74	3.21	3.76	3.26
18:2	0.67	1.61	3.27	2.87	3.39	2.88
18:3	0.21	0.41	0.28	0.33	0.37	0.38
20:4	—	—	0.16	—	—	—
20:5	—	—	—	—	—	—
22:5	—	—	—	—	—	—
22:6	—	—	—	—	—	—
P/S Ratio	0.10	0.25	0.32	0.26	0.33	0.31
Total lipid (%)	20.10	20.11	33.72	34.39	31.75	30.33
Cholesterol (mg/100 g)	60	65	—	79	68	71

Education and Welfare, 1980). In general, recommendations were made to reduce intakes of saturated fats, cholesterol, salt, and/or sugar, and to increase intakes of fiber and other complex carbohydrates (e.g., starch). Those recommendations could indirectly lead to decreased consumption of meats, eggs, and dairy products.

Those recommendations or goals have been a matter of some controversy. Assessment of their validity can be difficult (Ahrens, 1979). Frank disagreements have been published and alternative recommendations made (Food and Nutrition Board, 1980). The different opinions indicate that many issues involving the relationships between disease and diet remain unresolved. Useful discussions have been presented (Council for Agricultural Science and Technology, 1980; Franklin and Davis, 1981).

Thuringer	Turkey Breast meat	Turkey Ham	Turkey Roll Light and dark	Turkey Roll Light meat	Vienna sausage, canned
12.03	0.48	1.70	2.04	2.02	9.28
0.05	—	—	—	—	0.03
0.05	0.00	0.01	0.01	0.01	0.02
0.72	0.02	0.03	0.05	0.05	0.59
—	—	—	—	—	—
6.83	0.30	0.90	1.25	1.26	5.44
—	—	—	—	—	—
4.38	0.15	0.51	0.50	0.47	3.20
—	—	—	—	—	—
13.93	0.45	1.15	2.30	2.50	12.55
—	—	—	—	—	—
1.79	0.07	0.17	0.38	0.41	1.52
12.14	0.38	0.95	1.87	2.02	11.02
—	—	0.02	0.01	0.01	—
—	—	0.01	0.01	0.01	—
1.89	0.28	1.52	1.78	1.74	1.68
1.56	0.26	1.23	1.48	1.42	1.29
0.33	0.02	0.05	0.08	0.09	0.39
—	—	0.18	0.14	0.14	—
—	—	—	—	—	—
—	—	0.03	0.02	0.02	—
—	—	0.04	0.03	0.02	—
0.16	0.58	0.89	0.87	0.86	0.18
29.93	1.58	5.08	6.99	7.22	25.20
68	41	—	55	43	52

REFERENCES

Ahrens, E. H., Jr. (1979). *Am. J. Clin. Nutr.* **32,** Suppl., 2627–2631.

American Heart Association (1978). *Circulation* **58,** 762A–766A.

Anderson, R. A. (1985). *In* "Handbook of Nutrition in the Aged." (R. R. Watson, ed.), pp. 137–144. CRC Press, Boca Raton, Florida.

Bodwell, C. E., ed. (1977). "Evaluation of Proteins for Humans." Avi Publ. Co., Westport, Connecticut.

Bodwell, C. E., and Womack, M. (1978). *J. Food Sci.* **43,** 1543–1549.

Bodwell, C. E., Adkins, J. S., and Hopkins, D. T., eds. (1981). "Protein Quality in Humans: Assessment and *In Vitro* Estimation." Avi Publ. Co., Westport, Connecticut.

Bressani, R. (1977). *In* "Evaluation of Proteins For Humans" (C. E. Bodwell, ed.), pp. 204–232. Avi Publ. Co., Westport, Connecticut.

Chant, J. L., Day, L., Field, R. A., Kruggel, W. G., and Chang, Y.-O. (1977). *J. Food Sci.* **42,** 306–309.

C. E. Bodwell and B. A. Anderson

Table XVIII
Lipid Composition of Baby Foods[a]

Fatty acids (g/10 g)	Beef			Pork		
	Strained	Beef, junior	With beef heart, strained	Strained	Ham, strained	Ham, junior
Saturated, total	2.58	2.59	2.08	2.40	1.94	2.24
10:0	0.00	0.00	0.00	0.00	0.00	0.00
12:0	0.00	0.01	0.00	0.00	0.01	0.01
14:0	0.16	0.14	0.13	0.08	0.07	0.08
15:0	—	—	—	—	—	—
16:0	1.33	1.24	1.07	1.53	1.22	1.46
17:0	—	—	—	—	—	—
18:0	0.89	1.01	0.76	0.72	0.59	0.68
20:0	—	—	—	—	—	—
Monounsaturated, total	2.20	1.85	1.85	3.58	2.76	3.18
14:1	—	—	—	—	—	—
16:1	0.20	0.13	0.16	0.23	0.16	0.18
18:1	1.91	1.66	1.62	3.27	2.52	2.91
20:1	0.05	0.00	0.01	0.07	0.06	0.07
Polyunsaturated, total	0.24	0.16	0.21	0.78	0.78	0.91
18:2	0.09	0.11	0.16	0.69	0.71	0.82
18:3	0.05	0.03	0.02	0.03	0.03	0.03
20:4	0.05	0.02	0.03	—	0.04	0.05
20:5	—	—	—	—	—	—
22:5	—	—	—	—	—	—
22:6	—	—	—	—	—	—
P/S Ratio	0.09	0.06	0.10	0.33	0.40	0.41

[a] From U.S. Department of Agriculture (1978); no data available for chicken or turkey sticks, junior.

Council for Agricultural Science and Technology (1980). "Foods From Animals: Quantity, Quality and Safety." CAST, Ames, Iowa.

Cross, H. R., Ono, K., Johnson, H. K., and Moss, M. K. (1980). *40th Annu. Meet. Inst. Food Technol.* p. 158.

Food and Nutrition Board (1980). "Toward Healthful Diets." Nat. Acad. Sci., Washington, D.C.

Fox, J. D., and Ono, K. (1980). *40th Annu. Meet. Inst. Food Technol.* p. 157.

Franklin, K. R., and Davis, P. N. eds. (1981). "Meat in Nutrition and Health." National Livestock and Meat Board, Chicago, Illinois.

GAMMASTER (1981). "Food Irradiation Now." Martinus Nijhoff/Dr. W. Junk Publishers, Boston, Massachusetts.

Goldstrand, R. E. (1975). *Proc. Annu. Reciprocal Meat Conf.* **28,** 116–126.

Harris, R. S., and Karmas, E., eds. (1975). "Nutritional Evaluation of Food Processing," 2nd ed. Avi Publ. Co., Westport, Connecticut.

Harris, R. S., and von Loesecke, H., eds. (1960). "Nutritional Evaluation of Food Processing," Avi Publ. Co., Westport, Connecticut.

	Lamb		Veal		Chicken		Turkey	
	Strained	Junior	Strained	Junior	Strained	Junior	Strained	Junior
	2.31	2.56	2.29	2.39	2.03	2.47	1.91	2.31
	0.01	0.01	0.00	0.00	—	—	0.00	0.00
	0.00	0.00	0.02	0.01	0.00	0.00	0.03	0.04
	0.11	0.12	0.24	0.18	0.05	0.07	0.09	0.10
	—	—	—	—	—	—	—	—
	1.03	1.14	1.18	1.22	1.48	1.80	1.17	1.41
	—	—	—	—	—	—	—	—
	0.99	1.09	0.67	0.82	0.42	0.50	0.55	0.66
	—	—	—	—	—	—	—	—
	1.87	2.06	2.05	2.12	3.56	4.33	2.18	2.64
	—	—	—	—	—	—	—	—
	0.06	0.07	0.19	0.18	0.29	0.35	0.19	0.22
	1.78	1.96	1.79	1.88	3.25	3.95	1.96	2.37
	0.00	0.00	0.00	0.01	0.02	0.02	0.02	0.02
	0.20	0.22	0.16	0.17	1.92	2.33	1.45	1.76
	0.13	0.15	0.10	0.10	1.85	2.25	1.33	1.61
	0.05	0.06	0.03	0.03	0.04	0.05	0.06	0.08
	0.01	0.01	0.02	0.03	0.03	0.04	0.06	0.07
	—	—	—	—	—	—	—	—
	—	—	—	—	—	—	—	—
	—	—	—	—	—	—	—	—
	0.09	0.09	0.07	0.07	0.95	0.94	0.76	0.76

Hopkins, D. T. (1981). *In* "Protein Quality in Humans: Assessment and *In Vitro* Estimation" (C. E. Bodwell, J. S. Adkins, and D. T. Hopkins, eds.), pp. 169–194. Avi Publ. Co., Westport, Connecticut.

Josephson, E. S., Thomas, M. H., and Calhoun, W. K. (1978). *J. Food Process. Preserv.* **2,** 299–313.

Labuza, T. P. (1972). *CRC Crit. Rev. Food Technol.* **3,** 217–240.

Lawrie, R. A. (1979). "Meat Science," 3rd ed. Pergamon, Oxford.

McQuilkin, C., and Matthews, R. H. (1981). "Provisional Table on the Nutrient Content of Bakery Foods and Related Items." USDA, HNIS, Washington, D.C.

Mertz, W. (1985). *In* "NIH Workshop on Nutrition and Hypertension." (M. J. Horan *et al.,* eds.), pp. 271–276. Biomedical Information Corp., New York.

Monsen, E. R., Hallberg, L., Layrisse, M., Hegsted, D. M., Cook, J. D., Mertz, W., and Finch, C. A. (1978). *Am. J. Clin. Nutr.* **31,** 134–141.

Murphy, E. W., Brewington, C. R., Walter, B. W., and Nelson, M. A. (1979). "Health and Safety Aspects of the Use of Mechanically Deboned Poultry." U.S. Govt. Printing Office, Washington, D.C.

National Research Council (1980). "Recommended Dietary Allowances," 9th ed., Nat. Acad. Sci., Washington, D.C.

Table XIX
Examples of Amino Acid Scores Based on Individual Amino Acids for Meats and Meat Products[a]

Amino acid	Reference pattern (mg/g protein)	Cooked meats						Sausage and luncheon meat products				Baby foods			
		Beef round	Ham (fresh)	Ham (cured)	Chicken (dark with skin)	Chicken (light with skin)	Cod	Bologna (beef and pork)	Olive loaf	Turkey ham	Vienna sausage	Pork, ham, junior	Chicken sticks, junior	Turkey, strained	Turkey sticks, junior
Tryptophan	11	1.02	1.22	1.09	1.01	1.02	1.00	0.96	0.80	1.03	0.96	0.89	0.73	0.95	0.67
Threonine	35	1.25	1.34	1.27	1.18	1.18	1.25	1.23	1.14	1.27	0.99	1.24	1.12	1.26	1.14
Leucine	70	1.13	1.16	1.13	1.03	1.04	1.16	1.19	1.05	1.14	1.11	1.14	1.12	1.14	1.12
Isoleucine	42	1.07	1.15	1.04	1.18	1.20	1.10	1.21	0.84	1.24	1.29	1.13	1.19	1.19	1.09
Lysine	51	1.63	1.93	1.66	1.59	1.61	1.81	1.85	1.34	1.85	1.51	1.67	1.56	1.62	1.68
Methionine + cystine	26	1.42	1.45	1.59	1.53	1.54	1.55	1.58	1.45	1.51	1.64	1.46	1.19	1.66	1.20
Phenylalanine + tyrosine	73	1.00	1.04	1.04	0.97	0.98	0.99	1.10	0.93	1.08	1.02	0.98	1.10	1.05	1.09
Valine	48	1.01	1.12	0.90	1.01	1.01	1.07	1.18	0.89	1.11	1.16	1.07	1.09	1.06	0.99
Histidine	17	2.01	2.98	2.11	1.72	1.74	1.73	1.89	1.45	1.84	1.56	2.00	1.86	1.51	1.58
Total	373	1.22	1.36	1.24	1.19	1.20	1.26	1.32	1.07	1.30	1.21	1.24	1.21	1.24	1.19

[a] Scores calculated from amino acid data listed in Tables XIII–XV; reference pattern from the National Research Council (1980).

Table XX

Recommended Dietary Allowances (RDA) for Protein from a Mixed Diet and as Provided from Meats and Meat Products

| | Age (years) | Protein required (g/day) to meet the RDA | |
		Mixed diets[a]	Most meats and meat products[b]
Children (both sexes)	1–3	23	17.3
	4–6	30	22.5
	7–10	34	25.5
Males	11–14	45	33.8
	15–50	56	42.0
	51+	56	42.0
Females	11–18	46	34.5
	19–50	44	33.0
	51+	44	33.0

[a] Actual RDA values given by the National Research Council (1980); a 75% utilization level was assumed for a mixed diet compared to whole egg protein; whole egg protein was assumed to be utilized at only ~77% of its maximal value due to the loss of efficiency in utilization when consumed at requirement levels.

[b] Assumes that meat protein is utilized at levels equivalent to 95% of that for whole egg protein; the few data available from studies with humans suggest that meat protein is ~100% equivalent to egg protein.

Paul, A. A., and Southgate, D. A. T. (1978). "McCance and Widdowson's The Composition of Foods." Medical Research Council, London.

Price, J. F., and Schweigert, B. S., eds. (1971). "The Science of Meat and Meat Products," 2nd ed., Freeman, San Francisco, California.

Reiser, S. (1985). In "Nutrition Health." Vol. 3, pp. 203–216. Academic Publishers.

Reiser, S., Smith, J. C., Jr., Mertz, W., Holbrook, J. T., Scholfield, D. J., Powell, A. S., Canfield, W. K., and Canary, J. J. (1985). Amer. J. Clin. Nutr. 42(2), 242–251.

Reynolds, R. D., and Natta, C. L. (1985). Amer. J. Clin. Nutr. 41(4), 684–688.

Rice, E. E. (1971). In "The Science of Meat and Meat Products" (J. F. Price and B. S. Schweigert, eds.), 2nd ed., pp. 287–327. Freeman, San Francisco, California.

Robinson, J. R., Robinson, M. F., Levander, O. A., and Thomson, C. D. (1985). Amer. J. Clin. Nutr. 41(5), 1023–1031.

Sidwell, V. D., Foncannon, P. R., Moore, N. S., and Bonnet, J. C. (1974). Mar. Fish. Rev. 36, 21–35 (MFR Paper 1043).

Sidwell, V. D., Buzell, D. H., Foncannon, P. R., and Smith, A. L. (1977). Mar. Fish. Rev. 39, 1–11 (MFR Paper 1228).

Sidwell, V. D., Loomis, A. L., Loomis, K. J., Foncannon, P. R., and Buzell, D. H. (1978a). Mar. Fish. Rev. 40, 1–20 (MFR Paper 1324).

Sidwell, V. D., Loomis, A. L., Foncannon, P. R., and Buzell, D. H. (1978b). Mar. Fish. Rev. 40, 1–16 (MFR Paper 1355).

Table XXI
Percentage True Retention of Nutrients during Cooking[a]

| | Beef (separable lean) | | | Pork (separable lean) | | | Chicken | | | | Turkey, roasted | |
| | | | | | | | Roasted | | Stewed | | | |
	Broiled	Roasted	Braised	Broiled	Roasted	Braised	Flesh only	Flesh and skin	Flesh only	Flesh and skin	Flesh only	Flesh and skin
Water	63	65	50	58	65	45	63	59	70	70	65	64
Protein	99	99	98	99	102	99	101	99	101	98	99	100
Total Lipids	120	129	114	149	145	120	192	57	186	59	128	91
Ash	87	84	65	96	84	76	73	70	75	74	80	83
Calcium	103	91	84	74	95	—	98	92	87	80	129	130
Copper	100	100	100	97	102	102	95	91	93	91	68	69
Iron	94	99	102	81	103	101	97	87	94	88	96	96
Magnesium	86	83	66	96	74	—	77	76	66	67	79	82
Manganese	—	—	—	54	86	—	77	72	78	73	78	81
Phosphorous	83	84	81	—	—	73	80	79	69	69	81	83
Potassium	83	80	57	86	79	73	80	78	57	58	75	76
Sodium	85	83	59	88	82	72	82	76	71	69	76	78
Zinc	100	100	100	99	108	107	105	100	104	99	99	100
Ascorbic Acid	—	—	—	—	—	—	—	—	—	—	—	—
Thiamin	70	58	45	70	59	42	73	71	53	55	66	67
Riboflavin	92	98	86	102	94	73	90	89	95	93	82	85
Niacin	78	75	61	80	84	79	82	82	60	61	89	91
Pantothenic Acid	79	97	69	81	60	53	77	—	—	—	—	—
Vitamin B6	74	66	46	63	64	52	80	—	—	—	71	—
Folacin	87	88	72	—	—	—	—	—	—	—	61	—
Vitamin B12	75	72	67	91	82	60	65	—	48	—	64	—
Vitamin A	—	—	—	—	—	—	75	—	75	75	—	—
Cholesterol	106	103	103	115	114	108	95	—	—	—	88	88

[a] Values are true retentions (%); values for beef and pork are based on analyses of separable lean of raw cuts and lean separated from corresponding cuts after cooking; flesh only values for chicken and turkey are from analyses of flesh only after cooking with skin; values for beef based on analyses of 19 cuts from each of 15 carcasses, from U.S. Department of Agriculture (1986); values for pork based on analyses of seven cuts from each of 11 carcasses (spare ribs omitted), from U.S. Department of Agriculture (1983); values for chicken based on eight samples for all nutrients except pantothenic acid, vitamin B6, folacin, vitamin B12, vitamin A, and cholesterol (number of samples not specified); values for turkey based on 41 replications for all nutrients except vitamin B12 and vitamin A (number of samples not specified) and zinc, copper, and manganese (nine replications); values for chicken and turkey from U.S. Department of Agriculture (1979).

Slover, H. T., Thompson, R. H., Jr., Davis, C. S., and Merola, G. V. (1985). *J. Amer. Oil Chem. Soc.* **62**(4), 775–785.

Sommers, I. I., and Hagen, R. E. (1981). *In* "Meat in Nutrition and Health" (K. R. Franklin and P. N. Davis, eds.), pp. 19–34. National Livestock and Meat Board, Chicago, Illinois.

Thomas, M. H., Atwood, B. M., Wierbicki, E., and Taub, I. A. (1981). *J. Food Sci.* **46**, 824–828.

U.S. Department of Agriculture (USDA) (1944). "Family Food Consumption in the United States," Misc. Publ. No. 550. USDA, Washington, D.C.

U.S Department of Agriculture (USDA) (1954). "Food Consumption of Urban Families in the United States," Agric. Inf. Bull. No. 132. USDA, Washington, D.C.

U.S. Department of Agriculture (USDA) (1956). "Food Consumption of Households in the United States," Rep. No. 1. USDA, Washington, D.C.

U.S. Department of Agriculture (USDA) (1963). "Composition of Foods: Raw, Processed, Prepared," USDA Handb. No. 8. U.S. Govt. Printing Office, Washington, D.C.

U.S. Department of Agriculture (USDA) (1965). "Economics and Statistics Service," Stat. Bull. No. 364. USDA, Washington, D.C.

U.S. Department of Agriculture (USDA) (1968). "Food Consumption of Households in the United States, Spring, 1965," Rep. No. 1. USDA, Washington, D.C.

U.S. Department of Agriculture (USDA) (1969). "Pantothenic Acid, Vitamin B_6 and Vitamin B_{12} in Foods," Home Econ. Res. Rep. No. 36. U.S. Govt. Printing Office, Washington, D.C.

U.S. Department of Agriculture (USDA) (1976). "Composition of Foods: Dairy and Egg Products, Raw, Processed, Prepared," USDA, Agric. Handb. No. 8-1. Consumer and Food Economics Institute, U.S. Govt. Printing Office, Washington, D.C.

U.S. Department of Agriculture (USDA) (1978). "Composition of Foods: Baby Foods; Raw, Processed, Prepared," USDA Agric. Handb. No. 8-3. Consumer and Food Economics Institute, U.S. Govt. Printing Office, Washington, D.C.

U.S. Department of Agriculture (USDA) (1979). "Composition of Foods: Poultry Products; Raw, Processed, Prepared," USDA Agric. Handb. No. 8-5. Consumer and Food Economics Institute, U.S. Govt. Printing Office, Washington, D.C.

U.S. Department of Agriculture (USDA) (1980). "Composition of Foods: Sausage Products and Luncheon Meats: Raw, Processed, Prepared," USDA Agric. Handb. No. 8-7. Consumer and Food Economics Institute, U.S. Govt. Printing Office, Washington, D.C.

U.S. Department of Agriculture (USDA) (1981). "Economics and Statistics Service," Stat. Bull. No. 656. USDA, Washington, D.C.

U.S. Department of Agriculture (USDA) (1982a). "USDA Nationwide Food Consumption Survey, 1977–78, Food Consumption: Households in the 48 States, Spring 1977," Rep. No. H-1. USDA, Washington, D.C.

U.S. Department of Agriculture (USDA) (1982b). "Composition of Foods: Breakfast Cereals, Raw, Processed, Prepared," USDA Agric. Handb. No. 8-8. Consumer Nutrition Center, USDA, Washington, D.C.

U.S. Department of Agriculture (USDA) (1983). "Composition of Foods: Pork Products; Raw, Processed, Prepared," USDA Agric. Handb. No. 8-10. Consumer Nutrition Division, U.S. Govt. Printing Office, Washington, D.C.

U.S. Department of Agriculture (USDA) (1984). "USDA Nationwide Food Consumption Survey, 1977–78, Year 1977–78, Nutrient Intakes: Individuals in 48 States," Rep. No. I-2. USDA, Washington, D.C.

U.S. Department of Agriculture (USDA) Nutrition Monitoring Division (1986). "Composition of Foods: Beef Products; Raw, Processed, Prepared," USDA Agric. Handb. No. 8-13, U.S. Govt. Printing Office, Washington, D.C.

(USDA/HEW) U.S. Department of Agriculture and Department of Health, Education and Walfare (1980). "Nutrition and Your Health, Dietary Guidelines for Americans." U.S. Govt. Printing Office, Washington, D.C.

Table XXII

Nutritive Value of Meats, Meat Alternates, and Cereal Products per 1000 kcal[a]

	Beef, round, roasted		Pork, fresh ham, roasted, lean	Chicken				Cod fish, broiled, with margarine
				Baked, without skin		Fried (floured), with skin		
Food	Lean	Lean and fat		Light meat	Dark meat	Light meat	Dark meat	
Weight (g)	450	342	455	578	489	407	351	708
Proximate								
Protein (g)	130.5	89.60	128.7	178.7	133.7	123.8	95.64	143.3
Total lipids (g)	48.8	68.00	50.2	26.1	47.57	49.2	59.47	46
Carbohydrate (g)	0.0	0.00	—	—	—	7.4	14.3	2.83
Minerals								
Calcium (mg)	22	21	32	88.1	75	67	58	100
Copper (μg)	568	385	492	286	391	234	310	—
Iron (mg)	142	9.7	5.1	6.1	6.5	4.9	5.3	3.33
Magnesium (mg)	124	83	107	156	115	110	83	225
Manganese (μg)	71	49	166	95	104	105	136	—
Phosphorus (mg)	1075	732	1278	1252	874	866	620	1583
Potassium (mg)	1785	1212	1695	1429	1172	971	810	3142
Sodium (mg)	293	210	289	442	454	311	314	1042
Zinc (mg)	22.7	15.39	14.8	7.1	13.7	5.1	9.13	3.3
Vitamins								
Ascorbic acid (mg)	0	0	1.60	0	0	0	0	25
Thiamin (mg)	0.41	0.27	3.16	0.34	0.34	0.33	0.33	0.33
Riboflavin (mg)	1.17	0.79	1.60	0.68	1.15	0.53	.83	0.50
Niacin (mg)	18.45	12.69	22.46	71.84	32.01	49.00	24.00	15.16
Pantothenic acid (mg)	2.57	1.74	4.12	5.65	5.92	3.92	21.39	—
Vitamin B$_6$ (mg)	2.25	1.56	1.44	3.47	1.78	2.20	4.05	1.50
Folacin (μg)	53	35	53	20	40	14	29	—
Vitamin B$_{12}$ (μg)	12.6	8.96	3.26	1.97	1.55	1.34	1.07	4.92
Vitamin A (IU)	0	0	37	170	351	278	364	2083
Lipids								
Fatty Acids								
Saturated, total (g)	17.70	26.85	17.27	7.35	13.0	13.50	16.07	8.50
Monounsaturated, total (g)	22.74	31.19	22.57	8.91	17.42	19.52	23.38	19.16
Polyunsaturated, total (g)	2.04	2.70	6.10	5.65	11.03	10.96	13.71	14.83
Cholesterol (mg)	365	284	428	490	454	354	322	400

[a] Values for beef, cod, peanut butter, and dried beans calculated from USDA, Human Nutrition Information Service, Nutrient Data Research Branch (unpublished data); for pork from U.S. Department of Agriculture (1983); for chicken from U.S. Department of Agriculture (1979); for dairy and egg products from U.S. Department of Agriculture (1976); for oatmeal from U.S. Department of Agriculture (1982b).

[b] From McQuilkin and Matthews (1981).

U.S. Department of Health, Education, and Welfare (1979). "Healthy People — Surgeon General's Report on Health Promotion and Disease Prevention." U.S Govt. Printing Office, Washington, D.C.

U.S. Senate (1977a). "Dietary Goals for the United States." U.S. Govt. Printing Office, Washington, D.C.

U.S. Senate (1977b). "Dietary Goals for the United States," 2nd ed. U.S. Govt. Printing Office, Washington, D.C.

Whole milk	Skim milk	Cottage cheese	Cheddar cheese	Peanut butter	Dried Navy beans, cooked	Cereal, oatmeal, cooked	Bread[b] White	Bread[b] Whole wheat	Egg, cooked
1627	2849	968	246	169	658	1614	375	408	633
53.5	97.1	120.9	61.9	42.6	56.6	41.4	31.1	39.2	76.8
54.3	5.12	43.64	82.46	86.35	3.94	16.56	14.70	17.79	70.6
75.8	138.13	25.94	61.93	32.49	186.9	173.80	183.08	185.27	7.6
1940	3512	581	1789	58	446	174	473	294	354
—	—	—	—	778	2114	890	521	1395	
.80	1.2	1.3	1.7	3.07	16.58	11	10.65	13.95	13.17
220	326	51	70.18	296	399	386	79	379	76
—	—	—	—	3032	3316	9441	—	—	—
1520	2872	1276	1272	630	1067	1228	405	1061	1139
2467	4721	816	246	1180	2321	910	420	718	823
800	1465	3917	1544	799	5.18	7	1928	2595	873
6.20	11.40	3.60	7.72	4.92	7.25	7.93	2.33	6.85	9.12
13	23.2	Trace	0	0	0	0	Trace	Trace	0
0.62	1.05	.18	0.09	0.16	1.30	1.79	1.76	1.43	0.51
2.27	3.95	1.57	0.97	0.16	0.26	0.35	1.16	0.86	1.90
1.41	2.56	1.20	0.18	22.70	3.32	2.07	14.06	15.63	.38
5.13	9.42	2.07	1.05	1.85	1.97	3.24	1.61	3.01	10.89
0.67	1.16	0.65	0.18	0.69	0.67	0.34	0.13	0.76	0.76
80	151	120	44	175	—	62.1	131.3	224.4	304
5.81	10.81	6.04	2.02	0	0	0	0	0	9.75
2047	5814	1576	2631	0	0	262	Trace	Trace	3291
33.82	3.37	27.60	52.46	17.78	.47	3.03	3.49	3.10	21.14
15.67	1.40	12.44	23.33	39.79	.26	5.80	5.33	4.16	28.23
2.00	.23	1.34	2.37	25.56	2.28	6.69	4.95	7.55	9.12
887	47	143	263	0	0	0	4	0	3468

Van Arsdall, R. N. (1979). *In* "Foods From Animals: Quantity, Quality and Safety," Rep. No. 82, p. 49. Council for Agricultural Science and Technology, Ames, Iowa.

Veillon, C., Lewis, S. A., Patterson, K. Y., Wolf, W. R., and Harnly, J. M. (1985). *Anal. Chem.* **57**(11), 2106–2109.

Weihrauch, J. L., and Son, Y. (1983). *J. Am. Oil Chem. Soc.* **60**(12), 1971–1978.

Wiley, E. R., and Canary, J. J. (1985). *Nutr. Res.* **5**(6), 627–630.

Wolf, W. R., and Ono, K. (1980). *40th Annu. Meet. Inst. Food Technol.* p. 159.

World Health Organization (1981). "Wholesomeness of Irradiated Food," Tech. Rep. Ser. No. 659. WHO, Geneva.

Table XXIII

Nutritive Value of Meats, Meat Alternates, and Cereal Products in Specified Servings[a]

Food	Beef, round, roasted Lean	Beef, round, roasted Lean and fat	Pork, fresh ham, roasted, lean	Chicken Baked, without skin Light meat	Chicken Baked, without skin Dark meat	Chicken Fried (floured), with skin Light meat	Chicken Fried (floured), with skin Dark meat	Cod fish, broiled, with margarine
Serving size	3 oz (85 g)	3 oz (85 g)	3 oz (85 g)	3 oz (85 g)	3 oz (85 g)	3 oz (85 g)	3 oz (85 g)	3 oz (85 g)
Proximate								
Food energy (kcal)	189	248	187	147	174	209	242	120
Protein (g)	24.64	22.27	24.07	26.27	23.26	25.88	23.14	17.2
Total lipids (g)	9.21	16.90	9.38	3.83	8.27	10.28	14.37	5.52
Carbohydrate (g)	0	0	0	0	0	1.55	3.47	0.34
Minerals								
Calcium (mg)	4	5	6	13	13	14	14	12
Copper (mcg)	107	96	92	42	68	49	75	—
Iron (mg)	2.68	2.41	0.95	0.90	1.13	1.03	1.28	0.40
Magnesium (mg)	23	21	20	23	20	23	20	27
Manganese (μg)	13	12	31	14	18	22	33	—
Phosphorus (mg)	203	182	239	184	152	181	150	190
Potassium (mg)	337	301	317	210	204	203	196	377
Sodium (mg)	55	52	54	65	79	65	76	125
Zinc (mg)	4.30	3.83	2.77	1.05	2.38	1.07	2.21	0.40
Vitamins								
Ascorbic acid (mg)	0	0	0.3	0	0	0	0	3
Thiamin (mg)	0.08	0.07	0.59	0.05	0.06	0.07	0.08	0.04
Riboflavin (mg)	0.22	0.20	0.30	0.10	0.20	0.11	0.20	0.06
Niacin (mg)	3.40	3.15	4.20	10.56	5.57	10.23	5.81	1.82
Pantothenic acid (mg)	0.48	.43	0.77	0.83	1.03	0.82	0.98	—
Vitamin B_6 (mg)	0.43	0.38	0.27	0.51	0.31	0.46	0.27	0.18
Folacin (μg)	10	9	10	3	7	3	7	—
Vitamin B_{12} (μg)	2.38	2.23	0.61	0.29	0.27	0.28	0.26	0.59
Vitamin A (IU)	0	0	7	25	61	58	88	250
Lipids								
Fatty Acids								
Saturated, total (g)	3.34	6.67	3.23	1.08	2.26	2.82	3.89	1.02
Monounsaturated, total (g)	4.30	7.75	4.22	1.31	3.03	4.08	5.66	2.30
Polyunsaturated, total (g)	.39	.67	1.14	0.83	1.92	2.29	3.32	1.78
Cholesterol (mg)	69	71	80	72	79	74	78	48

[a] See footnote Table XXII.
[b] See footnote Table XXII.

| Whole milk | Skim milk | Cottage cheese | Cheddar cheese | Peanut butter | Dried Navy beans, cooked | Cereal, oatmeal, cooked | Bread[b] | | Egg, hard-boiled |
							White	Whole wheat	
1 cup (244 g)	1 cup (245 g)	1 cup (210 g)	1 oz (28 g)	¼ cup (32 g)	⅔ cup (127 g)	1 cup (234 g)	2 slices (50 g)	2 slices (50 g)	2 eggs (100 g)
150	86	217	114	189	193	145	134	123	158
8.03	8.35	26.23	7.06	8.06	10.92	6.00	4.14	4.81	12.14
8.15	.44	9.47	9.40	16.32	0.76	2.40	1.96	2.18	11.16
11.37	11.88	5.63	0.36	6.14	36.07	25.20	24.41	22.71	1.20
291	302	126	204	11	86	20	63	36	56
—	—	—	—	147	408	129	70	171	—
0.12	0.10	0.29	0.19	0.58	3.18	1.59	1.42	1.71	2.08
33	28	11	8	56	77	56	11	47	12
—	—	—	—	573	640	1369	—	—	—
228	247	277	145	119	206	178	54	130	180
370	406	177	28	223	448	132	56	88	130
120	126	850	176	151	1	1	257	318	138
0.93	0.98	0.78	0.88	0.93	1.40	1.15	0.31	0.84	1.44
3	2	Trace	0	0	0	0	Trace	Trace	0
0.09	0.09	0.04	0.01	0.03	0.25	0.26	0.24	0.18	0.08
0.34	0.34	0.34	0.11	0.03	0.05	0.05	0.16	0.11	0.30
0.21	0.22	0.26	0.02	4.29	0.64	0.30	1.88	1.92	0.06
0.77	0.81	0.45	0.12	0.35	0.38	0.47	0.22	0.37	1.72
0.10	0.10	0.14	0.02	0.13	0.13	0.05	0.02	0.09	0.12
12	13	26	5	33	—	9	18	28	48
0.87	0.93	1.31	0.23	0	0	0	0	0	1.54
307	500	342	300	0	0	38	Trace	Trace	520
5.07	0.29	5.99	5.98	3.36	0.09	0.44	0.47	0.38	3.34
2.35	0.12	2.70	2.66	7.52	0.05	0.84	0.71	0.51	4.46
0.30	0.02	0.29	0.27	4.83	0.44	0.97	0.66	0.93	1.44
133	4	31	30	0	0	0	1	0	548

9

Poultry Muscle as Food

P. B. ADDIS

Department of Food Science and Nutrition
University of Minnesota
St. Paul, Minnesota

The poultry meat industry is an increasingly important segment of the food industry. Steady increases in consumption of poultry, use of poultry meat as sausage ingredients, and the proliferation of new products have served to stimulate dynamic growth. Scientific and technological advances have paralleled those occurring in the mammalian meat industry. Indeed, the fundamental principles of muscle biology and meat processing science are the same in mammalian and avian species. Nevertheless, there are enough significant species differences in biology, postmortem physiology, processing, and storage problems to warrant a separate chapter on poultry muscle as a food.

Copyright © 1986 by Academic Press, Inc.
All rights of reproduction in any form reserved.

I. AVIAN MUSCLE BIOLOGY

In 1940, 28 weeks were required to produce a 17-lb turkey; in 1980, only 16 weeks were required. In the same period, feed conversion declined from 4.5 to 2.5. Remarkable advances in poultry production have been made, largely due to the shorter generation time of birds compared to swine, beef, and sheep. Additionally, Aves possess inherent advantages in body composition and efficiency that are due largely to their anatomic adaptations for flight. Characteristics of their muscle tissue are similarly influenced and are related to ultimate meat properties.

The following discussion of avian muscle biology reviews specific information on the avian musculoskeletal system and integrates into the discussion specific examples of relevance to ultimate meat properties.

A. Anatomic Adaptations for Flight

Domestic fowl have retained little of their ancestors' ability to fly but retain many of the anatomic characteristics that permit flight, including feathers, fused and pneumatic bones, air sacs, strong flight muscles, and homoiothermism. Avian bones are light because of their pneumatization (George and Berger, 1966). The strength of voluntary muscular activity is aided by a fused backbone. Birds contain air sacs, which are pulmonary extensions. The extreme elasticity of the air sacs aids in their ability to recoil and promote expiration (Fedde, 1976). The flight muscles constitute a major portion of the meat of domestic birds. The pectoralis (P), which depresses the wing, is the strongest muscle (George and Berger, 1966). Medial to the P is the supracoracoideus (S), or pectoralis minor, which elevates the wing.

Appreciation of the tremendous strength of the breast muscles may be obtained by considering the fact that carcass bones can be broken during electrical stunning or during the death struggle after exsanguination. Keel and wing bone breakages, attributed to excessive voltage and the resultant powerful contraction by the P, were a significant problem a decade ago (Ma and Addis, 1973).

The high yields of domestic chickens, *Gallus gallus,* and turkeys, *Meleagris gallopavo,* stem in part from evolutionary adaptations to flight. A 23-lb tom yields 4.8% skin, 2.2% separable fat, 8.3% bone, and 39% edible meat (Mountney, 1976), far superior to mammalian species. Breast, the largest portion of the carcass, commands the highest price. Figures for cooked birds range from 29% breast muscle as percentage of cooked chicken carcass to 40% in turkeys (Henrickson, 1978). Chickens and turkeys display the most intense differences in muscle color known to occur in a single animal. Breast meat approaches the whiteness of halibut; leg meat rivals pork and beef in redness. As will be seen, the significance of these variations extends far beyond the characteristic of color.

B. Red and White Muscle

Domestic birds are ideal for studies of differences between red and white muscle and an extensive literature exists. However, there is a danger in superficially classifying muscles based solely on their color and ignoring an equally important characteristic, namely, contraction speed. Paukul (1904) noted canine muscles that were classified as red but could be further subdivided into fast- and slow-contracting types. Thus, although color permits a gross estimate of percentage of muscle cells rich in myoglobin, it can be misleading in terms of speed of contraction.

1. HISTOCHEMISTRY

Histochemistry is an excellent tool for visualization of cellular details and categorization of muscle types. However, it is unable to provide quantitative biochemical information in the same manner as an enzyme assay.

Early systems of histochemical muscle fiber classification included the red–white system, which was too simplified to be accurate, and the red–intermediate–white system, which incorrectly categorized the intermediate fiber in terms of its ATPase activity. These problems were solved when Ashmore and Doerr (1971a,b) developed a system that was able to classify correctly the intermediate fiber in chickens and mammals. Two fundamental twitch types, α (fast-twitch) and β (slow-twitch) are superimposed on two color designations, R (red) and W (white), to produce three fundamental fiber types, βR, αR, and αW. α and β fibers were identified by ATPase staining, an example of which is shown in Fig. 1. Peter et al. (1972) independently developed a system exactly equivalent to that proposed by Ashmore and Doerr (1971a,b) by describing slow-twitch, oxidative (βR), fast-twitch, oxidative–glycolytic (αR), and fast-twitch, glycolytic (αW). Good correlation was also obtained by Burke et al. (1971) in a physiological study of motor units. Wiskus et al. (1976) demonstrated applicability of the βR, αR, αW system to turkeys and noted that few, if any, previous studies of poultry tenderness had included a muscle with βR fibers. Most investigators typically choose P (αW) and biceps femoris (αR) for tenderness studies. Addis (1978) concluded that the Ashmore–Doerr classification system was superior to all other systems for domestic animals and man.

The relative biological characteristics of the three fiber types (Table I) illustrate important differences and similarities. The key feature is the high adenosine triphosphatase activity in αR fibers. Previous systems classified the αR as

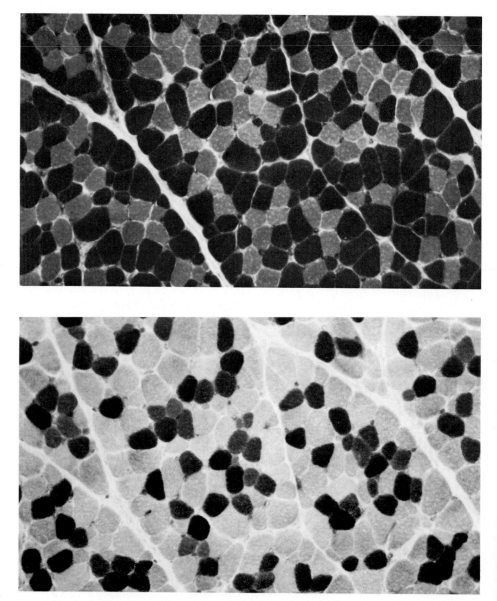

Fig. 1   Serial sections of *Sartorius* muscle from 4-week-old broiler (upper) and layer (lower) stained for ATPase after alkaline (left) and acid preincubation (right). Alkaline preincubation results in dark staining α fibers and light staining β fibers. Acid preincubation reverses the staining pattern. Weighted average fiber diameter was 17.1 μm. Line = 50 μm. Photomicrographs obtained with the help of E. D. Aberle from a study conducted in the author's laboratory, part of which was published by Aberle *et al.* (1979).

either slow, due to its red color, or intermediate, on the assumption that if its aerobic metabolism is intermediate, its myosin ATPase activity must also be intermediate between those found in red and white fibers.

2. PHYSIOLOGY

Physiological studies on muscle generally are concerned with contraction speed and its relation to mode of energy production. The overall activity of most avian muscles varies from tonic (slow, postural) to phasic (fast, locomotor) types and includes numerous mixed function muscles. The chicken latissimus dorsi (LD) complex consists of nearly purely tonic (anterior, ALD) and phasic (posterior, PLD) physiological types and has been closely studied.

The sarcoplasmic reticulum of the ALD is less well developed than the PLD, exhibiting fewer transverse tubules (Page, 1969). Rall and Schottelius (1973) noted that in isometric twitches PLD developed more tension and heat than ALD. The authors concluded that the PLD is designed for maximum power output, similar to the P; the ALD was best suited for prolonged tension maintenance.

3. BIOCHEMISTRY

Muscle is a fascinating tissue for studies of energy metabolism due to the existence of different cell types in the same muscle. Biochemical characteristics of fibers reflect not only their function (Table I) but also the nature, intensity, and duration of physical activity imposed. βR fibers are profusely supplied with mitochondria; αW fibers store much glycogen.

Table I
Relative Characteristics of the Three Fiber Types

Characteristic	βR	αR	αW
Color	Red	Red	White
Myoglobin content	High	High	Low
Capillary density	High	Intermediate	Low
Diameter	Small	Small-intermediate	Large
Number of mitochondria	High	Intermediate	Low
Succinic dehydrogenase	High	Intermediate	Low
Myofibrillar ATPase			
pH 4.00	+	−	−
pH 10.40	−	+	+
Contraction speed	Slow	Fast	Fast
Contractile action	Tonic	Phasic	Phasic
Lipase	High	High-intermediate	Low
Phosphorylase	Low	Intermediate	High

In terms of demands placed on muscles, few animals can match the capacity for locomotion by sustained muscular activity exhibited by birds. Migrant birds cover vast distances, 11,000 miles in the case of the Arctic Tern, often without stopping to eat, drink, or rest. Nonmigratory fowl such as the pheasant or grouse exhibit spectacular bursts of muscular activity but cannot sustain such performances for long periods of time. The domestic fowl flies little, if at all, but still displays powerful contractions of the breast muscles. In each case, fuels from different sources are required. An understanding of these differences is often instructive with regard to variations seen in the meat of all types of fowl used as food.

Premigratory fowl increase body weight by depositing large quantities of subcutaneous fat (McGreal and Farmer, 1956). During migration, fat and body weight decline. Fat is the chief fuel for migration, although liver and muscle glycogen are also used. George and Talesara (1961) noted that levels of succinic dehydrogenase in the P more than doubled prior to migration of *Sturnus roseus*. Studies on ducks and geese show similar results. Since contraction is a phasic process, αR fibers predominate in the P.

Undomesticated, nonmigratory fowl (pheasants, quail, partridge) do not require heavy accumulations of fat because flight distances are very short and, therefore, glycogen is the primary energy source. Lactate and CO_2 are formed. P is a mixed αR and αW, with αW the predominant type.

Domestic ducks and geese produce dark breast meat and heavy deposits of subcutaneous fat by virtue of the facts that they retain the capacity for premigratory preparation, like their wild counterparts, are fed *ad libitum,* and exercise very little. Domestic chickens and turkeys have undergone more intensive selection for muscle growth, a factor promoting leanness, are not descended from migratory fowl and, therefore, have less propensity to fatten. Breast muscles are almost entirely αW (Wiskus et al., 1976) and the fuel is glycogen. The type of metabolism in the P and S of domestic fowl explains their tendency toward severe postmortem myolactosis and concomitant pH decline, a significant factor in meat quality.

C. Growth and Development

1. EMBRYONIC DEVELOPMENT

Although early studies had considered embryonic muscle to be undifferentiated, Ashmore *et al.* (1973), by use of ATPase histochemical staining, detected two populations of fibers in chicks. Fertile eggs from normal and two dystrophic lines were examined at 2-day intervals from 10 days incubation until hatching. Samples of the "hatching" muscle (musculus complexus) were subjected to a battery of histochemical tests including acid- and alkaline-stable myofibrillar ATPase.

At 10 days, loosely organized myoblasts and myotubes are seen in areas destined to become muscles. Myoblasts and developing myotubes appear to cluster around larger myotubes. By 14 days, myoblasts are declining and myotubes are increasing in number. Large, centrally located myotubes stain darkly at pH 4.1, indicating that they are destined to become β-fibers in mature muscle. α-Fibers are formed by myoblast fusion on the surface of β-myotubes, which appear to be serving as a structural framework for the process. The findings of Ashmore et al. (1973) were consistent with electron microscopic data obtained by Kikuchi (1971), who suggested that larger myotubes, centrally located in developing fasciculi, served a directional role in development of second generation myotubes.

2. EX OVO DEVELOPMENT

After hatching, myofiber transformations occurring in chickens (Ashmore and Doerr, 1971a,b; Ashmore et al., 1973) are similar to those observed in mammals (Cassens and Cooper, 1969). At hatching, all fibers are red, therefore, either βR or αR. During a brief period after hatching, some αR fibers transform to αW. αW fibers possess superior capabilities for growth and Ashmore et al. (1972) proposed that increasing muscularity in animals, through genetic selection, is accomplished by transforming a greater percentage of αR to αW fibers. The implications for meat color are obvious.

3. INFLUENCE OF INHERITANCE

Inheritance may influence growth of muscle, and inadvertently the ultimate properties of meat, by effecting an increase in average size of fibers, an increased number of fibers, or both. The existence of populations of chickens selected for egg production or meat production allow for interesting comparisons.

Blunn and Gregory (1935) compared light White Leghorns to heavy Rhode Island Red chickens (adult weights 1561 and 3008 g, respectively). During embryonic development, hematoxylin staining revealed that Rhode Island Red chicks displayed more cells and mitotic figures but differences narrowed close to hatching. Ultimate embryo weights did not differ and were not highly correlated with cell number.

Smith (1963), in a similar heavy versus light breed comparison, concluded that hyperplasia is an embryonic event whereas hypertrophy occurs primarily after hatching. He noted more but smaller muscle cells in the heavy breed compared to the light. By 10 weeks of age, the heavy breed displayed larger fibers but since body weight was also heavier, one must question the validity of such a comparison. Measurement of mean cell diameter at equal body weight was not done but could be extrapolated, in which case cell diameter may have been slightly larger in the heavy breed. Further studies are needed on this point. Smith (1963) concluded that cell size was more important than cell number in determining muscle size.

Mizuno and Hikami (1971) noted more fibers in meat than egg chickens for some but not all muscles studied. Meat chickens displayed greater increase in DNA content after hatching than the egg breed.

From the foregoing investigations, it can be concluded that hypertrophy and, to a lesser degree, hyperplasia may be mechanisms whereby genes exert effects on muscle growth. However, none of these studies used histochemical fiber typing to determine details of muscle cell growth. Therefore, Aberle *et al.* (1979) studied *ex ovo* development in layer type and broiler type breeds but employed histochemical techniques as well as fiber diameter measurements. The results, obtained on 4-week chicks, indicated that broiler chickens had a greater proportion of β fibers and fewer α fibers in the sartorius (Fig. 1). However, of the α fibers present, more had transformed into αW fibers in broiler type chicks than in layer type chicks. The finding of significantly more β fibers in meat chickens would appear to be consistent with the findings of Ashmore *et al.* (1973), which suggested a directional role for β-myotubes in mixed fiber muscles. It is possible that genetic selection could involve selection for more β fibers to be developed in the early embryonic stages.

The differences noted by Aberle *et al.* (1979) for percentages in α fiber types were striking; broilers had 28% of their α fibers as αW compared to 11% for layer types. These findings are, therefore, consistent with the hypothesis of greater αR to αW transformation in heavily muscled animals (Ashmore *et al.*, 1972). Aberle *et al.* (1979) also noted that broilers had larger fibers than layers but did not make this comparison at an equivalent body weight (see Fig. 1).

There is potentially a great deal of information to be gained by studies on cellularity of undomesticated birds compared to their domesticated counterparts. Unfortunately, there is a paucity of information on this subject. Unpublished observations from our laboratory suggest that wild turkeys have smaller and more aerobic fibers than their domestic counterparts but comparisons at equivalent body weights were not obtained.

4. INFLUENCE OF HEREDITARY MYOPATHIES

Degenerative muscle diseases occur in domestic fowl and are either of basic scientific or practical significance. An example of the former is hereditary muscular dystrophy of the chicken, which serves as a model for the human dystrophies, including Duchenne. A very similar dystrophy in turkeys, marked by gross reduction in muscle size in breast and wing but not in red muscles and by fiber degeneration, has been reported (Harper and Parker, 1967). A different problem, although also genetically caused, is deep pectoral myopathy which has become a very significant economic problem. The S muscle only is affected and white turkeys are more frequently affected than bronze. Unlike dystrophies, which are simply transmitted as autosomal recessive traits, S muscle degenerative myopathy is a trait of variable penetrance and expressivity (Harper *et al.*, 1975), meaning that it is difficult to control by simple breeding plans. Therefore, the studies of Hollands *et al.* (1980) are of interest, since they suggested that serum creatine kinase (CK) is elevated in affected turkeys. Also relevant are studies from our laboratory on the development of a firefly luciferase method for blood CK that requires only a single drop of dried blood (Hwang *et al.*, 1977b) and has been successfully applied to genetic screening of the porcine stress syndrome (Hwang *et al.*, 1978).

5. INFLUENCE OF NUTRITION

Space does not permit a complete discussion of this topic. The reviews by Asghar and Pearson (1980) and Lin *et al.* (1980) are useful. A great deal of interest has been expressed in the production of leaner birds, based partly on consumer complaints about abdominal fat. Lin *et al.* (1980) listed the following factors that influence obesity in birds: breed, strain, genes, selection, diet, housing, environmental temperature, sex, and age. Selection for increased rate of growth or body weight results in selection for overly fat carcasses. Selection for feed efficiency is tantamount to selection for decreased body fat. Coon *et al.* (1981) demonstrated that a high-energy diet did not increase abdominal fat content of broilers over low-energy diets.

II. AVIAN MUSCLE POSTMORTEM

Biochemical events involved in the postmortem conversion of poultry muscle into meat affect several important quality attributes including tenderness, flavor, functional properties, and yield during processing. Tenderness is a key factor in consumer satisfaction. The occurrence of a small but significant number of extremely tough turkeys stimulated much research in this area. Although this issue is far from resolved, research has provided clues which have led to a reduction in the prevalence of tough turkeys.

Avian and mammalian muscles experience similar postmortem alterations but two important differences exist in poultry: the time period in which changes occur is compressed and the differences between red and white muscle are magnified. Pectoralis pH decline is faster in most turkeys than the most severely pale, soft, and exudative (PSE) porcine muscle (Ma and Addis, 1973), due in part to its nearly homogeneous αW fiber content (Wiskus *et al.*, 1976).

A. Antemortem Factors

Antemortem factors encompass environment during growth as well as immediately prior to exsanguination. Data on interaction of growing environment with stress response of birds at death are scarce but for pigs it has been shown that a comfortable growing environment results in animals that respond more acutely to heat stress immediately prior to death than animals acclimated to a warm environment (Addis *et al.*, 1967). Froning *et al.* (1978) studied antemortem (1) heat stress (42°C, 1 hr), (2) cold stress (4°C water, 20 min); (3) sodium pentabarbital anesthetization, and (4) free struggling; they noted that turkeys were tougher in groups (1) and (4) and displayed accelerated pH decline in group (4). Simpson and Goodwin (1975) noted that broiler tenderness was affected by plant, sex, and season with tenderness highest in the fall. In turkeys, breast muscle pH appears to be higher in winter, a factor that can beneficially influence both tenderness and water holding. Specific alterations in antemortem environment have been attempted so that birds would remain calm prior to exsanguination. The use of music, blue lights, and rubbing bars on the front of the breast have all had some measure of success.

Effect of age of bird on tenderness is probably mediated by collagen cross-link formation (Nakamura *et al.*, 1975). Highly cross-linked collagen displays elevated thermal stability and thereby increases meat toughness. Bawa *et al.* (1981) used papain and bromelain to attenuate age-induced toughness in spent fowl but noted off flavors associated with the use of 100 ppm tissue levels of papain.

B. Postmortem Factors

Both stress and struggling pre- and postexsanguination result in release of epinephrine and mobilization of glycogen. As a result, muscle lactic acid is produced. The causative role of lactate in toughness of breast muscle is clear (de Fremery, 1966; Khan and Nakamura, 1970). Heat stress causes lower ultimate pH values and tougher meat (Lee *et al.*, 1976). Khan (1971) demonstrated that toughness could be induced in broiler P by postmortem conditions similar to those that result in PSE meat.

In practical terms, struggling by birds on the processing line primarily influences biochemistry of P and S muscles. As discussed in Chapter 4, the completion of rigor mortis is a key event in postmortem muscle, marking the onset of aging–tenderization. The onset of rigor signals when a muscle may be chilled or frozen without potentially serious problems with toughness caused by thaw rigor or cold shortening. Several techniques have been tried as measures of degree of rigor completion or related factors important to tenderness. Aberle *et al.* (1971) studied the relationship of electrical impedance to turkey tenderness. At 45 min postmortem, P impedance at a 0.5-V potential was used to segregate birds into groups with high ($>600\ \Omega$) and low ($<450\ \Omega$) impedance. Low impedance birds experienced rapid depletion of ATP and CP and lactate accumulation. Differences in shear value were nonsignificant but favored the $>600\ \Omega$ group. Thigh muscles also were not significantly different in shear. Perhaps further studies will modify resistivity measurement to the degree that it can be used as a rapid, nondestructive tenderness measurement.

Khan and Frey (1971) developed a chemical measurement of degree of rigor for beef and chicken; an absorbance ratio of 258 nm/250 nm reflects the conversion of ATP to inosine monophosphate (IMP).

Ma et al. (1971) noted that measuring response to muscle stimulation provided a quick method for measurement of time course of rigor on excised strips of muscle (Table II). It was shown that muscles that will exhibit a long rigor time course display a long duration of electrical stimulation and a low threshold voltage. Thus, a parameter that requires only a few minutes (or even seconds if threshold is the only parameter measured) may be used to estimate another parameter of meat properties, rigor time course, which normally requires hours to determine. Ma et al. (1971) noted a wide diversity in rigor as 20 hens ranged from 25 to 391 min total time course. Postmortem aging, including sarcomere lengthening and myofibrillar fragmentation, was more rapid in fast than slow rigor carcasses. The results may also help explain the findings of Welbourne et al. (1968), who noted that shear values in turkey breast and thigh muscles were not correlated with sarcomere length measurements made at 3 hr postmortem and, therefore, probably made prior to rigor completion. Under such conditions, homogenization will trigger contraction of the sarcomeres and the measurement made will not reflect the true relationships between sarcomere length and shear value.

Electrical stunning, used to reduce the bruising and tearing that occurs as

Table II

Response to Electrical Stimulation and Time Course of Rigor Mortis of Turkey Hen P Muscle[a]

	Time Course of Rigor Mortis[a]		
Muscle parameter	Short $n = 4$	Intermediate $n = 8$	Long $n = 8$
Rigor completion (min)[b]	37 (25–48)[c]	93(84–107)	221(129–391)
Excitability threshold (volts)[d]	123(100–130)	58(25–130)	9 (2–25)
Contractility duration (sec)[e]	8 (0–30)	50 (0–102)	106 (48–156)

[a] Adapted from Ma et al. (1971).

[b] Time to complete the decrease in extensibility.

[c] Range for each subgroup in parenthesis.

[d] Minimum voltage that stimulated contraction.

[e] Time that muscle maintained ability to contract upon stimulation.

birds struggle on the exsanguination line, was curtailed in some plants due to the fact that it stimulated intense tetanic contractions, which often resulted in wing and keel bone breakage. Currently a modified stunning procedure is fairly widespread but not all plants have elected to use electrical stunning. Ma and Addis (1973) compared effects of physical restraint, struggling, and electrical stunning on P muscle properties. In general, struggling birds experienced more rapid loss of ATP, accumulation of lactate, and greater fluid losses after freezing and thawing. Significant correlations were noted between protein solubility and shear in P.

Observations that P toughening in turkeys may be similar to PSE musculature in pork, rather than sarcomere shortening, are consistent with observations made by Sayre (1968) and Khan (1968) on chicken muscle. Myosin extractability was shown to decrease rapidly during the first 4 hr of aging, coinciding with rigor, while actomysin levels remained low and constant. Actomyosin extractability began to increase 4–6 hr postmortem, whereas myosin continued to decline. The patterns appeared to reflect the time course of toughening and tenderization.

Both chronological age and postmortem aging time are important to tenderness and cooking time in turkeys. Mostert et al. (1966) noted that, for 12 to 24-week turkeys, increases in age and aging time resulted in decreased cooking time and Stadelman et al. (1966) showed decreased shear in the same birds with age and aging. In very old birds, such as spent turkey breeder hens, age-associated changes in connective tissue cause increased toughness. However, in younger birds, degree of fat finish explains why 24-week birds are more tender than 12-week birds.

Stadelman et al. (1966) concluded that a postmortem aging period is required to produce tender turkey meat. The authors noted that biceps femoris muscle tenderized more slowly than P. Not all researchers, however, agree that aging is necessary for acceptable tenderness in turkeys. Pickett and Miller (1967) studied effects of freezing in liquid nitrogen 40 min postmortem. No adverse thaw rigor effects were noted in the P and certain unspecified thigh muscles. Birds were frozen for a total of 3 weeks, which may have depleted ATP and reduced possibility of a rigor toughening upon thawing (Behnke et al., 1973a,b). It is also known that considerable postthawing tenderization is possible and can, therefore, replace the traditional prefreezing tenderization.

The early observations that tranquilized birds required a longer aging time than control birds (Dodge and Stadelman, 1960) and the finding of longer postmortem aging times required for red muscles than P (Stadelman et al., 1966) prompted Wiskus et al. (1973) to study rigor and aging–tenderization patterns in biceps femoris and peroneus longus. Rigor times for the biceps (163 min) and peroneus (241 min) suggested potential problems for cold shortening and/

or thaw rigor–induced toughening. Electrical stimulation measures significantly correlated with rigor time course in both muscles. Sarcomere length at 24 hr postmortem was significantly correlated to shear value in biceps (-0.45, $p < .05$) and peroneus (-0.68, $p < .01$) indicating that in red muscle, contraction was indeed a factor in postmortem toughening. Wiskus *et al.* (1973) noted evidence of thaw rigor toughening in gastrocnemius muscles. Birds used in this study had been sorted into rigor classifications by subjective means. To determine more accurately the effects of rigor state on response of muscle to freezing, the same authors injected Gecolate™ (glyceryl guaracolate), a muscle relaxant, into birds just prior to exsanguination. It was possible to produce pronounced differences in shear values between contralateral muscles in rigor-extended birds subjected to the following two sequences: (1) prerigor freezing, rapid thawing, and cooking; (2) aging in ice water 24 hr and cooking. Microscopic evidence of thaw rigor was obtained in samples from (1) (P. B. Addis, unpublished data).

Behnke *et al.* (1973a) demonstrated in chicken muscle that muscles frozen in a prerigor state and maintained at high subfreezing temperatures experience a rapid depletion of ATP and accumulation of lactate. A subsequent study (Behnke *et al.*, (1973b) indicated that holding prerigor muscle at $-3°C$ resulted in satisfactory tenderness. Therefore, in a commercial process, it would appear unlikely that prerigor freezing causes toughness via the classical thaw rigor mechanism. Although ATP depletion would occur at a slower rate at $-20°C$ (typical of commercial freezer storage), compared to $-3°C$, the extended storage times would probably permit significant ATP depletion in most carcasses and an attenuation of thaw rigor potential.

Thus a satisfactory explanation for the small number of very tough turkeys remains elusive. Nevertheless, from the facts reviewed in the previous paragraphs, it is possible to postulate an explanation. It is known that red muscles are affected more severely than white muscle. Also, the problem is rarely if ever noted where the traditional chilling–aging period is observed. Birds with the potential for a long rigor time course appear to be most severely affected. If thaw rigor is prevented by ATP depletion during storage, the longer-than-normal rigor patterns may delay onset of aging tenderization and, by this means, cause toughness.

In summary, the following combination of events may be involved in the development of unusual toughness in turkey meat: (1) prerigor freezing of birds with inherently long rigor time courses; (2) brief storage during marketing and distribution; and (3) rapid thawing by consumer or, worse yet, cooking from the frozen state. The third factor is perhaps the key since significant postthawing tenderization can occur and if this factor is not permitted, severe toughness could result.

III. PRESERVATION AND STORAGE

This discussion emphasizes scientific principles used in prevention of some unique problems associated with poultry spoilage. Detailed practical information is available (Mountney, 1976; Henrickson, 1978).

A. Packaging

Most chicken is marketed fresh; most turkey as frozen. Yet both industries seek greater balance. Marketing frozen chicken greatly reduces microbial spoilage. Fresh turkeys provide consumer with improved convenience and could hold the key to making turkey more acceptable as a year-round product.

Cut-up poultry is usually tray-packed and overwrapped with any one of a number of films, each with its own set of properties. The film is chosen to accomplish a given shelf life objective at a minimum cost. Spencer *et al.* (1956) demonstrated the advantages of anaerobic packaging: fryer halves packed in cellophane spoiled more quickly at $0°C$ than those in heat-shrinkage polyvinylidine. The results were significant in that aerobes account for most spoilage problems in poultry. Debevere and Voets (1973) attributed the superiority of heat-shrinkable film to thickening during shrinkage. Fick's Law dictates reduced diffusion of oxygen as the film thickness is increased.

The most important new development in packaging of poultry products, controlled atmosphere storage (CAS), provides an opportunity to decrease costs and thereby increase profits for the processor (Sander and Soo, 1978). CAS involves developing a specific gaseous environment around fresh meat within a package. Carbon dioxide (CO_2) is most frequently used. CO_2 is effective at inhibiting psychrophiles, including *Pseudomonas,* as well as *Staphylococcus aureus, S. sporogenes, Clostridium botulinum, Salmonella,* and some *Enterococcus.*

CAS has been used for poultry since about 1978. CAS eliminates ice, water spillage, and associated problems and increases payloads by 20 to 25% for chicken. CAS extends shelf life dramatically to 18 to 21 days compared with 7 to 9 days for ice pack.

CAS is accomplished by evacuating the air, backflushing with CO_2, and then sealing. The packaging material should be resistant to transmission of air, CO_2, and water. The ultimate spoilage occurring in CAS is not aerobic microbial but rather rancidity or souring.

A variation of CAS is hypobaric storage, which involves low pressure, low temperature, and high humidity and which has been shown to increase the shelf life of broilers (Restaino and Hill, 1981).

B. Preservation

1. REFRIGERATION

Crushed ice, with associated problems of water uptake and microbiological growth, remains the primary method for refrigeration of broiler carcasses. In the absence of ice, refrigerated broilers must be packaged to prevent dehydration. "CO_2 snow-pack" was compared to ice-pack by Landes (1972) who noted that iced birds tended to be more tender and produce slightly lesser drip losses. Solid CO_2 is lighter than ice, which is advantageous in lowering transportation costs.

2. FREEZING

Poultry meat is frozen in a variety of ways including liquid nitrogen, propylene glycol immersion, blast air, still air, or plate. Cryogenic methods produce a slightly superior product but general application of quick freezing methods has been limited by cost (Sebranek, 1982). In addition, much of the early literature was ambivalent concerning proposed advantages of cryogenics, a fact that prompted Streeter and Spencer (1973) to conduct a thorough comparison of quality in broilers refrigerated, cryofrozen, and conventionally frozen. Paired halves were used with left half used as a nonfrozen control and right half subjected to one of four different treatments: (1) conventional, air blast ($-29°C$, 100 fpm); (2) cryogenic immersion in liquid dichlorodifluoromethane at $-30°C$; (3) cryogenic, spray of liquid nitrogen ($-195.8°C$); and (4) cryogenic, combination of air blast and liquid nitrogen spray. Samples were stored 7 days, thawed, cooked, and evaluated. Of the freezing methods, liquid nitrogen treatment caused formation of the smallest amount of drip but values (2.2%) were higher than control (0%). Air blast–frozen halves experienced the greatest cooking losses; unfrozen, the least. Shear values and bone darkening were essentially unaffected by freezing treatments. Bone darkening is a consistent quality problem in frozen broiler carcasses. No significant differences were noted for organoleptic quality among the five treatments.

Cunningham (1975) published a comprehensive review on consumer acceptability of frozen poultry. Consumers apparently detect differences in quality that go undetected in organoleptic studies in which panel participants were "blinded" from knowing which birds were frozen. Cooking prior to freezing results in a less tender product. Temperature fluctuations during frozen storage are very damaging to meat quality.

Drip loss is an important determinant for both yield and quality of meat and results from microscopic damage to muscle caused by formation and growth of ice crystals. Size and location of crystals are key factors governing tissue damage. Fast freezing produces small intracellular crystals; slow freezing, large extracellular crystals. Gerrits and Jansonius (1975) studied recrystallization during storage in chicken semitendinosus and demonstrated that recrystallization occurs more rapidly early in storage. Cryofrozen muscle exhibited recrystallization earlier during $-10°C$ storage than muscle frozen at $-22°C$. The results are a relevant consideration for fast freezing since meat which is cryofrozen is seldom cryostored, meaning that ample opportunity exists for crystal growth during the inevitable temperature increase that occurs after cryofreezing is completed.

3. THERMAL PROCESSING

Heat processing of poultry meat, once limited to canning and home cooking, now includes an extensive commercial enterprise encompassing an impressive array of fully cooked, convenience products. The industrial processor must fulfill several product requirements, including color, flavor, tenderness, and wholesomeness, while attempting to maximize product yield.

Practical guides to home cooking of turkeys are presented by Mountney (1976) and Henrickson (1978). Briefly, the use of a thermometer in the deepest part of the thigh is recommended. End point temperature ranges are $82–85°C$.

Concerning commercial cooking of turkey products, long-time, low-temperature (LTLT) processing of the product in a heat-shrinkable package produces greater yield than high-temperature, short-time (HTST) processing, and/or dry heating. Vacuum-packaged meat is cooked in a moist environment, either a high-humidity smokehouse (operated without smoke) or a water bath. Some difficulties in uniformity of temperature may occur in water cookers. Vacuum-packaged meat should not be processed in a dry oven, as temperatures typically used will severely overcook meat. Interestingly, there have been no valid scientific studies comparing moist and dry systems for cooking the same type of product because there is no overlapping of the temperature ranges used in the two methods. Dry ovens must be operated at $122°C$ or higher to insure against microbial growth. Meat in a vacuum bag is subjected to oven temperature ranges of 60 to $93°C$. The reasons that packaged meat can be efficiently cooked at much lower temperatures than dry-cooked meat concern two factors: namely, heat capacity of the heating medium, which increases as humidity level increases, and the prevention of surface evaporation and its concomitant surface cooling of the dry cooked meat. Godsalve et al. (1977a,b) designed a controlled convection oven and with it demonstrated the importance of the surface cooling as a barrier to heat transfer into meat. In dry cooking, heat is transferred into meat as moisture is transferred out. The surface temperature cannot exceed the boiling point until the meat is cooked beyond organoleptic doneness. In contrast, the surface of meat in an overwrap is free to assume the temperature of the heating medium immediately. As a result, pouch-cooked meat may be cooked efficiently at very low environmental temperatures. In addition, evaporative cooking losses are eliminated or held inside the pouch where, in time, some resorption of fluid can occur.

Numerous factors influence product yield during heat processing. Ultimate pH and protein solubility of poultry muscle may increase during colder seasons and an improvement in cooking yield may be seen at these times. Other factors, however, are more important and are under control of the processor. End-point temperature, rate of heat transfer, and post-oven temperature increase are critical factors that must be under close control if the processor is to maximize yields. All three factors are more easily controlled by using a LTLT – moist heat system because temperature gradients are lower.

USDA regulations require an internal temperature of 71°C for uncured and 68°C for cured poultry, as a means of reducing risk from *Salmonella*. Thigh meat is often cooked to 74°C to decrease redness. To maximize yields, poultry should be cooked to the lowest possible internal temperature (71°C for un-cured) at the lowest possible rate of heat transfer to accomplish the process in a reasonable time. Hwang *et al.* (1977a) demonstrated that for cooking times as long as 14 hr that shrink loss and protein solubility were temperature-dependent and time-independent phenomena.

As a specific example of a cooking procedure used for turkey rolls that contain both light and dark meat, an oven temperature of 76°C could be used throughout the cook cycle with the meat removed at close to 73°C. Postoven temperature increases would probably elevate the final internal temperature to the target 74°C, depending on meat mass. Using higher temperature gradients will increase shrink loss by causing an increased gradient within the meat, overcooking the outer portion, and creating excessive post-oven temperature increases, overcooking the interior. Stepwise or incremental increases in oven temperature are not advantageous with regard to shrink and could create mi-crobiological problems. Incremental temperature programs are necessary, however, for emulsion products.

C. Chemical Deterioration during Storage

Chemical deterioration during storage of poultry meat primarily involves development of rancidity during refrigeration or frozen storage or damage to protein components during frozen storage. Triglyceride rancidity, warmed-over flavor, and even autoxidation of muscle lipids in nitrite-treated products such as turkey franks and bologna and turkey ham may result in reduced acceptance of poultry meat. Protein solubility and functionality losses during storage are particularly significant in poultry meat.

1. RANCIDITY

Rancidity represents perhaps the largest area of research on poultry meat products. The tendency of birds to deposit highly unsaturated fat, the new uses to which poultry meat is put (precooked frozen convenience products), and the methods for recovering poultry meat (mechanical deboning) all tend to make rancidity a great concern. Economic implications are enormous. Off flavors due to rancidity are the key source of consumer dissatisfaction or lack of enthusiastic acceptance by consumers. Rancidity is also responsible for loss of vitamins, protein functionality, and desirable color of poultry meat. Findings that acute and chronic toxicity effects may be associated with the consumption of rancid foods are also significant.

In reviews, Pearson *et al.* (1977) and Allen and Foegeding (1981) explain that it is the high content of unsaturated fatty acids in both lipid fractions, triglycerides and phospholipids, which result in increased susceptibility to rancidity in birds. Turkeys are especially susceptible because they store less vitamin E than chickens. Therefore, although feeding increased levels of vitamin E retards tissue oxidation, it is not totally effective. Vitamin E is stored effectively by adipocytes but less efficiently in muscle phospholipids, the oxidation of which leads to warmed-over flavor (WOF). In recent years, WOF rancidity has been a more significant problem than autoxidation of triglycerides.

Of the four known forms of vitamin E $(\alpha, \beta, \gamma, \delta)$, α-tocopherol possesses the most biological activity. α-Tocopherol is usually added to feed in the esterified form (α-tocopherol acetate) to protect the hydroxyl from oxidation. When consumed by animals, the vitamin is quickly de-esterified to the active form. The use of α-tocopherol acetate supplementation has been reviewed by Marusich (1978a,b, 1980). Numerous studies have demonstrated improved tissue stability in birds supplemented with vitamin E. Webb *et al.* (1972) studied BHT, ethoxyquin, and D,1-α-tocopherol acetate supplementation and noted that the latter two compounds were effective in stabilizing pre-fried frozen broiler parts.

The expense of vitamin E is such that researchers have attempted to determine optimum methods and levels of application. Uebersax *et al.* (1978) compared 100 I.U./kg feed/day with biweekly subcutaneous injections. Breast, thigh, and mechanically deboned turkey meat (MDTM), and turkey loaf, formulated with breast plus 25% MDTM, were evaluated. Tocopherol supplementation resulted in a lowering of thiobarbituric acid (TBA) values over controls. Feeding was more effective than injection of E. TBA values did increase with storage time with breast, MDTM and thigh ranking low, intermediate, and high, respectively. Hens displayed lower TBA values than toms. Vitamin E also improved stability of loaves.

Bartov and Bornstein (1977) reported that α-tocopherol acetate and ethoxyquin improved stability of both fat and muscle of broilers, whereas BHT improved fat stability only. If broilers were given a fat supplement, BHT and ethoxyquin were more able to stabilize fat than α-tocopherol acetate, but only ethoxyquin stabilized meat. Dietary ethoxyquin stabilized carcass fat logarithmically and thigh linearly. Bartov and Bornstein (1978) compared vitamin E (20 and 40 mg/kg) and ethoxyquin (125 and 300 mg/kg) in fat-supplemented diets. A linear relationship was noted between E content of fat tissue and fat stability. Duration of vitamin E feeding was correlated with fat stability.

The results have suggested that dietary components other than vitamin E may influence stability by interacting with E. Thus, feeding fat diverts some E to the increased depot fats and away from muscle. Few studies on turkeys have been conducted to determine the interaction between vitamin E levels and other dietary components. Therefore, Rethwill et al. (1981) observed muscle stability in turkeys fed a variety of fat sources and vitamin E levels. A four-by-three full factorial design was used including four diets (corn–soybean with no added fat, 8% hydrolyzed animal and vegetable fat, 8% animal fat, or 8% soybean oil) and three levels of E supplementation (none, the usual level of 10 I.U./kg, and 300 I.U./kg). Birds were processed at 14 weeks of age, divided in half and vacuum-packaged. One half was held at 3°C for 3 weeks to simulate storage conditions experienced by a turkey carcass marketed as a fresh bird. The other half was stored at $-30°C$ for 6 months, thawed, cooked and stored at 4°C for 48 hr.

Data obtained on fresh meat demonstrated the efficacy of vitamin E supplementation as control carcasses reached a TBA value of 2.84, considered very oxidized, whereas at 300 I.U./kg E, TBA reached only 0.14. Animal fat and animal–vegetable fat diets elevated TBA at the control vitamin E level. Samples thawed after 6 months storage were all low in TBA value before cooking but developed significant warmed-over flavor rancidity after cooking; source of dietary fat had virtually no influence on rancidity. Supplementation with α-tocopherol acetate at 300 I.U./kg decreased TBA values but did not completely protect against WOF. Supplementation was significantly more effective in control-diet birds than in the other three treatments.

Attempts to characterize the chemical nature of WOF have been reasonably successful. Ruenger et al. (1978) compared gas chromatograms of freshly cooked and reheated turkey and identified several different peaks. Mass spectrometry was used to identify two key peaks as heptaldehyde and n-nona-3,6-dienal, both known products of autoxidation.

Einerson and Reineccius (1977) studied inhibition of WOF in turkey by retorting at 121°C for 50 min ($F_o = 17$, $Z = 10°C$). Compared to freshly cooked or retorted meat, reheated turkey developed significant WOF. Water-soluble, antioxidant substances, determined by dialysis to have low molecular weights, were believed to be involved. The antioxidants are probably formed by the browning reaction during retort treatment. The authors hypothesized that low molecular weight reductones donate two hydrogen atoms and inhibit the free radical phase of lipid oxidation. Reductones of this type display a propensity to give up two hydrogens and form dehydroreductones.

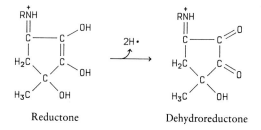

| Reductone | Dehydroreductone |

Further characterization of retort-induced antioxidants was accomplished by Einerson and Reineccius (1978) using Sephadex G-50 and G-15 column chromatography. Of the fractions collected from G-50 columns, those providing the greatest antioxidant activity appeared to be <500 MW. The authors concluded that the active fraction has a molecular weight of 2500, displays antioxidant activity as low as 200 ppm, is nonvolatile, very hygroscopic, and a strong reducing agent.

The growing popularity of ground turkey, a product highly susceptible to lipid oxidation, has stimulated searches for ways to inhibit rancidity in this product. Dawson and Schierholz (1976) noted that TBA values were increased by cooking, grinding, and storage. Younathan et al. (1980) inhibited rancidity with hot-water extracts of eggplant and peels of yellow onions, potatoes, and sweet potatoes. Spencer and Verstrate (1975) noted that dietary supplementation as well as direct addition of vitamin E protected ground turkey. Olson and Stadelman (1980) demonstrated antioxidant effects in ground turkey for BHA, BHT, and silicone combinations.

Perhaps the most labile lipid oxidation system known is MDTM. CO_2 chiller–mixers have traditionally been used to reduce the risk of microbiological problems. Noble (1976) noted that solid CO_2 chilling of mechanically deboned chicken meat (MDCM) resulted in a drop in meat pH of 0.5 units that persisted 24 hr but did not appear to influence peroxide value. In contrast, Uebersax et al. (1977) found that "CO_2 snow" resulted in rapid increases in TBA values during frozen storage. Subsequently, Uebersax et al. (1978) compared CO_2 chilling to alternative methods for MDCM and MDTM. CO_2 cooling increased TBA values in both types of meat. It therefore appears likely that CO_2 increases susceptibility to rancidity by lowering pH and thereby increasing lipid autoxidation. Alternative cooling procedures are currently being developed. Also, limiting the frozen storage period to <3 months appears to reduce rancidity in MDTM.

Lipid autoxidation can be catalyzed by numerous agents including trace metals, light, enzymes, oxygen, exogenous rancid fat, and heme pigments. Copper and iron are common catalysts that may contaminate curing salts, water used in processing, and equipment. Salts containing antioxidants or treated with chelating agents to remove prooxidants are useful for preventing rancidity in products. Classical antioxidants include BHA, BHT, and PG. Citrate is a useful secondary antioxidant or synergist. Nitrite and ascorbate are effective antioxidants, nitrite at >50 ppm, ascorbate at >500 ppm. Ascorbate is a prooxidant below 500 ppm. Phosphates and smoke compounds are effective antioxidants.

Emphasis has been placed on natural products with antioxidant properties. Protein hydrolyzates are synergistic with phenolic antioxidants. Autolyzed yeast protein provided protection equivalent to 0.02% BHA alone (Bishov and Henick, 1975). Similar results were obtained with hydrolyzed vegetable protein.

Other natural antioxidants include those isolated from bacteria (Smith and Alford, 1970), flavanoids (Pratt, 1972), mold-fermented soybean products such as dried tempeh (György et al., 1974), ferulic acid (Yagi and Ohishi, 1979), and quercetin (Hudson and Mahgoub, 1980).

Kawashima et al. (1981) studied antioxidant properties of various combinations of peptides and sugar. Peptides with branched chain amino acids, if combined with xylose, resulted in significant antioxidant activity. Unfortunately, the sugar–amino acids studied were inactive in aqueous systems.

Obviously, the processor faced with rancidity has a formidable array of chemicals with which to attack the problem. Further studies are needed to assist in prevention of the specific problem of WOF and other types of phospholipid autoxidation.

2. PROTEIN DAMAGE

Khan (1966) studied changes occurring in proteins during frozen storage of chicken P muscle. Actomyosin extractability decreased during frozen storage, as did ATPase activity and sulfhydril content.

IV. PROCESSING AND FABRICATION

A vast new array of poultry products have been developed in recent years. Although the general principles involved in the manufacture of poultry products are similar to their counterpart in "red meat," some important differences exist. The growth of further processed poultry has been phenomenal. Poultry frankfurters account for nearly 10% of the United States market for franks.

Numerous other products have been formulated from poultry meat, especially turkey, which include ham, bologna, pastrami, sausages of various types including salami, ground meat, and bacon-like strips. More basic products are also becoming popular including parts: wings, drumsticks, roasts, breasts, thighs, hindquarters, rolls (precooked roasts), roasts, steaks, and cutlets. Poultry is also used in the food service and canning industries in pies, sandwiches, soups, and casseroles. Spent hens are often used as sources of meat for further processing but, due to age-induced toughness, this can result in meat of reduced tenderness and palatability, especially juiciness. Finally, vast quantities of mechanically deboned poultry meat (MDPM) are used in fabrication of further processed turkey products. As of 1980, the annual production of mechanically deboned meat from poultry exceeded 300 million lb in the United States.

Mountney (1976) has itemized detailed steps in fabrication of many products. Therefore, this review will emphasize application of scientific principles to the production of further processed turkey products.

A. Traditional Products

Upon arrival at the plant, birds commence the following sequence (abbreviated) of processing steps: shackling, electrical stunning, bleeding, scalding ($53-58°C$ for chickens, $60-61°C$ for turkeys for 30 to 60 sec), washing, evisceration, cleaning, and chilling. USDA requires chilling to $4°C$ within 4 hr, 6 hr, and 8 hr for <4 lb, $4-8$ lb, and >8 lb carcasses, respectively. Lillard (1982) has reviewed pros and cons of chilling systems available. Carcasses proceed into marketing channels as packaged or iced whole birds or are cut up for parts or boned out for further processing. Yield is a critical factor and is influenced by genes, size, age, nutrition, preslaughter holding time and temperature, all processing steps, and chilling. Ice-water chilling results in water uptake and an improved yield (Anonymous, 1975). Phosphates in chill water tend to reduce water uptake but increase subsequent retention (Schermerhorn et al., 1963). A loss of 0.5% will occur during cutting for production of parts and meat used in further processing.

B. Cured Products

Curing involves use of sodium nitrite to fix myoglobin into a relatively stable pink pigment, nitric oxide hemochrome. Salt and spices add preservative and flavoring effects. Mountney (1976) cautions against attempting to adapt cure formulas for red meat products to poultry. Delicate turkey flavor can easily be masked by cures with a high salt and spice content. Curing is effective at retarding microbial growth and rancidity. Sodium nitrate has no influence on curing per se but may influence residual nitrite levels during long-term storage if micrococci are active.

A problem that is as common as, if not more common than, insufficient color in cured products is development of cured color in products that are not meant to be cured. Cured color can be developed by contaminated nitrite, by carbon monoxide and nitrogenous gasses from ovens and exhaust fumes, and by a combination of nitrate and metabisulfite or thiosulfite, the latter mixture usually resulting from nitrate in vegetables and the reducing agent found in vegetable fresheners. Fabrication of uncured and cured items in the same plant can lead to unexpected coloration of uncured products.

C. Smoked Products

Excellent reviews of smoking of meat have been published (Draudt, 1963; Wistreich, 1977). Smoked products are usually brined (but not always cured with nitrite) with a salt–sugar–spice mixture prior to smoking. To achieve good yield, it is important to select carcasses that are not excessively fat or lean. Smoking deposits compounds that are bactericidal, bacteriostatic and strongly antioxidative. Traditional, liquid-atomized, and liquid-regenerated processes are available (Wistreich, 1977). The type of smoke process used is an extremely critical factor affecting product quality and, therefore, should be selected with care.

Breclaw and Dawson (1970) prepared boneless light and dark meat rolls from spent fowl and studied various smoke treatments: (1) control, (2) liquid brine smoke, (3) concentrated liquid smoke, (4) smokehouse, (5) oil-base smoke, (6) oil-base smoke plus cure, and (7) dry smoke flavoring. Smoking method had a significant effect on tenderness of both types of rolls; treatments (2), (3) and (4) adversely affected tenderness. Organoleptic analysis demonstrated that all light products were favored over dark meat.

Oblinger *et al.* (1977) showed that brining and smoking increased tenderness in dark meat of spent fowl.

Smoking and brining can also be applied to broilers. Stubblefield and Hale (1977) used response surface statistics to determine procedures for smoked chicken and recommended 4–5% NaCl in cooked meats and sucrose levels not to exceed 20–30% of NaCl level. $NaNO_3$ was found to be unnecessary. A short equilibrium time (6–12 hr) was effective.

D. Cured and Smoked Products

Properly manufactured, cured–smoked turkey is extraordinarily palatable and possesses exceptional keeping qualities. Wisniewski and Mauer (1979) investigated for methods of curing–smoking compared to a conventional method of curing plus roasting. A dry-curing–smoking procedure and conventional process reduced bacterial counts but elevated TBA values. Cured, liquid-smoked turkey received the lowest sensory scores.

Turkey ham has competed well with traditional ham. Made from boneless, skinless turkey thigh meat, additives include water, salt, ham seasoning, phosphate, and curing agents. Meat and additives are mixed or tumbled until meat becomes tacky and the mixture is stuffed, smoked, and cooked 60 to 82°C for 1 hour at 40% relative humidity. Internal temperature should reach 68°C.

E. Loaves and Rolls

Technology has developed methods to bind together separate pieces of meat into a loaf, either precooked (roll) or uncooked (roast), which appears similar to intact muscle. Chicken and turkey rolls are manufactured as thermoplastic or thermoset products (Bauerman, 1979). Thermoplastic products are fabricated by cooking poultry meat, skins, and broth to 71°C, mixing in additives and binders, extruding mixture into moisture-proof casings, pasteurizing at 71°C, and chilling. Typical thermoset products are fully cooked turkey breasts and rolls. Meat may be massaged or tumbled to cause surfacing of salt-soluble protein. For rolls, additives are added and mixture is extruded and cooked. After cooking, the bag may be stripped off, juices drained, and roll repackaged; however, this is not usually necessary, since the additives used minimize shrink. In contrast, turkey breasts cannot be made with binders and do produce free juices during cooking. Visible shrink is not accepted in this product, necessitating a repackaging step.

Rolls often are sliced for sandwich meat or other uses. During slicing, meat pieces previously bound by salt-soluble protein may separate and constitute a serious product defect. A common cause of incomplete binding of meat pieces is insufficient time of tumbling and, therefore, insufficient migration of salt-soluble protein to the meat surface. Use of salt and polyphosphate during tumbling is also important. Use of frozen–thawed meat sometimes results in incomplete binding due to limitations on binding protein. Emulsion and egg albumen are effective binders. If emulsion is used, a stepwise smokehouse schedule should be selected to prevent emulsion breakdown.

Kardouche *et al.* (1978) demonstrated the efficacy of soy protein isolate in binding of cubes of P muscle, pre- and postrigor. Prerigor processing and isolate increased yield 3 and 8%, respectively.

Acton (1972) demonstrated that as meat particle size decreased, there was an increase in binding strength, decrease in cooking loss, and an increase in salt- and water-soluble protein extractability in chicken loaves made from leg and thig meat.   Total extractable protein was correlated with cooking loss (−0.90) and binding strength (0.90).

F. Comminuted Products

The rapid development of poultry sausage has been stimulated by a combination of price advantages and nutritional superiority.   Cost of production of poultry frankfurters is reduced primarily by availability of low-cost MDPM. Nutritional advantages stem from the lack of USDA moisture and fat regulations for poultry sausage that enables processors to add more water, and therefore less fat, as a method for minimizing costs.   As a result, turkey franks sell for as much as one-third less than traditional wieners.   USDA permits the use of up to 15% poultry meat in comminuted red meat products such as franks and bologna.

Poultry franks may display an inferior bite texture due to the high water/ protein ratio.   Firmness may be increased by elevating protein and/or decreasing moisture, but these alterations cause increased formulation costs.   Baker *et al.* (1970) adjusted pH of chicken from 4.6 to 8.6 and noted optimum firmness at 6.1 (normal) and a precipitous decline in firmness above and below normal pH.

Cunningham and Froning (1972) reviewed the process of poultry meat emulsification.   Although some disagreement exists, light turkey meat possesses greater ability as an emulsifier, both for capacity to emulsify fat and stability of emulsion formed, than dark meat.   Emulsion temperature appears to be inversely related to tensile strength of processed meat (Hargus *et al.,* 1970). Extended chopping decreased protein solubility.

Froning and Neelakantan (1971) showed that prerigor turkey is superior to rigor muscle as a fat emulsifier.   The authors also studied fat-emulsifying capacity and emulsion stability of actomyosin, myofibrils, actin, myosin, meat, and sarcoplasmic protein at pH 7.0 and pH 6.0, reflecting pre- and postrigor muscle, respectively.   pH 7 generally resulted in greater emulsifying capacity than pH 6 but, for sarcoplasmic protein, the reverse was true.   Myosin produced emulsions of lower stability than actin and sarcoplasmic proteins, most notably at pH 6.0.   Actin appears to be little affected by salt.   The authors concluded that capacity and stability are independently important factors that need consideration in production of comminuted turkey products.

Studies by Galluzo and Regenstein (1978a,b,c) have provided theoretical information on poultry meat emulsions.   Chicken breast myosin was used in a "timed" emulsification system that, coupled with results obtained by SDS electrophoresis, permitted detailed evaluation of functionality of various protein fractions.   It was noted that emulsifying capacity decreased as pH approached the pI of myosin. For pH $>$ pI, myosin assumes a net negative charge and increases in hydration and fat emulsification are evident. As emulsification proceeds, all four myosin chains appear to participate on an equal basis, indicating that myosin behaves as an intact protein at the fat globule – water interface. Myosin exceeded actin in emulsifying capacity and also differed from actin in characteristics of the emulsion it formed. Myosin was more rapidly removed from solution and incorporated into the interface than actin and formed finer-grained and thicker emulsions than actin. Actomyosin behaved much like myosin alone as an emulsifier. Under conditions such that actomyosin was dissociated by ATP, actin remained in the aqueous phase during emulsification, whereas mysosin was preferentially used. Similar results were obtained on isolated myofibrils. It was also noted that tropomyosin tended to remain in the aqueous phase, whereas troponin was used in interfacial membrane formation.

Divalent cations have been shown by Whiting and Richards (1978) to influence emulsification in chicken. In a 3% NaCl breast extract, 10 mM of $CuCl_2$ decreased, $FeCl_2$ slightly increased, and $ZnCl_2$ greatly increased emulsifying capacity. $CaCl_2$ had no effect.

In the manufacture of poultry franks two basic ingredients, MDPM and skin, are used. The naturally occurring proportion of skin on the carcass constitutes the USDA limit. Chickens, being smaller than turkeys and possessing a greater surface-area-to-volume ratio, provide greater quantities of fat to chicken franks than do turkeys. As a result, turkey franks are leaner.

A typical turkey frank formulation includes a mixture of 88% MDTM and 12% skin in the meat block and a fat content of 18 to 20%. Many processors will preblend MDTM. The formulation also includes salt, cure, water, and corn syrup solids. Alkaline phosphates and caseinates are permitted in poultry sausage, in contrast to red meat sausage. MDPM has a high water content, necessitating addition of little or no added water at mixing. Salt, cure, and water are added to meat and mixed 2 – 3 min or preblended 24 hr. Spices, corn syrup solids, and additives are added and meat mixed 2 more min. If rework is to be added, it should be limited to 5% if possible. The mixture is then emulsified, extruded, and smoked.

As is true in red meat emulsions, overchopping can result in emulsion breakdown from fat coalescence. Poultry meat emulsions may be even more sensitive than red meat emulsions due to the fact that MDPM may be in a highly comminuted state depending upon the type of deboner used. Some mechanical deboners produce coarsely comminuted meat; others, finely comminuted. Thus, the sausage maker must be familiar with the types of MDPM being supplied at any one time and adjust amount of emulsification according to fineness of incoming MDPM. In this way, overchopping and underchopping may both be avoided. Chopping temperature ranges of 10 to 13°C are maximum if preblending is used.

Stable emulsions of very low fat content are easily made with MDPM. Unfortunately, a rubbery texture precludes extra-lean wieners from general acceptance by consumers.

G. Fermented Sausage

Little published information is available on fermented poultry sausage but Keller and Acton (1974) investigated production of fermented, semi-dry turkey sausage. After grinding, skin and meat were mixed for 2 min; curing agents, seasoning, dextrose, and salt were added and the blend mixed for 8 min. The starter culture, *Pediococcus cerevisiae*, was added and the blend mixed for 2 min, after which the initial temperature, 5°C, climbed to 12°C. The mix was stuffed into a 52-mm fibrous casing and hung in a fermentation chamber at 38°C and 95% relative humidity. The authors noted that frozen starter culture required only 24 hr to ferment turkey sausage; freeze-dried culture required 48 hours. Following fermentation, sausages were heated at 66°C for 30 min and ultimately cooked t 71°C. At this stage, the product was sprayed with a cold water rinse and placed in a 7–8°C drying room with 10–15 air changes/hr. The relative humidity was 88–92%. Lyophilized cultures required sausage to be dried 10 days; frozen, 12 days.

The procedure used by Keller and Acton (1974) differs somewhat from those used for dry and semi-dry sausages in the red meat industry. Traditional processes for semi-dry and dry sausages specify that these are mix-type sausages, not comminuted in the usual sense, so that a definite particle size is obtained. Salt is added at the end of the chop and temperatures are kept low, near freezing, so that extraction of salt-soluble protein will not occur. Of course, preblending with salt would never be practiced. As with franks, the producer of fermented poultry sausage should consider the fineness of MDPM available.

Results obtained by Keller and Acton (1974) demonstrated that lyophilized culture exhibited a longer lag phase than frozen culture, necessitating a 48-hr process for lyophilized cultures compared to 24 hr for frozen ones. pH declined from 6.0 to 4.6 while lactate was increasing to 0.8%. The pI of meat in this study was found by the authors to be 5.0. At 4.6, many proteins have assumed a positive charge and have established a degree of hydration by electrostatic–dipolar interactions between positive proteins and dipolar water molecules. Results demonstrated that a 10–12-day drying period was required to attain the moisture/protein ratio consistent with a semi-dry sausage classification.

H. Poultry Meat as a Product Ingredient

Trimmings from the boning process have long been a source of poultry meat for products. However, hand deboning muscles such as the P and S for roasts, rolls, and other products leaves a considerable amount of meat on the bones. Mechanical deboning is much more effective at removal of protein from bone than hand deboning. Mechanical deboning as a process began in the fish industry. It was first used in poultry in 1960 and later adapted for mammalian species. An excellent review of MDPM has been published by Froning (1976) who lists the following products in which MDPM is used: bologna, pickle bologna, pimento bologna, salami, salami sausage, turkey treats, frankfurters, combination rolls, ground raw patties, cutlets, and diced turkeys.

MDPM has relatively good emulsion-forming ability (stability) up to 12.8°C but not higher (Froning, 1976). Compared to hand-deboned meat, MDPM has similar water- and fat-binding properties. However, as skin content increases, fat emulsion capacity and stability both decline.

Froning and Johnson (1973) developed a method of improving MDPM by centrifugation. Fowl meat was deboned either immediately after chilling or following 2 months of frozen storage at −29°C. High-speed centrifugation improved protein content, decreased fat content, improved emulsifying capacity and stability, and decreased TBA values. The effects on TBA values are suggestive that heme pigments from bone marrow are a source of the instability problems associated with MDPM.

Lyon et al. (1981) studied the functionality of MDCM and isolated soy protein (ISP) in the production of poultry rolls. ISP significantly reduced shrink. Rolls containing ISP and MDCM (10 or 20%) exhibited less shrink than those with ISP but without MDCM. Color intensity increased as percentage of MDCM increased.

Lipid oxidation has been a key quality consideration in MDPM. Oxidative deterioration of MDPM may result in several problems in the finished product including rancidity, color loss, reduced protein functionality, and reduced nutritional value. Rancidity increases with storage time so recent efforts to market MDPM more rapidly have reduced the rancidity problem considerably. Studies by Moerck and Ball (1974) indicated that for MDCM, muscle lipids, including fatty acids with as many as six double bonds, were the major contributors to the development of rancidity during 4°C storage.

It is likely that the high heme iron content, contributed by marrow, stimulates oxidation. MDPM has the characteristic of a finely ground meat or emulsion and, as such, it does not resemble muscle or anything with an intact texture. Several areas of research have developed to overcome this problem. Use of extruders, soy protein and heat- and freeze-texturization are some important developments, according to Mauer (1979).

Although much research has been conducted on MDPM, fewer studies have been completed on the residue from the mechanical deboning process. Actually, a considerable quantity of high quality protein is to be found in bone residue. Unfortunately, most of this product is rendered. Studies have, however, been conducted on the recovery and utilization of protein from MDCM residue. Lawrence (1981) studied the recovery of alkaline-extracted protein from mechanically deboned poultry residues. Optimal recovery was obtained by a combination of 22°C, pH 10.5, and 30 min mixing time. No lysinoalanine was formed. Lawrence (1981) also demonstrated the freeze-texturization of recovered protein.

V. UNIQUE ASPECTS OF POULTRY MICROBIOLOGY

In considering the problems of poultry meat microbiology, it is important to differentiate between fresh and further processed products and whether the organisms are a public health or meat spoilage problem. Most bacteria found on the poultry carcass are harmless to man; a proportion, but still fewer than one in 100, are involved in spoilage (Barnes, 1976). Food poisoning bacteria in poultry products are primarily *Salmonella* sp. and *Clostridium perfringens (welchii)* whereas *Pseudomonas* sp. are responsible for much of the spoilage. Barnes (1976) presented a review of the spoilage organisms of fresh poultry meat, including a number that are able to grow at refrigerator temperatures. Bailey *et al.* (1979) investigated shelf life of broilers as influenced by packaging system used. Vacuum-packed, CO_2-packed (65 and 20% CO_2) and ice-packed broilers are compared. CO_2 at 65% increased shelf life by 1 day over 20% CO_2 and 5 days over ice pack. CO_2 packaging resulted in spoilage by *Lactobacilli;* ice-packed birds were spoiled by *Pseudomonas* sp.

Zottola and Busta (1971) surveyed a variety of further processed turkey products for microbiological quality. Of 35 raw products examined, all were found to contain coliforms including 19 *Escherichia coli*, 25 *Staphylococcus aureus,* and 7 *Clostridium perfringens. Salmonellae* were found in only three samples. Of 38 cooked samples, 16 contained coliforms (6 *C. perfringens*) and only one *S. aureus.*

VI. SENSORY QUALITIES

It is believed that the protein fraction of meat is responsible for the "general meat flavor." The lipid fraction, including carbonyls, forms the "species" flavor. Sulfur compounds are involved in general meat flavor. In spite of intense efforts, however, our understanding of meat flavor, including poultry, is far from clear.

Off flavor problems, on the other hand, have yielded some information of practical significance. Reineccius (1979) presented a useful review of off flavors in meat and fish. "Fishy" flavor can result from diets high in polyunsaturated fatty acids such as fish oil or linseed oil. Vitamin E in the diet reduces the severity of "fishy" flavor, suggesting that an oxidative mechanism is involved. The other major off flavors include WOF in cooked meats and rancid flavor of both triglycerides and MDCM, all three of which respond at least partly to treatment by antioxidants.

VII. NUTRITIONAL ASPECTS OF POULTRY MEAT PRODUCTS

The dramatic increase in consumption of poultry meat and poultry meat products has been stimulated in part by the public's awareness of nutritional advantages of these products over beef, pork, and lamb. Most of the interest centers on the question of fat and cholesterol and, unfortunately, some confusion exists. Undoubtedly, poultry meat has three specific advantages, namely, a more unsaturated fat, a lower fat content (especially in franks and bologna) and fewer calories per serving. The confusion occurs when the consumer, and even some people in the poultry production and processing industry, assume that a leaner product may be equated with a low cholesterol product. In fact, turkey meat is slightly higher than beef in cholesterol content precisely because it is leaner than beef. More cholesterol is found in the lean portion of meat than the fat portion.

Poultry muscle is considerably lower in iron than beef and lamb. Pork exceeds poultry meat in thiamine content.

A key reference is Demby and Cunningham (1980), a review of the literature on factors affecting composition of chicken meat.

VIII. SUMMARY AND FUTURE RESEARCH NEEDS

The purpose of this chapter has been to attempt to recapitulate previous chapters and demonstrate the applicability of this information to processing of poultry meat products. Certainly, those researchers who have worked on the same types of problems in different species would attest to the synergism that is experienced in interspecies studies.

In terms of future research needs, most discussion is limited to processing. Although much has been learned about rancidity, the problem is still a significant one. Further investigations are needed on inhibiting WOF by nutritional methods, additives, or processing procedures. Nutritional research should be conducted on the potential antioxidant capabilities of β-carotene, riboflavin, selenium, methionine, and cysteine to name a few. Newer sources of antioxidants that could be added to meat directly should be sought. Continued studies on processing procedures such as vacuum mixing, chopping, and tumbling are needed.

Poultry white meat is in great demand. Therefore, more product development work is needed to utilize the excess of available dark meat. Studies should be conducted to determine more accurately the basis for the preference of most consumers for a light meat over dark meat.

Studies should be conducted on the development of freeze-dried, prerigor poultry meat (see Chapter 6) for use as a sausage ingredient or in precooked ground or comminuted product. Meat from turkey breeder hens would appear to be ideally suited for use as freeze-dried, prerigor meat.

Space does not permit a thorough discussion of all areas of research that could lead to important advances in the rapidly evolving field of poultry products technology.

REFERENCES

Aberle, E. D., Stadelman, W. J., Zachariah, G. L., and Haugh, C. G. (1971). *Poult. Sci.* 50, 743.
Aberle, E. D., Addis, P. B., and Shoffner, R. N. (1979). *Poult. Sci.* 58, 1210.
Acton, J. C. (1972). *J. Food Sci.* 37, 240.
Addis, P. B. (1978). *In* "Malignant Hyperthermia" (J. A. Aldrete and B. A. Britt, eds.), p. 109. Grune & Stratton, New York.
Addis, P. B., Johnson, H. R., Thomas, N. W., and Judge, M. D. (1967). *J. Anim. Sci.* 26, 466.
Allen, C. E., and Foegeding, E. A. (1981). *Food Technol. (Chicago)* 35, 253.
Anonymous (1975). *North Cent. Reg. Res. Publ.* No. 226; *S.D., Agric. Exp. Stn. [Bull.]* B630.
Asghar, A., and Pearson, A. M. (1980). *Adv. Food Res.* 26, 53.
Ashmore, C. R., and Doerr, L. (1971a). *Exp. Neurol.* 30, 431.
Ashmore, C. R., and Doerr, L. (1971b). *Exp. Neurol.* 31, 408.
Ashmore, C. R., Tompkins, G., and Doerr, L. (1972). *J. Anim. Sci.* 34, 37.
Ashmore, C. R., Addis, P. B., Doerr, L., and Stokes, H. (1973). *J. Histochem. Cytochem.* 21, 266.
Bailey, J. S., Reagan, J. O., Carpenter, J. A., Schuler, G. A., and Thomson, J. E. (1979). *J. Food Prot.* 42, 218.
Baker, R. C., Darfler, J., and Vadehra, D. V. (1970). *J. Food Sci.* 35, 693.
Barnes, E. M. (1976). *J. Sci. Food Agric.* 27, 777.
Bartov, I., and Bornstein, S. (1977). *Br. Poult. Sci.* 18, 47.
Bartov, I., and Bornstein, S. (1978). *Br. Poult. Sci.* 19, 129.
Bauerman, J. F. (1979). *Food Technol. (Chicago)* 33, 42.
Bawa, A. S., Orr, H. L., and Usborne, W. R. (1981). *Poult. Sci.* 60, 744.
Behnke, J. R., Fennema, O., and Cassens, R. G. (1973a). *J. Agric. Food Chem.* 21, 5.
Behnke, J. R., Fennema, O., and Haller, R. W. (1973b). *J. Food Sci.* 38, 275.
Bishov, S. J., and Henick, A. S. (1975). *J. Food Sci.* 40, 345.

Blunn, C. T., and Gregory, P. W. (1935). *J. Exp. Zool.* **70**, 397.

Breclaw, E. W., and Dawson, L. E. (1970). *J. Food Sci.* **35**, 379.

Burke, R. E., Levine, D. N., Zajac, F. E., III, Tsairis, P., and Engel, W. K. (1971). *Science* **174**, 709.

Cassens, R. G., and Cooper, C. C. (1969). *Adv. Food Res.* **19**, 1.

Coon, C. N., Becker, W. A., and Spencer, J. V. (1981). *Poult. Sci.* **60**, 1264.

Cunningham, F. E. (1975). *World's Poult. Sci. J.* **31**, 136.

Cunningham, F. E., and Froning, G. W. (1972). *Poult. Sci.* **51**, 1714.

Dawson, L. E., and Schierholz, K. (1976). *Poult. Sci.* **55**, 618.

Debevere, J. M., and Voets, J. P. (1973). *Br. Poult. Sci.* **14**, 17.

de Fremery, D. (1966). *In* "The Physiology and Biochemistry of Muscle as a Food, 1" (E. J. Briskey, R. G. Cassens, and J. C. Trautman, eds.), p. 205. Univ. of Wisconsin Press, Madison.

Demby, J. H., and Cunningham, F. E. (1980). *World's Poult. Sci. J.* **36**, 25.

Dodge, J. W., and Stadelman, W. J. (1960). *Poult. Sci.* **39**, 672.

Draudt, H. N. (1963). *Food Technol. (Chicago)* **17**, 1557.

Einerson, M. A., and Reineccius, G. A. (1977). *J. Food Process. Preserv.* **1**, 279.

Einerson, M. A., and Reineccius, G. A. (1978). *J. Food Process. Preserv* **2**, 1.

Fedde, M. R. (1976). *In* "Avian Physiology" (P. D. Sturkie, ed.), 3rd ed., p. 123. Springer-Verlag, Berlin and New York.

Froning, G. W. (1976). *Food Technol. (Chicago)* **30**, 50.

Froning, G. W., and Johnson, F. (1973). *J. Food Sci.* **38**, 279.

Froning, G. W., and Neelakantan, S. (1971). *Poult. Sci.* **50**, 839.

Froning, G. W., Babji, A. S., and Mather, F. B. (1978). *Poult. Sci.* **57**, 630.

Galluzzo, S. J., and Regenstein, J. M. (1978a). *J. Food Sci.* **43**, 1757.

Galluzzo, S. J., and Regenstein, J. M. (1978b). *J. Food Sci.* **43**, 1761.

Galluzzo, S. J., and Regenstein, J. M. (1978c). *J. Food Sci.* **43**, 1766.

George, J. C., and Berger, A. J. (1966). "Avian Myology." Academic Press, New York.

George, J. C., and Talesara, C. L. (1961). *Comp. Biochem. Physiol.* **3**, 267.

Gerrits, A. R., and Jansonius, F. A. T. (1975). *Qual. Poult. Meat, Proc. Eur. Symp., 2nd, 1975* pp. 16(1)–16(9).

Godsalve, E. W., Davis, E. A., Gordon, J., and Davis, H. T. (1977a). *J. Food Sci.* **42**, 1038.

Godsalve, E. W., Davis, E. A., and Gordon, J. (1977b). *J. Food Sci.* **42**, 1325.

György, P., Murata, K., and Sugimoto, Y. (1974). *J. Am. Oil Chem. Soc.* **51**, 377.

Hargus, G. L., Froning, G. W., Mebus, C. A., Neelakantan, S., and Hartung, T. E. (1970). *J. Food Sci.* **49**, 1625.

Harper, J. A., and Parker, J. E. (1967). *J. Hered.* **58**, 189.

Harper, J. A., Bernier, P. E., Helfer, D. H., and Schmitz, J. A. (1975). *J. Hered.* **66**, 362.

Henrickson, R. L. (1978). "Meat, Poultry and Seafood Technology." Prentice-Hall, Englewood Cliffs, New Jersey.

Hollands, K. G., Grunder, A. A., Williams, C. J., and Gavora, J. S. (1980). *Br. Poult. Sci.* **21**, 161.

Hudson, B. J. F., and Mahgoub, S. E. O. (1980). *J. Sci. Food Agric.* **31**, 646.

Hwang, P. T., Addis, P. B., Rosenau, J. R., Nelson, D. A., and Thompson, D. R. (1977a). *J. Food Sci.* **42**, 590.

Hwang, P. T., Addis, P. B., Rempel, W. E., and Antonik, A. (1977b). *J. Anim. Sci.* **45**, 1015.

Hwang, P. T., McGrath, C. J., Addis, P. B., Rempel, W. E., Thompson, E. W., and Antonik, A. (1978). *J. Anim. Sci.* **47**, 630.

Kardouche, M. B., Pratt, D. E., and Stadelman, W. J. (1978). *J. Food Sci.* **43**, 882.

Kawashima, K., Itoh, H., and Chibata, I. (1981). *Agric. Biol. Chem.* **45**, 987.

Keller, J. E., and Acton, J. C. (1974). *J. Food Sci.* **39**, 836.

Khan, A. W. (1966). *Cryobiology* **3**, 224.

Khan, A. W. (1968). *J. Inst. Can. Sci. Technol. Aliment.* **1**, 86.

Khan, A. W. (1971). *J. Food Sci.* **36**, 120.
Khan, A. W., and Frey, A. R. (1971). *Can. Inst. Food Technol. J.* **4**, 139.
Khan, A. W., and Nakamura, R. (1970). *J. Food Sci.* **35**, 266.
Kikuchi, T. (1971). *Tohoku J. Agric. Res.* **22**, 1.
Landes, D. R. (1972). *Poult. Sci.* **51**, 1765.
Lawrence, R. A. (1981). M.S. Thesis, University of Alberta, Alberta, Canada.
Lee, Y. B., Hargus, G. L., Hagberg, E. C., and Forsythe, R. H. (1976). *J. Food Sci.* **41**, 1466.
Lillard, H. S. (1982). *Food Technol. (Chicago)* **36**, 58.
Lin, C. Y., Friars, G. W., and Moran, E. T. (1980). *World's Poult. Sci. J.* **36**, 103.
Lyon, C. E., Lyon, B. G., and Hudspeth, J. P. (1981). *Poult. Sci.* **60**, 584.
Ma, R.T.-I., and Addis, P. B. (1973). *J. Food Sci.* **38**, 995.
Ma, R. T.-I., Addis, P. B., and Allen, C. E. (1971). *J. Food Sci.* **36**, 125.
McGreal, R. D., and Farner, D. S. (1956). *Northwest Sci.* **30**, 12.
Marusich, W. L. (1978a). *Feedstuffs* **50**, 20.
Marusich, W. L. (1978b). *Feedstuffs* **50**(51), 25.
Marusich, W. L. (1980). *In* "Vitamin E — A Comprehensive Treatise" (L. J. Machlin, ed.), p. 445. Dekker, New York.
Maurer, A. J. (1979). *Food Technol. (Chicago)* **33**, 48.
Mizuno, T., and Hikami, Y. (1971). *Jpn. J. Zootech. Sci.* **42**, 526.
Moerck, K. E., and Ball, H. R., Jr. (1974). *J. Food Sci.* **39**, 876.
Mostert, G. C., Harrington, R. B., and Stadelman, W. J. (1966). *Poult. Sci.* **45**, 359.
Mountney, G. J. (1976). "Poultry Products Technology." Avi Publ. Co., Westport, Connecticut.
Nakamura, R., Sekoguchi, S., and Sato, Y. (1975). *Poult. Sci.* **54**, 1604.
Noble, A. C. (1976). *J. Inst. Can. Sci. Technol. Aliment.* **9**, 105.
Oblinger, J. L., Janky, D. M., and Koburger, J. A. (1977). *J. Food Sci.* **42**, 1347.
Olson, V. M., and Stadelman, W. J. (1980). *Poult. Sci.* **59**, 2733.
Page, S. G. (1969). *J. Physiol (London)* **205**, 131.
Paukul, E. (1904). *Arch. Physiol.* **5**, 100–200.
Pearson, A. M., Love, J. D., and Shorland, F. B. (1977). *Adv. Food Res.* **23**, 1.
Peter, J. B., Barnard, R. J., Edgerton, V. R., Gillespie, C. A., and Stempel, K. E. (1972). *Biochemistry* **11**, 2627.
Pickett, L. D., and Miller, B. F. (1967). *Poult. Sci.* **46**, 1148.
Pratt, D. E. (1972). *J. Food Sci.* **37**, 322.
Rall, J. A., and Schottelius, B. A. (1973). *J. Gen. Physiol.* **62**, 303.
Reineccius, G. A. (1979). *J. Food Sci.* **44**, 12.
Restaino, L., and Hill, W. M. (1981). *J. Food Prot.* **44**, 535.
Rethwill, C. E., Bruin, T. K., Waibel, P. E., and Addis, P. B. (1981). *Poult. Sci.* **60**, 2466.
Rueanger, E. L., Reineccius, G. A., and Thompson, D. R. (1978). *J. Food Sci.* **43**, 1198.
Sander, E. H., and Soo, H. M. (1978). *J. Food Sci.* **43**, 1519.
Sayre, R. N. (1968). *J. Food Sci.* **33**, 609.
Schermerhorn, E. P., Adams, R. L., and Stadelman, W. J. (1963). *Poult. Sci.* **50**, 1456.
Sebranek, J. G. (1982). *Food Technol. (Chicago)* **36**, 120.
Simpson, M. D., and Goodwin, T. L. (1975). *Poult. Sci.* **54**, 275.
Smith, J. H. (1963). *Poult. Sci.* **42**, 283.
Smith, J. L., and Alford, J. A. (1970). *Lipids* **5**, 795.
Spencer, J. V., and Verstrate, J. A. (1975). *Poult. Sci.* **54**, 1820.
Spencer, J. V., Ecklund, M. W., Sauter, E. H., and Hard, M. M. (1956). *Poult. Sci.* **35**, 1173.
Stadelman, W. J., Mostert, G. C., and Harrington, R. B. (1966). *Food Technol. (Chicago)* **20**, 110.
Streeter, E. M., and Spencer, J. V. (1973). *Poult. Sci.* **52**, 317.
Stubblefield, J. D., and Hale, K. K., Jr. (1977). *J. Food Sci.* **42**, 1349.

Uebersax, K. L., Dawson, L. E., and Uebersax, M. A. (1977). *Poult. Sci.* **56**, 707.
Uebersax, M. A., Dawson, L. E., and Uebersax, K. L. (1978). *Poult. Sci.* **57**, 924.
Webb, J. E., Brunson, C. C., and Yates, J. D. (1972). *Poult. Sci.* **51**, 1601.
Welbourne, J. L., Harrington, R. B., and Stadelman, W. J. (1968). *J. Food Sci.* **33**, 450.
Whiting, R. C., and Richards, J. F. (1978). *J. Food Sci.* **43**, 312.
Wiskus, K. J., Addis, P. B., and Ma, R. T.-I. (1973). *J. Food Sci.* **38**, 313.
Wiskus, K. J., Addis, P. B., and Ma, R.T.-I. (1976). *Poult. Sci.* **55**, 562.
Wisniewski, G. D., and Mauer, A. J. (1979). *J. Food Sci.* **44**, 131.
Wistreich, H. (1977). *Proc. Meat Ind. Res. Conf.* p. 37.
Yagi, K., and Ohishi, N. (1979). *J. Nutr. Sci. Vitaminol.* **25**, 127.
Younathan, M. L., Marjan, Z. M., and Arshad, F. B. (1980). *J. Food Sci.* **45**, 274.
Zottola, E. A., and Busta, F. F. (1971). *J. Food Sci.* **36**, 1001.

10

Fish Muscle as Food

W. D. BROWN

Institute of Marine Resources
Department of Food Science and Technology
University of California, Davis
Davis, California

Copyright © 1986 by Academic Press, Inc.
All rights of reproduction in any form reserved.

I. INTRODUCTION

To a biologist, a fish may be a cold-blooded aquatic vertebrate animal, having fins and gills.   However, to muscle food scientists and consumers the term fish takes on quite a different and much broader connotation.   It may well be applied to any sort of aquatic animal used for food, ranging from a bivalve mollusk residing in an estuarine inlet (an oyster) to a free-swimming, broad-ranging finfish, such as a salmon.   Certainly the term seafood, if not fish, is used to encompass a variety of animal muscle food resources far in numerical excess of the few land-based creatures discussed elsewhere in this volume.

As an example of this diversity, the fish family Scombridae alone is composed of some 15 genera including about 48 species of epipelagic oceanic fishes (Collette, 1978).   This group includes the mackerels, bonitos, and tunas commonly used for human food.   From the food scientist's viewpoint, it is an important group of fish because of economic considerations based on the value of canned products and because of the fact that these are the fish primarily involved in scombroid poisoning, a food intoxication problem discussed in Section IV,B,3,a.

A shopper buying canned tuna fish will be concerned with whether the product is solid pack or chunk style, or perhaps with whether the tuna is prepared in oil or water, rather than with biological classification of the fish. Nonetheless, if the product is called "white," it contains albacore tuna *(Thunnus alalunga)* whereas if it bears the more common "light" designation, it is probably yellowfin *(T. albacares)* or skipjack *(Katsuwonus pelamis)* tuna. While these fish share in common voracious appetites, rapid swimming speeds, and a migratory nature, they differ significantly in size, and capture and handling methods may vary greatly (Joseph *et al.*, 1980).

Other fish, such as flounder and sole, are content to lie on the bottom of the ocean and their activities, and presumably their metabolic rates, differ markedly from those of tuna.   Still other fish, such as members of the important cod family, fall between these two extremes.

The term shellfish commonly is used to include groups of animals that, from a biological standpoint, are themselves widely divergent.   Frequently, such a designation includes both crustacea and mollusks, including shrimp, crab, lobster, oyster, mussels, clams, and others.   However even this type of classification is subject to exception, since, for example, squid are indeed members of the phylum Mollusca, but have no shell.   Crabs falling into this category range from the soft-shell varieties found in the Gulf and the eastern United States, to the Dungeness variety of the West Coast and their distant relative, the monstrous King crab from Alaska.

Abalone, a mollusk highly prized in California, Japan, and elsewhere, comes in many forms: so-called reds and pinks, together with green, white, and black members of this group.   Abalone found throughout the world are called by dozens of other names; in Greece, for example, these delicacies are known as Venus' ears while Australians use the far less romantic designation of mutton fish (Cox, 1962).

Even the popular and appealing lobster offers itself in several guises. The so-called clawed lobsters, which include the American (or Maine lobster) together with the European and Norwegian versions (scampi), are only one among several major families of decapod crustacea constituting "lobsters." Another is that of the spiny lobster, which in some versions is called the Pacific, and which in others is often sold as rock lobster, or lobster tails. Slipper lobsters are yet another group and, while of excellent quality, are of limited economic significance (Phillips et al., 1980).

Because of the diversity of aquatic animals used as food by man, the reader is advised to exercise caution when confronted with any attempt on the part of a researcher or writer to group together scientifically considerations of the properties of muscles from "fish." Nevertheless, seafoods tend to be grouped together by the food distributor if not by the food scientist. Consequently, some effort will be made in this chapter to deal with such generalizations as are possible, with the full realization that limitations are real.

Fortunately, muscle from aquatic animals does share in common with muscle from terrestrial animals many properties, and general features of muscle structure, function, and conversion to food have been treated in detail by others in this book. It would be inappropriate in this chapter to duplicate the earlier coverage. Rather, an attempt will be made here to emphasize those aspects of fish muscle as food that are unique or unusual, or that take on some degree of significance greater than they might in other animal tissues.

II. COMPARATIVE BIOCHEMISTRY OF FISH MUSCLE

Fish clearly differ significantly from land animals in that they live in an aqueous environment, with the majority of food fish being marine in origin and therefore requiring some mechanism for osmoregulation. The need for energy for heat in poikilothermic animals obviously is less than that of homoiotherms; however, some aquatic animals can go for astonishingly long periods of time without eating (in some cases during periods of substantial activity) and therefore must have means of maintaining metabolic energy reserves. Some, but not all, have swim bladders to help with buoyancy, and many may alter body composition to make flotation a more likely possibility. Thus, there are many necessary adaptations undergone by these animals. In spite of this, many aspects of their biochemistry, including features of muscle structure and function, seem to parallel quite closely those seen in land animals.

A. Myofibrillar Proteins

J. J. Connell and associates at the Torry Research Station have done extensive work with cod muscle proteins, especially myosin (Connell and Howgate, 1959; Connell, 1960, 1963; Connell and Olcott, 1961; Mackie and Connell, 1964). While their findings are obviously of interest from a comparative standpoint, they have more general applicability in terms of myosin biochemistry as well. In general, the properties of cod myosin resemble those of myosins in mammalian muscle, such as the rabbit model. The fish myosin, however, exhibits considerable instability and tends to aggregate readily.

Skipjack tuna myosin was also found to have properties similar to those of rabbit myosin, including amino acid composition, ATPase activity, pH activity maxima, and sedimentation properties. Tuna myosin was more stable than cod, but less stable than rabbit myosin (Chung et al., 1967).

Studies on myosin of spiny lobster have, in general, indicated that this protein resembles those of other animals (Tomioka et al., 1974, 1975). Carp myosin was one of the earliest of the myosins studied (Hamoir et al., 1960); it has been found to have properties much in common with those of the rabbit protein (Tsuchiya and Matsumoto, 1975). The latter authors have done considerable work in detailing properties of the proteins of striated muscle of squid mantle, including studies of the ATPase activity of actomyosin and myosin of squid (Horie et al., 1975; Tsuchiya et al., 1978). Their work has included elucidation of the physiochemical properties of squid paramyosin and tropomyosin (Tsuchiya et al., 1980a,b).

There are some indications, based on thermal stability, resistance to denaturation and tryptic digestion, and properties of ATPase, that muscle proteins from fish in warmer waters are more stable than those from fish in colder environments (Hasnain et al., 1976; Arai et al., 1976; Yabe et al., 1978).

B. Connective Tissue Proteins

Perhaps the most important feature of connective tissue proteins in fish muscle is their low concentration. This is one of the main reasons that tenderness is not a particular concern with regard to most fresh seafood products. Matsumoto (1980) has suggested that the total amount of stroma proteins in fish muscle is only about 3 to 5%.

Perhaps because of the traditional use of fish preparations from fish swim bladders (originally those of sturgeon) for isinglass, there have been some studies of fish collagens used for such purposes. Of more interest to considerations of fish muscle as food are the studies of Love's group dealing with connective tissues of fish. They have shown that there are, in fact, differences in the strength of connective tisues holding together musculature of various species of fish (Yamaguchi *et al.*, 1976).

C. Muscle Proteases

In recent years, there has been considerable interest in autolytic degradation of fish tissue. It is generally speculated that proteolytic enzymes (cathepsins), present as normal tissue constituents, play a role in postmortem changes in muscle that may well result in undesirable flavor and texture alterations.

Makinodan and co-workers have been engaged in a series of studies dealing with muscle proteases isolated from carp. These workers have isolated cathepsins A and D as well as an alkaline and neutral protease from this muscle source and have reported on various properties of these enzymes (Makinodan and Ikeda, 1976a,b, 1977; Makinodan *et al.*, 1979, 1980). Muscle cathepsins have also been studied in marine fish. Reddi *et al.* (1972) have described in some detail the physical and chemical properties of a catheptic enzyme from flounder muscle. This cathepsin had its highest relative specific activity in the lysosomal fraction, as might be expected. Geist and Crawford (1974) examined muscle cathepsin in three species of Pacific sole. Of interest is the fact that their sensory evaluation measurements did not support a significant degradative role for these proteases. Eitenmiller (1974) demonstrated a cathepsin D type enzyme activity in the muscle of white shrimp. He concluded that the absence of cathepsins A, B, and C suggests that shrimp muscle has a proteolytic enzyme profile different from that of mammalian tissue. The same group (Bauer and Eitenmiller, 1974) purified an arylamidase from shrimp muscle. Sakai and Matsumoto (1981) have found several proteases in squid mantle muscle to be active at neutral and alkaline, as well as in acid, pH ranges.

Deng (1981) has suggested that, during the heating of mullet muscle, alkaline proteases may play a role in tenderization. He feels that the texture of this fish is influenced by protein–protein interaction and by enzyme hydrolysis, with the predominant influence of the two factors being dependent on time–temperature relationships.

D. Functional Properties

Consideration of the functional properties of fish muscle is an area deserving of more attention than it has received to date. Part of the reason for the apparent lack of interest in this area is that production of comminuted fish products, even on a relative basis, is far less than that involving red meats. The potential production and use of fish protein concentrate received a great deal of attention for many years; however, this product has yet to attain any significant degree of commercial importance. While there are a multitude of reasons for this, one is undoubtedly the fact that most fish protein concentrate preparations were devoid of functionality. There have been suggestions for means of producing this material to yield a product with dehydration and emulsifying capabilities (e.g., Cobb and Hyder, 1972); however, this is not yet a common practice.

On a world scale, there are important products that depend on some functional properties of fish muscle, such as emulsifying capacity and water-binding ability. So-called kamaboko, a spiced gel product made from fish, is enormously popular in Japan as are several types of fish sausages. There are indications that the use of minced fish products (discussed later herein) is on the increase in the United States and elsewhere.

Groninger, Spinelli, and their associates have done extensive work in developing functional properties in fish proteins. They have noted that emulsifying capacity of fish protein isolates is closely related to moisture content and water activity (Koury and Spinelli, 1975). One of their key contributions is the preparation of acylated fish protein, usually with acetyl of succinyl groups, and the demonstration that such modified proteins are potentially useful as functional ingredients in food systems (Groninger and Miller, 1975; Spinelli *et al.,* 1977; Lee *et al.,* 1981).

Childs (1974) found little variation in the emulsifying capacity of white muscle from several species of Oregon groundfish. Shimizu *et al.* (1981) have provided a rather extensive summary of the temperature gelation curves of meat paste prepared from 49 species of fish. They found species to vary in their capacity for involvement in two phases of the gel-forming process, these authors call a structure-setting reaction and a structure disintegration reaction.

E. Nonprotein Nitrogenous Constituents

The free amino acids profile of fish muscle is influenced by many factors, including species, habitat, and perhaps even position of sampling. Mackie and Ritchie (1974) have examined free amino acids of muscle from several species of fish. They find taurine, glycine, alanine, and lysine to be present generally in relatively large amounts. Konosu *et al.* (1974) have measured distribution of various nitrogenous constituents in the muscle of eight species of fish. They found that 90% or more of the total consisted of amino acids, small peptides, trimethylamine oxide, trimethylamine, creatine, creatinine, and nucleotides. The pattern of distribution, however, varied considerably from species to species, except for nucleotides.

Two types of nonprotein nitrogen compounds appear to be particularly important in fish muscle. The first is that of the imidazole compounds, particularly histidine, carnosine (β-alanyl-histidine), and anserine (β-alanyl-1-methyl-histidine). These substances are sometimes found in large amounts in a variety of fish (Lukton and Olcott, 1958). Histidine is an important constituent because it can be decarboxylated to histamine by a variety of bacteria and therefore may be involved in scombroid poisoning. This topic is discussed in Section IV,B,3,a the role of anserine and carnosine in producing histamine or histaminelike substances is not clear.

The second type of compound of particular interest is that of the amines frequently found in fish. Perhaps the most important of these is trimethylamine oxide (TMAO). It can be converted to trimethylamine, which contributes to the characteristic fishy odors of fish undergoing spoilage. The assay of these materials as a measure of quality is discussed in Section VI,C.

TMAO appears to have an essentially ubiquitous occurrence in aquatic animals of marine origin. It is not commonly found in fresh water fish. Two useful earlier summaries dealing with this substance in fish are those of Dyer (1952) and Groninger (1959). Harada and associates have examined TMAO distribution in a variety of marine animals, including mollusks, arthropods, echinoderms, elasmobranchs, and teleosts (Harada et al., 1970, 1971, 1972; Harada and Yamada, 1973). The role of TMAO is still obscure; it has been suggested that it may be involved in osmoregulation. The distribution of TMAO in marine animals is uneven and seems to be affected by species, season, size and habitat. Levels of 100 mg nitrogen (from TMAO) per 100 g fish are not uncommon.

F. Myoglobins

The muscle respiratory protein myoglobin has received much attention following the determination of the primary structure of sperm whale myoglobin. Subsequent work largely has been with mammalian myoglobins, but a reasonable amount of work has been done with fish myoglobins as well. Part of this is because of the general interest in using comparative biochemistry to help elucidate aspects of protein structure – function relationships, but part is also due to the fact that various derivatives of myoglobin are responsible for color in a variety of fish products.

Myoglobins from several different fish have been prepared and various properties of these myoglobins determined, including amino acid composition, terminal amino acids, chromatographic and electrophoretic properties, spectra, denaturation behavior, oxygen affinity, relative ease of autoxidation, and molecular weights. In general, fish myoglobins have been found to have amino acid compositions substantially different from those of mammalian myoglobins, and, unlike many other myoglobins, they contain a cysteine residue. The

N-terminus usually was found to be blocked, and the fish myoglobins, relatively speaking, are somewhat more unstable. Oxygen binding and many other properties were found to resemble closely those of mammalian myoglobins. Examples of species studied include carp (Hamoir, 1953); several tunas (Matsuura and Hashimoto, 1955; Rossi-Fanelli and Antonini, 1955; Konosu et al., 1958); salmon and tunas (Brown et al., 1962); yellowfin tuna (Hirs and Olcott, 1964); shark, mackerel, skipjack, and bigeye tuna (Tomita and Tsuchiya, 1971); several species of trout and salmon (Amano and Tsuyuki, 1975); dolphin fish (Bannister and Bannister, 1976); bluefin tuna (Balestrieri et al., 1978); and mackerel and sardine (Yamaguchi et al., 1979).

The complete amino acid sequence of yellowfin tuna myoglobin has been reported by Watts et al. (1980). This particular tuna myoglobin previously had been shown to have a blocked N-terminus (Rice et al., 1979), as does myoglobin from another tuna species (Amano et al., 1976), and to have a myoglobin fold, or three-dimensional structure, similar to that of sperm whale myoglobin as determined by x-ray crystallography (Lattman et al., 1971). The tuna protein, however, is known to differ from sperm whale myoglobin in hydrodynamic behavior and stability to denaturation (Fosmire and Brown, 1976) and to be more susceptible to autoxidation (Brown and Mebine, 1969).

The sequence analysis of yellowfin tuna myoglobin showed an aminoacetylated chain of 146 amino acid residues. From 79 to 85 amino acid substitutions were noted between this myoglobin and those of mammals, birds, and shark. Two external regions (containing six and seven residues), which had been highly conserved in other myoglobins, were found to be greatly or totally altered in the tuna protein (see Table I). Significant differences in the extent of electrostatic bonding between tuna and other myoglobins were also noted.

Many of the amino acid substitutions found in the tuna protein have also been noted in carp myoglobin (A. Romero-Herrera, personal communication, 1980). Myoglobins from two species of shark have also been isolated and their amino acid sequences determined (Fisher and Thompson, 1979; Fisher et al., 1980). Examination of these sequences in comparison with those of tuna and

Table I

Amino Acid Sequence of Conserved Region in Myoglobins

Myoglobin	Residue number[a]						
	133	134	135	136	137	138	139
Mammals	Lys	Ala	Leu	Glu	Leu	Phe	Arg
Chicken and penguin	Lys	Ala	Leu	Glu	Leu	Phe	Arg
Turtle and lizard	Lys	Ala	Leu	Glu	Leu	Phe	Arg
Yellowfin tuna	Asn	Val	Met	Gly	Ile	Ile	Ile

[a] Based on numbering system used for sperm whale myoglobin.

carp, and the many mammalian myoglobins studied, will allow for more meaningful evolutionary considerations.

G. Seasonal Variation

Some may question the appropriateness of ascribing a comparative feature to seasonal variation. Nonetheless, no seasonal variation in cattle, sheep, swine, or poultry comes near the extremes seen in fish. Love (1980) has treated this topic in detail; he points out that all of the attributes that govern the desirability of fish muscle as food (flavor, odor, texture, color, and surface appearance) can be said to vary with season. As an example, the technological problem of *gaping*, the falling apart of fish muscle tissue due to weakness in sheets of connective tissue binding muscle cells together, is seasonal in nature. Love feels that the variation in postmortem pH in fish (Love, 1979) is the most important factor influencing gaping in fish.

In the Unites States, seasonal effects in salmon are well known because of the intensive interest in sport fishery. Many salmon become unfit for human consumption at or near the time of their spawning, particularly those species for whom the time of reproduction is also the end of their life cycle. During their migration from the ocean back to spawning grounds, many biochemical changes occur; one of interest to the food scientist is the variation in fat content of these animals, with the lipid deposits furnishing a source of energy to the salmon during the long swim and consequently decreasing significantly in quantity during the course of the run.

H. Lipids

Marine oils are well known to contain high levels of polyunsaturated fatty acids, and to include, in many cases, significant levels of classes of lipids not ordinarily found in appreciable quantities in terresterial animals. These include wax esters, alkyldiacylglycerols, and hydrocarbons. To put the importance of waxes into perspective, it should be noted that copepods (enormously abundant small marine crustaceans that are important in the food chain) synthesize and store waxes. It has been estimated that half of the organic matter synthesized by the primary producers in the ocean is converted for a time into waxes (Benson and Lee, 1975). It should be noted, however, that some have questioned the degree to which wax esters are persistent dietary survivors (Kayama and Nevenzel, 1974).

While details of the roles of these lipid classes are not always well understood, it is clear that they can perform a variety of functions. One is certainly to serve as a metabolic energy reserve. These lipid deposits also provide a substantial

degree of buoyancy and require no osmoregulation, factors of obvious importance to a swimming animal living in a saltwater environment. Sargent (1976)
has summarized aspects of the structure, metabolism, and function of lipids in
marine organisms, and Ackman (1980) has provided a useful treatment of fish
lipids that includes rather extensive tabulations of the composition of various
fish oils and a lengthy bibliography.

I. Pressure

Marine animals, including some of those used for food, inhabit may regions
of the world's oceans and may, therefore, find themselves subject to widely
varying thermal and pressure regimes. It is conceivable that adaptation to such
habitats may have involved alterations at the macromolecular level. Such
adaptations, if in fact they do occur, are of obvious interest to the evolutionary
and/or comparative biochemist, and perhaps to the food scientist as well.
Efforts are now being made to study some of these interesting problems,
particularly those dealing with effects of temperature and pressure on enzyme
function (Somero, 1978). A treatment of the theory of pressure effects on
enzymes has been provided by Morild (1981).

J. Red versus White Muscle

Consideration of comparative biochemical aspects of red versus white muscle is a topic unto itself, and no attempt will be made here to treat it in
comprehensive fashion. In general, properties of red and white muscle in fish
are quite similar to those of such muscle types in land animals. Hamoir and
associates have examined extracts of red and white muscle from carp, with
particular emphasis on lower molecular weight components (Hamoir and Konosu, 1965; Konosu et al., 1965; Pechere and Focant, 1965; Hamoir et al., 1972).
They concluded that criteria of differentation based on protein analysis that
distinguish anaerobic and aerobic types of muscle in higher vertebrates also are
valid in the case of fish. Workers in the same laboratory also reported differences in digestibility, stability, and ATPase activity of myosins isolated from the
two types of carp muscle (Syrovy et al., 1970).

The two muscle types are found in many fish, and even lightly pigmented fish,
such as halibut and sole, have small bands of red muscle. In more active fish the
amount of red muscle increases, and in very active fish, such as the tunas, red
muscle may contribute a significant percentage of the total muscle tissue.
Because of this, and because of the ability of tuna to maintain temperatures well
above ambient, there has been a substantial amount of research dealing with
properties of red and white muscle in tuna. An overview of some of this work is

given below. It should also be noted that red muscle recovered from tuna during the processing of white (light) muscle for human food is an important constituent of many pet foods. For this purpose, fish are first cooked, then the muscle types cleaned and divided by hand. Separation of the red and white muscle in raw fish is not done commonly on an industrial scale, although it has been suggested that, due to differences in lipid content, the muscle types could be separated by differential flotation (Matsumoto, 1980).

Tuna red muscle, with its capacity for high rates of aerobic metabolic activity, has attracted the interest of many investigators, including those interested in temperature regulation. It is of considerable interest that, of the fishes, tuna are among the fastest swimmers, have the highest metabolic rates, and have the warmest bodies, being capable of producing temperatures in the propulsive musculature of up to 20°C warmer than the surrounding seawater under extreme conditions. Body temperatures up to 10°C higher than ambient have been noted frequently.

This topic has been treated in detail by Stevens and Neill (1978), who have concluded that aerobic metabolism occurring primarily in the red muscle of skipjack tuna can account for a 10°C excess temperature effect. These authors point out that there is also an anatomical basis for warm-bodiedness in tunas based on short-circuiting heat from venous to arterial sides of the circulatory system before it reaches the gills. Because of the potential for significant temperature elevation, and data suggesting that some tunas are capable of thermoregulation, it may well be that tuna are close to being homoiotherms.

Hochachka *et al.* (1978) have summarized the work of others, as well as their own, and noted that data on skeletal muscle of skipjack tuna support a partitioning between red and white muscle function but also suggest that there is a considerable overlap. Red muscle is highly vascularized and has large amounts of fat and glycogen, high levels of enzymes of aerobic metabolism, and a profusion of mitochondria. The contribution of red muscle to swimming at all speeds seems to be dependent on aerobic metabolism (carbohydrate or fat oxidation, or both). Skipjack tuna red muscle seems fairly typical for teleost fish, while tuna white muscle has some novel features. There are large amounts of glycogen and high glycolytic enzyme activities; however, there are also enzymes of aerobic metabolism present and quite active. Mitochondria are much more abundant than in most teleost white muscle. White muscle contribution to rapid swimming is supported almost totally by one of the most intense anaerobic glycolysis known in nature. However, during periods of intense feeding at sea, with a concomitant increase of muscle temperature of up to 10°C, white muscle work has been found to be supported mostly by an aerobic, glucose-based metabolism.

As swimming velocity increases there is probably a change from the predominant use of pure red muscle to that of predominantly white muscle. White muscle apparently is used during acceleration and for short-term, rapid swimming. Its use at some threshold velocity may be species and size dependent (Sharp and Pirages, 1978). Localization of red musculature is also species dependent, and whether such muscle is located near the surface or is found in internal arrays may be an adaption related to habitat, specifically having to do with temperature maintenance. Thus, it is postulated by Sharp and Pirages (1978) that species preferring cooler water show maximum internalization of red muscle as an adaptation to allow maintenance of a "tropical" red muscle temperature.

Interesting color plates detailing relative distribution of red muscle in scombroid fish may be found in the preface of a monograph edited by Sharp and Dizon (1978).

III. NUTRITIONAL QUALITIES OF FISH MUSCLE

As previously noted, there is great diversity among animals we call fish. It is therefore not surprising that there is substantial variation in composition of fish tissues. There are vast differences between one species and another, between individuals, and even between different parts of the same fish. Stansby (1976) has called attention to many factors involved in such differences in chemical composition in his report dealing with analyses of fish from the Northeast Pacific Ocean.

In Section II,G the tremendous change in fat content of salmon during their migration to the spawning grounds was mentioned. For this and many other reasons, including species differences, it would be wise to view with skepticism any food composition tables that purport to give the amount of fat in salmon. This is but one somewhat dramatic example; however, similar reservations are in order when evaluating single or mean values for many nutrients in seafoods.

Chapter 8 of this book covers nutritional values of meat, including fish. Therefore, the present treatment is brief and highly selective, being restricted to certain topics particularly relevant to seafood products.

Sidwell et al. (1974, 1977, 1978a,b) have provided a series of papers summarizing composition of the edible portion of raw (fresh or frozen) crustaceans, finfish, and mollusks. This collection contains a rather comprehensive review of literature values of essentially all important nutrients. Many sections of the book edited by Heen and Kruezer (1962) are also of interest, as is one in particular of the series of books by Borgstrom (1962) dealing with fish as food.

A. Proteins

Protein from seafood sources is of high quality, containing all of the essential amino acids in good quantity and in balanced amounts. The amount of protein present in most fish products used for human consumption ranges from ~ 15 to 20% on a wet weight basis. As is true for almost all food products, such values are generally based on nitrogen determination, usually by Kjeldahl analyses, and therefore may not truly represent "protein." This might well be particularly true for some species of fish. In tunas, for example, values of protein content > 20% are frequently noted; this may be due in part to the fairly large contribution to total nitrogen of the nonprotein nitrogenous materials present in substantial amounts in these fish.

Most literature values deal with marine fish. However, Mai *et al.* (1980) studied several species of freshwater fish and found no major differences between amino acid compositions of freshwater fish and those of marine fish.

There has not been a great deal of work done dealing with changes in protein quality as a result of processing of fish. Tanaka *et al.* (1980) studied available lysine losses in fish proteins as a result of heating; such losses were minimal, except when reducing sugars were purposely added. Poulter and Lawrie (1977) examined samples of several species of fish stored frozen for extended periods of time. They found no nutritionally significant decrease in lysine availability or *in vitro* digestibility of the muscle proteins in their samples. Groninger and Miller (1979) reported that there was some lowering of the protein efficiency ratio of fish protein samples that had been acylated.

B. Lipids

Some features of the general makeup of fish lipids have been noted earlier. Fish oils from those aquatic animals ordinarily used as food consist for the most part of triglycerides and phospholipids. The total amount of fat in fish is extraordinarily variable. Stansby (1976) found a range of 0.2 to 23.7% in total oil content of the samples included in his study. He noted that oil content has been measured as high as 85% in the edible flesh of some fish. This, of course, is unusually high. The sum of the oil and water content in fish tends to approximate 80%. Fatty deposits found as such frequently are located immediately beneath the skin of fish and along the abdominal wall.

1. FATTY ACIDS

As a generality, fish oils contain large amounts of highly unsaturated fatty acids. They differ from other natural sources of unsaturated fatty acids, such as vegetable oils, in the variety of fatty acids present. Fish oils may contain 16 or more fatty acids in relatively major amounts with many more occurring in smaller quantities. They are also more unsaturated than other oils, commonly having substantial quantities of fatty acids with four, five, or even six double bonds. Longer chain length fatty acids, that is, those having 22 carbons, are also frequent constituents of fish oils (see Table II for examples). Linoleic acid,

Table II
Polyunsaturated Fatty Acids in Canned Fish[a]

Fatty acid	Percentage of total fatty acids	
	Red salmon	Yellowfin tuna[b]
$18:3\omega3$	6.75	0.63
$20:3\omega9$	0.33	0.37
$20:3\omega6$	0.54	0.49
$20:3\omega3$ and $20:4\omega6$	7.78	6.69
$18:4\omega3$	1.24	1.00
$20:4\omega3$	5.37	0.84
$22:4\omega6$	0.81	1.52
$20:5\omega3$	10.09	12.59
$20:5\omega6$	2.67	3.33
$22:5\omega3$	3.51	2.66
$22:6\omega3$	16.44	32.34

[a] For complete analyses of fatty acids in these samples, see Bonnett et al. (1974).
[b] Canned in brine.

the so-called essential fatty acid found in large quantities in vegetable oils, is present in rather insignificant quantities ($<1\%$) in fish oils. Bonnet et al. (1974) have reported fatty acid composition of 32 finfish, crustaceans, and mollusks. Ackman (1980) has provided a useful summary article dealing with fish lipids.

2. CHOLESTEROL

Shellfish usually are considered to have relatively high levels of cholesterol, and their use in dietary regimes in which cholesterol levels are regulated is sometimes restricted. One of the problems is that earlier literature values may be erroneous because of limitations in the methodology available at the time of analysis. Idler and Wiseman (1972) addressed this issue in their review of molluscan sterols. Kritchevsky and DeHoff (1978) have analyzed several sea foods for cholesterol content, using different analytical methods. They concluded that variation in sterol content obtained by different methods may be due to the presence of sterols other than cholesterol.

Pearson (1977, 1978) has provided cholesterol contents for a variety of Australian seafoods. Her data are derived from a wide variety of animals, such as several different species of prawns, lobsters, and crabs, but suffer to some extent from the fact that the number of animals in each sample was usually very small. In broad ranges, using rounded figures, she found ~ 160–200 mg cholesterol per 100 g wet weight in prawns; similar figures for crab and lobsters were 40–100 mg/100 g, and for mollusks, including oysters, 40–70 mg/100 g.

There is considerable evidence suggesting that seafood in the diet, probably via its lipid fraction, has a cholesterol-lowering effect. Carpenter (1980) has provided an analysis of some of these reports. It is interesting that crustaceans such as shrimp and lobster have a dietary requirement for cholesterol, or at least for a preformed sterol nucleus.

C. Minerals and Vitamins

For the most part, mineral and vitamin compositions of seafood muscle present few unusual features. The content of mercury may sometimes be troublesome, and there may be loss of thiamine in fish. Both of these problems are treated in subsequent sections. Because of the general concern regarding dietary sodium, some tuna packers have stopped the practice of adding salt to the cans.

Sidwell *et al.* (1977, 1978a) have summarized literature values for sodium, potassium, chlorine, calcium, phosphorus, and magnesium (1977); for a number of microelements (1978a); and for vitamins (1978b) in crustaceans, finfish, and mollusks. In general, fish provide reasonable amounts of these nutrients. Fish liver oils are rich in vitamins A and D, hence the traditional use of cod liver oil for this purpose. The amount of niacin is particularly high in tuna (Gordon *et al.*, 1979), on the order of five times the amount found in less active bottom fish. It seems likely that this is due to the presence of relatively large amounts of NAD/NADP in tuna muscle because of the high metabolic rates and swimming speeds of these animals.

D. Thiaminase

An antithiamine factor has long been recognized as being present in fish. So-called Chastek's paralysis was first noted in the early 1930s in foxes fed a diet rich in raw carp. This disorder was subsequently shown to be due to thiamine deficiency. This factor is generally called thiaminase and, indeed, much of the early work (summarized by Higashi, 1961) indicated that there was such an enzyme found in a variety of fish, both marine and freshwater. Concentrations are highest in viscera, but the enzyme is found in muscle tissue as well (Tsuzimura *et al.*, 1972; Ishihara *et al.*, 1972, 1973). It is heat labile, although the extent of loss of this antithiamine activity on heating is variable (Luna *et al.*, 1968; Tang and Hilker, 1970; Ishihara *et al.*, 1972). There is evidence for the existence of an antithiamine factor involving a heme group that may not be enzymatic in nature (Porzio *et al.*, 1973). The latter workers refer to this activity as a thiamine-modifying property of heme proteins.

IV. MICROBIAL PROBLEMS WITH FISH MUSCLE

A. Spoilage

Fish, in general, usually spoil more rapidly than other muscle foods, particularly when mishandled, and such spoilage is primarily bacterial in nature. Organisms normally present on the skin and in the intestinal tract of living animals become ready contaminants of seafood products. Bacteria living in ocean waters obviously must be somewhat halophilic in nature, and considerable attention has been given to marine microbiology and to the possibility that significantly different bacterial populations may be found in marine fish. However, it appears that organisms relatively common to other muscle foods are responsible for fish spoilage as well. Bacteria native to fish and added adventitiously during landing and processing include *Acinetobacter, Arthrobacter, Flavobacter/Cytophaga, Microccoccus, Moraxella,* and *Pseudomonas*. Those primarily responsible for spoilage include *Pseudomonas, Acinetobacter,* and *Moraxella* species. General treatment of this topic may be found in another chapter in this book and elsewhere (Chichester and Graham, 1973, Rheinheimer, 1974, Shewan, 1977; Skinner and Shewan, 1977; Liston, 1980).

One spoilage characteristic found in fish and not in other muscle foods is that of trimethylamine formation. This very odoriferous amine, generally considered to be a major constituent of the fishy smell associated with spoiling seafoods, is produced by a number of bacteria capable of reducing trimethylamine oxide, a natural constituent of marine fish as discussed in Section II,E. The formation of trimethylamine is so common that measurement of this substance is used as an index of quality deterioration. Trimemethylamine oxide can also be degraded enzymatically during cold or frozen storage of seafoods; however, this normally yields dimethylamine and formaldehyde. Production of these materials may indicate undesirable changes, but these are not usually considered to play any major role in spoilage *per se*.

Ammonia is also found in spoiling fish. Because of its high concentration in elasmobrachs, urea is a good substrate for bacterial action, producing ammonia in these animals. Substantial quantities of ammonia may also be found as a result of microbial deterioration of other seafoods, such as shrimp.

B. Foodborne Infections and Intoxications

The center for Disease Control in the United States recently reported that fish, mollusks, marine crustaceans, and marine mammals were involved in ~11% of the reported foodborne disease outbreaks in the Unites States during the period 1970–1978 (Bryan, 1980). Some of the more important infections and intoxications are noted in the following sections. In addition to these, there are outbreaks of diseases such as typhoid, cholera, and hepatitis as a result of ingestion of uncooked shellfish harvested from polluted waters. This is more a matter of proper sewage treatment than of factors associated with the normal conversion to fish to food. Similarly, there is always the danger of contamination of seafoods due to mishandling during preparation, leading to staphylococcal intoxication, salmonellosis, and/or other types of human illness.

1. VIBRIO PARAHAEMOLYTICUS

Vibrio parahaemolyticus, along with Clostridium botulinum type E, is one of the few pathogenic bacteria known to occur naturally on fish. Others may be involved somewhat indirectly, such as those that produce histamine in scombroid fish, as discussed below.

Vibrio parahaemolyticus is a halophilic organism that is widely distributed in warm waters, particularly in inshore marine and estuarine areas. It is a major causative agent of gastroenteritis in Japan. That it is relatively less important in the United States and elsewhere is thought to be due to its lack of recognition by public health authorities, as well as to the fact that consumption of raw or lightly cooked seafoods is less. Fortunately, the organism is highly sensitive to heat ($> 48°C$) and cold ($0-5°C$) and does not survive normal food processing.

It is also killed on exposure to fresh water. There have been problems of human illness resulting from consumption of precooked products that were subsequently reexposed to the organism, however. Liston has prepared summaries dealing with this organism (1973, 1980).

2. CLOSTRIDIUM BOTULINUM, TYPE E

This organism is widely distributed in marine and lake sediments, although its occurrence is by no means uniform and many apparently good natural habitats lack significant populations of this organism. It is a frequent contaminant of fish, and as such is usually thought of as the fish botulism organism. Fortunately, it does not appear to grow and produce toxin in living animals. It can grow at relatively low temperatures ($\sim 3°C$) but, from a practical standpoint, has been a health hazard only in processed fish, not in fresh or frozen products. This is generally a result of inadequate heat treatment that kills competing organisms but allows C. botulinum spores to survive. A review dealing with C. botulinum in fishery products is that of Hobbs (1976).

3. TOXINS

From a worldwide standpoint, the problem of oral fish intoxicants, that is, fish that are poisonous to eat, is significant and puzzling. A variety of toxins are known. Human responses range from discomfort to death, and an aquatic animal that is safe to eat in one location may be dangerous in another, or at another time of the year. A series of books by Halstead (1965, 1967, 1970, 1978) offers a comprehensive treatment of this topic. The same author has provided a briefer volume with treatment of a selected number of problems (Halstead, 1980). Pertinent reviews are those of Scheuer (1970, 1977). Brief attention is given below to a few of the toxins that are of most significance to those primarily concerned with muscle as food.

a. Scombroid Poisoning

This type of poisoning involves the ingestion of poorly handled scombroid fish including saury, bonito, seerfish, mackerel, and tunas, which contain relatively large amounts of histamine. These fish normally contain large amounts of free histidine and other imidazole compounds in their muscle tissue. Many bacteria have the capability of decarboxylating histidine to produce histamine. If the levels of histamine are sufficiently high, for example, 100 mg/100 g, eating the fish can produce symptoms of scombroid poisoning. These include a flushing of the face and neck, headache, cardiac palpitation, dizziness, faintness, itching and burning of the mouth and throat, rapid and weak pulse, and in some victims, a rash and gastrointestinal upset. Practical problems include the fact that histamine is heat stable and that toxic levels of histamine may be reached before the fish appear spoiled by the usual criteria.

While many bacteria can decarboxylate histidine, *Proteus* strains are among the most commonly implicated. It is not yet known whether decarboxylation of the related peptides carnosine and anserine results in any toxic product (see Fig. 1). Whether histamine acting alone is solely responsible for the toxicity is somewhat confusing, inasmuch as there are many reports indicating that histamine taken orally by human subjects is not toxic. Also, morbidity rates vary greatly from 0.07 to 100%. Such variation is undoubtedly due in part to the

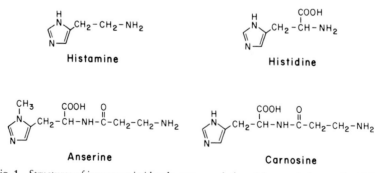

Fig. 1   Structures of important imidazole compounds found frequently in scombroid fish.

fact that fish in the same lots differ in their histamine content; the latter may also vary in different locations in the same fish. Motil and Scrimshaw (1979) have demonstrated, however, that human volunteers, studied under controlled conditions, do respond adversely to oral ingestion of histamine. A summary of the general problem of histamine toxicity from fish products is available (Arnold and Brown, 1978). Lerke *et al.* (1978) have an interesting report of an outbreak of scombroid poisoning from consumption of sashimi (raw tuna) in the San Francisco area that discusses many of the common features and perplexing questions associated with this problem.

Since histidine is a normal constituent of scrombroid fish, and since very many types of bacteria can decarboxylate it to histamine, there are almost always low levels of histamine present in processed fish from this group. In canned tuna, such levels almost always are < 10 mg/100 g. Proper refrigeration of raw material prevents formation of excessive amounts of histamine and in fact, it appears that tuna that has spoiled at temperatures < 10°C has never been implicated in scrombroid posioning (Lerke *et al.,* 1978).

b. Ciguatera

This is a serious and fairly common seafood intoxication caused primarily by tropical shorefishes. More than 400 species of fish have been incriminated. Food fishes previously found to be safe to eat may suddenly, and for no obvious reason, become ciguatoxic. It is now believed that the toxicity is due to the fact that animals in question have been feeding on toxic materials. It appears that a dinoflagellate found on the surface of brown seaweed may be the primary producer of ciguatoxin. A complex food web may be involved that includes filter-feeding invertebrates, plankton-feeding fishes, herbivorous fishes, and carnivorous fishes that feed on herbivorous fishes.

Paresthesia (tingling or burning of the skin) is the classic symptom of ciguatera toxicity and is commonly used for diagnostic purposes. Gastrointestinal upsets may be noted together with neurological disorders ranging from sensory disturbances to serious convulsions and muscular paralysis. Contrary to some early reports, the fatality rate is low. Poisonous fish cannot be detected by appearance, and animal tests are required for confirmation, inasmuch as simple chemical tests are not yet available. The toxin is not destroyed by cooking (Halstead, 1980; Higerd, 1980).

c. Puffer Fish Poisoning

Of all the marine fish, puffer fish are among the most poisonous, containing in their viscera (especially in ovaries and liver) a potent nerve poison, tetrodotoxin. This toxin may produce violent death in short order. In spite of this, puffer fish is a popular food among connoisseurs in Japan who consume this fish (called *fugu*) in restaurants employing specially trained cooks. These cooks remove the poisonous parts; skeletal muscle ordinarily is nontoxic. Although it is unusual, diners in such resturants have been known to die. In such cases, the chef is subject to legal action on the basis of professional negligence.

d. Paralytic Shellfish Poisoning

The poison (saxitoxin) is produced by dinoflagellates, namely species of *Gonyaulax*. Bivalve mollusks such as clams, mussels, oysters, and scallops feed by filtering dinoflagellates and other organisms from ocean water. The mollusks themselves are not harmed by the toxic dinoflagellates but concentrate saxitoxin. When it is present in large amounts, other animals, including man, can become ill as a result of eating toxic shellfish. The public frequently associates this problem with red tides, caused by a bloom, or heavy increase, in the population of dinoflagellates that causes a red to brown water color, depending on the density of the organisms. This correlation is not based on fact. In California, for example, other nontoxic species of *Gonyaulax* are responsible for most red tides. The time of year is more critical, and there is a quarantine on mussels in California from May 1 to October 21. In spite of this quarantine, and in spite of the fact that signs are posted along virtually all beaches, there are periodic outbreaks of paralytic shellfish poisoning. In California in 1980, there were 98 reported cases and two deaths. On a statistical basis, mussels were most often involved; abalone, crab, and shrimp are not filter feeders, and therefore present no danger.

The toxin is not destroyed by heating. One of the great difficulties in monitoring areas for dangerous populations of *Gonyaulax* species is that, to date, only animal (mouse) assays are thought to be reliable. These are expensive and time-consuming. Efforts have been made to develop more rapid chemical assays. However, workers in the field still question the applicability of existing chemical assays for routine monitoring of saxitoxin levels in shellfish (White and Maranda, 1978). A useful general treatment of red tide and paralytic shellfish poisoning is that of Dale and Yentsch (1978).

V. STORAGE, PRESERVATION, AND PROCESSING OF FISH MUSCLE

The processing of seafoods has as its objective the provision of means for widespread distribution from point of harvest of products that are safe, palatable, and nutritious. One of the major differences between fish and other muscle foods is that point of harvest may well be uncertain, and the amount of harvestable product will almost surely be unknown. The advent of successful aquaculture ventures may circumvent this problem, at least to some extent, but at present the oceans are human's source of supply for the great majority of fish.

Preservation, and perhaps processing as well, starts on the vessel in which fish are caught. This may range from some sort of chilling involving ice, refrigeration, or freezing on board, to actual final processing such as is done on Russian and Japanese factory ships. In any case, the crucially important factor is the extent to which microbial spoilage is retarded until the product either reaches the consumer or is converted into some type of shelf stable form. Chemical alterations such as the development of rancidity and undesirable texture and color changes, while of obvious importance, must always be considered after proper concern has been given to inhibition of bacterial spoilage.

There have been changes in recent years in the means by which fish are distributed. On a world scale, more fish is marketed in the fresh form than in any other although this amount decreased to 40% or less in the mid-1970s. The relative amount going into frozen and canned products is increasing, to ~26% in 1976, while that going into cured products (smoked, salted) decreased to ~15% (Steinberg, 1980). In the United States, the amount of fish marketed that is truly fresh, that is, has not been frozen and subsequently thawed and sold as "fresh," is quite small. Figures are suprisingly difficult to obtain, but the amount is on the order of a few per cent.

The U.S. Department of Commerce via its National Marine Fisheries Service (NOAA) publishes a yearly document detailing statistical information concerning fisheries of the United States (Anonymous, 1980). The latter document reveals, for example, that, in 1980, the landings of menhaden in the United States were 2.5 billion pounds. On a poundage basis, the menhaden fishery is by far the largest in this country; however, virtually all of this catch is reduced to fish meal, oil, and solubles. Such products, while of great economic importance, are beyond the scope of the present chapter. The following sections concentrate on means of preservation of fish muscle for direct use as human food.

A. Chilling

The chilling of fish is a vitally important means of preservation. It is applied to a host of products, including many that subsequently will be distributed in other than fresh form, for example, canned.

A number of means are employed to chill fish. Ordinary ice (wet ice) is perhaps the most important, and its use continues to be widespread. It is effective because of its high latent heat of fusion. There have been efforts to incorporate inhibitory substances such as antibiotics and nitrite into ice. Many of these are effective, but because of potential human health hazards and resulting regulatory considerations, they have not attained widespread use.

Ice-chilled or mechanically refrigerated seawater (RSW) is also employed. RSW provides good control, but does require onboard refrigeration equipment. In boats that are equipped to utilize this technology, it is an effective means of cooling. Saturation of salt water with carbon dioxide has been used to further enhance the extent of microbial inhibition. The use of gaseous carbon dioxide in a variety of gas mixtures in refrigerated holds, vans, or other containers has also been employed. This technology is noted in Section V,F,1 on the use of modified atmospheres. Some consideration has been given to the use of superchilling, lowering the temperature of fish to about -1 to $-3\,°C$. At these temperatures, some of the water in the product will be converted to ice. However, the product is not frozen in the usual sense of the word.

Whatever the means utilized, the objective in chilling fish is simple: take the product to the lowest possible temperature (without freezing) as quickly as feasible and maintain this temperature until the product is consumed or subjected to some type of processing. Inattention to proper chilling undoubtedly is responsible for more loss of product, or for inferior quality, than is any other factor. Ronsivalli and Baker (1981) have reviewed the low-temperature preservation of seafoods.

In addition to the type of chilling employed, other features of the handling of fish at sea are important. These include to what extent the product is cleaned (eviscerated, heads removed), and how it is stored (large or small holds, stacked or packed in boxes) and even how the fish is caught (hook and line or nets, type of net). Consideration of such factors is beyond the scope of the present treatment.

The handling of landed fish that is to enter the fresh market is relatively straightforward. It is cleaned, if this was not already done on a boat, and cut into portions or left in the round. Usually, this is a hand operation, but machines may be employed. Crustaceans such as crab and shrimp may be precooked before distribution. The technology primarily involves good sanitation practices and continued proper chilling of the product throughout the distribution chain and at the retail level.

Generally speaking, few changes take place in fresh fish that are related to the conversion of muscle to food, except those associated with microbial spoilage. Exceptions to this might include surface discoloration due to pigment oxidation in some fish, and the development of off odors. Usually in fresh products, the latter are associated with early stages of bacterial spoilage rather than with strictly chemical changes, such as oxidative rancidity.

B. Freezing

Properly frozen and stored seafoods retain their quality for extended periods of time. Generally, products held at or below $0\,°F$ (about $-18\,°C$) do not deteriorate for months. They may be held for a year or more at still lower temperatures (e.g., −40°C) but such storage is not common.

Most seafood products are frozen in processing facilities rather than on boats, although there are important exceptions. Tuna fish, for example, are frequently frozen at sea by immersion in chilled brine. Large purse seiners may freeze hundreds, or even thousands, of tons of tuna for delivery to canneries. Direct immersion in brine or other coolants may be used for a variety of products at fish-freezing plants. Freezing solutions that can be used for this purpose include liquid nitrogen, and some of the series of Freon compounds. In more common use are conventional freezing techniques including plate freezing, spray freezing, and blast freezing. Freezing in refrigerated air (sharp freezers) is also still done, although this process is relatively slow. Ronsivalli and Baker (1981) offer descriptions of the various types of freezers. They also call attention to the importance of thawing, and its effect on product quality. This aspect receives too little attention. A number of problems can result; for example, it takes longer to thaw a product than to freeze it under similar heat transfer conditions. Surface microorganisms may well be provided with good growing conditions for much of the thaw period.

Frozen products may undergo a number of undesirable changes. Some examples of these are included in the following sections. General features of freezing preservation of seafoods have been described for finfish (Slavin, 1968a,b; Banks and Waterman, 1968) and for shellfish (Dassow, 1968a,b; Fieger and Novak, 1968a,b; Peters, 1968a,b). Bramsnaes (1969) has summarized aspects of quality and stability of frozen seafood.

1. PROTEIN DENATURATION AND MICROSTRUCTURAL CHANGES

Freezing and frozen storage of seafoods can result in undesirable textural changes. Muscle may become tougher, drier, and lose its water-binding capability. These changes are undoubtedly due in part to protein denaturation. A number of factors are thought to be involved, including a decrease in the amount of liquid water available, an increase (perhaps localized to a degree) in tissue salt concentration, and mechanical damage resulting from ice crystal formation. There are changes during the storage period that go considerably beyond those induced by the freezing process. Chu and Sterling (1970), for example, measured solubility of whole muscle protein as well as intrinsic viscosity, isoelectric point, ATPase activity, UV absorbance, and optical rotatory dispersion (ORD) of purified myosin extracts from Sacramento blackfish. Virtually all measures indicated that there was more denaturation in the frozen stored samples (30 days at −5°C) than found just after freezing. As expected, both of these types of samples showed more evidence of denaturation than was found in fresh (unfrozen) muscle.

There is an overall decrease in protein extractability during frozen storage that is attributed primarily to alternations in the myofibrillar fraction, particularly the myosin – actomyosin system. There apparently are fewer changes in sarcoplasmic proteins. The nature of the reactive groups responsible for protein aggregation is not well understood. Sulfhydryl groups have been implicated (Mao and Sterling, 1970b), but details remain to be elucidated. The fact that myofibrillar proteins of stored frozen fish can be solubilized in sodium dodecyl sulfate solutions has been taken as an indication that many of the linkages formed in proteins in frozen fish must be of a secondary nature (i.e., hydrogen or hydrophobic bonds). Since formaldehyde is a product of decomposition of trimethylamine oxide (as noted in Section V,B,3) and since formaldehyde is highly reactive with a variety of functional groups in proteins, it has been widely suspected as being an agent involved in covalent cross-linking of proteins in frozen stored fish. Connell's work (1975) on the role of formaldehyde as a protein cross-linking agent during frozen storage of cod throws doubt on this concept, although the addition of formaldehyde does decrease the extractable protein content of cod muscle (Castell et al., 1973a,b). The loss in water-binding capacity is probably due to an increase in cross-linkages between myofibrillar proteins. Such loss influences the functional properties of minced fish but may also have rather pronounced effects on texture in ordinary fish muscle.

A number of workers have studied the release of free fatty acids from lipids (probably phospholipids) in fish that occurs simultaneously with protein denaturation during frozen storage (see Sikorski et al., 1976). There apparently is no simple relationship between lipid hydrolysis and protein denaturation, but loss in protein extractability would appear to be, at least in part, due to interaction with free fatty acids. If there is accompanying lipid oxidation, then it is reasonable to assume that some polymerization and aggregation of proteins results from the presence of oxidized lipids since it is generally accepted that lipid – protein polymer formation results from a reaction between oxidized fatty acids and proteins.

Matsumoto and colleagues (1980) have done extensive studies with regard to the cryoprotective effect of a variety of substances, using a carp actomyosin model system. One of their earliest findings (Noguchi and Matsumoto, 1970) was that sodium glutamate was effective in preventing the denaturation of actomyosin. Another paper (Ohnishi et al., 1978) summarizes some of their earlier work and reports details with regard to the effect of a variety of amino acids on actomyosin filaments in frozen muscle, using electron microscopy. They found great variations in protective effect, with glutamate again being most effective.

Efforts have been made to describe changes in freezing and frozen storage at the ultrastructural level. Childs (1973) found that during frozen storage of cod, the extractability of whole myofibrils decreased more rapidly than did the combined extractability of component myofibrillar proteins. Oguni *et al.* (1975) used electron microscopy to show that as a result of freezing and frozen storage of carp muscle, actomyosin filaments were aggregated in side-to-side and crosswise fashion and also entangled into netlike masses. Mao and Sterling (1970a) examined frozen Sacramento blackfish muscle for changes in crystallinity by X-ray diffraction and water vapor sorption. They concluded that changes in the amount of crystallinity could not account for all observed textural changes. Aitken and Connell (1977) used low-angle X-ray diffraction to study structural changes in muscle of frozen stored cod. They concluded that severe toughening and loss of water-holding capacity could occur without measurable decrease in interfilament distance and that an increase in hydrophobicity of the proteins as a result of denaturation probably had more of an effect.

Giddings and Hill (1976, 1978) have used scanning electron microscopy to study the effects of freezing on crustacean muscle. These investigators were using blue crab and shrimp; such products are cooked before freezing. They found that ice crystal growth and tissue disruption occurs largely during postfreezing frozen storage. One of the interesting sidelight observations in their 1978 paper was the fact that so-called double freezing of shrimp did not result in additional tissue disruption. This involves thawing frozen raw shrimp, cooking, and refreezing for distribution.

Sikorski *et al.* (1976) have provided a useful review dealing with protein changes in frozen fish. Another by Matsumoto (1980) deals generally with chemical deterioration of muscle proteins during frozen storage but emphasizes fish. Both reviews include important references to the earlier literature. Avoidance or minimization of changes in frozen fish such as enhanced drip loss, decrease in water-binding capacity, pigment oxidation, and adverse changes in texture await better understanding of the basic phenomena involved.

2. LIPID OXIDATION AND HYDROLYSIS

Virtually all fish contain highly unsaturated fatty acids as major components of their lipids. However, the total amount of fat may vary greatly. Obviously fatty fish such as herring, mackerel, and salmon will be more susceptible to appreciable oxidation. In spite of the fact that the low temperatures of frozen storage inhibit the lipid oxidation reactions, such reactions do take place. They may be enhanced by tissue disruption associated with freezing and almost always are increased in products that become partially dehydrated during frozen storage. It also appears that vacuum packaging of fatty fish such as salmon in a low oxygen–permeable film is advantageous (Yu *et al.*, 1973).

A second deteriorative change in fish lipids in frozen muscle is that of lipolysis. This takes place rather rapidly in some products, depending on temperature. Hardy *et al.* (1979) have reported, for example, that during frozen storage of cod, a type of fish low in fat, the major change is that of lipolysis. One interesting feature of fish muscle is that there is an acceleration of the decomposition of phospholipids as a result of freezing (e.g., Hanaoka and Toyomizu, 1979). Hardy (1980) has provided a review article dealing with postmortem changes in fish muscle lipids that includes discussion of lipolysis and autoxidation in frozen tissues.

3. DECOMPOSITION OF TRIMETHYLAMINE OXIDE

As noted earlier, trimethylamine oxide (TMAO) is a normal constituent of the muscle of many marine fish. Bacterial reduction of TMAO to trimethylamine (TMA) results in a fishy odor associated with spoilage. There is little or no production of TMA from TMAO during frozen storage; however, there is significant decomposition of TMAO to dimethylamine (DMA) and formaldehyde (Fig. 2). The latter may be involved in cross-linking of proteins as discussed in Section V,B,1. Castell *et al.* (1973a,b) found that large amounts of TMA were produced in iced fish, but none was formed in frozen fish of the same species in 60 days. Cod, pollock, and cusk were found to produce more DMA in frozen than in iced fillets. This appears to be a relatively general observation, although there is significant variation among species (Castell *et al.*, 1973a,b; Mackie and Thomson, 1974).

It is generally assumed that the production of DMA and formaldehyde during frozen storage is enzymatic in nature and there is considerable literature in support of this conclusion. Shenouda (1980) provides considerable detail concerning the present state of knowledge of TMAOase. However, Spinelli and Koury (1979, 1981) have provided evidence suggesting that the formation of DMA does not necessarily result from enzymatic activity. They found that it was formed in heat-dried muscle, for example. Furthermore, these workers have shown specifically that DMA can be formed by the reaction of TMAO with catabolites of cysteine or with ferrous iron. With regard to the latter

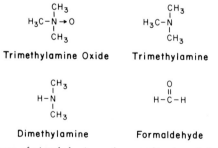

Fig. 2   Structure of trimethylamine and some of its degradation products.

point, it is of interest that Castell *et al.* (1971) had previously shown that formation of DMA during frozen storage of gadoid species was reduced by removal of dark lateral muscle from these fish. Such muscle contains considerable myoglobin, and it is conceivable that the iron present in that protein may participate in the reaction proposed by Spinelli and Koury. The fact that endogenous ionic constitutents or substances formed postmortem may promote the formation of DMA may have some bearing on the observation that production of DMA is usually found to be greater in minced fish.

4. MYOGLOBIN

Most fish muscles contain relatively small amounts of the pigment myoglobin. Red muscle found in tuna and other fish is obviously an exception to this rule. In spite of the usually low concentration of myoglobin, the bright, slightly pink surface appearance of most cut surfaces of fish (other than salmonids) is due entirely or in part to derivatives of this pigment. Of these, oxymyoglobin is the form normally present. When it undergoes oxidation to metmyoglobin the result is a change in surface appearance to a dull tan or brown color, which is considered less desirable. This oxidation rate decreases with temperature, but for reasons not yet fully understood, the oxidation rate increases sharply at the time pigment solutions are frozen (Brown and Dolev, 1963). Oxidation of oxymyoglobin in frozen tuna has been studied in some detail by Bito (1964, 1965a,b). Once freezing has taken place, this undesirable oxidation is minimized by storage at the lowest possible temperature. Matsumoto and Matsuda (1967) showed that freezing in liquid nitrogen minimized the problem. However, this technology is not yet in common use with fish.

C. Canning

The canning of fish is an important means of preservation. The extensive heat treatment involved in the sterilization process substantially alters the nature of the raw product so that, in effect, a new product is formed. This technology is quite standardized and the principles involved are obviously the same as those that apply to all canned muscle foods. Ordinarily, there are relatively few deteriorative changes that take place in the canned products. A few technological problems that sometimes are found with canned products are treated in Section VI of this chapter.

Examples of canned finfish include tuna, salmon, sardines, anchovy, and mackerel while canned shellfish include shrimp, clams, and crab. Two important commercial products are tuna and salmon. There are significant differences in the manner in which these fish are handled. Tuna are normally processed from the frozen state. Following thawing, they are eviscerated by hand and the heads may be removed from larger fish. They are then precooked in large chambers and cooled, usually overnight. The white muscle components are removed by hand to go into the canning line, while red muscle is processed for pet food and skin and bones are saved for fish meal production. Cans may be labeled in a variety of styles (solid pack, chunk, flaked, or grated) depending on the type of pieces contained. Traditionally, vegetable oil and salt were added prior to sterilization. However, in recent years the amount of tuna packed in water is increasing substantially, and in many cases salt is no longer routinely added.

Salmon canning, however, usually employs unfrozen fish that are mechanically cleaned. Can-length portions of skin-on cleaned fish are prepared and the cans filled mechanically. The product is then retorted. The heat applied during sterilization is sufficient to soften bones and render them edible.

D. Drying, Salting, and Smoking

These older methods frequently are used in combination to insure the production of a safe product. The trend obviously is away from utilization of these approaches for preservation of significant quantities of product except in those parts of the world where more advanced technologies may not readily be applied. In other areas, these methods may be used for specialized products important to national or ethnic populations, such as pickled herring (salted) or lutefisk (dried cod). The use of smoking at the present time is primarily to impart a pleasant taste, and this method commonly is used in conjunction with some other to provide preservation. Smoked, salted products familiar to most readers include kippered herring, finnan haddie, and smoked salmon. The relative importance of these types of processing is outlined by Steinberg (1980); the book edited by Burgess *et al.* (1967) includes various sections detailing aspects of the technologies.

E. Minced Fish Products

Minced fish products are becoming much more common. They may be based on a fresh or frozen preparation made either from whole or parts of fish, including muscle residues removed mechanically from so-called fish frames after filleting. In Japan, minced fish is called *surimi* and the latter is the primary constitutent of an important jellylike product called *kamaboko*. In Asia, a variety of fish sausages are popular. In the United States, increased interest in minced fish was stimulated in part by a desire to utilize muscle tissue remaining on filleted fish, and in part by interest in underutilized fish species. Realization of the potential for economic development of these resources depended at first on the development of efficient mechanical deboning machines, and at present on the successful incorporation of the resultant minced fish into acceptable products. For many purposes, minced fish can be frozen successfully prior to its use. Obviously, this would be no problem with respect to the production of fish sticks, which was the first major use of minced fish in the United States. With other products, however, there may be more reason for concern. Thus, the functional properties of frozen minced fish are affected, primarily due to lower water-holding capacity, loss of gel-forming ability, and lowering of emulsifying capability. The extent of denaturation is thought to be related to rheological properties of fish sausages and kamaboko. Successful preparation of kamaboko is closely related to properties of the myofibrillar protein in fish muscle. This apparently is particularly true with regard to water-holding capacity, which probably explains the concern with denaturation (Suzuki, 1981).

Other potential problems with minced fish (frozen or not) include the possibility of enhanced lipid oxidation and discoloration problems. The latter may involve increasing the rate of autoxidation of oxymyoglobin and the mechanical inclusion of melanin particles from skin of fish into minced product. Also, the increased handling makes possible the introduction of an added microbial load.

Various aspects of the utilization of minced fish may be found in the articles of Keay (1976), Baker *et al.* (1977), King (1977), and Lanier and Thomas (1978).

F. New Technology

The fish processing industry, in general, is rather resistant to change. Improved procedures in many instances reflect only updating or modernization, such as a new machine or device to do more efficiently precisely what the old device accomplished. Engineering developments have taken place, however, examples being the present use of pumps by some fisherman for unloading fish from holds, the manufacture of mechanical deboners, and the occasional design of a highly specialized piece of apparatus for labor savings, such as an automatic squid-cleaning machine (Singh and Brown, 1981).

In addition to improved or new mechanical devices, there have been some new approaches to fish preservation that are worthy of note. A few of these of potential importance are outlined briefly below.

1. MODIFIED ATMOSPHERES

Modified atmospheres are employed by placing a product in a gas or mixture of gases for some period of time for storage and/or distribution. The product usually is sealed in a container, ranging from a rail car or sea van to individual pallel loads, or even retail packages. There may be changes in gas composition due to leakage, absorption, or respiratory activity. In controlled atmosphere systems, selected gases or percentages are maintained during the storage period by periodic or continuous addition. Modified atmospheres are simpler and less expensive to create and are quite satisfactory for most practical purposes.

It has been known for decades that the use of modified atmospheres, particularly those containing high levels of carbon dioxide (CO_2), was effective in inhibiting spoilage of muscle foods. One of the problems with red meats is the fact that in high concentrations of CO_2, oxymyoglobin oxidizes rapidly to metmyoglobin for reasons that are obscure at this time. This oxidation creates a severe practical problem, which, however, is much less pronounced in most fish due to their low myoglobin concentration. In spite of this, the use of modified atmospheres with fish has not been widespread.

In recent years, more attention has been given to this technology. For example, a modified atmosphere of 80% CO_2, balance air, was found to be effective in extending shelf life of fresh rockfish fillets. Aerobic plate counts and trimethylamine levels were significantly lower in fillets held in modified atmospheres than in those held in air (Fig. 3). The reason for the effectiveness of carbon dioxide is not known. There was a decline in surface pH on fillets held in such atmospheres, and measurement of oxidation – reduction potentials suggested a more aerobic environment on the surface of such fillets (Parkin *et al.*, 1981). Similar findings were made in an earlier study in which lower levels of carbon dioxide (20 and 40%) were employed (Brown *et al.*, 1980). Care must be given to possible differences among classes of product stored, but in general the higher levels of carbon dioxide are more effective.

Salmon has been shipped successfully under commercial conditions in 60% carbon dioxide (Vernath and Robe, 1979). Other recent studies deal with the potential for CO_2 atmosphere preservation of fish sealed in individual packages

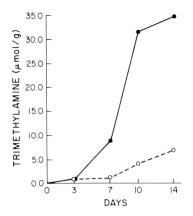

Fig. 3   Formation of trimethylamine in rockfish fillets stored in modified atmospheres at 2°C. O, 80% carbon dioxide, 20% air; ●, air control.

(Banks *et al.*, 1980) and for the use of NaCl in combination with CO_2 for extending shelf life of fish fillets (Mitsuda *et al.*, 1980).

The use of modified atmospheres with fish appears to have great potential. Some concern has been expressed about possible growth of *Clostridium botulinum* type E. However, this can be avoided by proper attention to refrigeration (which should be done in any case); also, the use of anaerobic atmospheres is not required. Wolfe (1980) has discussed the general use of modified atmospheres for meats, fish, and produce. Temperature effects, particularly those having to do with microbial growth in modified atmospheres, have been considered in detail by Ogrydziak and Brown (1981).

2. HYPOBARIC STORAGE

Hypobaric storage is a type of environmental control that involves a combination of low pressure, low temperature, high humidity, and ventilation, the latter to provide for gas exchange. In the case of hypobaric storage, no gas other than air is supplied. While this technology has been applied more to fruits and vegetables than to muscle foods, it has justifiably attracted the attention of workers interested in meats and fish. Restaino and Hill (1981) have discussed the microbiology of meats held in a hypobaric environment. Haard *et al.* (1979) suggested, based on their work with Atlantic herring and cod, that the hypobaric process might be an effective means of extending the storage life of eviscerated fish. Varga *et al.* (1980) studied fillets of cod, herring, and mackerel held in a hypobaric chamber at several temperatures. They found a 10–15% extension in keeping times of fillets stored hypobarically at 0°C. Summaries dealing with the use of hypobaric conditions include those of Mermelstein (1979) and Jamieson (1980).

3. FLAME STERILIZATION

An alternate to conventional thermal processing is flame sterilization. This is a method of commercially sterilizing canned foods that involves heating cans by direct contact with a burner flame while the cans are being rotated to insure proper convection (Leonard *et al.*, 1975). Stèriflamme is the name given to a commercial sterilizer that has been used primarily with fruits and vegetables. The process essentially uses a high-temperature, short-time process for microbial destruction with the potential for better maintainence of product quality (color, texture, and flavor). It can also be employed as a high-vacuum flame sterilization process, which is surprisingly energy-efficient (Carroad *et al.*, 1980). The potential for use of this technique with seafood is yet to be fully evaluated; however, it is certainly worthy of exploration, and some efforts are presently underway.

VI. TECHNOLOGICAL PROBLEMS AND
ORGANOLEPTIC CONSIDERATIONS

Technological problems with seafood products generally fall into two categories. The first includes quality features that have public health implications and are therefore of concern not only to the industry, but to regulatory agencies as well. Some of these are inherent in the raw material and may not be amenable to control by the processor other than by removal of product from the distribution chain, usually following some type of analysis of the constituent in question. The second category has more to do with product quality in the usual sense and with attributes that bear on the organoleptic properties of the product.

The development of methods for the assessment of fish quality is a topic that has received much attention. Two approaches appear to have been utilized most frequently. The first is the measure of nitrogenous substances such as trimethylamine (and its breakdown products), total volatile bases, volatile basic nitrogen, amino nitrogen, or ammonia. The second is the measure of degradation products of nucleotides, with the substance most commonly assayed being hypoxanthine. Other investigators have measured volatile acids, total reducing substances, surface pH, and alcohols as indices of quality deterioration. Histamine may be determined specifically because of its role in scombroid toxicity.

Microbiological assays of quality ordinarily involve some type of total plate count or determination of populations of classes of microbes, such as psychrophiles. Considerable effort has been given to development of methods to accomplish these objectives. Other microbiological determinations have had as their objective some understanding of the nature of the spoilage involved.

There are even physical means of measuring fish quality. The Torry Fish Freshness Meter (Torrymeter), for example, measures the parallel capacitance and resistance of the sample being tested. A log function of these readings has been found to correlate with sensory evaluations of fish freshness.

Ultimately, determination of fish quality must be based on organoleptic considerations. Such determinations are reported, albeit infrequently; however, with some exceptions, the level of sensory science employed in studies of fish quality has been marginal at best. There is a need for systematic comparison of sensory measurements with chemical and microbial parameters.

Jhaveri et al., (1978) have provided an informative summary of abstracts of methods used to assess fish quality. Martin et al. (1978) have discussed means of quality assessment of fresh fish. Connell's book (1980) discusses applicability of various methods and also offers a general treatment of control of fish quality. The following sections include brief discussions of some specific technological problems with seafoods that are of particular concern.

A. Chlorinated Hydrocarbon Residues

There is less concern now than there was some years ago with the contamination of fishery products with the insecticide DDT and its metabolites and with the industrial chlorinated hydrocarbon group called polychlorinated biphenyls (PCBs). This is primarily because the use of DDT has now been banned in many countries, including the United States, and the manufacture and use of PCBs is quite restricted. Nevertheless, these compounds are highly resistant to degradation and tend to accumulate in fatty tissues of organisms high in the food chain, including fish used for human consumption. Consequently, there is continuing interest in monitoring levels of these compounds. Stout and Beezhold (1981) have measured chlorinated hydrocarbon levels in fishes and shellfishes of the northeastern Pacific Ocean, while Stout et al. (1981) have surveyed chlorinated hydrocarbon residues in menhaden products. Finfishes from the northeastern Pacific rarely contained residues in excess of 1 ppm, and shellfishes measured never exceeded 0.5 ppm. The menhaden study involved samples collected over a 9-year period from 1969 to 1977. In that case, the findings only partially confirmed the hypothesis that chlorinated hydrocarbon levels have decreased with time. However, the problem clearly is becoming less acute, at least in the United States.

B. Mercury

In 1953, there occurred the now famous tragic episode of mercury poisoning of Japanese villagers living near Minamata Bay. Organic mercury compounds discharged in industrial waste fed into the bay were concentrated by fish. People consuming the affected fish ultimately became ill, and many fatalities resulted. This incident was responsible for much attention being focused on levels of mercury in seafoods generally. Extensive analyses revealed appreciable levels of mercury only in swordfish and tuna, among commercially popular fish. That these fish are affected is due to the facts that they occupy an upper echelon in the food chain and that individual fish live for many years and have a prolonged time for accumulation of mercury.

In 1969 the U.S. Food and Drug Administration (FDA) established a 0.5 ppm guideline as a maximum safe limit for mercury in fish. It subsequently developed that mercury in fish apparently is derived for the most part from natural sources rather than from industrial contamination. Samples of tuna taken from museum specimens dated as early as around 1900 were found to have levels of mercury as high as those found today. In 1979, the FDA raised the action level for mercury to 1 ppm in fish. Such levels are those at or above which FDA will have products removed from distribution and marketing. Virtually all tuna fish analyses reveal amounts below this level.

The toxic form of mercury is organic alkyl mercury, usually methyl mercury. Bacteria can convert inorganic mercury to this form, which is then transmitted through the food chain. There is evidence that selenium protects against mercury toxicity (Ganther and Sunde, 1974) and levels of selenium, particularly in marine fish, tend to follow to some degree the levels of mercury (Luten *et al.*, 1980). There are also reports that mercury levels can be reduced in comminuted fish products by washing such preparations with cysteine solutions (Spinelli *et al.*, 1973; Teeny *et al.*, 1974). A review article dealing with mercury in fish is available (Newberne, 1974).

C. Fishiness

A fishy smell certainly is present in seafoods of poor quality. It is unfortunate that much of the world's population associates such odors with all fish. The compounds responsible are usually volatile amines, and it appears that trimethylamine is one of the most likely culprits. As already noted, this compound is formed readily by a host of bacteria, using as a substrate the odorless trimethylamine oxide found normally in fish. The problem is minimized by handling procedures that reduce bacterial spoilage.

Considerable research effort has been given to analytical procedures for measuring TMA levels. Such levels are used frequently as a chemical index of quality in technological studies of fish products. The most common method of analysis involves Dyer's test, formation of the picrate salt of TMA and its determination by spectrophotometric assay. Attention has been given to extraction procedures (Murray and Gibson, 1972a,b; Bullard and Collins, 1980). Castell *et al.* (1974) have described a procedure that allows for simultaneous measurement of TMA and dimethylamine (DMA) in fish. A simple method for TMA analyses that should be suitable for routine quality control work has been reported by Chang *et al.* (1976). This method employs a TMA-specific electrode. Assay of TMA in canned products is not common. However, Tokunaga (1975) has suggested that the ratio of DMA to TMA may be a useful quality index for canned albacore tuna.

D. Rancidity

Since many fish contain highly unsaturated fatty acids, sometimes in fairly large quantity, it is not surprising that the development of oxidative rancidity is an important technological problem with seafoods. Inasmuch as bacterial spoilage usually limits the storage life of fresh fish to a few days, or at most a few weeks, rancidity is more likely to be a problem in frozen products. Other types of products in which lipid oxidation is likely include those prepared by traditional salting and drying methods (in which case, however, rancid flavors may be considered by some populations as desirable) and by-products such as fish meal, where rapid lipid oxidation in the heated, dried final product may actually present a fire danger if not properly controlled.

The nature of the oxidation is well understood to be a free radical chain reaction. It may be an autoxidation reaction or may involve catalysts such as metal ions or heme compounds. For example, the more rapid oxidation usually noted in red muscle is probably due to its content of myoglobin, which can act as a catalyst, as well as to the larger amounts of lipid ordinarily present. The freezing process may enhance the possibility of oxidation by altering cell integrity and making contact of reactants more likely. Dehydration, if not controlled, can have the same effect, and lipid oxidation in unduly dehydrated frozen products can be severe.

Several means may be employed to minimize oxidation. In frozen products, the most obvious is the use of the lowest possible temperature for storage. What is ideal from a chemical standpoint may not, of course, be practical. Packaging of product is important. This may range from proper glazing, which minimizes dehydration and creates a barrier for oxygen access, to vacuum packaging. Chemical additives, usually introduced as dipping agents, may include antioxidants and chelating agents. In those products used for human foods, examples of such additives include ascorbate (or erythorbate), EDTA, and citric acid. Ethoxyquin frequently is added to fish meal. A vitamin E preparation is added to cat food made from tuna red meat. Otherwise a deficiency of this nutrient may be induced in cats that feed exclusively on such products and, as a result, have an unusually large intake of highly unsaturated fatty acids.

E. Texture

Texture is usually not a problem with fresh seafood products. There is no particular concern with toughness, as in the case of red meats, and certainly aging would not be recommended. As noted earlier, there is relatively little connective tissue in fish.

There are, however, problems with adverse texture changes in some seafoods as a result of freezing and frozen storage. Usually, in the case of finfish, such changes are described as resulting in dryness, and, sometimes, toughness. In shellfish, products may also become tough and dry and may be described as "cardboardy." Some of these problems are serious enough to influence the technology employed. Thus, lobsters are almost invariably cooked before freezing, since some of those frozen raw may be unpredictably tough when subsequently thawed and cooked.

Changes in water-binding capacity, properties of muscle proteins, and aggregation undoubtedly effect texture. Features of such changes are included in Section V,B on freezing of seafoods.

One of the more unusual problems with texture in seafoods is that found in abalone. The muscle from this animal that is used for food (the "foot") is extraordinarily tough. The fresh or frozen tissue must be subjected to vigorous pounding prior to cooking, or the resulting product may be so tough as to be virtually inedible. Tenderizers, marinades, and the like are without their usual effect. Olley and Thrower (1977) have described properties of abalone, which they understandably label an esoteric food.

F. Struvite

Struvite is a hard, clear crystal of magnesium ammonium phosphate ($MgNH_4PO_4 \cdot 6H_2O$) sometimes found in canned fish, such as tuna, and canned shellfish, such as crab and shrimp. Consumers are particularly bothered by this problem inasmuch as the crystals closely resemble fragments of glass. Struvite readily dissolves in dilute acid, such as vinegar, and this fact may be used to convince purchasers that the product is not harmful. However, it is best avoided entirely. The formation of struvite in tuna is related to pH and is more likely to occur in fish having pH values > 6.0. Mixing of different fish is one means of alleviating the problem. A more direct approach is to add sodium acid pyrophosphate (0.5% by weight) to the cans prior to heat sterilization. This functions as a sequestrant and removes magnesium ions, thereby preventing struvite formation.

G. Discolorations

One of the most important quality attributes of seafood products is that of color. Surface appearance is noted by consumers prior to any evaluation of odor, taste, or texture. A variety of discolorations are sometimes encountered in fish. Some appear to be unique to seafoods while others have been noted with other products as well. Among the latter would be sulfide blackening of canned products and nonenzymatic surface browning; such browning in fish may be associated with lipid oxidation, which can provide an ample supply of carbonyl compounds to react with amino groups which are always present. There may be fading of carotenoid type colors in some fish; this too may be related to lipid oxidation and concomitant oxidation of these pigments. Some specific discoloration problems with fish that are of commercial importance are listed below.

1. BLACK SPOT (MELANOSIS) IN SHRIMP

This is an enzymatic problem caused by the formation of melanin pigments by polyphenoloxidases. It may be seen in a variety of crustaceans. In industrial practice, sodium bisulfite or sodium metabisulfite is employed to inhibit the reaction in shrimp.

2. BLUEING IN CRAB

This is seen as a blue or blackish surface discoloration in cooked crab. Melaninlike compounds may be involved in blueing, but the exact cause is not clear. Copper-containing hemocyanin is the respiratory protein in crabs, and involvement of this compound has been suspected. Babbitt et al. (1973) have suggested that the reaction(s) may involve oxidation of phenols that may be enzymatically initiated, but which then proceeds further nonenzymatically in the presence of metals under alkaline conditions. The use of some sort of acid to lower the pH seems to prevent the blue discoloration. Boon (1975) has reviewed general aspects of discoloration in processed crabmeat.

3. GREENING IN TUNA

This term has several meanings within the industry. It is sometimes used to indicate discoloration phenomena in a general sense but is more appropriately applied to a greenish pigment sometimes seen in canned tuna. This pigment is produced when tuna myoglobin is denatured in the presence of trimethylamine oxide (TMAO) and cysteine. The denaturation apparently exposes the sulfhydryl group in the protein, which then reacts with free cysteine to form a disulfide bond between this amino acid and the denatured myoglobin. TMAO presumably acts as a mild oxidizing agent to promote the disulfide bond formation (Koizumi and Matsuura, 1967; Grosjean et al., 1969). Khayat (1978) has reported that the production of this green pigment could be inhibited by addition of cysteine, homocysteine, and antioxidants such as ascorbic acid and propyl gallate.

Another important discoloration reaction in tuna occurs when the normal, pink denatured globin ferrohemochrome of cooked tuna flesh has undergone oxidation to the ferric derivative, which is tan or brown. This problem is sometimes included in the general industry designation of greening although the pigment is not green. Various of these off colors in tuna have been discussed by Tomlinson (1966).

Yet another discoloration problem is that of orange tuna. This, however, apparently does not involve heme groups but is due to nonenzymatic browning involving glycolytic intermediates, particularly glucose 6-phosphate.

4. OXIDATION OF OXYMYOGLOBIN

Oxymyoglobin is responsible for the bright, light pink surface color of most cut or skinned fish surfaces. When its ferrous iron undergoes oxidation to the ferric state, metmyoglobin is formed. The latter pigment is brown and lends an undesirable tan or brownish cast (depending on concentration) to the fish. The reaction is faster at lower pH values, is markedly slowed by lowering the temperature, and may be more rapid in fish than in other myoglobins (Brown and Mebine, 1969). The oxidation apparently proceeds via deoxymyoglobin as an intermediate; hence the presence of moderate concentrations of oxygen actually helps prevent it from taking place by shifting the equilibrium involved to oxymyoglobin. A review dealing with the role of myoglobin pigments in muscle foods, including fish, is available (Livingston and Brown, 1981).

At present, the best means of inhibiting this oxidation is the maintainance of low temperatures, ideally just above freezing. Both nonenzymatic and enzymatic reducing systems are known to exist in fish as well as mammalian systems (Brown and Snyder, 1969; Shimizu and Matsuura, 1971; Al-Shaibani et al., 1977a,b). It is not yet clear how best to utilize such systems in a practical sense. A recent detailed investigation of metmyoglobin reductase from beef heart muscle (Hagler et al., 1979) may provide a stimulus for similar investigations utilizing fish sources of the enzyme.

H. Myths

In the last section of this chapter, it seems appropriate to note briefly some persistent myths and misconceptions associated with the use of fish muscle as food. Such ideas are not limited, of course, to seafood products, but their counterparts with regard to the consumption of red meats and poultry may conjure up less interesting philosophical, not to say scientific, contemplation. Price (1979) has provided a summary of some of these misconceptions, from which the following list is taken.

1. SHELLFISH SHOULD BE EATEN ONLY IN "R" MONTHS

This idea was based originally on factors related to reproduction of European oysters; baby oyster shells in the brood chamber of adults are crunchy and definitely not oysterlike. Oysters in the United States do not brood young, but the public has mistakenly transferred this myth to a belief associated with paralytic shellfish poison.

2. MAHI-MAHI IS PORPOISE OR DOLPHIN MEAT

Mahi-Mahi is the name frequently given in restaurants to dolphin fish, a teleost caught in tropical waters. The mammalian dolphin is not even a distant relative.

3. EATING SEAFOOD WITH MILK WILL MAKE YOU SICK

This idea has been around for many years and was a notion firmly planted in the author's mind at an early age. There is no known basis in fact.

4. FISH IS BRAIN FOOD

Fish frequently contains high levels of phosphorus. The idea is related to this and dates back to the nineteenth century, when it was found that compounds containing phosphorus are abundant in the brain. It would be smart not to believe this misconception.

5. RAW OYSTERS ARE AN APHRODISIAC

The concept presumably has to do with (1) the oyster's content of cholesterol and (2) the sterol structure of sex hormones. This myth probably has done wonders for the sale of raw oysters. It is tempting to suggest it as an area for future research.

VII. SUMMARY AND RESEARCH NEEDS

One important aspect that differentiates aquatic animals from others whose muscle is used for food is that of diversity. An extraordinarily broad collection constitutes seafoods, and the differences among these animals cannot be ignored. Consideration must be given to important comparative biochemical aspects. These include occurrence and properties of red and white muscle, and the fact that red muscle in some fish has critical metabolic significance. Fish, in general, contain highly unsaturated fatty acids, more so than land animals. Their myofibrillar and sarcoplasmic proteins have properties similar to those of mammals, but may be somewhat less stable, and the myofibrillar proteins may have a greater tendency to aggregate. The amount of connective tissue protein is small. There are usually present many low molecular weight muscle constituents. Among those of particular interest are imidazole compounds, including histidine, anserine, and carnosine, and trimethylamine oxide.

From a nutritional standpoint, seafoods provide a good source of high-quality protein and have relatively generous amounts of vitamins and minerals. The fats, while highly unsaturated, do not include large quantities of essential fatty acids. The cholesterol content is generally thought to be high in shellfish, but early values may be erroneous. Some fish contain an antithiamine factor which is usually designated as thiaminase.

Fish undergo rather rapid microbial spoilage involving organisms similar to those found in other muscle foods. There are some unique problems with seafoods, including the potential for microbial decarboxylation of histidine, resulting in toxic levels of histamine, and the reduction of trimethylamine oxide to trimethylamine, a major contributor to the fishy smell of spoiling fish. *Vibrio parahemolyticus* is a potential hazard with fish, as is *Clostridium botulinum* type E. A variety of toxins are important, including those involved in ciguatera, puffer fish poisoning, and paralytic shellfish poisoning.

Temperature control is particularly important for the storage and preservation of fish, and temperatures near freezing are ideal. Frozen fish may be of high quality, but there may be problems with the use of this technology involving protein denaturation, aggregation, and alterations of texture and functional properties. Fats may undergo hydrolysis as well as oxidation. The heme pigments are particularly susceptible to enhanced oxidation in frozen products.

New methods showing promise include the use of modified atmospheres, hypobaric storage, and flame sterilization. Many technological problems remain. Some, such as contamination of seafoods with chlorinated hydrocarbons and mercury, relate to the raw material and their importance seems to be diminishing. Others, such as oxidative rancidity and undesirable changes in texture and color, are yet to be fully controlled.

A number of research needs may be listed.

1. Better understanding of protein denaturation, particularly in frozen fish, including elucidation of cross-linking phenomena and the role of formaldehyde, and the effects of these changes on water-binding capacity in particular and functional properties in general.

2. Development of means of enhancing functional properties in fish, such as in minced fish products.

3. Studies of the enzymatic and/or nonenzymatic production of formaldehyde and dimethylamine from trimethylamine oxide.

4. Understanding of problems with texture, particularly in crustaceans, with abalone as a special case.

5. The role of pressure on enzymatic and other reactions.

6. More comprehensive assays of nutrients in seafoods, using sampling procedures adequate to control for species and seasonal variation, both of which are extremely critical in fish.

7. Evaluation of the potential for expanded use of new freezing technologies, including cryogenics, and the effectiveness of cryoprotective agents.

8. Studies of microbial growth on products held in modified atmospheres, including establishing the nature of the organisms involved in posttreatment spoilage, means of further inhibition of the growth of such organisms, and the relative potential for growth of *Clostridium botulinum* type E.

9. Determination of the roles of anserine and carnosine in fish, and their contribution, if any, to the scombroid toxicity problem.

10. The development of improved means for measuring lipid oxidation and of new and more effective antioxidants and packaging materials.

11. Investigations of factors influencing oxidation of oxymyoglobin, including the relationship of metmyoglobin reductases to such oxidation *in vivo* and the basis for the enhancement of this oxidation by carbon dioxide.

Finally, it should be noted that much work is needed dealing generally with comparative biochemistry of fish. Such studies will not only aid in understanding the biology of these animals as well as basic biochemical and physiological phenomena but may materially aid the scientist interested in the conversion of fish muscle to food.

REFERENCES

Ackman, R. G. (1980). *In* "Advances in Fish Science and Technology" (J. J. Connell, ed.), pp. 86–103. Fishing News Books Ltd., Farnham, Surrey, England.
Aitken, A., and Connell, J. J. (1977). *Bull. Inst. Int. Froid, Annex* **1**, 187–191.
Al-Shaibani, K. A., Price, R. J., and Brown, W. D. (1977a). *J. Food Sci.* **42**, 1013–1015.
Al-Shaibani, K. A., Price, R. J., and Brown, W. D. (1977b). *J. Food Sci.* **42**, 1156–1158.
Amano, H., and Tsuyuki, H. (1975). *Bull. Jpn. Soc. Sci. Fish.* **41**, 885–894.
Amano, H., Hashimoto, K., and Matsuura, F. (1976). *Bull. Jpn. Soc. Sci. Fish.* **42**, 577–581.
Anonymous (1981). "Fisheries of the United States, 1980." U.S. Dept. of Commerce, Washington, D.C.
Arai, K., Hasnain, A., and Takano, Y. (1976). *Bull. Jpn. Soc. Sci. Fish.* **42**, 687–695.
Arnold, S. H., and Brown, W. D. (1978). *Adv. Food Res.* **24**, 113–154.
Babbitt, J. K., Law, D. K., and Crawford, D. L. (1973). *J. Food Sci.* **38**, 1101–1103.
Baker, R. C., Regenstein, J. M., Raccach, M., and Darfler, J. M. (1977). "Frozen Minced Fish." New York Sea Grant Institute, Albany.
Balestrieri, C., Colonna, G., Giovane, A., Irace, G., Servillo, L., and Tota, B. (1978). *Comp. Biochem. Physiol. B* **60B**, 195–199.
Banks, A., and Waterman, J. J. (1968). *In* "The Freezing Preservation of Foods" (D. K. Tressler, W. B. Van Arsdel, and M. J. Copley, eds), Vol. 3, pp. 250–265. Avi Publ. Co., Westport, Connecticut.
Banks, H., Nickelson, R., II, and Finne, G. (1980). *J. Food Sci.* **45**, 157–162.
Bannister, J. V., and Bannister, W. H. (1976). *Comp. Biochem. Physiol. B* **53B**, 57–60.
Bauer, B. A., and Eitenmiller, R. R. (1974). *J. Food Sci.* **39**, 10–14.
Benson, A. A., and Lee, R. F. (1975). *Sci. Am.* **232**, 76–86.

Bito, M. (1964). *Bull. Jpn. Soc. Sci. Fish.* 30, 847–857.

Bito, M. (1965a). *Bull. Jpn. Soc. Sci. Fish.* 31, 534–539.

Bito, M. (1965b). *Bull. Jpn. Soc. Sci. Fish.* 31, 540–545.

Bonnett, J. C., Sidwell, V. D., and Zook, E. G. (1974). *Mar. Fish. Rev.* 36(2), 8–14.

Boon, D. D. (1975). *J. Food Sci.* 40, 756–761.

Borgstrom, G., ed. (1962). "Fish as Food," Vol. 2 Academic Press, New York.

Bramsnaes, F. (1969). *In* "Quality and Stability of Frozen Foods" (W. B. Van Arsdel, M. J. Copley, and R. L. Olson, eds.), pp. 217–236. Wiley, New York.

Brown, W. D., and Dolev, A. (1963). *J. Food Sci.* 28, 211–213.

Brown, W. D., and Mebine, L. B. (1969). *J. Biol. Chem.* 244, 6696–6701.

Brown, W. D., and Snyder, H. E. (1969). *J. Biol. Chem.* 244, 6702–6706.

Brown, W. D., Martinez, M., Johnstone, M., and Olcott, H. S. (1962). *J. Biol. Chem.* 237, 81–84.

Brown, W. D., Albright, M., Watts, D. A., Heyer, B., Spruce, B., and Price, R. J. (1980). *J. Food Sci.* 45, 93–96.

Bryan, F. L. (1980). *J. Food Prot.* 43, 859–876.

Bullard, F. A., and Collins, J. (1980). *Fish. Bull.* 78, 465–473.

Burgess, G. H. O., Cutting, C. L., Lovern, J. A., and Waterman, J. J., eds. (1967). "Fish Handling and Processing," Chem. Publ. Co., New York.

Carpenter, K. J. (1980). *In* "Advances in Fish Science and Technology" (J. J. Connell, ed.), Fishing News Books Ltd., Farnham, Surrey, England.

Carroad, P. A., Leonard, S. J., Heil, J. R., Wolcott, T. K., and Merson, R. L. (1980). *J. Food Sci.* 45, 696–699.

Castell, C. H., Smith, B., and Neal, W. (1971). *J. Fish. Res. Board Can.* 28, 1–5.

Castell, C. H., Smith, B., and Dyer, W. J. (1973a). *J. Fish. Res. Board Can.* 30, 1205–1213.

Castell, C. H., Neal, W. E., and Dale, J. (1973b). *J. Fish. Res. Board Can.* 30, 1246–1248.

Castell, C. H., Smith, B., and Dyer, W. J. (1974). *J. Fish. Res. Board Can.* 31, 383–389.

Chang, G. W., Chang, W. L., and Lew, K. B. K. (1976). *J. Food Sci.* 41, 723–724.

Chichester, C. O., and Graham, H. D., eds. (1973). "Microbial Aspects of Fishery Products." Academic Press, New York.

Childs, E. A. (1973). *J. Food Sci.* 38, 718–719.

Childs, E. A. (1974). *J. Fish. Res. Board Can.* 31, 1142–1144.

Chu, G. H., and Sterling, C. (1970). *J. Text. Stud.* 2, 214–222.

Chung, C.-S., Richards, E. G., and Olcott, H. S. (1967). *Biochemistry* 6, 3154–3161.

Cobb, B. F., III, and Hyder, K. (1972). *J. Food Sci.* 37, 743–750.

Collette, B. B. (1978). *In* "The Physiological Ecology of Tunas" (G. D. Sharp and A. E. Dizon, eds.), pp. 7–39. Academic Press, New York.

Connell, J. J. (1960). *Biochem. J.* 75, 530–538.

Connell, J. J. (1963). *Biochim. Biophys. Acta* 74, 374–385.

Connell, J. J. (1975). *J. Sci. Food Agric.* 26, 1925–1929.

Connell, J. J. (1980). "Control of Fish Quality." Fishing News Books Ltd., Farnham, Surrey, England.

Connell, J. J., and Howgate, P. F. (1959). *Biochem. J.* 71, 83–86.

Connell, J. J., and Olcott, H. S. (1961). *Arch. Biochem. Biophys.* 94, 128–135.

Cox, K. W. (1962). "California Abalones, Family Haliotidae." Resources Agency of California, Dept. of Fish and Game, Sacramento.

Dale, B., and Yentsch, C. M. (1978). *Oceanus* 21, 41–49.

Dassow, J. A. (1968a). *In* "The Freezing Preservation of Foods" (D. K. Tressler, W. B. Van Arsdel, and M. J. Copley, eds.), Vol. 2, pp. 197–208. Avi Publ. Co., Westport, Connecticut.

Dassow, J. A. (1968b). *In* "The Freezing Preservation of Foods" (D. K. Tressler, W. B. Van Arsdel, and M. J. Copley, eds.), Vol. 3, pp. 266–275. Avi Publ. Co., Westport, Connecticut.

Deng, J. C. (1981). *J. Food Sci.* **46**, 62–65.
Dyer, W. J. (1952). *J. Fish Res. Board Can.* **8**, 314–324.
Eitenmiller, R. R. (1974). *J. Food Sci.* **39**, 6–9.
Fieger, E. A., and Novak, A. F. (1968a). *In* "The Freezing Preservation of Foods" (D. K. Tressler, W. B. Van Arsdel, and M. J. Copley, eds.), Vol. 2, pp. 209–215. Avi Publ. Co., Westport, Connecticut.
Fieger, E. A., and Novak, A. F. (1968b). *In* "Freezing Preservation of Foods" (D. K. Tressler, W. B. Van Arsdel, and M. J. Copley, eds.), Vol. 3, pp. 276–288. Avi Publ. Co., Westport, Connecticut.
Fisher, W. K., and Thompson, E. O. P. (1979). *Aust. J. Biol. Sci.* **32**, 277–294.
Fisher, W. K., Koureas, D. D., and Thompson, E. O. P. (1980). *Aust. J. Biol. Sci.* **33**, 153–167.
Fosmire, G. J., and Brown, W. D. (1976). *Comp. Biochem. Physiol. B* **55B**, 293–299.
Ganther, H. E., and Sunde, M. L. (1974). *J. Food Sci.* **39**, 1–5.
Geist, G. M., and Crawford, D. L. (1974). *J. Food Sci.* **39**, 548–551.
Giddings, G. G., and Hill, L. H. (1976). *J. Food Sci.* **41**, 455–457.
Giddings, G. G., and Hill, L. H. (1978). *J. Food Process. Preserv.* **2**, 249–264.
Gordon, D. T., Roberts, G. L., and Heintz, D. M. (1979). *J. Agric Food Chem.* **27**, 483–490.
Groninger, H. S., Jr. (1959). *U.S., Fish Wildl. Serv., Spec. Sci. Rep.-Fish.* **333**.
Groninger, H. S., Jr., and Miller, R. (1975). *J. Food Sci.* **40**, 327–330.
Groninger, H. S., Jr. and Miller, R. (1979). *J. Agric. Food Chem.* **27**, 949–955.
Grosjean, O. K., Cobb, B. F., Mebine, B., and Brown, W. D. (1969). *J. Food Sci.* **34**, 404–407.
Haard, N. F., Martins, I., Newbury, R., and Botta, R. (1979). *J. Inst. Can. Sci. Technol. Aliment.* **12**, 84–87.
Hagler, L., Coppes, R. I., Jr., and Herman, R. H. (1979). *J. Biol. Chem.* **254**, 6505–6514.
Halstead, B. W. (1965). "Poisonous and Venomous Marine Animals of the World," Vol. 1. U.S. Govt. Printing Office, Washington, D.C.
Halstead, B. W. (1967). "Poisonous and Venomous Marine Animals of the World," Vol. 2. U.S. Govt. Printing Office, Washington, D.C.
Halstead, B. W. (1970). "Poisonous and Venomous Marine Animals of the World," Vol. 2 (cont'd.). U.S. Govt. Printing Office, Washington, D.C.
Halstead, B. W. (1978). "Poisonous and Venomous Marine Animals of the World." Darwin Press, Inc., Princeton, New Jersey.
Halstead, B. W. (1980). "Dangerous Marine Animals," 2 ed. Cornell Maritime Press, Centreville, Maryland.
Hamoir, G. (1953). *Nature (London)* **171**, 345–346.
Hamoir, G., and Konosu, S. (1965). *Biochem. J.* **96**, 85–97.
Hamoir, G., McKenzie, H. A., and Smith, M. B. (1960). *Biochim. Biophys. Acta* **40**, 141–149.
Hamoir, G., Focant, B., and Disteche, M. (1972). *Comp. Biochem. Physiol. B* **41B**, 665–674.
Hanaoka, K., and Toyomizu, M. (1979). *Bull. Jpn. Soc. Sci. Fish.* **45**, 465–468.
Harada, K., and Yamada, K. (1973). *J. Shimonoseki Univ. Fish.* **22**, 77–94.
Harada, K., Takeda, J., and Yamada, K. (1970). *J. Shimonoseki Univ. Fish.* **18**, 287–295.
Harada, K., Yamamoto, Y., and Yamada, K. (1971). *J. Shimonoseki Univ. Fish.* **19**, 105–114.
Harada, K., Deriha, T., and Yamada, K. (1972). *J. Shimonoseki Univ. Fish.* **20**, 249–264.
Hardy, R. (1980). *In* "Advances in Fish Science and Technology" (J. J. Connell, ed.), pp. 103–111. Fishing News Books Ltd., Farnham, Surrey, England.
Hardy, R., McGill, A. S., and Gunstone, F. D. (1979). *J. Sci. Food Agric.* **30**, 999–1006.
Hasnain, A., Tamura, H., and Arai, K. (1976). *Bull. Jpn. Soc. Sci. Fish.* **42**, 783–791.
Heen, E., and Kreuzer, R., eds. (1962). "Fish in Nutrition." Fishing News (Books) Ltd., London.
Higashi, H. (1961). *In* "Fish as Food" (G. Borgstrom, ed.), Vol. 1, pp. 441–442. Academic Press, New York.

Higerd, T. B. (1980). *Proc. Annu. Trop. Subtrop. Fish. Technol. Conf. Am., 5th, 1980* pp. 62–63.

Hirs, C. H. W., and Olcott, H. S. (1964). *Biochim. Biophys. Acta* **82**, 178–180.

Hobbs, G. (1976). *Adv. Food Res.* **22**, 135–185.

Hochachka, P. W., Hulbert, W. C., and Guppy, M. (1978). *In* "The Physiological Ecology of Tunas" (G. D. Sharp and A. E. Dizon, eds.), pp. 153–154. Academic Press, New York.

Horie, N., Tsuchiya, T., and Matsumoto, J. J. (1975). *Bull. Jpn. Soc. Sci. Fish.* **41**, 1039–1045.

Idler, D. R., and Wiseman, P. (1972). *J. Fish. Res. Board Can.* **29**, 385–398.

Ishihara, I., Yasuda, M., and Morooka, H. (1972). *Bull. Jpn. Soc. Sci. Fish.* **38**, 1281–1287.

Ishihara, T., Kinari, H., and Yasuda, M. (1973). *Bull. Jpn. Soc. Sci. Fish.* **39**, 55–59.

Jamieson, W. (1980). *Food Technol. (Chicago)* **34**(3), 64–71.

Jhaveri, S., Montecalvo, J., Jr., Karakoltsidis, P., and Howe, J. (1978). Mar. Tech. Rep. No. 69. University of Rhode Island, Kingston.

Joseph, J., Klawe, W., and Murphy, P. (1980). "Tuna and Billfish — Fish Without a Country." Inter-American Tropical Tuna Commission, La Jolla, California.

Kayama, M., and Nevenzel, J. C. (1974). *Mar. Biol.* **24**, 279–285.

Keay, J. N., ed. (1976). "Proceedings of a Conference on Production and Utilization of Mechanically Recovered Fish Flesh." Torry Research Station, Aberdeen, Scotland.

Khayat, A. (1978). *J. Food Technol.* **13**, 117–127.

King, F. J. (1977). *Mar. Fish. Rev.* **39**(4), 1–4.

Koizumi, C., and Matsuura, F. (1967). *Bull. Jpn. Soc. Sci. Fish.* **33**, 839–842.

Konosu, S., Hashimoto, K., and Matsuura, F. (1958). *Bull. Jpn. Soc. Sci. Fish.* **24**, 563–566.

Konosu, S., Hamoir, G., and Pechere, J. F. (1965). *Biochem. J.* **96**, 98–112.

Konosu, S., Watanabe, K. and Shimizu, T. (1974). *Bull. Jpn. Soc. Sci. Fish.* **40**, 909–915.

Koury, B. J., and Spinelli, J. (1975). *J. Food Sci.* **40**, 58–61.

Kritchevsky, D., and DeHoff, J. L. (1978). *J. Food Sci.* **43**, 1786–1787.

Lanier, T. C., and Thomas, F. B. (1978). "Minced Fish: Its Production and Use," UNC Sea Grant Publ. North Carolina State University, Raleigh.

Lattman, E. E., Nockolds, C. E., Kretsinger, R. H., and Love, W. E. (1971). *J. Mol. Biol.* **60**, 271–277.

Lee, K., Groninger, H. S., and Spinelli, J. (1981). *Mar. Fish. Rev.* **43**(3), 14–19.

Leonard, S., Merson, R. L., Marsh, G. L., York, G. K., Heil, J. R., and Wolcott, T. (1975). *J. Food Sci.* **40**, 246–249.

Lerke, P. A., Werner, S. B., Taylor, S. L., and Guthertz, L. S. (1978). *West. J. Med.* **129**, 381–386.

Liston, J. (1973). *In* "Microbial Safety of Fishery Products" (C. O. Chichester and H. D. Graham, eds.), pp. 203–213. Academic Press, New York.

Liston, J. (1980). *In* "Advances in Fish Science and Technology" (J. J. Connell, ed.), Fishing News Books Ltd., Farnham, Surrey, England.

Livingston, D. J., and Brown, W. D. (1981). *Food Technol. (Chicago)* **35**(5), 244–252.

Love, R. M. (1979). *J. Sci. Food Agric.* **30**, 433–438.

Love, R. M. (1980). "The Chemical Biology of Fishes," Vol. 2, pp. 350–387. Academic Press, New York.

Lukton, A., and Olcott, H. S. (1958). *Food Res.* **23**, 611–618.

Luna, Z. G., Marzan, A. M., Montilla, A. A., and Caasi, P. I. (1968). *Philipp. J. Sci.* **97**, 145–151.

Luten, J. B., Ruiter, A., Ritskes, T. M., Rauchbaar, A. B., and Riekwel-Body, G. (1980). *J. Food Sci.* **45**, 416–419.

Mackie, I. M., and Connell, J. J. (1964). *Biochim. Biophys. Acta.* **93**, 544–552.

Mackie, I. M., and Ritchie, A. H. (1974). *Proc. Int. Congr. Food Sci. Technol., 4th, 1974* Vol. I, pp. 29–38.

Mackie, I. M., and Thomson, B. W. (1974). *Proc. Int. Congr. Food Sci. Technol., 4th, 1974* Vol. I, pp. 243–250.

Mai, J., Shetty, J. K., Kan, T., and Kinsella, J. E. (1980). *J. Agric. Food Chem.* **28**, 884–885.
Makinodan, Y., and Ikeda, S. (1976a). *Bull. Jpn. Soc. Sci. Fish.* **42**, 239–247.
Makinodan, Y., and Ikeda, S. (1976b). *Bull. Jpn. Soc. Sci. Fish.* **42**, 665–670.
Makinodan, Y., and Ikeda, S. (1977). *J. Food Sci.* **42**, 1026–1033.
Makinodan, Y., Hirotsuka, M., and Ikeda, S. (1979). *J. Food Sci.* **44**, 1110–1117.
Makinodan, Y., Hirotsuka, M., and Ikeda, S. (1980). *Bull. Jpn. Soc. Sci. Fish.* **46**, 1507–1510.
Mao, W., and Sterling, C. (1970a). *J. Text. Stud.* **1**, 338–341.
Mao, W., and Sterling, C. (1970b). *J. Text. Stud.* **1**, 484–490.
Martin, R. E., Gray, R. J. H., and Pierson, M. D. (1978). *Food Technol. (Chicago)* **32**(5), 188–192.
Matsumoto, J. J. (1980). *In* "Chemical Deterioration of Proteins" (J. R. Whitaker and M. Fujimaki, eds.), pp. 95–124. Am. Chem. Soc., Washington, D.C.
Matsumoto, J. J., and Matsuda, E. (1967). *Bull. Jpn. Soc. Sci. Fish.* **33**, 224–228.
Matsumoto, J. J., Hosoda, H., and Tsuchiya, T. (1980). *Bull. Jpn. Soc. Sci. Fish.* **46**, 1131–1135.
Matsuura, F., and Hashimoto, K. (1955). *Bull. Jpn. Soc. Sci. Fish.* **20**, 946–950.
Mermelstein, N. H. (1979). *Food Technol. (Chicago)* **33**(7), 32–40.
Mitsuda, H., Nakajima, K., Mizuno, H., and Kawai, F. (1980). *J. Food Sci.* **45**, 661–666.
Morlid, E. (1981). *Adv. Protein Chem.* **34**, 93–166.
Motil, K. J., and Scrimshaw, N. S. (1979). *Toxicol. Lett.* **3**, 219–223.
Murray, C. K., and Gibson, D. M. (1972a). *J. Food Technol.* **7**, 35–46.
Murray, C. K., and Gibson, D. M. (1972b). *J. Food Technol.* **7**, 47–51.
Newberne, P. M. (1974). *CRC Crit. Rev. Food Technol.* **4**, 311–335.
Noguchi, S., and Matsumoto, J. J. (1970). *Bull. Jpn. Soc. Sci. Fish.* **36**, 1078–1087.
Ogrydziak, D. M., and Brown, W. D. (1982). *Food Technol. (Chicago)* **36**(5), 86–96.
Oguni, M., Kubo, T., and Matsumoto, J. J. (1975). *Bull. Jpn. Soc. Sci. Fish.* **41**, 1113–1123.
Ohnishi, M., Tsuchiya, T., and Matsumoto, J. J. (1978). *Bull. Jpn. Soc. Sci. Fish.* **44**, 755–762.
Olley, J., and Thrower, S. J. (1977). *Adv. Food Res.* **23**, 143–186.
Parkin, K. L., Wells, M. J., and Brown, W. D. (1981). *J. Food Sci.* **47**, 181–184.
Pearson, J. A. (1977). *CSIRO Food Res. Q.* **37**, 33–39.
Pearson, J. A. (1978). *CSIRO Food Res. Q.* **38**, 62–64.
Pechere, J. F., and Focant, B. (1965). *Biochem. J.* **96**, 113–118.
Peters, J. A. (1968a). *In* "The Freezing Preservation of Foods" (D. K. Tressler, W. B. Van Arsdel, and M. J. Copley, eds.), Vol. 2, pp. 216–223. Avi Publ. Co., Westport, Connecticut.
Peters, J. A. (1968b). *In* "The Freezing Preservation of Foods" (D. K. Tressler, W. B. Van Arsdel, and M. J. Copley, eds.), Vol. 3, pp. 289–294. Avi Publ. Co., Westport, Connecticut.
Phillips, B. F., Cobb, J. S., and George, R. W. (1980). *In* "The Biology and Management of Lobsters" (J. S. Cobb and B. F. Phillips, eds.), Vol 1, pp. 1–82. Academic Press, New York.
Porzio, M. A., Tang, N., and Hilker, D. M. (1973). *J. Agric. Food Chem.* **21**, 308–310.
Poulter, G. R., and Lawrie, R. A. (1977). *J. Sci. Food Agric.* **28**, 701–709.
Price, R. J. (1979). "Marine Advisory Publication," Leafl. No. 21120, Div. Agric. Sci., University of California.
Reddi, P. K., Constantinides, S. M., and Dymsza, H. A. (1972). *J. Food Sci.* **37**, 643–648.
Restaino, L., and Hill, W. M. (1981). *J. Food Prot.* **44**, 535–538.
Rheinheimer, G. (1974). "Aquatic Microbiology." Wiley, New York.
Rice, R. H., Watts, D. A., and Brown, W. D. (1979). *Comp. Biochem. Physiol. B* **62B**, 481–487.
Ronsivalli, L. J., and Baker, D. W., II (1981). *Mar. Fish. Rev.* **43**(4), 1–15.
Rossi-Fanelli, A., and Antonini, E. (1955). *Arch. Biochem. Biophys.* **58**, 498–500.
Sakai, J., and Matsumoto, J. J. (1981). *Comp. Biochem. Physiol. B* **68B**, 389–395.
Sargent, J. R. (1976). *In* "Biochemical and Biophysical Perspectives in Marine Biology" (D. C. Malins and J. R. Sargent, eds.), Vol. 3, pp. 149–212. Academic Press, New York.
Scheuer, P. J. (1970). *Adv. Food Res.* **18**, 141–161.

Scheuer, P. J. (1977). *Acc. Chem. Res.* **10**, 33–39.

Sharp, G. D., and Dizon, A. E., eds. (1978). "The Physiological Ecology of Tunas." Academic Press, New York.

Sharp, G. D., and Pirages, S. (1978). *In* "The Physiological Ecology of Tunas" (G. D. Sharp and A. E. Dizon, eds.), p. 74. Academic Press, New York.

Shenouda, S. Y. K. (1980). *Adv. Food Res.* **26**, 275–311.

Shewan, J. M. (1977). *In* "The Bacteriology of Fresh and Spoiling Fish and the Biochemical Changes Induced by Bacterial Action." *Proc. Trop. Prod. Inst. Conf. Handl. Process. Mark. Trop. Fish, London,* pp. 51–66.

Shimizu, C., and Matsuura, F. (1971). *Agric. Biol. Chem.* **35**, 468–475.

Shimizu, Y., Machida, R., and Takenami, S. (1981). *Bull. Jpn. Soc. Sci. Fish.* **47**, 95–104.

Sidwell, V. D., Foncannon, P. R., Moore, N. S., and Bonnett, J. C. (1974). *Mar. Fish. Rev.* **36**(3), 21–35.

Sidwell, V. D., Buzzell, D. H., Foncannon, P. R., and Smith, A. L. (1977). *Mar. Fish. Rev.* **39**(1), 1–11.

Sidwell, V. D., Loomis, A. L., Loomis, K. J., Foncannon, P. R., and Buzzell, D. H. (1978a). *Mar. Fish Rev.* **40**(9), 1–20.

Sidwell, V. D., Loomis, A. L., Foncannon, P. R., and Buzzell, D. H. (1978b). *Mar. Fish. Rev.* **40**(12), 1–16.

Sikorshi, Z., Olley, J., and Kostuch, S. (1976). *CRC Crit. Rev. Food Sci. Nutr.* **8**, 97–129.

Singh, R. P., and Brown, D. E. (1981). *Calif. Agric.* **35**(7 and 8), 4–6.

Skinner, F. A., and Shewan, J. M., eds. (1977). "Aquatic Microbiology." Academic Press, New York.

Slavin, J. W. (1968a). *In* "The Freezing Preservation of Foods" (D. K. Tressler, W. B. Van Arsdel, and M. J. Copley, eds.), Vol. 2, pp. 179–196. Avi Publ. Co., Westport, Connecticut.

Slavin, J. W. (1968b). *In* "The Freezing Preservation of Foods" (D. K. Tressler, W. B. Van Arsdel, and M. J. Copley, eds.), Vol. 3, pp. 233–249. Avi Publ. Co., Westport, Connecticut.

Somero, G. N. (1978). *Biochem. Biophys. Perspect. Mar. Biol.* **4**, 1–27.

Spinelli, J., and Koury, B. (1979). *J. Agric. Food Chem.* **27**, 1104–1108.

Spinelli, J., and Koury, B. J. (1981). *J. Agric. Food Chem.* **29**, 327–331.

Spinelli, J., Steinberg, M. A., Miller, R., Hall, A., and Lehman, L. (1973). *J. Agric. Food Chem.* **21**, 264–268.

Spinelli, J., Koury, B., Groninger, H., Jr., and Miller, R. (1977). *Food Technol. (Chicago)* **31**, 184–187.

Stansby, M. E. (1976). *Mar. Fish. Rev.* **38**(9), 1–11.

Steinberg, M. A. (1980). *In* "Advances in Fish Science and Technology" (J. J. Connell, ed.), pp. 34–48. Fishing News Books Ltd., Farnham, Surrey, England.

Stevens, E. D., and Neill, W. H. (1978). *In* "Fish Physiology" (W. S. Hoar and D. J. Randall, eds.), Vol. 7, pp. 315–356. Academic Press, New York.

Stout, V. F., and Beezhold, F. L. (1981). *Mar. Fish. Rev.* **43**(1), 1–12.

Stout, V. F., Houle, C. R., and Beezhold, F. L. (1981). *Mar. Fish. Rev.* **43**(3), 1–13.

Suzuki, T. (1981). *In* "Water Activity: Influences on Food Quality" (L. B. Rockland and G. F. Stewart, eds.), p. 743. Academic Press, New York.

Syrovy, I., Gaspar-Godfroid, A., and Hamoir, G. (1970). *Arch. Int. Physiol. Biochim.* **78**, 919–934.

Tanaka, M., Okubo, S., Suzuki, K., and Taguchi, T. (1980). *Bull. Jpn. Soc. Sci. Fish.* **46**, 1539–1543.

Tang, N. Y., and Hilker, D. M. (1970). *J. Food Sci.* **35**, 676–677.

Teeny, F. M., Hall, A. S., and Gauglitz, E. J., Jr. (1974). *Mar. Fish. Rev.* **36**(5), 15–19.

Tokunaga, T. (1975). *Bull. Jpn. Soc. Sci. Fish.* **41**, 547–553.

Tomioka, H., Yamaguchi, K., Hashimoto, K., and Matsuura, F. (1974). *Bull. Jpn. Soc. Sci. Fish.* **40**, 1269–1275.
Tomioka, H., Yamaguchi, K., Hashimoto, K., and Matsuura, F. (1975). *Bull. Jpn. Soc. Sci. Fish.* **41**, 51–58.
Tomita, H., and Tsuchiya, Y. (1971). *Tohoku J. Agric. Res.* **22**, 228–238.
Tomlinson, N. (1966). *Bull., Fish. Res. Board Can.* **150**.
Tsuchiya, T., and Matsumoto, J. J. (1975). *Bull. Jpn. Soc. Sci. Fish.* **41**, 1319–1326.
Tsuchiya, T., Yamada, N., Mori, H., and Matsumoto, J. J. (1978). *Bull. Jpn. Soc. Sci. Fish.* **44**, 203–207.
Tsuchiya, T., Fukuhara, S., and Matsumoto, J. J. (1980a). *Bull. Jpn. Soc. Sci. Fish.* **46**, 197–200.
Tsuchiya, T., Shinohara, T., and Matsumoto, J. J. (1980b). *Bull. Jpn. Soc. Sci. Fish.* **46**, 893–896.
Tsuzimura, M., Michinaka, K., and Watanabe, S. (1972). *J. Jpn. Soc. Food Nutr.* **25**, 533–537.
Varga, S., Keith, R. A., Michalik, P., Sims, G. G., and Reiger, L. W. (1980). *J. Food Sci.* **45**, 1487–1491.
Vernath, M. F., and Robe, K. (1979). *Food Process.* **40**, 76–79.
Watts, D. A., Rice, R. H., and Brown, W. D. (1980). *J. Biol. Chem.* **255**, 10916–10924.
White, A. W., and Maranda, L. (1978). *J. Fish. Res. Board Can.* **35**, 397–402.
Wolfe, S. K. (1980). *Food Technol. (Chicago)* **34**(3), 55–58.
Yabe, K., Nakamura, K., Suzuki, M., and Ito, Y. (1978). *Bull. Jpn. Soc. Sci. Fish.* **44**, 1345–1350.
Yamaguchi, K., Lavéty, J., and Love, R. M. (1976). *J. Food Technol.* **11**, 389–399.
Yamaguchi, K., Takeka, N., Ogawa, K., and Hashimoto, K. (1979). *Bull. Jpn. Soc. Sci. Fish.* **45**, 1335–1339.
Yu, T. C., Sinnhuber, R. O., and Crawford, D. L. (1973). *J. Food Sci.* **38**, 1197–1199.

Index